CARBOHYDRATES — Polyhydroxy aldehydes, or polyhydroxy ketones, or substances that can be hydrolyzed to such aldehydes or ketones.

CARBONYL GROUP — The reactive group to be found in aldehydes, ketones, and acyl compounds, i.e., $\diagup\!\!\!\!C\!=\!O.$

CARBOXYL GROUP — The functional group characteristic of the carboxylic acids,

i.e., $-C\diagup^{O}_{\diagdown OH}$.

CHIRALITY — The left-handed or right-handed property possessed by compounds that makes possible the existence of enantiomers.

CODON — Coding units for specific amino acids made up of triplet combinations of nucleotides in DNA.

CONFIGURATION — The spatial arrangement of atoms in a molecule that may be changed only by breaking and rearranging bonds.

CONFORMATION — A spatial arrangement of groups in a molecule permitted by their rotation about single bonds.

CONJUGATED SYSTEM — A system of alternate single and double bonds through which chemical activity may be transmitted from one part of the molecular structure to the other.

COVALENCE — Bonding through electron-pair sharing.

DEXTROROTATORY — A term applied to optically active compounds that rotate plane polarized light clockwise or to the right.

DIASTEREOISMERS — Stereoisomers that are not mirror images of each other.

ELECTROPHILIC REAGENT — A reagent that seeks a site of high electron density, e.g., an acid.

ENANTIOMERS — Optical isomers that are nonsuperimposable mirror images.

ESTERS — A general class of acid derivatives in which the ionizable hydrogen of the acid has been replaced by a hydrocarbon group.

ETHERS — A general class of compounds in which two hydrocarbon groups are joined through an oxygen atom.

Continued on back endpapers.

ORGANIC
CHEMISTRY

ORGANIC CHEMISTRY

A Brief Course

FOURTH EDITION

Walter W. Linstromberg
University of Nebraska at Omaha

Henry E. Baumgarten
University of Nebraska—Lincoln

D. C. HEATH AND COMPANY
Lexington, Massachusetts   Toronto

Consulting Editor, **Jacob Kleinberg**

Preface

As numerous teachers of organic chemistry have observed, the learning of organic chemistry is very similar to the learning of a foreign language. The alphabet is short and simple: C, H, N, O, Cl, Br, I, F, S, P, and a few other inorganic "letters." But there are 2×10^6 words! To be able to use these words, first in simple phrases (one-step reactions and syntheses), then later in sentences (multistep reactions and syntheses), and finally in paragraphs (bio-organic sequences and pathways), the student must learn a good deal of new chemistry. The study of this new language by means of the chemistry of the various functional groups appears to be the simplest, most natural, and most effective way of easing the burdens of the learning process.

Intended for use in a brief introductory course in organic chemistry, the fourth edition, like its predecessors, is organized around organic functional groups. Other organizational schemes based, for example, on reaction types or reaction mechanisms have certain advantages for the *advanced* student, but none of the organizational schemes experimented with over the past 25 years seems to meet the special requirements of the average *beginning* student of organic chemistry.

Although there are some pedagogical advantages in the separation of aliphatic and aromatic chemistry, time generally does not permit this separation in a brief course. The unfortunate result is inadequate attention to the aromatic compounds. Therefore, in this text the two classes of compounds are treated together with due regard for their chemical differences.

The "rules of grammar" for the language of organic chemistry include not only the all-embracing principles of chemistry (as taught in general and in physical chemistry), but also the more specific concepts of contemporary organic reaction mechanisms. Even a limited understanding of these important principles helps in learning the facts of organic chemistry and unifies the apparently (but not truly) diverse organic reactions and syntheses by showing the definite interrelationships among them. Obviously, in a brief course time will permit neither an extensive nor an intensive review of physical organic chemistry or of organic reaction mechanisms, nor allow an exposition of the experimental basis for current theoretical beliefs and an indication of the uncertainties in these beliefs. In this text theoretical organic chemistry and organic reaction mechanisms are used as adjuncts to the study of the properties, syntheses, and reactions of organic compounds. An attempt has been made to keep the discussions of these topics terse and succinct. We have aimed at simplicity while retaining accuracy and avoiding unnecessary mechanistic rigor. Unfortunately, beginning students tend to accept reaction mechanisms as factual material, particularly when time does not permit a detailed discussion of their derivations and limitations. The instructor will want to remind the students occasionally that organic reaction mechanisms are rarely, if ever, proved and represent at best our current rationalizations of how reactions take place.

In the preparation of the fourth edition, the entire text was reviewed for conformity with our best current understanding of both fact and theory (reduced to a level consistent with a brief course). About one-fifth of the text represents new or revised material: corrections, additions, supplementation. The revisions will be found throughout the text with perhaps greatest concentration in the first ten chapters. A few new reactions have been added, largely in response to specific requests from users (for example, the Diels-Alder reaction, the Wittig reaction, conjugate addition to α, β-unsaturated carbonyl systems, quinone syntheses and reactions, carbenes). Chapter 1 (General Principles) has been augmented by sections on rules for writing Lewis structures; reaction mechanisms, reaction intermediates, and transition states; acids and bases (electrophiles and nucleophiles); and resonance structures. More extensive use has been made of Fischer projections in Chapter 5 (Stereoisomerism), as experience has shown that students (and many experienced chemists as well) find it easier to handle the two-dimensional structures than to visualize structures in three dimensions. A short section on the chemistry of bivalent sulfur compounds, those most likely to be encountered in biochemistry or in the life sciences, has been appended to Chapter 13 (now titled Amines and Other Nitrogen Compounds; Sulfur Compounds). The chapter on the chemistry of petroleum (formerly Chapter 5) has been moved to the end of the text, where it joins the other special-topics chapters.

The principal addition to the fourth edition is a new chapter (Chapter 6) on spectroscopy (nuclear magnetic resonance, infrared, and ultraviolet-visible (electronic)). Probably no recent development in chemistry has had as great an effect on the growth and development of our knowledge of organic chemistry during the past half century as the addition of spectroscopic techniques to the armamentarium of the organic chemist. The chapter on spectroscopy is somewhat more extensive than those describing spectral techniques usually found in brief texts. With so little time available in beginning courses, it is essential that the text provide sufficient detail to support and supplement the lectures, for the student is largely on his own in mastering material of this type. For this reason the discussion of spectroscopy is intended to be at a relatively low level but more informative than the highly simplified narratives often found in texts for brief courses.

The number of exercises for the student has been increased substantially; and, again in response to requests from users, several more demanding exercises have been added to most of the chapters. Instructors and students are reminded that solutions to all of the exercises as well as numerous additional exercises (largely relatively simple exercises of a review nature) may be found in the student manual (*Study Guide with Problems and Solutions for Organic Chemistry, a Brief Course*) that has been prepared by Professor Robert L. Zey.

The authors are indebted to students and colleagues at the University of Nebraska who have kindly contributed in one way or another to the preparation of this edition. Dr. Naba K. Gupta and Dr. Bandanna Chatterjee provided valuable suggestions for the improvement of Chapter 15 on the amino acids, peptides, and proteins and Chapter 16 on the nucleic acids. The ^{1}H nmr, ir, and uv-vis spectra in Chapter 6 were run by Mr. Paul Y.-N. Chen. The ^{13}C spectrum of sucrose was run by Mr. Alan Sopchik. The Ortep drawing of 3-carbomethoxybicyclo[1.1.0]butane-1-carboxylic acid was provided by Dr. Victor W. Day from his research collection of structures established by X-ray crystallography.

The authors are also indebted to the many instructors who have used previous editions of this text and who have suggested additions, changes, or other improvements that have been incorporated into the fourth edition. Especially helpful were the suggestions of our colleague, Dr. D. N. Marquart. We like to think that this text is written to serve the needs of instructors of brief courses in organic chemistry, and we urge all users and potential users to share with us their ideas for the improvement of future editions.

Walter W. Linstromberg
Henry E. Baumgarten

Contents

3 *The Unsaturated Hydrocarbons—Alkenes and Alkynes*

6 *Determination of Molecular Structure; Spectroscopy*   149

7 *Organic Halogen Compounds*   205

10 The Carboxylic Acids and Their Derivatives 305

11 Fats, Oils, Waxes; Soaps and Detergents; Prostaglandins 337

14 *Carbohydrates*   415

Introduction

Organic chemistry according to its original definition dealt only with those substances of natural *organic* origin—that is, products of plants or animals. It was thought by early chemists that the production of an organic compound involved the "vital force" of a living organism. The science of chemistry thus was separated into two broad divisions—**inorganic** and **organic.** In 1828, however, the German chemist Friedrich Wöhler[1] heated some ammonium cyanate, NH_4OCN, and obtained urea.

$$NH_4OCN \xrightarrow{\text{heat}} \begin{array}{c} H \\ H-N \\ \quad\quad C=O \\ H-N \\ H \end{array}$$

For the first time the vital force theory began to be questioned. Certainly nothing could be more organic than urea for it is excreted in the urine of mammals, and ammonium cyanate was prepared from two inorganic substances, potassium cyanate (KOCN) and ammonium sulfate (($NH_4)_2SO_4$). However, the ammonium cyanate that Wöhler used had been prepared by the calcination of animal bones and for this reason some of his contemporaries still held the vitalistic theory to be valid. It was not until some time later when Kolbe[2] succeeded in converting chloroacetic acid into acetic acid by heating it with zinc that the vital force theory was completely abandoned by chemists and the relationship of the two branches of chemistry clearly recognized. As is often the case, names persist even though they are no longer descriptive. Today, **organic chemistry** is recognized as the study of most **carbon containing compounds.** Over two million of these compounds of carbon have been synthesized or studied, and the list of known organic compounds continues to grow daily. In many instances the organic chemist not only has recreated structures found in nature but has improved upon them as well. Moreover, he has synthesized many compounds which have no counterparts in nature. Indeed, the synthesis of new and useful compounds is one of the most important aspects of organic

[1] Friedrich Wöhler (1800–1882). Professor of Chemistry, University of Göttingen.

[2] Adolph Wilhelm Kolbe (1818–1884). Professor of Chemistry, University of Leipzig.

chemistry. It was thus that we obtained the synthetic fibers nylon and dacron, the local anesthetics novocaine and xylocaine, the various elastomers we use as substitutes for rubber, the films and molded articles of plastic, the sulfa drugs, to mention but a few of the more familiar products. Other synthetics that you will read about in this text would make a long and impressive list. The credits we accord the research chemist for the many products which make our lives more healthful, more comfortable, and more enjoyable certainly are well deserved. However, in fairness it must be said that occasionally a synthetic product proves to be a mixed blessing. DDT is a case in point. While the use of this powerful insecticide has eradicated legions of insect pests costly and dangerous to man, it has at the same time found its way into the ecosystem with deleterious results. Each new synthetic product, therefore, may have its price, and man, in his technological and economic zeal, must be ever vigilant to preserve a natural balance in his environment. To do otherwise would threaten the existence of all living things—himself included.

It is highly desirable that we have a knowledge of organic chemistry, if for no other reason than to enjoy a greater appreciation of the world around us. Our foods, clothing, fuels, and medicines are, for the most part, organic in nature. There is hardly a phase of our daily lives that is not related to or dependent upon this tremendously important and fascinating science.

A working knowledge of organic chemistry is a necessary prerequisite to a clearer understanding of those studies of which it is an integral part—i.e., medicine, nursing, pharmacy, home economics, agriculture, and all their related fields.

The need for getting off to a good start in an organic chemistry course cannot be overemphasized. It is in order, therefore, to offer to the student a few suggestions which will make the study of organic chemistry orderly, meaningful, and interesting. Read your assignments carefully. Learn each new concept and term as you come to it. Use the reviews, exercises, and problems which you will find within the body of the text and at the end of each chapter. Most important, by the extensive use of scratch pad and pencil apply through practice what you have learned. *This is the only way you will remember the material you study.* Finally, ask questions both in class and in the laboratory.

In the beginning it may appear confusing that the laboratory preparation of one class of compounds frequently requires the use of another class not yet studied. Cheer up! It soon will become obvious that in this manner the synthesis of one substance automatically employs a reaction of the other. Inasmuch as the study of organic chemistry involves principally learning the **names, structures, preparation,** and **reactions** of many compounds, the whole scheme of presentation is one of integrating these constituents. As in a jig-saw puzzle, the pieces will fit into place as you proceed through the course.

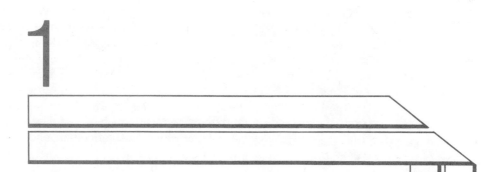

1

General Principles

Introduction

The analysis of any plant or animal product reveals that it always contains carbon and hydrogen and usually also oxygen and/or nitrogen. In many organic compounds sulfur, phosphorus, and the halogens are also to be found. In some cases even a metallic element makes up part of a naturally occurring organic structure.[1] It would be correct to say that a very small number of elements (less than a dozen) comprise most organic compounds. This is small indeed when compared to the large number of different elements found in inorganic substances. How is it possible for such a small number of elements to

---

[1] Iron is present in the hemoglobin of blood to the extent of 0.34%, magnesium comprises about 2.7% of the chlorophyll found in green plants, and cobalt is present in Vitamin B_{12} to the extent of 4.34%.

account for over two million organic compounds? To obtain the answer to this question we need to examine the manner in which the atoms in compounds are held together. It is not enough to know the number and kind of atoms an organic compound contains. We must be able to translate each molecular formula into a structure. When we have done this, we will see that carbon is nearly unique among the elements. Not only does the carbon atom bond to each of the elements cited above, but it shows a predilection for bonding to other carbon atoms. The unique role of the carbon atom in the formation of organic compounds will be understood better after a brief review of atomic structure.

1.1   Atomic Structure

In your introductory course in chemistry you learned that an atom consists of a positively charged nucleus surrounded by negative electrons equal in number to the charge on the nucleus. You also learned that the electrons are distributed around the nucleus in energy levels called **principal quantum shells** designated by the letters K, L, M, N, O, P, and Q, or by numbers 1–7, in each case starting with the shell nearest the nucleus. If n is the number of the principal shell, the number of electrons needed to completely fill that shell with electrons is $2n^2$. Thus, the first shell (K shell) can hold a maximum of two electrons, the second shell (L shell) a maximum of eight electrons, etc. Electrons in the first shell have the lowest energy, the energy increasing with shell number. In addition to principal shells, there are still other spatial categories for electrons.

Principal shells are divided into **subshells** designated by letters in order of filling as s, p, d, and f. The first principal shell has only one subshell, designated as the $1s$. The second principal shell has two subshells designated $2s$ and $2p$ subshells and the third principal shell consists of three subshells, namely $3s$, $3p$, and $3d$. Inasmuch as our present study is limited principally to compounds of carbon formed with elements of low atomic numbers, we need not concern ourselves at this time with higher subshells. The subshells are further divided into **atomic orbitals.** An orbital is not easily defined but may be described as the region in the space surrounding an atomic nucleus which is most likely to be occupied by an electron. Imagine an electron in the s orbital as a point of light capable of being photographed. A time exposure on a photographic film then would reveal a circular cloud, dense toward the center and diffuse at its outer boundary. Next imagine a *surface* encompassing part of the electron cloud with the following properties: (1) the probability of finding the electron at any point on the surface is everywhere the same, and (2) the electron will be found within or on (but not outside) the surface 90% (or any other arbitrarily chosen percentage) of the time. For an s orbital the surface will be a sphere; therefore, we will use the sphere as a geometrical representation for the s orbital. All s orbitals, or orbitals in s subshells, are spheres, with their centers at the nucleus of the atom (Fig. 1.1). However, the shapes of orbitals vary. The K shell has

FIGURE 1.1 *The s Orbital*

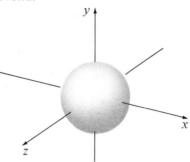

only the $1s$ orbital. The L shell has, in addition to the $2s$ orbital, which also is spherical but somewhat larger than the $1s$ orbital, three p orbitals. The p orbitals are of equal energy and are dumbbell-shaped with a lobe on either side of the atomic nucleus (Fig. 1.2). The axes of the p orbitals are perpendicular to each other and lie along the coordinate axes with the nucleus at the origin. The orbitals are differentiated as $2p_x$, $2p_y$, or $2p_z$, where the small subscripts refer to the x, y, and z axes. Each $2p$ orbital has a **node,** which in this instance is a nodal plane defined by the two coordinate axes perpendicular to the orbital axis. Thus, the $2p_x$ orbital has the yz plane as a node (or nodal plane). The probability of finding an electron at the node is zero. The total number of nodes is always one less than the principal quantum number, n. Each orbital, regardless

FIGURE 1.2 *Atomic $2p_x$, $2p_y$, and $2p_z$ Orbitals with Axes Mutually Perpendicular*

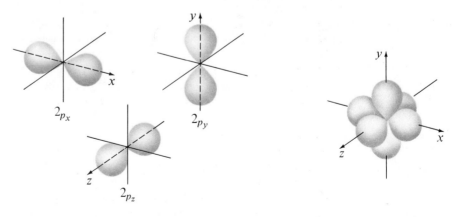

TABLE 1.1 *Schematic Representation of Electron Orbitals for the First Ten Elements of the Periodic Table**

Element	At. No.	Normal K shell 1s	Normal L shell 2s	2p_x	2p_y	2p_z	Excited K shell 1s	Excited L shell 2s	2p_x	2p_y	2p_z	Valence
H	1	↓										1
He	2	↑↓										0
Li	3	↑↓	↓									1
Be	4	↑↓	↑↓	○	○	○	↑↓	↓	↓	○	○	2
B	5	↑↓	↑↓	↓	○	○	↑↓	↓	↓	↓	○	3
C	6	↑↓	↑↓	↓	↓	○	↑↓	↓	↓	↓	↓	4
N	7	↑↓	↑↓	↓	↓	↓						3
O	8	↑↓	↑↓	↑↓	↓	↓						2
F	9	↑↓	↑↓	↑↓	↑↓	↓						1
Ne	10	↑↓	↑↓	↑↓	↑↓	↑↓						0

* The spin of an electron is shown by a small arrow. Since the spins of two electrons occupying the same orbital must be opposite (positive and negative), the arrows representing such electrons are shown pointing in opposite directions. The empty (broken line) circles in the diagram represent orbitals potentially available to electrons of higher energy. The normal state is the lowest energy state for the atom; thus, all electrons are in the lowest energy orbitals allowed by the rules. An excited state is a high energy state in which one or more electrons have been promoted to orbitals of higher energy.

of its designation, has a maximum capacity of two electrons, and, when the orbital is filled, the two electrons must be oriented in opposite directions or must have **opposite spin.** Further, in any principal shell electrons occupy orbitals of lower energy first—that is, they occupy *s* orbitals before *p* orbitals, *p* before *d*, etc. Moreover, an orbital in any subshell is not occupied by a pair of electrons until each orbital of the subshell is occupied by one electron. Table 1.1 shows graphically how electrons are distributed around the nuclei of the first ten elements.

EXERCISE 1.1

Expand Table 1.1 to include the M, or third principal quantum shell, and show the electron distribution of the next eight elements in the Periodic Table.

Chemical Bonding

The only elements which have a completely filled outermost, or valence, shell belong to the family of elements referred to as the noble gases in Group O of the periodic table.[2] A completely filled outermost shell appears to represent a stable arrangement, and many elements in their reactions with each other exhibit a tendency to fill their outermost shells through a transfer or sharing of electrons. Either transfers or sharing result in the formation of chemical bonds.

1.2 *Electrovalent, or Ionic, Bonds*

When sodium and chlorine enter into chemical combination the product, sodium chloride, is a solid compound in which both sodium and chlorine appear as ions, or charged particles. By contributing its one outermost (valence) electron to the chlorine atom, sodium is left deficient by one electron and becomes a sodium ion with a charge of plus one. Chlorine, by acquiring the extra electron from sodium, becomes a chloride ion with a charge of minus one.

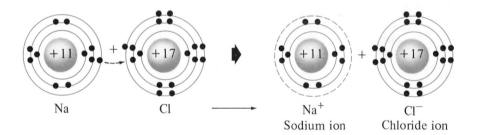

| Na | Cl | → | Na⁺ | Cl⁻ |

$$\text{Na} \qquad \text{Cl} \longrightarrow \underset{\text{Sodium ion}}{\text{Na}^+} \qquad \underset{\text{Chloride ion}}{\text{Cl}^-}$$

After the transfer of the electron is completed, both sodium and chloride ions have filled outer shells of electrons; the former ion has an electronic configuration equivalent to that of the noble gas (Ne) preceding it in the periodic table, and the chloride ion has the configuration of the noble gas (Ar) immediately following (Table 1.2).

In the sodium chloride crystal each sodium ion is surrounded by six chloride ions and each chloride ion is surrounded by six sodium ions in a cubic

---

[2] The outermost (valence) shell is usually regarded as completed when it has eight electrons, that is, filled *s* and *p* subshells even though the shell can potentially hold more than eight electrons. Exceptions: hydrogen and helium have a filled outermost shell when the shell contains two electrons—the 1*s* subshell.

TABLE 1.2 *An Abbreviated Periodic Table*

	1	2	3	4	3	2	1	0
1	^{1}H							^{2}He
2	^{3}Li	^{4}Be	^{5}B	^{6}C	^{7}N	^{8}O	^{9}F	^{10}Ne
3	^{11}Na	^{12}Mg	^{13}Al	^{14}Si	^{15}P	^{16}S	^{17}Cl	^{18}Ar
4	^{19}K	^{20}Ca					^{35}Br	^{36}Kr
5	^{37}Rb	^{38}Sr					^{53}I	^{54}Xe
6	^{55}Cs	^{56}Ba					^{85}At	^{86}Rn

(The numbers across the top are the usual valences of the elements in that column.)

structure (Fig. 1.3). These ions of opposite charge are held together by strong, attractive electrostatic forces, or **electrovalent bonds.**

Another example of an ionic compound is lithium fluoride. The lithium atom, like sodium, also gives up the single electron in its outermost shell. The lithium ion, with two electrons remaining in its K shell, thus has the same spatial arrangement of electrons (called **configuration**) as helium. Fluorine, on the other hand, attains the configuration of neon by accepting an electron. The compound has the formula Li^+F^-.

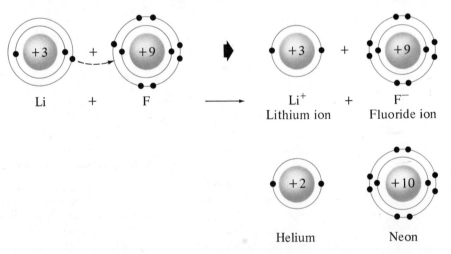

Thus, many inorganic compounds are formed by transfers of electrons between the elements to produce, not molecules, but aggregates of charged particles. Crystals of inorganic salts usually are hard substances with high melting points, because the electrostatic forces that hold the ions together in the crystal are great and not easily broken. When inorganic salts dissolve in water, the ions present in the crystals combine with water molecules, dissociate, and produce solutions which conduct an electric current. These properties are not generally characteristic of organic compounds.

FIGURE 1.3 *The Spatial Arrangement of Sodium and Chloride Ions in a Crystal of Sodium Chloride (Common Salt)*

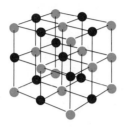

(a) The black circles represent sodium ions, the color circles chloride ions;

(b) The small spheres represent sodium ions and the large spheres chloride ions.

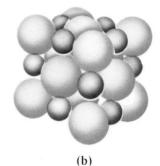

(a)

(b)

1.3   *Covalent Bonds*

A covalent single bond forms between two atoms through the *sharing* of a pair of electrons. Two atomic orbitals, one from each atom, overlap and form *two* new orbitals called **molecular orbitals,** which encompass the nuclei of both atoms. The two molecular orbitals are quite different in energy, and only the lower energy, **bonding orbital** is used in the formation of covalent bonds. We will not need to use the higher energy orbital, or **antibonding orbital,** in most of our study of organic chemistry but will require it in the discussion of ultraviolet spectra (Chapter 6). In the most common case, one electron from each bonded atom occupies the bonding molecular orbital to form the two-electron covalent bond. As in the case of a filled atomic orbital, the electrons in the pair occupying the molecular orbital must have opposite spins. Now each electron, which before bonding was attracted only to its own nucleus, is also subject to the attraction of a second nucleus. This additional attractive force gives the bond strength and makes for a stable arrangement. Ions are not formed in this type of union since there is no transfer of electrons.

The simplest example of the covalent single bond is found in the hydrogen molecule. Each hydrogen atom with a single electron in its $1s$ orbital can supply one of the shared electron pair which fills the molecular orbital. The shape of the resulting molecular orbital is no longer spherical but, as might be imagined, more sausage-shaped. The bond formed between the two hydrogen atoms is cylindrically symmetrical about its axis and is called a **σ** (**sigma bond**). It is a very strong bond and results in a very stable molecule. In Fig. 1.4 the approximate shapes of both the atomic and molecular orbitals and the energy relationships between them are shown.

A number of atoms share more than one pair of electrons. Nitrogen, for example, is capable of sharing three pairs of electrons to form three covalent bonds. In the nitrogen molecule, N_2, each nitrogen atom supplies three electrons

FIGURE 1.4 *Atomic and Molecular Orbitals of Hydrogen and Their Energy Relationships*

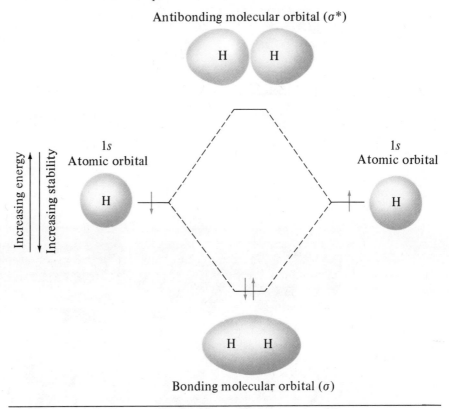

to a common linkage. In ammonia, NH_3, nitrogen shares three unpaired electrons with those of three hydrogen atoms. Oxygen is capable of providing two electrons for two shared electron pairs, as in H_2O and CO_2.

Nitrogen	Ammonia	Carbon dioxide	Water

Structures drawn in the above manner are called **Lewis structures.** Each shared pair of electrons is in a bonding molecular orbital. Each unshared pair (one pair on each nitrogen atom, two pairs on each oxygen atom) is in a **non-bonding** molecular orbital. The nature of the orbitals involved in multiple bonds will be discussed in later chapters (Secs. 3.2 and 3.15).

Since electrons are the same regardless of their origin, we can use a short line (—) to represent the shared electron pair (or covalent single bond), double lines for two shared pairs (covalent double bonds), and triple lines for three shared pairs (covalent triple bonds).

$$\ddot{N}\equiv\ddot{N} \qquad \overset{\displaystyle H}{\underset{\displaystyle H}{\overset{|}{\underset{|}{\ddot{N}-H}}}} \qquad \overset{+}{\underset{+}{O}}=C=\overset{+}{\underset{+}{O}} \qquad \overset{\displaystyle \overset{oo}{O}\colon}{\underset{\displaystyle H}{\overset{|}{\underset{|}{H-O}}}}$$

Nitrogen	Ammonia	Carbon dioxide	Water

EXERCISE 1.2

Using dots and small asterisks to represent electrons, draw the Lewis structures for: (1) Hydrogen peroxide, H_2O_2, (2) Chlorine, Cl_2, (3) Phosgene, $COCl_2$.

In our discussion of the covalent bond we have emphasized the idea of electron pair sharing as a give-and-take proposition in which each atom sharing in the bond makes an equivalent contribution. Frequently structures are formed in which both electrons comprising the shared pair of a covalent bond come from only one of the atoms. Such an arrangement is possible when one atom has an empty valence orbital and the orbital of the other contains an unshared pair of electrons. This type of bond is called a **coordinate covalent,** or **semipolar bond.** An excellent illustration of the coordinate covalent bond is found in the boron fluoride–ammonia complex. Using dots to represent the valence electrons of boron and hydrogen, small circles for the five outermost electrons of nitrogen, and small x's for the seven electrons of fluorine, we can show how B, N, F, and H are bonded to each other in this structure.

Boron fluoride	Ammonia	Boron fluoride–ammonia complex

$$\overset{\displaystyle F}{\underset{\displaystyle F}{F-B\leftarrow N-H}}\overset{\displaystyle H}{\underset{\displaystyle H}{}} \qquad \text{or} \qquad \overset{\displaystyle F}{\underset{\displaystyle F}{F-\overset{-}{B}-\overset{+}{N}-H}}\overset{\displaystyle H}{\underset{\displaystyle H}{}}$$

The bond with an arrowhead indicates a coordinate covalent, or semipolar, bond between nitrogen and boron, with nitrogen as the donor of the electron pair and boron as the acceptor. Since nitrogen has, in effect, bestowed upon boron a charge almost equivalent to one electron and lost a corresponding amount, the donation also may be indicated by placing a plus sign on nitrogen and a minus sign on boron. All other bonds shown in the structure are of the ordinary covalent type.

EXERCISE 1.3

Trimethylamine oxide has the electron configuration shown. Redraw this structure using a different color or symbolism to indicate the origin of all electrons. Are any bonds of the coordinate covalent type?

$$
\begin{array}{c}
\text{H} \\
\overset{\circ\circ}{\text{H}\!:\!\text{C}\!:\!\text{H}} \\
\text{H} \qquad \overset{\circ\circ}{\quad} \\
\text{H}\!:\!\text{C}\ \ :\ \text{N}\ :\ \text{O}\!: \\
\text{H} \qquad \overset{\circ\circ}{\quad} \\
\text{H}\!:\!\text{C}\!:\!\text{H} \\
\text{H}
\end{array}
$$

The mastery of the use of Lewis structures is an almost essential element in the study of organic chemistry. A Lewis structure for almost any molecule or ion can be drawn by the following procedure.

1. Determine the total number of valence electrons for the entire molecule or ion by counting the valence electrons that can be contributed by each neutral atom. Add one electron if the species is a negative ion; subtract one election if the species is a positive ion.

N_2 Total valence electrons $= 2 \cdot \ddot{N} : \ = 2 \times 5 = 10$

NH_3 Total valence electrons $= \ \cdot \ddot{N} : \ + 3\,H \cdot \ = 5 + 3 = 8$

$NH_4{}^+$ Total valence electrons $= \ \cdot \ddot{N} : \ + 4\,H \cdot \ - e = 5 + 4 - 1 = 8$

HO^- Total valence electrons $= :\ddot{O} : \ + H \cdot \ + e = 6 + 1 + 1 = 8$

2. Draw the atomic symbols arranged in approximately the correct geometric pattern. Fill in the single, double, and triple bonds using a short line (—) for each pair of electrons in the bond. Add the unshared electrons (pairs or single electrons) using one dot for each electron.

3. Each atom should, to the greatest extent possible, have a filled valence shell: 8 electrons for first-row atoms and 8, 10, or 12 for second-row atoms. (*Note:* Recall that a covalent bond counts as 2 electrons for *each* atom in determining filled valence shells.)

$:N{\equiv}N:$ (8 + 8 valence electrons) rather than $:\ddot{N}{=}N:$ (8 + 6 electrons)

$:\ddot{O}{=}C{=}\ddot{O}:$ (8 + 8 + 8 electrons) rather than $:\ddot{O}{-}C{=}\ddot{O}:$ (8 + 6 + 8 electrons)

4. Maximize the number of bonds and minimize the number of unpaired electrons. (See examples in (3).)
5. Find the **formal charge** on each atom. First, assign to each atom all of its unshared electrons and *half* of the electrons in its covalent bond pairs. Then substract this total number of electrons from the nuclear charge for the atom.

Sulfate ion (8, 10, or 12 electrons permissible in sulfur's valence shell)

Nearly all bonds which bind atoms together to form organic compounds are of the covalent type. The unit particles in most organic compounds are molecules, not ions of opposite charge as in inorganic salts. Since the attractive forces between molecules are much weaker than those between ions in salts, many organic substances are liquids with low boiling points. They ordinarily have low melting points if crystalline solids. Organic substances, for the most part, are only slightly soluble or insoluble in water. If slightly soluble, their aqueous solutions seldom conduct an electric current, unless they are the salts of organic acids or salts of organic bases. Reactions between organic molecules are usually slow and take place only when molecules collide with sufficient kinetic energy to break, rearrange, or form new bonds. For this reason organic reactions frequently require long periods of heating and the presence of a catalyst before any appreciable changes occur. For example, a covalently bound halogen atom may show no tendency to precipitate as silver halide when merely shaken with a silver nitrate solution. On the other hand, the reaction of an ionic halide with silver nitrate is immediate.

The Covalence of Carbon

1.4 *Hybridized Orbitals*

Since the covalence which an element generally exhibits usually is determined by the number of unpaired electrons in half-filled orbitals in its valence shell, the case of carbon deserves special mention. In nearly all organic compounds, carbon exhibits a covalence of four. How can this be when Table 1.1 shows carbon has but two unpaired electrons in its outer shell? The answer is that, in the first excited state of carbon attained by promoting one of the $2s$ electrons to the higher energy, vacant $2p_z$ orbital (as shown in Table 1.1), there are four singly-occupied orbitals, one $2s$ and three $2p$. If these four orbitals are mixed (mathematically) an excited atom with four equivalent orbitals results.

FIGURE 1.5   sp³-*Hybridized Orbitals:* (*a*) *An* sp³-*Hybridized Orbital,* (*b*) *Conventional Symbol Used to Represent an* sp³ *Orbital,* (*c*) *Four Tetrahedral* sp³ *Orbitals of the Carbon Atom*

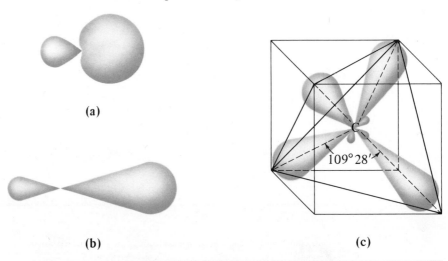

(a)

(b)

(c)

The energies of the four equivalent orbitals are equal and intermediate in value between the energy of a $2s$ orbital and that of a $2p$ orbital. The total energy (for all four) is the same as the sum of the energies of the $2s$ and three $2p$ orbitals. These blended orbitals are said to be *hybridized* and are called sp^3 (pronounced s-p-three) orbitals. Chemical theory predicts that the four sp^3 hybrid orbitals should be directed from the carbon atom toward the corners of a tetrahedron (Fig. 1.5).

    If the four hybrid sp^3 orbitals are allowed to overlap with the $1s$ orbitals of four hydrogen atoms, four bonding molecular orbitals of the sigma type will be formed (as well as four antibonding orbitals). The resultant molecule, CH_4, should have the carbon atom at the center and four hydrogen atoms at the vertices of a regular tetrahedron. Experimental data are in agreement with this prediction. Thus, for example in methane, CH_4, the principal constituent of natural gas, the four covalent C—H bonds are directed toward the vertices of a regular tetrahedron (Fig. 1.6).[3] With carbon in the center, these four sigma bonds make angles of 109° 28′ with each other. They are of the same length (1.09 Å).[4] The energy required to break any one of them is the same (102 kcal).[5] The bonds, therefore, *must* be equivalent.

---

[3] A regular tetrahedron is easily constructed by drawing a cube and connecting all nonadjacent corners.

[4] The Angstrom (Å) is equal to 10^{-8} centimeter.

[5] The *kilocalorie* (kcal) is equal to 1,000 calories.

FIGURE 1.6 *Structural Representations of Methane: (a) Ball and Stick Model in Regular Tetrahedron, (b) Tetrahedral Structure Showing Electrons Involved in Bonding and Bond Angles, (c) Stuart Model*

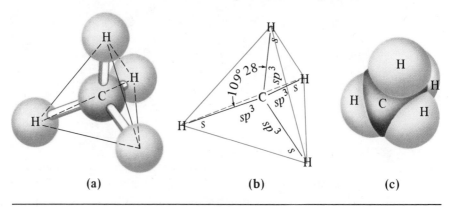

(a)   (b)   (c)

What have we gained by the process of hybridization, which required the use of the higher energy excited state orbitals? First, we were able to form four *equivalent* bonds from carbon to hydrogen, rather than two. Thus, we lowered the energy of the molecule more by forming two extra bonds than we lost by using the higher energy orbitals. This alone would not justify hybridization, for we could have used the $1s$ and three $2p$ excited state orbitals without hybridization to form four non-equivalent covalent bonds. However, the use of hybridized orbitals permitted the formation of *stronger* bonds because of the better overlapping properties of sp^3 orbitals. Furthermore, the tetrahedral geometry of the sp^3 orbitals allows the maximum *separation in space* of the four bonds. Since the electron pairs in one bond repel those in the other bonds, when the bonds are farther apart, the energy of the system will be lower and the molecule will be more stable.

Hybridization likewise can be considered to take place in the case of boron to give three hybridized sp^2 orbitals, and with beryllium to yield two hybridized sp orbitals. Were it not for such hybridization, one would expect the covalence of carbon to be two and that of boron one; beryllium would have no electrons available for chemical combination. This, we know, is contrary to fact.

EXERCISE 1.4

If the boron atom uses three sp^2 hybridized orbitals for bonding with fluorine, and each B—F bond is separated from the other by a maximum distance, what is the value of the bond angle? What is the geometry of the boron trifluoride molecule?

1.5   *Polar Covalent Bonds*

In molecules in which the bonded atoms are identical, as in the hydrogen or chlorine molecules (H—H, Cl—Cl), the bonding electrons are equally shared. However, were hydrogen to be substituted for either of the chlorine atoms or chlorine for either of the hydrogen atoms in the above examples, then the shared pair of electrons binding the two atoms in each instance would be drawn somewhat closer to the chlorine atom. Such displacement, however small, results in a polar bond and confers upon the more electronegative (electron-attracting) element—chlorine, in this case—a partial negative charge (delta minus) and leaves the other part of the molecule with a partial positive charge (delta plus).

$$H \overset{..}{\underset{..}{\times} Cl} : \qquad \overset{\delta+ \quad \delta-}{\underset{\longmapsto}{H\text{—}Cl}}$$

This slight shift in electron density is indicated by an arrow directed toward the more electronegative element. Opposite the arrowhead the shaft is crossed by a short vertical bar as in a plus sign.

A rough determination of the relative electronegativities of elements may be obtained from their position in the periodic chart. Electronegativities increase as we proceed from left to right or from bottom to top through the periodic arrangement. This establishes fluorine as the most electronegative of the elements. The polarity of a molecule, as you will discover later, frequently determines its behavior in the course of a chemical reaction. For this reason you should have some knowledge of electronegativities.

EXERCISE 1.5

In which of the hydrogen halides will the bond between the hydrogen and the halogen atom be most polar? Least polar?

Analysis of Organic Compounds and Calculation of Formulas

1.6   *Empirical Formulas*

The symbols for the elements present in a substance expressed in the simplest atomic ratio in whole numbers is called the **empirical formula.** The ratio in which the atoms of elements are present in an organic compound is found from the percentage composition of the compound. Since most organic substances contain carbon, hydrogen, and oxygen, they usually will burn to produce carbon dioxide and water. These gases may be collected and weighed.

The presence of nitrogen, sulfur, and the halogens in an organic substance is revealed by fusion of a sample with molten sodium: nitrogen yields sodium cyanide, $NaCN$; sulfur yields sodium sulfide, Na_2S; and the halogens yield the sodium halides, $NaCl$, $NaBr$, or NaI. An analysis for these inorganic, water-soluble, fusion products easily shows whether nitrogen, sulfur, or the halogens are present in the sample.

Let us determine the empirical formula of a white, crystalline, water-soluble compound which burns but gives negative tests for nitrogen, sulfur, and the halogens. If a carefully weighed sample is heated in a combustion tube while a stream of oxygen is passed over it, the hydrogen will be oxidized to H_2O and the carbon to CO_2. These gases may be collected in two separate absorption tubes. One tube is filled with anhydrous magnesium perchlorate, $Mg(ClO_4)_2$, which collects the water, and the other is filled with sodium hydroxide which collects the carbon dioxide. These absorption tubes are carefully weighed before and after combustion and their gain in weight is taken equal to the amount of each gas collected.

Let us suppose that the sample taken weighed 0.1800 g, the CO_2 collected 0.264 g, and the H_2O 0.108 g.

$$\text{Weight of sample} = 0.1800 \text{ g}$$
$$\text{Weight of } CO_2 = 0.264 \text{ g}$$
$$\text{Weight of } H_2O = 0.108 \text{ g}$$

Since carbon comprises 12 g of the total gram molecular weight of carbon dioxide and hydrogen accounts for 2 g of the gram molecular weight of water, the percentages of carbon, hydrogen, and oxygen in the sample may be calculated (using rounded atomic weights) as follows:

$$\% \text{ Carbon} = \frac{0.264 \times 12/44}{0.1800} \times 100 = 40$$

$$\% \text{ Hydrogen} = \frac{0.108 \times 2/18}{0.1800} \times 100 = 6.6$$

$$\% \text{ Oxygen} = 100 - (40 + 6.6) = 53.4$$

The simplest gram-atom ratio is found by dividing the percentage of each constituent by its gram atomic weight:

$$C_{40/12}H_{6.6/1}O_{53.4/16}, \text{ which may be reduced to } C_{3.3}H_{6.6}O_{3.3}$$

Further reduction to a simple whole number gives an atomic ratio of CH_2O, which is the empirical formula for the unknown.

The only substance with this simple ratio in its molecular structure is formaldehyde—a gas; the sample was a solid. In order to determine the **molecular formula** of the sample we must know its molecular weight.

1.7 *Molecular Formulas*

For many years the molecular weights of organic molecules were determined by simple physical methods such as measurement of the freezing point depression of a suitable solvent or measurement of the volume, pressure, and temperature of a given weight of gas. Although these methods are still employed to a limited extent, at present it is more common to use spectroscopic methods such as (1) nuclear magnetic resonance spectroscopy (Chapter 6) to *estimate,* or (2) mass spectroscopy[6] to determine precisely (to several decimal places) the molecular weights of organic compounds. Let us assume that the molecular weight of the solid described above, whose empirical formula was found to be CH_2O, is determined to be 180. To find the molecular formula we proceed as follows:

$$M. W. = 180$$
$$CH_2O \text{ formula weight} = 30$$

thus, $\qquad\qquad (CH_2O)_n = 180; \ n = 6$

The correct molecular formula for the solid is $C_6H_{12}O_6$.

1.8 *Structural Formulas and Isomerism*

As was stated in the introductory section, it is not enough simply to determine the number and kinds of elements comprising an organic compound. Let us illustrate this point. We have already encountered methane whose molecular formula is CH_4 and have learned that in methane the four hydrogens are bonded to a central carbon atom. Now the gaseous hydrocarbon butane has the formula C_4H_{10}. How can the four carbon and ten hydrogen atoms be arranged so that each carbon has four bonds and each hydrogen only one? We discover that there are two possibilities. In one structure, no carbon would be bonded to more than *two* others, whereas in the second structure, one carbon could be attached to *three* others:

Normal butane, B.P., 0.5°C
(a continuous "straight" chain structure)

Isobutane (Methylpropane) B.P., −11.7°C
(a branched chain structure)

In order to draw these two structures, we simply have allowed the *sp*³ hybrid orbitals of carbon atoms to overlap those of other carbons, as well as with *s*

---

[6] From a high resolution mass spectrum run on a few micrograms of an organic compound it is often (but not always) possible to determine not only a precise molecular weight but also the molecular formula.

FIGURE 1.7 *Ball and Stick Models of (a) n-Butane and (b) Isobutane (Carbon Atoms Are Indicated in Black, Hydrogen Atoms in Brown)*

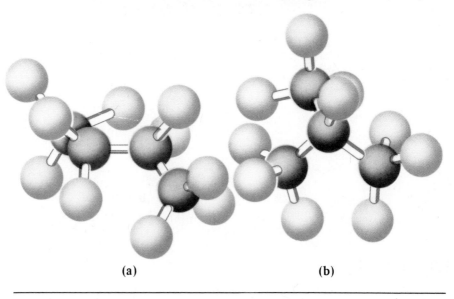

(a)   (b)

FIGURE 1.8 *Molecular and Scale Models of (a) n-Butane and (b) Isobutane*

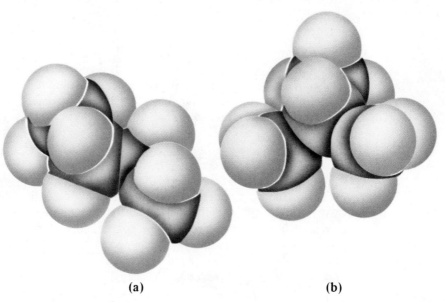

(a)   (b)

orbitals of hydrogen atoms, to form the four sigma bonds that each carbon is entitled to. (See also Figs. 1.7 and 1.8 on page 17.)

Although both compounds have the *same molecular formula,* they have *different structures* and are called **isomers.** (Gr. *isos,* same; *meros,* part.)

EXERCISE 1.6

Is there more than one structure possible for dichloromethane, CH_2Cl_2, if its geometry is tetrahedral? Would more than one structure be possible if its structure were pyramidal with carbon at the apex and hydrogen and chlorine atoms at the corners of the base?

Another pair of gaseous hydrocarbons has the molecular formula C_4H_8. By following the rules of valence for these two elements and bonding the carbon atoms in a continuous **cyclic** arrangement we are able to draw two different C_4H_8 structures.

Cyclobutane, B.P., 13.0°C   Methylcyclopropane, B.P., 4.0°C

In Chapter 3 you will discover that this same molecular formula will permit four additional structures. While the bonds involved in these structures will not all be of the sigma type as in the previous examples, the six compounds represented by the molecular formula will all be isomers, nevertheless.

You can see from the preceding examples that carbon is capable of sharing an electron-pair not only with hydrogen and other elements, but also with another carbon atom. You also have seen that it is capable not only of forming "straight" and branched chain structures, but cyclic ones as well. The tetrahedral nature of the carbon atom and its ability to bond to other carbons makes possible a great number of organic compounds by the combination of a relatively few different atomic species. If the number of carbon atoms is increased from four to ten as in the compound $C_{10}H_{22}$, we could draw *seventy-five* different structures for this molecular formula. It has been calculated that the molecular formula of tetracontane, $C_{40}H_{82}$, would permit the formation of 62,491,178,805,831 different structures! Of course, this number of compounds with the molecular formula $C_{40}H_{82}$ are not all known, and it is doubtful that more than a few even exist. The point is exaggerated simply to show that the structural formula, or graphic representation, of a compound is necessary to show the geometric arrangement of the atoms.

The determination of a structure for an unknown organic compound may, at first thought, appear to be a formidable task. Quite often it is. However, in recent years the availability of sophisticated instruments for measuring certain physical properties has made it much easier for the organic chemist to classify a compound and to determine its structure. In addition he also obtains from these physical measurements some idea as to its chemical properties, and a clue to its synthesis, as well.

Among the most useful instrumental methods employed by present-day organic chemists in the determination of structure are several methods based on absorption spectroscopy. These will be described in Chapter 6.

1.9 *Functional Groups*

Chemical reactions involving organic compounds center, for the most part, about some unique structural feature known as a **functional group.** This group not only bestows a characteristic behavior upon the molecule, but also identifies it as belonging to a certain family of compounds or **homologous series.** For example, methyl alcohol, or methanol (used extensively at one time as radiator antifreeze), has the structure

$$
\begin{array}{c}
\text{H} \\
| \\
\text{H}-\text{C}-\text{O} \\
| \qquad \diagdown \\
\text{H} \qquad \text{H}
\end{array}
$$

The next higher member, or **homolog,** in this series is ethyl alcohol, the alcohol of beverages, which has the structure

$$
\begin{array}{c}
\text{H} \quad \text{H} \\
| \quad\ | \\
\text{H}-\text{C}-\text{C}-\text{O} \\
| \quad\ | \quad \diagdown \\
\text{H} \quad \text{H} \quad \text{H}
\end{array}
$$

It is clear from their structures that a feature common to both alcohols is the **hydroxyl group,** —OH, which is the functional group that characterizes the alcohol family. Simply by increasing the length of the hydrocarbon chain each time by an increment of $-CH_2-$ we can write formulas for other members of the series. The size and geometry of the unreactive portion of the structure may modify, but usually does not alter, the reactions characteristic of the functional group. This greatly simplifies the study of organic chemistry. In the following chapters you will learn how to build into a structure the desired functional group, how to replace it by another, and in some cases, how to eliminate it entirely. Table 1.3 lists the principal families, or classes, of compounds encountered in organic chemistry, the corresponding functional groups, and a common example within each class.

TABLE 1.3 *Classes of Compounds and Their Functional Groups*

Class Name	Functional Group	General Formula*	Common Example
Alcohols	—O—H	R—O—H	Ethyl alcohol
Ethers	C—O—C	R—O—R	Diethyl ether
Aldehydes			Acetaldehyde
Ketones	C=O		Acetone
Carboxylic Acids			Acetic acid
Amines			Ethylamine

*The R used in a general formula refers to that portion of the molecule other than the functional group. In most cases $R = C_n H_{2n+1}$. R is called the alkyl group.

Example: methyl   ethyl

EXERCISE 1.7

The molecular formula of a compound is $C_4H_{10}O$. See if you can incorporate these fifteen atoms into seven different structures. What classes of compounds are represented by these structures?

EXERCISE 1.8

Write Lewis structures for each of the classes of organic compounds shown in Table 1.3. Start with the structures in the General Formula column and add electrons (using one dot for each electron) until you have accounted for all of the valence electrons.

1.10 Reaction Mechanisms. Reaction Intermediates. Transition States.

Although ionic compounds may undergo reactions by dissociation into component ions and recombination to yield products, as stated in Sec. 1.3, reactions of organic molecules generally take place when the molecules collide with sufficient kinetic energy to break, rearrange, or form bonds. Although we will often write a single, simple equation for an organic reaction, in reality the reaction may be quite complex, requiring several steps as the various bonds are broken or new bonds are formed. For example, the reaction that we might write as

$$A + B \longrightarrow C + D$$

could involve the following typical two-step sequence:

(1) $A + B \longrightarrow$ Transition State I $\longrightarrow$ Reaction Intermediate
(2) Reaction Intermediate $\longrightarrow$ Transition State II $\longrightarrow$ C + D

For such a sequence we can draw a **reaction profile** or **reaction coordinate diagram** of the type shown in Fig. 1.9. The change in the potential energy of the system is plotted (along the ordinate) against the progress of the reaction (along the abcissa) or the **reaction coordinate.** As the reaction begins and the reactants, A and B, come together, bonds *begin* to break or form; and the energy increases to a first maximum at **transition state** I. The energy then falls as the initial breakage and formation of bonds is completed to give a **reaction intermediate.** The energy then rises again as further bond breakage and formation begins to a new maximum at transition state II. Finally, the energy falls to the level of that of the products, C and D, as all bond breakage and formation is completed.

In these equations and drawings the transition states (or activated complexes) are transient species of high energy with definite geometries (but partially broken or formed bonds). Usually we will not be able to specify their

FIGURE 1.9   *Reaction Coordinate Diagram*

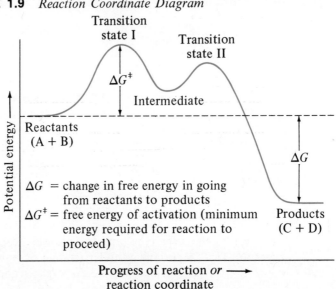

exact structures, except by inference; nor will we be able to make physical measurements on them. Reaction intermediates are found at energy minima (between transition states) and have definite structures (fully formed bonds). Sometimes they can be isolated, and, even if not, they can often be observed by physical methods (usually by spectroscopy (Chapter 6)) under carefully chosen reaction conditions. Not all reaction profiles will look exactly like Fig. 1.9; they may be simpler or more complex, but the general features will be similar.

The detailed description of a reaction in which all transition states and reaction intermediates are specified is called a **reaction mechanism.** In other words, the equation for a reaction tells us *what* takes place, and the reaction mechanism tells us *how* it takes place.

Among the reaction intermediates involved in organic reactions are four species, often called reactive intermediates because they are usually difficult to isolate. These are the **carbonium ion** (or carbocation), **carbanion, free radical, and carbene.** The relationship between these species is illustrated in the following schematic equation:

Each of these reactive intermediates is a high energy species, having only a transitory existence, although some related substances of higher molecular weight may be of greater stability.

Both methyl radical and methyl cation are electron-deficient in that they lack a full octet of valence electrons, although only the cation carries a positive charge. The state of hybridization in methyl cation is sp^2 (see Sec. 1.4 and Fig. 3.1), precisely what is needed to permit maximum separation of the electrons in the three bonds. Therefore, methyl cation (like most other carbonium ions) is planar. The state of hybridization in methyl radical is less certain, but we can assume that it is approximately sp^2 with the *odd* (unpaired) electron in a $2p$ orbital, and that the radical is approximately planar. The negatively charged methyl anion is approximately sp^3-hybridized and has a pyramidal geometry (similar to that of ammonia, NH_3:) with the unshared pair of electrons in an sp^3-like orbital. The state of hybridization of carbene is more complex and will not be discussed in this text. Although these reactive intermediates are short-lived and cannot be isolated, physical measurements have been made on carbonium ions, carbanions, and free radicals. Reactive intermediates of these types are very important in understanding organic reactions and will be encountered frequently in your study of organic chemistry.

| Methyl cation[7] | Methyl radical[7] | Methyl anion[7] |

EXERCISE 1.9

Assume that carbene can exist in two forms, one of which has sp^2 hybridization (Fig. 3.1) and the other, sp hybridization (Sec. 3.15). Draw structures similar to those given for methyl cation, methyl radical, and methyl anion which will show the bonds and the two unshared (not necessarily paired) electrons in the two forms of carbene. If correctly done, your structures will probably not be far from those being used for carbene at present.

---

[7] In an effort to show three dimensions in two, chemists resort to the following conventions: a solid line represents a bond in the plane of the paper; a dotted line represents a bond to an atom behind the plane of the paper; and a wedge represents a bond to an atom in front of the plane of the paper.

1.11 *Acids and Bases. Electrophiles and Nucleophiles.*

Examination of the presently accepted mechanisms of organic reactions shows that many of them are best described as acid-base reactions. Since acids and bases can be defined in several ways, we will review the definitions of interest to us.

As defined by Brønsted and Lowry, an acid is a substance that supplies protons (to a base), and a base is a substance that accepts protons (from an acid). Examples of Brønsted-Lowry acids that are familiar to you include the strong acids, hydrochloric acid (HCl) and sulfuric acid (H_2SO_4), as well as the weak acid, acetic acid (CH_3—CO_2H). Typical Brønsted-Lowry bases are the strong base, hydroxide ion (OH^-) and the weak base, ammonia (NH_3). The strengths of Brønsted-Lowry acids and bases are measured by the position of the equilibrium

$$HA + B: \overset{K}{\rightleftarrows} A:^- + BH^+$$

acid base conjugate conjugate
base of HA acid of B:

where the conjugate acid and conjugate base are defined as the products resulting from the proton exchange, and K is the equilibrium constant. When we compare the strengths of organic acids and bases, we will be using the Brønsted-Lowry definition.

The most general and most useful definition of acids and bases is that of Lewis, which is based on the sharing of an electron pair between acid and base. A Lewis acid is a species that accepts an electron pair (from a Lewis base), and a Lewis base is a species that donates an electron pair (to a Lewis acid). Typical Lewis acids are the proton (H^+), silver ion (Ag^+), boron trifluoride (BF_3), aluminum chloride ($AlCl_3$), and carbonium ions (such as CH_3^+)—that is, positive ions and species containing atoms with unfilled octets or electron-deficient atoms. Lewis bases are essentially the same as Brønsted-Lowry bases, and typical Lewis bases are hydroxide ion (OH^-), ammonia ($:NH_3$), chloride ion (Cl^-), water (H_2O), and carbanions (such as $CH_3:^-$)—that is, negative ions, species having unshared electron pairs, or, in some cases, species with loosely bound shared pairs. A typical Lewis acid-base reaction is the formation of the boron trifluoride-ammonia complex (p. 9).

In organic chemistry reagents that are electron-seeking (Lewis acids) and can form a covalent bond by accepting an electron pair *from a carbon atom* are called **electrophiles** (electron-loving). Reagents that are electron-donating (Lewis bases) and can form a covalent bond to carbon by donating a pair of electrons *to a carbon atom* are called **nucleophiles** (nucleus-loving). It is important not to confuse the terms basicity and nucleophilicity. As stated above, basicity is measured by the position of an *equilibrium* and is the affinity of a base for a proton. **Nucleophilicity** is the affinity of a base for a carbon atom and is measured by the *rate* of a reaction in which the nucleophile forms a bond to a carbon atom in a standard substrate.[8]

---

[8] The term *substrate* refers to the organic compound undergoing a reaction with a given reagent.

EXERCISE 1.10

Classify the following species as Lewis acids or Lewis bases:

(1) $:\ddot{\underset{..}{F}}:^-$, (2) $CH_3:\ddot{\underset{..}{O}}:CH_3$, (3) $:\ddot{\underset{..}{Br}}^+$, (4) $H:\underset{H}{\underset{..}{C}}::\underset{H}{\underset{..}{C}}:H$, (5) $MgBr_2$.

1.12 Resonance Structures

There are many molecules, ions, and radicals whose physical and chemical properties are not satisfactorily represented by a single Lewis structure. A familiar example is nitrate ion, NO_3^-.

I II III IV

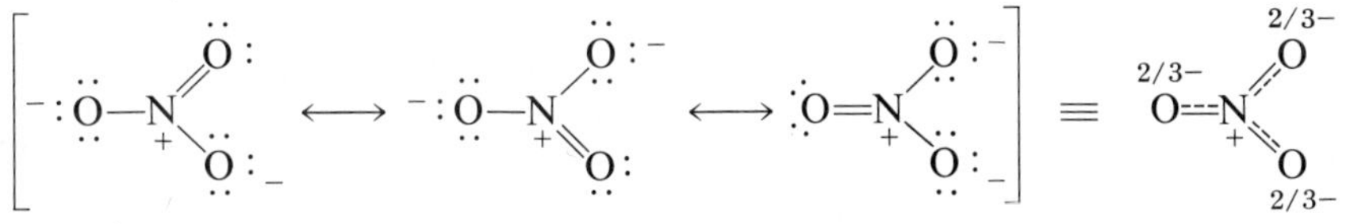

As shown, we can draw three equivalent Lewis structures with two negatively charged, singly-bonded oxygen atoms and a third doubly-bonded oxygen. In the actual ion the negative charges are distributed equally over all oxygen atoms; that is, there is a unit positive charge on nitrogen and two-thirds of a negative charge on each oxygen. This fractional charge distribution cannot be shown by a single Lewis structure. We can, and do, draw structures such as IV, which approximates the actual structure; however, structures such as IV are not as useful as Lewis structures in most chemical applications. In general, whenever two or more Lewis structures can be written for a molecule, ion, or radical, which differ only in the distribution of electrons, the structure of that species *may* not be satisfactorily represented by any one of the Lewis structures, but may be better described as a **resonance hybrid** of them all. In such situations all of the reasonable Lewis structures (called **contributing forms** or canonical forms) are written as shown, connected with a *double-headed* arrow and enclosed in a bracket. The meaning of this symbolism [↔] is that the actual structure is a hybrid, or some blend or mix, of all the structures shown. It is *not* to be confused with an equilibrium (⇄) between several independently existing species. None of the species, I, II, or III exists—what exists is a species intermediate between them all.

The reason that the resonance hybrid exists (rather than an equilibrium mixture of I, II, and III) is that the hybrid is of lower energy and more stable than any of the contributing forms. A factor that contributes to the increased stability of many hybrids is the dispersal of electron surpluses or deficiencies over as many atoms as are able to accept them. In the hybrid nitrate ion, the net negative charge is distributed over all three electronegative oxygen atoms, rather than only two.

Resonance structures will be used freely and frequently in this text; therefore, a few rules will be given to guide you in drawing such structures.

1. The positions of the *nuclei* of the atoms must be the same in all contributing forms (resonance structures); the only changes are in electron assignments.

2. All resonance structures must have the same number of *paired* electrons.

3. In deciding which of two (or more) resonance structures is the more important (more stable, lower energy) the following guidelines should be applied *in the order given* ((a) is more important than (b), etc.).
    (a) Structures with filled octets in first row atoms are more important than those with unfilled octets. This is true even if it causes a positive charge to appear on an electronegative atom.
    (b) Structures of neutral molecules with no separation or minimum separation of charge are more important than those with considerable charge separation. Charge separation is more favorable when negative charges are on electronegative elements, and positive charges are on electropositive elements.
    (c) A structure with a greater number of bonds is usually more important than one with a lesser number.

4. Resonance stabilization of a species is greatest when there are two (or more) *equivalent* structures which can be said to be of greatest importance (that is, of lowest energy). Conversely, a single Lewis structure is least likely to be satisfactory when two or more equivalent structures of lowest energy can be written.

5. If there is only one contributing form of low energy, to a first approximation, the resonance hybrid may be expected to have the properties expected of that form, and we will say that resonance is not important.

The application of these rules may be illustrated with the following examples of species you will meet later in the text: (1) ethene, (2) formaldehyde, (3) protonated formaldehyde, (4) allyl cation.

	More important	Less important	Comment
(1)	$\underset{H}{\overset{H}{\diagdown}}C{=}C\underset{H}{\overset{H}{\diagup}}$	$\underset{H}{\overset{H}{\diagdown}}{}^{+}C{-}C^{-}\underset{H}{\overset{H}{\diagup}}$	Charge separation on atoms of same electronegativity. Resonance not important.
(2)	$\underset{H}{\overset{H}{\diagdown}}C{=}\ddot{O}:$	$\underset{H}{\overset{H}{\diagdown}}{}^{+}C{-}\ddot{O}:^{-}$	Charge separation on atoms of unlike electronegativity. Rule 3(a), (b), (c) favor form on left, but form on right is reasonable. Resonance important.

	More important	Less important	Comment
(3)			No charge separation. Rule 3(a), (c) favor form on left, but form on right is reasonable. Resonance important.
(4)			No charge separation. Two forms have same structure, are equivalent. Resonance is very important.

Note the use of the small, curved arrows. These curved arrows are employed in two ways: as sort of a bookkeeping device to keep track of electrons as we construct resonance structures and as a means of showing the general flow of electrons in some reaction mechanisms.

To save time and conserve space we will not write all of the resonance forms of a species unless it is necessary to do so. Instead we will use the most important resonance structure to represent the structure of the species.

Summary

1. Carbon, which has four electrons available for bonding, has a great tendency to bond to itself in straight chain, branched chain, and cyclic structures.

2. The covalent bond involves a pair of shared electrons between two atoms. This type of bond predominates in organic compounds.

3. When carbon is bonded to four other atoms or groups, it does so through four sp^3 hybridized orbitals. The bonds so formed are directed to the vertices of a tetrahedron. (If the bonded groups are identical, it will be a regular tetrahedron.)

4. Empirical and molecular formulas are determined by qualitative tests and quantitative measurements.

5. Formulas of organic compounds reveal little unless written as structures.

6. Compounds having the same molecular formula, but different structures, are isomers.

7. Organic reactions nearly always involve some structural feature known as a functional group.

8. A reaction mechanism is a detailed description of a reaction which includes all transition states and intermediates.

9. Carbonium ion, carbanions, and free radicals are important intermediates in organic reactions.

10. Electrophiles are reagents that can form a covalent bond by accepting an electron pair from a carbon atom. Nucleophiles are reagents that can form a covalent bond by donating a pair of electrons to a carbon atom.

11. Whenever two or more Lewis structures may be written for a molecule, ion, or radical, the actual structure may be a hybrid of all contributing structures.

New Terms

atomic orbitals	isomers
carbanion	molecular formula
carbonium ion	molecular orbital
covalence	nucleophile
electrophile	polar bond
electronegativity	reaction intermediate
empirical formula	reaction mechanism
free radical	resonance hybrid
functional group	sigma bond
homologous series	tetrahedral angle
hybridized orbitals	transition state

Supplementary Exercises

EXERCISE 1.11 Using dots, asterisks or any other symbolism, write Lewis structures for the following.

(a) H_2S
(b) HCN
(c) ethane, C_2H_6
(d) methyl alcohol
(e) formaldehyde, CH_2O

(f) chloroform, $CHCl_3$
(g) chloromethane, CH_3Cl
(h) dimethyl ether, C_2H_6O
(i) BF_3
(j) methylamine, CH_5N

EXERCISE 1.12 Contrast the physical properties of common salt with those of naphthalene (moth flakes), $C_{10}H_8$, with specific reference to: (a) solubility in water; (b) solubility in gasoline; (c) formation of electrolytic solutions; (d) flammability; (e) melting points; (f) vapor pressure. Which of the above properties makes naphthalene a good larvacide?

EXERCISE 1.13 What unique characteristics of carbon make possible the formation of the great number of organic compounds?

EXERCISE 1.14 The number in parentheses accompanying the following is the number of isomeric structures permitted for that molecular formula. Can you draw them?

    (a) C_3H_8 (1)   (e) C_5H_{12} (3)

    (b) C_3H_7Cl (2)   (f) C_3H_6O (6)

    (c) C_3H_8O (3)   (g) C_2H_3N (2)

    (d) $C_2H_4Cl_2$ (2)   (h) C_3H_7N (6)

EXERCISE 1.15 In the structure CCl_4 what is the maximum number of chlorine atoms which may lie in the same plane? (*Hint: See Fig. 1.6.*)

EXERCISE 1.16 In which of the following substances does the molecule contain polar bonds? Which ones are polar molecules?

    (a) CH_4 (f) C_2H_6

    (b) CH_3Br (g) CCl_2F_2

    (c) CCl_4 (h) BF_3

    (d) HI   (i) CH_3OH

    (e) CO_2 (j) diethyl ether, $C_2H_5OC_2H_5$

EXERCISE 1.17 Calculate the percentage composition of each element in the following.

    (a) CH_4O (d) CH_4N_2O

    (b) CH_5N (e) $C_{12}H_{22}O_{11}$

    (c) CH_3I

EXERCISE 1.18 Combustion of a 0.060 g sample of an organic substance containing C, H, N, and O yielded 0.044 g of CO_2 and 0.036 g of H_2O. In a separate determination a sample of the same size yielded 22.4 ml of nitrogen gas when corrected to STP. What is the empirical formula of the compound?

EXERCISE 1.19 The organic dye, indigo, gave an analysis of 73.35% carbon, 3.8% hydrogen, and 10.7% nitrogen. Molecular weight determinations gave values in the 250–275 range, and qualitative tests showed that no other elements, except oxygen, were present. Calculate the molecular formula for indigo.

EXERCISE 1.20 For a series of molecules in which there are two atomic nuclei *lying along the x-axis,* draw sketches illustrating the overlapping of the following pairs of orbitals on the two nuclei. Use the simplified symbols: circles for *s* orbitals, figure-eight shapes for *p* orbitals, and tear-drop shapes for hybrid orbitals. Ignore the other orbitals on the atoms, if any.

    (a) hydrogen 1*s* and fluorine $2p_x$ (c) carbon $2p_y$ and carbon $2p_y$

    (b) carbon $2p_x$ and carbon $2p_x$ (d) carbon sp^3 and carbon sp^3

EXERCISE 1.21 Label the following pairs of structures as isomers or resonance forms.

(a) 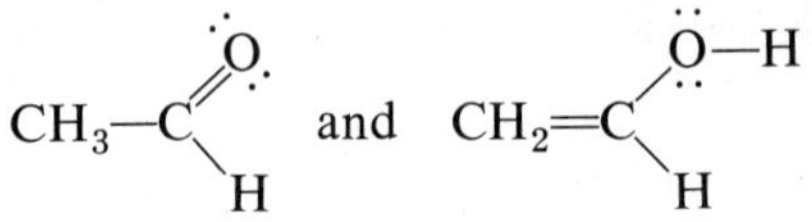 $CH_3-\overset{\overset{\displaystyle ..\,O\,..}{\|}}{C}\diagdown_H$ and $CH_2=\overset{\overset{\displaystyle ..\,O-H}{}}{C}\diagdown_H$

(b) $CH_3-\overset{\overset{\displaystyle :O:}{\|}}{C}-CH_3$ and $CH_3-\overset{\overset{\displaystyle :\ddot{O}:^-}{|}}{\underset{+}{C}}-CH_3$

(c) $CH_3-\overset{\overset{\displaystyle }{}}{\underset{\underset{\displaystyle :\ddot{O}:_-}{|}}{C}}=\ddot{O}$ and $CH_3-\overset{\overset{\displaystyle }{}}{\underset{\underset{\displaystyle :O:}{\|}}{C}}-\ddot{O}:^-$

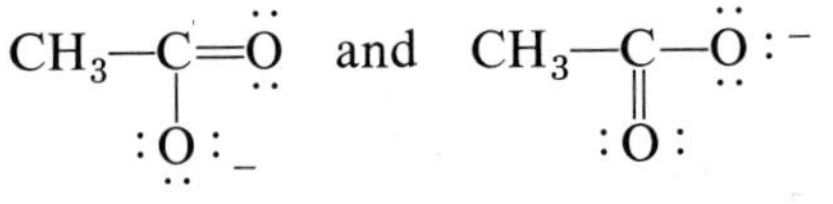

(d) 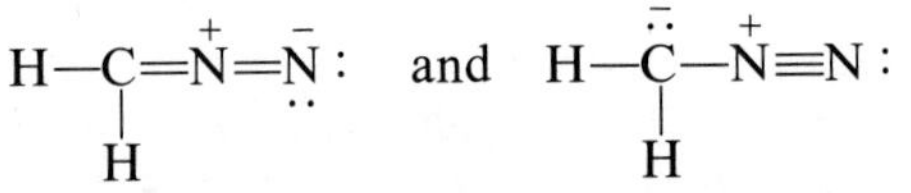 $H-\overset{\overset{\displaystyle }{}}{\underset{\underset{\displaystyle H}{|}}{C}}=\overset{+}{N}=\overset{-}{\ddot{N}}:$ and $H-\overset{\overset{\displaystyle \cdot\cdot\,-}{}}{\underset{\underset{\displaystyle H}{|}}{C}}-\overset{+}{N}\equiv N:$

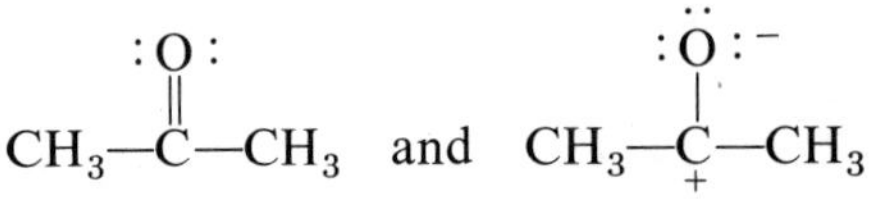

(e)

$$\begin{array}{c}H\\ |\\ C\end{array}$$

and

benzene structures

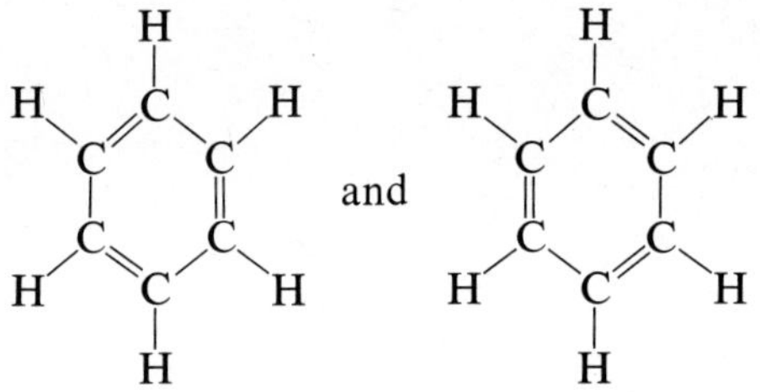

(f) $CH_2=C=CH_2$ and $CH_3-C\equiv C-H$

(g) $CH_3-C\equiv N:$ and $CH_3-\overset{+}{N}\equiv\overset{-}{C}:$

(h) 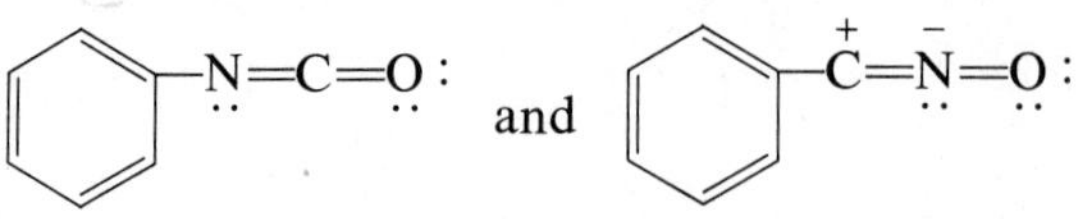 (phenyl)$-N=C=O:$ and   (phenyl)$-\overset{+}{C}=\overset{-}{N}=O:$

EXERCISE 1.22 Calculate the formal charge on each atom in the following substances.

(a) 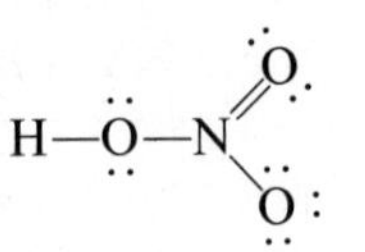 $H-\ddot{O}-N\diagup^{\displaystyle ..\,O\,..}_{\diagdown\, \ddot{O}:}$

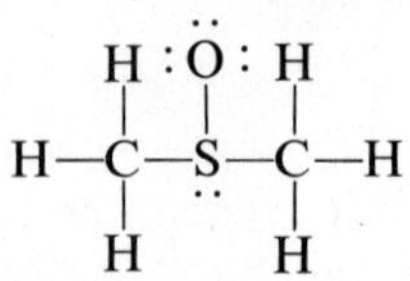

(b) 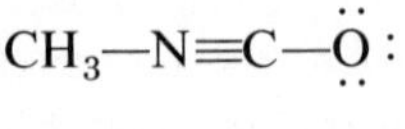 $CH_3-N\equiv C-\ddot{O}:$

(c) $H-\overset{\overset{\displaystyle H}{|}}{\underset{\underset{\displaystyle H}{|}}{C}}-\overset{\overset{\displaystyle :\ddot{O}:}{|}}{\underset{\underset{\displaystyle }{|}}{S}}-\overset{\overset{\displaystyle H}{|}}{\underset{\underset{\displaystyle H}{|}}{C}}-H$

(d) 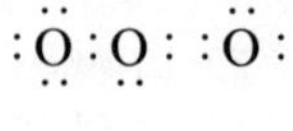 $:\ddot{O}:\overset{..}{\underset{..}{O}}::\ddot{O}:$

2

The Alkanes, or Saturated Hydrocarbons

Introduction

The hydrocarbons, as the name reveals, are compounds containing only carbon and hydrogen. The various structural forms, which combinations of these two elements permit, provide us with the simplest introduction possible to the study of organic chemistry. These useful compounds are isolated from petroleum, our principal fossil fuel, and therefore comprise a great energy source. However, through reactions whereby the hydrogen atoms are substituted by other groups a large number of other useful compounds result. Thus, the hydrocarbons, by name and structure, have a fundamental relationship to other organic substances.

2.1 *Structure and Formulas of the Alkanes*

The **alkanes,** or **saturated hydrocarbons,** are compounds composed of carbon and hydrogen in which each carbon atom is covalently linked to four other atoms by single electron pair bonds. The alkanes also are frequently referred to as the **paraffins** (L. *parum,* little; *affinis,* affinity) because of their relative inertness.

The molecular formulas for the alkanes are easily obtained from the general formula C_nH_{2n+2}, where *n* is the number of carbon atoms present. Each member of a family, or homologous series, differs from its immediate relatives by a methylene group, $-CH_2$. However, the structures for isomers

TABLE 2.1 *Formulas of the Alkanes*

n	No. of Isomers	Molecular Formula	Projection Formula and Name	Condensed Formula
1	None	CH_4	Methane	CH_4
2	None	C_2H_6	Ethane	CH_3CH_3
3	None	C_3H_8	Propane	$CH_3CH_2CH_3$
4	2	C_4H_{10}	Normal butane	$CH_3(CH_2)_2CH_3$
			Isobutane	$(CH_3)_2CHCH_3$

TABLE 2.1 *Continued*

n	No. of Isomers	Molecular Formula	Projection Formula and Name	Condensed Formula
5	3	C_5H_{12}	H—C—C—C—C—C—H (each C with H above and below) **Normal pentane**	$CH_3(CH_2)_3CH_3$
			Isopentane	$(CH_3)_2CHCH_2CH_3$
			Neopentane	$C(CH_3)_4$

which have the same molecular formula, as we have already noted (Sec. 1.8), are written in different ways. Table 2.1 lists the formulas and structures of the alkanes containing one to five carbon atoms.

2.2 *Nomenclature of the Alkanes*

Names of the first four members of the alkane family have no simple derivation and must be memorized. However, naming the higher members of the series does follow a system, and learning them is relatively easy. The number of carbon atoms present in a continuous chain is indicated by a Greek prefix which is followed by the suffix *-ane:*

Penta = 5; C_5H_{12} is **pentane**
Hexa = 6; C_6H_{14} is **hexane,** etc. as on p. 43

The unbranched, or continuous, chain is called **normal** and must be indicated in the name as a prefix, *n-*, to differentiate it from its branched chain isomers:

n-Pentane   Isopentane, or 2-Methylbutane   Neopentane, or 2,2-Dimethylpropane

In the sections which follow we will learn how to name highly branched structures, but before this can be done we first must learn the names of some of the "branches."

2.3   *Alkyl Groups*

Alkyl groups, as such, have no independent existence but are simply structural derivatives useful in the naming of compounds.

Removal of one of the hydrogen atoms of an alkane produces an alkyl group. The residual group then is named by replacing the *-ane* suffix of the parent hydrocarbon by *-yl:*

$$\underset{\text{Methane}}{H-\overset{\overset{\displaystyle H}{|}}{\underset{\underset{\displaystyle H}{|}}{C}}-H} \quad \text{becomes} \quad \underset{\text{Methyl}}{H-\overset{\overset{\displaystyle H}{|}}{\underset{\underset{\displaystyle H}{|}}{C}}-} \quad \text{or more simply written as } CH_3-$$

$$\underset{\text{Ethane}}{H-\overset{\overset{\displaystyle H}{|}}{\underset{\underset{\displaystyle H}{|}}{C}}-\overset{\overset{\displaystyle H}{|}}{\underset{\underset{\displaystyle H}{|}}{C}}-H} \quad \text{becomes} \quad \underset{\text{Ethyl}}{H-\overset{\overset{\displaystyle H}{|}}{\underset{\underset{\displaystyle H}{|}}{C}}-\overset{\overset{\displaystyle H}{|}}{\underset{\underset{\displaystyle H}{|}}{C}}-} \quad \text{or simply } C_2H_5-$$

For the third member of the series, propane, there is only one structure as a saturated hydrocarbon, but there are two different **propyl groups.** This is possible because the two end carbons are attached to only *one* other carbon while the middle carbon is attached to *two* others.

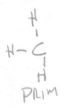

$$H-\overset{\overset{\displaystyle H}{|}}{\underset{\underset{\displaystyle H}{|}}{C}}-\overset{\overset{\displaystyle H}{|}}{\underset{\underset{\displaystyle H}{|}}{C}}-\overset{\overset{\displaystyle H}{|}}{\underset{\underset{\displaystyle H}{|}}{C}}-H$$

Propane

A carbon bonded to only one other carbon is called **primary,** one bonded to two carbons is called **secondary** (*sec*) and to three carbons, **tertiary** (*tert*). Hydrogens bonded to these carbons are designated according to the types of carbons to which they are attached. Thus, a primary carbon has bonded to it *three* primary hydrogens and *one* alkyl group. A secondary carbon has *two* secondary hydrogens and *two* alkyl groups. A tertiary carbon is bonded to only *one* tertiary hydrogen, but to *three* alkyl groups.

The six primary hydrogens shown bounded by solid lines in the formula for propane are all equivalent. Removal of any one of these six hydrogen atoms gives the **normal propyl group**

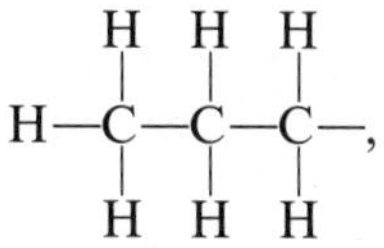

$$\underset{\underset{\text{H}}{|}}{\overset{\overset{\text{H}}{|}}{\text{C}}}-\underset{\underset{\text{H}}{|}}{\overset{\overset{\text{H}}{|}}{\text{C}}}-\underset{\underset{\text{H}}{|}}{\overset{\overset{\text{H}}{|}}{\text{C}}}-, \quad \text{or} \quad CH_3-CH_2-CH_2-, \quad \text{or} \quad n\text{-}C_3H_7-$$

Note that for brevity the normal propyl group can be shown simply as $n\text{-}C_3H_7-$. Removal of one of the secondary hydrogens shown within the broken circles in the formula for propane gives the **isopropyl** group

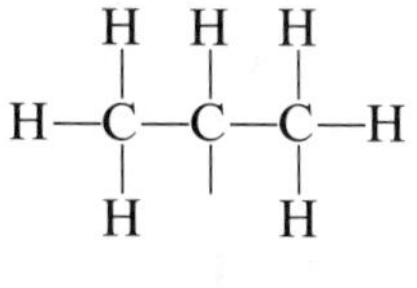

$$\text{H}-\underset{\underset{\text{H}}{|}}{\overset{\overset{\text{H}}{|}}{\text{C}}}-\underset{\underset{\text{H}}{|}}{\overset{\overset{\text{H}}{|}}{\text{C}}}-\underset{\underset{\text{H}}{|}}{\overset{\overset{\text{H}}{|}}{\text{C}}}-\text{H} \quad \text{or} \quad (CH_3)_2CH- \quad \text{or} \quad iso\text{-}C_3H_7-$$

Obviously the general formula for an alkyl group is C_nH_{2n+1}. The number of different alkyl groups that can be produced from any alkane depends upon how many different primary, secondary, or tertiary hydrogen atoms can be replaced. We can illustrate this very easily using the two isomeric butanes.

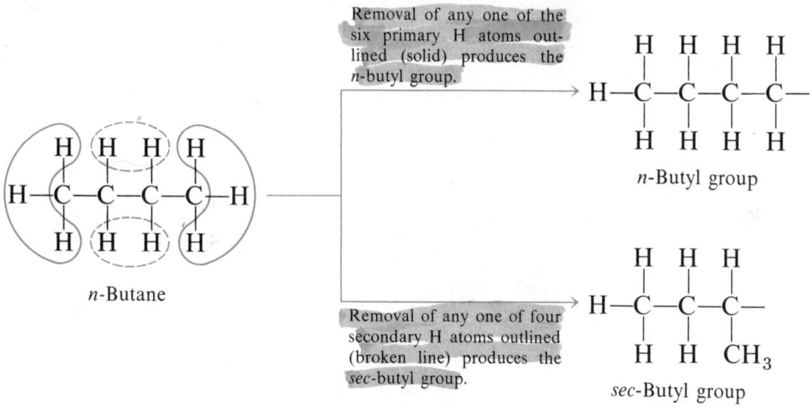

Removal of any one of the six primary H atoms outlined (solid) produces the *n*-butyl group.

n-Butyl group

Removal of any one of four secondary H atoms outlined (broken line) produces the *sec*-butyl group.

sec-Butyl group

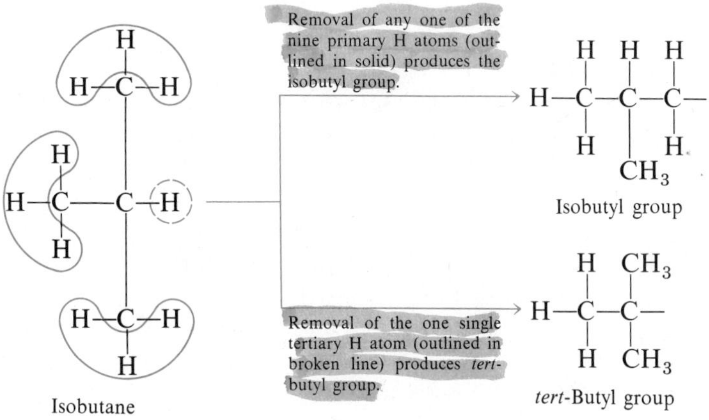

n-Butane

Removal of any one of the nine primary H atoms (outlined in solid) produces the isobutyl group.

Isobutyl group

Removal of the one single tertiary H atom (outlined in broken line) produces *tert*-butyl group.

tert-Butyl group

Isobutane

TABLE 2.2 *Commonly Used Alkyl Groups*

Hydrocarbon	Group	Example of Common Usage

Methane

$$H-\overset{\displaystyle H}{\underset{\displaystyle H}{C}}-H$$

Methyl

$$H-\overset{\displaystyle H}{\underset{\displaystyle H}{C}}-$$

CH_3I
Methyl iodide
(Iodomethane)

Ethane

$$H-\overset{\displaystyle H}{\underset{\displaystyle H}{C}}-\overset{\displaystyle H}{\underset{\displaystyle H}{C}}-H$$

Ethyl

$$H-\overset{\displaystyle H}{\underset{\displaystyle H}{C}}-\overset{\displaystyle H}{\underset{\displaystyle H}{C}}-$$

C_2H_5OH
Ethyl alcohol (Ethanol)

Propane

$$H-\overset{\displaystyle H}{\underset{\displaystyle H}{C}}-\overset{\displaystyle H}{\underset{\displaystyle H}{C}}-\overset{\displaystyle H}{\underset{\displaystyle H}{C}}-H$$

n-Propyl

$$H-\overset{\displaystyle H}{\underset{\displaystyle H}{C}}-\overset{\displaystyle H}{\underset{\displaystyle H}{C}}-\overset{\displaystyle H}{\underset{\displaystyle H}{C}}-$$

$CH_3CH_2CH_2Br$
n-Propyl bromide
(1-Bromopropane)

Isopropyl

$$CH_3-\overset{\displaystyle H}{\underset{\displaystyle \,}{C}}-CH_3$$

$$CH_3-\overset{\displaystyle H}{\underset{\displaystyle OH}{C}}-CH_3$$

Isopropyl alcohol
(2-Propanol)

---

n-Butane

$$H-\overset{\displaystyle H}{\underset{\displaystyle H}{C}}-\overset{\displaystyle H}{\underset{\displaystyle H}{C}}-\overset{\displaystyle H}{\underset{\displaystyle H}{C}}-\overset{\displaystyle H}{\underset{\displaystyle H}{C}}-H$$

n-Butyl

$CH_3CH_2CH_2CH_2-$

$CH_3CH_2CH_2CH_2-Cl$
n-Butyl chloride
(1-Chlorobutane)

---

sec-Butyl

$$CH_3-\overset{\displaystyle H}{\underset{\displaystyle H}{C}}-\overset{\displaystyle H}{\underset{\displaystyle \,}{C}}-CH_3$$

$$CH_3-\overset{\displaystyle H}{\underset{\displaystyle H}{C}}-\overset{\displaystyle H}{\underset{\displaystyle CH_3}{C}}-Br$$

sec-Butyl bromide
(2-Bromobutane)

Isobutane

$$H-\overset{\displaystyle H}{\underset{\displaystyle H}{C}}-\overset{\displaystyle H}{\underset{\displaystyle CH_3}{C}}-\overset{\displaystyle H}{\underset{\displaystyle H}{C}}-H$$

Isobutyl

$$CH_3-\overset{\displaystyle H}{\underset{\displaystyle CH_3}{C}}-CH_2-$$

$$CH_3-\overset{\displaystyle H}{\underset{\displaystyle CH_3}{C}}-CH_2-Br$$

Isobutyl bromide
(1-Bromo-2-methylpropane)

---

tert-Butyl

$$CH_3-\overset{\displaystyle CH_3}{\underset{\displaystyle CH_3}{C}}-$$

$$CH_3-\overset{\displaystyle CH_3}{\underset{\displaystyle CH_3}{C}}-Br$$

tert-Butyl bromide
(2-Bromo-2-methylpropane)

An *iso* compound is one that has a methyl substituent on the next to last carbon atom. The isopropyl group is the only example of an *iso* group which at the same time is also a secondary group.

Branched-chain hydrocarbons which contain six or more carbon atoms have so many isomeric structures that to attempt to characterize them by simple prefixes such as iso, *sec, tert,* and the like is hopeless. It soon became apparent to chemists throughout the world that with the growing number of compounds some better system of naming was necessary.

In order to cope with the problem of standardizing nomenclature an international committee of chemists met at Geneva, Switzerland in 1892 to develop and adopt a systematic method for the naming of organic compounds. A commission from the International Union of Chemists (IUC) met in Liege, Belgium in 1930 and again as the International Union of Pure and Applied Chemistry (IUPAC) at Amsterdam in 1949. Rules originally adopted and subsequently modified by these groups of chemists have given us an orderly system of nomenclature known first as the **Geneva system,** or the **IUC system,** but more recently as the **IUPAC system.** With the increasing use of the computer in preparing indices and searching the chemical literature, some time-honored rules have been changed recently. The use of common names is being increasingly discouraged although they are still used for relatively simple structures and must be learned. However, the name of any compound, and most certainly that of a complex structure, is unmistakable if the IUPAC system is followed.

2.4 *Rules for Naming the Alkanes*

A. Select the *longest continuous chain* of carbon atoms. If there are two chains of equal length, take the one with the larger number of branches. This is the **parent chain.**

B. Count the number of carbon atoms in the parent chain and give the chain the name of the normal alkane (called the **parent hydrocarbon**) having that number of carbon atoms. The compound will be named as a derivative of the parent hydrocarbon. (Note: The prefix *n-* is not used in this system.)

C. Number the parent chain from one end to the other so that the **substituents** (side chains or groups) can be located by number (see Step E!).

D. Arrange all of the substituent group names in alphabetical order and place them in front of the parent hydrocarbon name, indicating the *number* of each type of group present by prefixes and the location (position) of each group by an Arabic numeral.

Prefixes:	di = 2	hexa = 6	deca = 10
	tri = 3	hepta = 7	undeca = 11
	tetra = 4	octa = 8	dodeca = 12
	penta = 5	nona = 9	

Ignore the above prefixes and the prefixes, *sec-* and *tert-*, in establishing the alphabetical order (i.e., ethyl before dimethyl). The unitalicized prefixes, cyclo-, iso-, and neo-, are considered part of the substituent name and are included in establishing alphabetical order. If two groups are attached to the same carbon atom of the parent chain, repeat the Arabic numeral.

E. The substituents must be given the lowest numbers possible. In comparing two numbering schemes, use the scheme which has the lower *first* number for a substituent. If the first numbers are the same in both schemes, use the lower second number, etc. If the numbers are the same in both schemes, use the scheme that gives the lower first number to the substituent that will appear first in the name (lower in the alphabet).

F. Each homologous series of compounds has a characteristic *suffix*. The suffix for the alkanes is *-ane*.

G. Write the name as one word, separating Arabic numerals by commas and numerals from alkyl or other group names by hyphens.

Let us apply these rules of nomenclature to a few examples:

$$\begin{array}{c} CH_3 \\ | \\ CH_3-C-CH_3 \\ | \\ CH_3 \end{array} \qquad \begin{array}{c} CH_3 \\ | \\ CH_3-C-CH_2-CH_3 \\ | \\ H \end{array}$$

(1)   (2)   (3)   (4)

2,2-Dimethylpropane   2-Methylbutane
(Neopentane)   (Isopentane)

$$\begin{array}{c} CH_3 \qquad CH_3 \\ | \qquad\quad | \\ CH_3-C-CH_2-C-CH_3 \\ | \qquad\quad | \\ H \qquad CH_3 \end{array}$$

(5)   (4)   (3)   (2)   (1)

2,2,4-Trimethylpentane
(*Not* 2,4,4-Trimethylpentane)

$$\begin{array}{c} CH_3 \quad H \qquad\qquad\qquad\qquad\qquad CH_3 \\ | \qquad | \qquad\qquad\qquad\qquad\qquad | \\ CH_3-CH_2-C-\!-C-CH_2-CH_2-CH_2-CH_2-C-CH_3 \\ | \qquad | \qquad\qquad\qquad\qquad\qquad | \\ H \quad CH_3 \qquad\qquad\qquad\qquad\qquad H \end{array}$$

2,7,8-Trimethyldecane
(*Not* 3,4,9-Trimethyldecane)[1]

---

[1] The older rule required that the sum of the numbers be as small as possible. Many names based on this rule are in the chemical literature. The new rules (Rule E) require that the lower *first* number (2 rather than 3) be used. To determine the lower first number, arrange all of the numbers (regardless of where they come *in the name*) in a sequence and compare the sequences (1,1,3 rather than 1,2,2; 1,1,3,4 rather than 1,2,2,3).

EXERCISE 2.1

Five different structures may be drawn for the molecular formula C_6H_{14}. One isomer has a continuous chain of six carbons, two have continuous chains of five carbons, and two have continuous chains of four carbons. Draw structures for all five compounds and name them according to IUPAC rules.

Examples of substituents other than (or in addition to) alkyl groups:

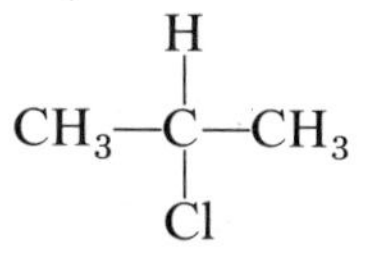

$$
\begin{array}{ccc}
\text{H} & \text{CH}_3 & \text{H} \\
| & | & | \\
\text{CH}_3\!-\!\text{C}\!-\!\text{CH}_3 & \text{CH}_3\!-\!\text{C}\!-\!\text{CH}_2\!-\!\text{Cl} & \text{CH}_3\!-\!\text{C}\!-\!\text{CH}_2\!-\!\text{Cl} \\
| & | & | \\
\text{Cl} & \text{CH}_3 & \text{Cl}
\end{array}
$$

2-Chloropropane   1-Chloro-2,2-   1,2-Dichloropropane
(Isopropyl chloride)   dimethylpropane   (Propylene chloride)
(Neopentyl chloride)

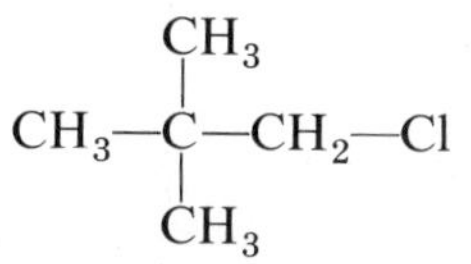

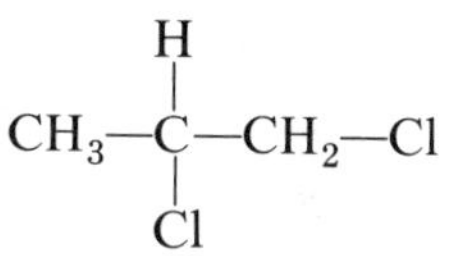

EXERCISE 2.2

A student named the structure below incorrectly as 1-chloro-2-ethyl-2-methylpropane.

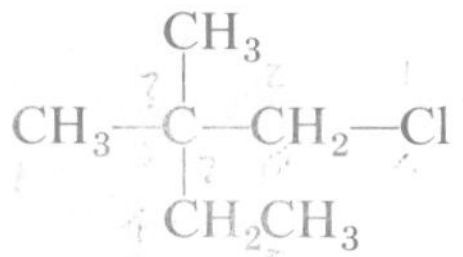

$$
\begin{array}{c}
\text{CH}_3 \\
| \\
\text{CH}_3\!-\!\text{C}\!-\!\text{CH}_2\!-\!\text{Cl} \\
| \\
\text{CH}_2\text{CH}_3
\end{array}
$$

Tell why this name is objectionable and assign a correct name to the compound.

2.5 *Conformational Isomers*

The single bond between carbon atoms in a molecule permits rotation of these atoms as if the shared pair of electrons were a pivot point between two tetrahedra (Fig. 2.1).

Frequently, this ability to revolve around the line joining two carbons is called **free rotation.** How free or easy it is depends upon the nature of the atoms or groups occupying the other three corners of each tetrahedron. If these groups are large or bulky, free rotation not only may be inhibited, but prevented entirely. Again, if groups on adjacent carbons are of high electron density such as chlorine or bromine, they will tend to repel each other and therefore will occupy positions as far from each other as possible. The different spatial arrangements made possible by rotation about a single bond are called **conformations** and a small energy barrier must be overcome in order for adjacent carbon atoms to rotate from one conformation to another. The conformational

FIGURE 2.1   *Ball and Stick Model of Ethane Showing Arrangement of Atoms in (a) Staggered and (b) Eclipsed Conformations*

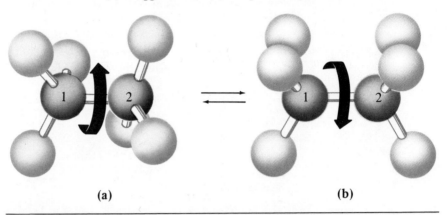

(a)   (b)

isomers possible for ethane and the potential energy changes required when carbon atom 2 is rotated through successive 60° angles may be shown more clearly using the representations first devised by Professor M. S. Newman of Ohio State University. In the **Newman projections** of the ethane molecule (Fig. 2.2), the viewer is looking along the axis connecting the two carbon atoms. The conformational isomers or **conformers** of 1,2-dichloroethane are shown in Figure 2.3. Note the three conformations shown and the **dihedral angles** between the chlorine atoms in each conformation: (a) **anti** (180°), (b) **gauche,** or skew (60°), and (c) **eclipsed** (0°).

θ = dihedral angle

It is important that we distinguish between such structural arrangements made possible by a rotation about bonds and those made possible by breaking and rearranging bonds. Regardless of the number of different conformations a molecule may assume, it still retains the same identity and all its physical and chemical properties. This is not always the case when bonds are broken and rearranged. For example, were we to break the bonds connecting the chlorine and the hydrogen atoms to the number 2 carbon of 2-chlorobutane and exchange the positions of the chlorine and the hydrogen, the result would be a 2-chlorobutane but, as we shall see in Chapter 5, a different 2-chlorobutane than the one we altered.

FIGURE 2.2 *Potential Energy Changes Required to Rotate Carbon Atoms of Ethane Through Successive 60° Angles*

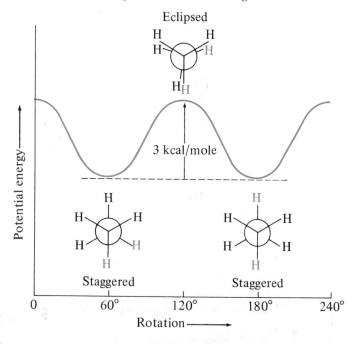

FIGURE 2.3 *Scale Models Showing (a) Anti, (b) Gauche, and (c) Eclipsed Conformations of 1,2-Dichloroethane*

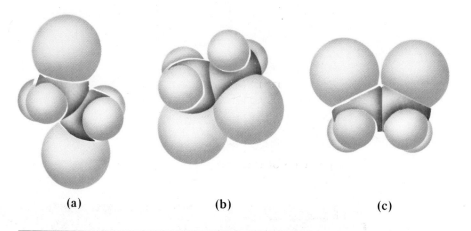

(a)   (b)   (c)

FIGURE 2.4   *Ball and Stick Models of* n-*Pentane**

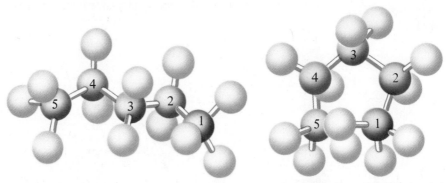

*The model at right shows the "head-to-tail" arrangement possible because of singly-bonded carbons.

Free rotation about singly-bonded carbons permits hydrocarbons in a continuous chain to be anything but "straight." They may zig-zag, turn corners, spiral, and even cyclize in a head-to-tail arrangement. Note the near cyclic form normal pentane (Fig. 2.4) can assume when carbon atoms numbered (2) and (3) are each rotated through an angle of 109°28′. For this reason we must inspect a carbon chain carefully before numbering it. The longest chain is not necessarily the horizontal one.

EXERCISE 2.3

Conformational isomers often are shown by pseudo-three-dimensional or perspective formulas in which the dotted line is the bond below, the wedge the bond above, and the solid line the bond within the plane of the paper. Consider the structure below a representation of *n*-butane.

$$x \diagdown \underset{\underset{z}{y}}{C} - \underset{\underset{z'}{}}{C} \diagup^{x'}_{y'}$$

Draw both the pseudo-three-dimensional and Newman projections of (a) the anti conformation; (b) the gauche conformation. Which conformation is most likely for *n*-butane? Why?

2.6   *Physical Properties of the Alkanes*

The first four members of the alkane family are gases. Those containing five to seventeen carbon atoms are liquids at room temperature, while those members having eighteen or more carbon atoms are solids. All hydrocarbons are

TABLE 2.3   *Physical Properties of Some Normal Saturated Hydrocarbons*

Name	Formula	M.P., °C	B.P., °C	Specific Gravity	Normal State
Methane	CH_4	−182.6	−161.4	—	gas
Ethane	C_2H_6	−172.0	−88.3	—	gas
Propane	C_3H_8	−187.1	−44.5	—	gas
n-Butane	C_4H_{10}	−135.0	−0.5	—	gas
n-Pentane	C_5H_{12}	−129.7	36.2	0.6264	liquid
n-Hexane	C_6H_{14}	−94.0	69	0.6594	liquid
n-Heptane	C_7H_{16}	−90.5	98.4	0.6837	liquid
n-Octane	C_8H_{18}	−56.8	124.6	0.7028	liquid
n-Decane	$C_{10}H_{22}$	−32	175	0.730	liquid
n-Pentadecane	$C_{15}H_{32}$	10	271	0.772	liquid
n-Octadecane	$C_{18}H_{38}$	28	308	0.77	solid

insoluble in water but dissolve easily in most organic solvents. They are colorless and tasteless. When pure they are odorless. The odor of natural gas is not that of methane, its principal constituent, but that of a contaminant purposely added to allow for the detection of leaks. Table 2.3 summarizes properties of some representative members of the alkane series.

2.7 Preparation of the Alkanes

The alkanes are isolated from petroleum and usually are consumed as mixtures. It is seldom necessary to synthesize a hydrocarbon. Rather, if one is required alone, it is separated from its homologs by one of several laboratory techniques. A number of organic compounds, when properly treated, do react to produce hydrocarbons. In order to illustrate the preparation of an alkane, however, it will be necessary for us to use compounds we have yet to meet in subsequent chapters.

For our illustration we will use a relatively old reaction (*ca.* 1900) that provides a product in which the number of carbon atoms is the *same* as that in the starting material and a relatively new reaction (*ca.* 1965) in which the number of carbon atoms in the product is the sum of those in the reactants. An important clue to the choice of which, among many, synthetic reactions you should use to prepare a given compound is the change, if any, in the number of carbon atoms in the product compared to that in the reactant. In each of our examples we will start from an alkyl halide, R-X,[2] and convert it into an **organometallic compound,** a compound in which the carbon atom is bonded to a metal by a bond that may be covalent or ionic or somewhere in between, depending on the metal.

---

[2] R is a general designation for an alkyl group, C_nH_{2n+1}; X is a general designation for a halogen. These symbols are used throughout the text.

A. Reduction of Alkyl Halides (via the Hydrolysis of a Grignard Reagent). In 1901 Victor Grignard[3] discovered one of the most useful of all chemical reactions. He prepared methylmagnesium iodide, an organometallic compound, by treating methyl iodide with fresh magnesium turnings in anhydrous ether (diethyl ether). In the organometallic compound formed, the alkyl group is united to the metal.

$$
\begin{array}{ccc}
\quad H & & \quad H \\
\mid & \xrightarrow[\text{ether}]{\text{anhydrous}} & \mid \\
H-C-I + Mg & & H-C-MgI \\
\mid & & \mid \\
\quad H & & \quad H
\end{array}
$$

Methyl   Methylmagnesium
iodide   iodide

The overall equation for the preparation of alkylmagnesium halides, or Grignard reagents, as they came to be known, can be written

$$R-X + Mg \xrightarrow{\text{ether}} RMgX$$

The reaction also proceeds satisfactorily with higher molecular weight alkyl iodides, bromides, or chlorides. Grignard reagents are extremely reactive and, when hydrolyzed (reacted with water), are easily converted into alkanes.[4] Methylmagnesium iodide, on hydrolysis, yields methane; ethylmagnesium bromide, when hydrolyzed, gives ethane.

$$CH_3MgI + H_2O \longrightarrow CH_4 + HO-Mg-I$$
$$C_2H_5MgBr + H_2O \longrightarrow C_2H_6 + HO-Mg-Br$$

In practice, hydrolysis usually is accomplished by the use of dilute hydrochloric acid which converts the basic magnesium halide, $HO-Mg-X$, to the water-soluble magnesium halide, MgX_2.

B. Coupling of Alkyl Halides. Alkyllithium compounds can be prepared by a procedure similar to that used for Grignard reagents.

$$R-X + 2 Li \xrightarrow{\text{ether}} R-Li + LiX$$

$$CH_3-CH_2-CH_2-CH_2-Br + 2 Li \xrightarrow{\text{ether}} CH_3-CH_2-CH_2-CH_2-Li + LiBr$$

n-Butyl bromide   *n*-Butyllithium

Both Grignard reagents and organolithium reagents are widely used in the laboratory and in industry. The preparation of other organometallic compounds generally requires special techniques beyond the scope of this text.

---

[3] Victor Grignard (1871–1935). French chemist and winner of the Nobel Prize in chemistry (1912) for the discovery and development of the reaction that bears his name.

[4] Because Grignard reagents are easily hydrolyzed, glassware used in their preparation must be dry and reagents used must be anhydrous. Unless these precautions are taken, Grignard reagents will not form.

Organolithium compounds react with cuprous iodide to yield a different class of organometallic compounds known as **cuprates** (or lithiumdialkyl copper compounds).

$$2 \, CH_3—Li + CuI \longrightarrow Li(CH_3)_2Cu + LiI$$

Methyllithium   Lithiumdimethyl
          cuprate

These relatively new organometallic compounds have been found to react with alkyl halides to give alkanes in which the total number of carbon atoms is the sum of those in the alkyl halide plus those in *one* of the two alkyl groups in the cuprate as shown in the following example.

$$CH_3—CH_2—CH_2—CH_2—Br + Li(CH_3)_2Cu \longrightarrow CH_3—CH_2—CH_2—CH_2—CH_3 + CH_3Cu + LiBr$$

n-Butyl bromide   Lithiumdimethyl   *n*-Pentane
           cuprate
(4 carbons)   (1 carbon)   (5 carbons)

2.8   *Chemical Properties of the Alkanes*

The only reactions the saturated hydrocarbons are capable of undergoing without a rupture of carbon–carbon bonds are reactions in which hydrogen of the parent hydrocarbon is replaced by some other atom or group. Such replacements are called **substitution reactions.** Generally, the hydrocarbons are unreactive toward most reagents but are attacked by halogens and oxygen under certain conditions. Reactions with halogens are substitution reactions; attack by oxygen results in the breaking of carbon–carbon and carbon–hydrogen bonds. When alkanes are heated to extreme temperatures in the absence of air (a treatment referred to as *pyrolysis,* or *cracking*), the carbon–carbon and the carbon–hydrogen bonds are also ruptured to give a variety of fragmentary products. Catalytic cracking of long chain hydrocarbons as carried out in the petroleum industry (Chapter 18) makes possible the production of great quantities of motor fuel.

A. Oxidation (Combustion). Alkanes are resistant to ordinary chemical oxidizing reagents such as potassium permanganate, $KMnO_4$, and potassium dichromate, $K_2Cr_2O_7$, but may be oxidized to carbon dioxide and water when ignited in the presence of an excess of oxygen. Great quantities of heat energy called *heats of combustion* are released per mole of hydrocarbon burned, and it is this reaction which makes hydrocarbons so useful as fuels. As may be seen from the following three reaction equations, each CH_2 increment in the hydrocarbon chain adds approximately 156 kcal. to the heat of combustion.

$$CH_4 + 2 \, O_2 \longrightarrow CO_2 + 2 \, H_2O + 211 \, kcal$$
$$C_2H_6 + 3\tfrac{1}{2} \, O_2 \longrightarrow 2 \, CO_2 + 3 \, H_2O + 368 \, kcal$$
$$C_3H_8 + 5 \, O_2 \longrightarrow 3 \, CO_2 + 4 \, H_2O + 526 \, kcal$$

EXERCISE 2.4

The boiling point of *n*-pentane is 36.2°, and the heat of combustion for the liquid is given in the literature as 833 kcal/mole. Why is this value *less* than that obtained by the simple addition of CH_2 heat equivalents to any of the values given in the preceding three examples?

Incomplete combustion of alkanes results in the formation of carbon (soot) and the very dangerous carbon monoxide.

$$2 CH_4 + 3 O_2 \longrightarrow 2 CO + 4 H_2O$$
$$CH_4 + O_2 \longrightarrow C + 2 H_2O$$

Although both of the preceding reactions are to be avoided if we are to heat our homes safely and economically, the reaction that produces soot (carbon black) is of vital importance to our tire and rubber industry.

B. Substitution (Halogenation).   At ordinary temperatures, and in the absence of light, chlorine does not react with saturated hydrocarbons. At elevated temperatures and in the presence of sunlight or ultraviolet light, the hydrogen atoms are replaced by one or more chlorine atoms. A substitution of hydrogen by chlorine is known as a **chlorination** reaction. The reaction pathway by which chlorine substitutes for hydrogen in a hydrocarbon has been well established as one involving the formation of **free radicals.** A free radical in its reaction with another molecule can lead each time to the production of another free radical. The reaction is thus a self-propagating chain reaction which continues until all materials are consumed or until two free radicals meet and unite. A reaction pathway followed in this manner is referred to as a **free radical chain mechanism.** The first steps in the chlorination of methane may be shown as follows:

$$Cl:Cl \xrightarrow{\text{sunlight}} 2\,Cl\cdot$$

H—C—H + Cl· ⟶ HCl + H—C· (with H atoms above and below each carbon)

Methyl radical

H—C· + Cl₂ ⟶ H—C—Cl + Cl·; etc.

Chloromethane
(Methyl chloride)

The chlorination of methane is not restricted to the exclusive conversion of all methane to methyl chloride but proceeds until all four hydrogen atoms (on at least some) of the methane molecules have been substituted. The chlorination of a hydrocarbon therefore may result in a mixture of products.

$$
\begin{array}{c}
\quad\ \text{H} \\
\ | \\
\text{H}-\text{C}-\text{Cl} \quad + \text{Cl}_2 \longrightarrow \text{HCl} + \quad \text{H}-\text{C}-\text{Cl} \\
\ | \\
\quad\ \text{H}
\end{array}
\qquad
\begin{array}{c}
\text{Cl} \\
| \\
\text{C}-\text{Cl} \\
| \\
\text{H}
\end{array}
$$

Chloromethane   Dichloromethane
(Methyl chloride)   (Methylene chloride)

$$
\begin{array}{c}
\text{Cl} \\
| \\
\text{H}-\text{C}-\text{Cl} + \text{Cl}_2 \longrightarrow \text{HCl} + \quad \text{H}-\text{C}-\text{Cl} \\
| \\
\text{H}
\end{array}
\qquad
\begin{array}{c}
\text{Cl} \\
| \\
\ \\
| \\
\text{Cl}
\end{array}
$$

Trichloromethane
(Chloroform)

$$
\begin{array}{c}
\text{Cl} \\
| \\
\text{H}-\text{C}-\text{Cl} + \text{Cl}_2 \longrightarrow \text{HCl} + \quad \text{Cl}-\text{C}-\text{Cl} \\
| \\
\text{Cl}
\end{array}
\qquad
\begin{array}{c}
\text{Cl} \\
| \\
\ \\
| \\
\text{Cl}
\end{array}
$$

Tetrachloromethane
(Carbon tetrachloride)

The chlorination of either methane or ethane in which only one hydrogen atom has been replaced can result in only one monochlorinated product because all hydrogen atoms in these two alkanes are of the primary type. Propane, on the other hand, can yield two different monochlorinated products because it has secondary hydrogen as well as primary hydrogen atoms which may be replaced (Sec. 2.3). This being the case, a question arises as to which propyl chloride would we obtain if propane is chlorinated. Actually, both *n*-propyl chloride and isopropyl chloride are produced but not in equal amounts.

$$
\begin{array}{c}
\text{H H H} \\
| \ | \ | \\
\text{H}-\text{C}-\text{C}-\text{C}-\text{H} + \text{Cl}_2 \xrightarrow{\text{uv}} \\
| \ | \ | \\
\text{H H H}
\end{array}
\begin{array}{c}
\text{H H H} \\
| \ | \ | \\
\text{H}-\text{C}-\text{C}-\text{C}-\text{Cl} + \\
| \ | \ | \\
\text{H H H}
\end{array}
\begin{array}{c}
\text{H H H} \\
| \ | \ | \\
\text{H}-\text{C}-\text{C}-\text{C}-\text{H} \\
| \ | \ | \\
\text{H Cl H}
\end{array}
$$

Propane   1-Chloropropane   2-Chloropropane
                (*n*-Propyl chloride)   (Isopropyl chloride)
                    45%   55%

Generally speaking, a tertiary hydrogen is more readily replaced than a secondary one, and a secondary more readily than a primary hydrogen.

Bromine also reacts with the alkanes when the reaction is catalyzed by high frequency radiation, but does so much more slowly than chlorine. Iodine fails to react when catalyzed in this manner, but fluorine reacts with explosive violence. The reasons for this difference in the reaction behavior among members of the same family is a matter of energy relationships between reactants and products. To understand these relationships we need to know what energy requirements must be met in order for molecules to meet and react.

2.9 *Bond Dissociation Energies and Heats of Reaction*

In order for two molecules to react they must collide in a certain way with sufficient energy to break existing bonds and to form new ones. The breaking of a chemical bond requires energy. Conversely, the formation of a new bond yields energy. It would seem, then, that if the formation of bonds in the products were to yield a greater amount of energy than that expended in the breaking of bonds in the reactants, that is, if the reaction were **exothermic,** the reaction should proceed spontaneously. This is not the case, however. Consider the following analogy: Your automobile is on a plateau on the western side of the continental divide, at a higher elevation than Missouri, and thus possesses potential energy. But the ride to Kansas City is not simply one of coasting downhill. Some energy must be expended to climb over the divide as well as to overcome the intervening hills. Thus it is with chemical reactions. The energy necessary to get over the "hump" is called the **activation energy.** The net heat change for the overall reaction in going from reactants to products is called the

TABLE 2.4 *Bond Dissociation Energies (kcal/mole at 25°)*

H—H	104	F—F	38	$O{=}O$	119
H—F	135	Cl—Cl	58	O—H	111
H—Cl	103	Br—Br	46	C$=$O[a]	175
H—Br	87	I—I	36	H_3C—H	104
H—I	71				
H_3C—F	108	$CH_3\overset{\text{H}}{\underset{\text{H}}{\text{C}}}$—Cl	81.5	$CH_3\overset{\text{H}}{\underset{\text{H}}{\text{C}}}$—H	98
H_3C—Cl	83.5	$CH_3CH_2\overset{\text{H}}{\underset{\text{H}}{\text{C}}}$—Cl	81	$CH_3CH_2\overset{\text{H}}{\underset{\text{H}}{\text{C}}}$—H	98
H_3C—Br	70	$(CH_3)_2\overset{\text{H}}{\text{C}}$—Cl	81	$(CH_3)_2\overset{\text{H}}{\text{C}}$—H	95
H_3C—I	56	$(CH_3)_3C$—Cl	80	$(CH_3)_3C$—H	92

[a] $CH_2{=}O$

heat of reaction and is symbolized by ΔH, the capital Greek delta signifying *change*. The heat of reaction for many chemical reactions can be calculated from a knowledge of bond dissociation energies. The dissociation energies for a number of different bonds are given in Table 2.4.

The following example illustrates how Table 2.4 may be used.

Problem: Calculate ΔH for the monochlorination of methane.

Solution:

(1) Write the equation to show exactly which bonds are to be broken and which new bonds are to be formed.

$$\underset{104}{H-\overset{\displaystyle H}{\underset{\displaystyle H}{C}}\!\!+\!\!H} + \underset{58}{Cl\!\!+\!\!Cl} \longrightarrow \underset{83.5}{H-\overset{\displaystyle H}{\underset{\displaystyle H}{C}}\!\!+\!\!Cl} + \underset{103\ \ (kcal)}{H\!\!+\!\!Cl}$$

(2) Add bond energies on both sides of the equation. Breaking the C—H bond and the Cl—Cl bond requires 162 kcal; formation of the C—Cl and the H—Cl bonds yields a total of 186.5 kcal.

(3) Subtract the heat liberated through bond formation from the heat expended in bond dissociation. The difference is ΔH for the reaction.

$$\Delta H = 162 - 186.5 = -24.5\ kcal$$

The sign is negative when the net result is *exothermic*.
The sign is positive when the net result is *endothermic*.

Potential energy changes involved in chemical reactions may be illustrated graphically by means of reaction coordinate diagrams (Fig. 1.9). In such diagrams the height of the **energy barrier** (the "hump") represents the **activation energy** or the minimum energy needed to get the reaction to "go". If the activation energy is high, at any instant only a few of the many molecules present will have the energy needed to get over the barrier; and the reaction will be slow. Thus, the activation energy of a reaction affects its *rate*. The principal reason we heat organic reaction mixtures is to increase the number of molecules with energies greater than the energy barrier and, thereby, speed up the reaction. Because changes in potential energy are not easily calculated, as a first *approximation* we will substitute changes in heat content (ΔH) for the difference in free energy between reactants and products and **heat of activation** ($\Delta H^{\ddagger}$) for the activation energy. With these substitutions we can draw the diagrams shown in Fig. 2.5, which correspond to the two chain propagation steps.[5]

---

[5] In multistep reactions, such as the chlorination of methane, a separate reaction coordinate diagram may be required for each step. Sometimes, as in Fig. 1.9, two diagrams can be combined to represent a two-step reaction, but this is not desirable for reactions in which a new reactant is introduced in the second step.

FIGURE 2.5   *Reaction Profiles for the Chain Propagation Steps in the Chlorination of Methane*

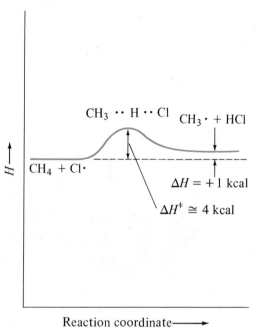

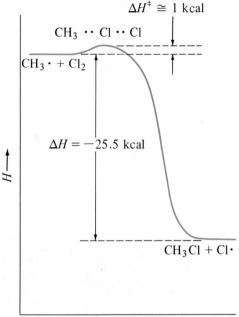

EXERCISE 2.5

Calculate ΔH for the following reaction:

$$CH_3-H + Br-Br \longrightarrow CH_3-Br + H-Br$$

Answer: -7 kcal

EXERCISE 2.6

It is not possible to calculate $\Delta H^{\ddagger}$ from bond dissociation energies; however, the calculation of ΔH is straightforward for the reactions shown in Fig. 2.5. Show how the values shown may be calculated.

2.10   *Cycloalkanes*

The cycloalkanes are saturated hydrocarbons that have ring structures. The general formula for this class of hydrocarbons is C_nH_{2n}.

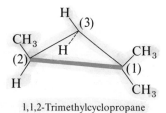

Cyclopropane   Cyclobutane   Cyclopentane

All bond
angles $= 109°28'$

Cyclohexane

Only the ring carbons are counted in naming the cycloalkanes and, as in open chain structures, they are numbered in order to designate substituents by the smallest numbers. The alkyl substituent also becomes one word with the name of the parent cycloalkane.

1,1,2-Trimethylcyclopropane

Cyclopropane, which is used as an inhalation anesthetic, may be prepared by a coupling reaction of 1,3-dibromopropane. Methods for the preparation of other cycloalkane derivatives are described in later sections.

1,3-Dibromopropane   Cyclopropane

Although the general formula of the cycloalkanes is the same as that of the alkenes (next chapter), their chemical behavior generally is more like that of the alkanes than that of the alkenes. Cyclopropane and cyclobutane are exceptions to this general rule, for reasons explained in the next section.

EXERCISE 2.7

How many cyclic structures of formula C_5H_{10} can you draw and name?

2.11 *Ring Structure and Strain Theory*

In order to bring the terminal carbons of propane together and form a bond between them, the tetrahedral angle of 109°28′ between carbon-carbon bonds (Sec. 2.5) must be changed to 60°, which is the value for the angles in an equilateral triangle. Each valence bond in cyclopropane thus is distorted from the normal tetrahedral angle by 1/2(109°28′ − 60°) or 24°44′. A bond angle deviation of this magnitude places the ring structure of cyclopropane under considerable strain and makes it more reactive than propane. With bromine, substitution does not occur. Instead, the ring is opened and the bromine adds at each end of the chain.

$$\underset{\substack{H_2 \\ C \\ H_2C \longrightarrow CH_2}}{} + Br_2 \xrightarrow{CCl_4} Br-CH_2-CH_2-CH_2-Br$$

1,3-Dibromopropane

Other reagents that react with carbon-carbon double bonds (Secs. 3.6–3.8) may also cause the ring to open.

$$\underset{\substack{H_2 \\ C \\ H_2C \longrightarrow CH_2}}{} + HBr \longrightarrow CH_3-CH_2-CH_2-Br$$

1-Bromopropane
(*n*-Propyl bromide)

$$\underset{\substack{H_2 \\ C \\ H_2C \longrightarrow CH_2}}{} + H_2 \xrightarrow[80°]{Ni} CH_3CH_2CH_3$$

Propane

In 1885, von Baeyer[6] advanced the theory that cycloalkane rings of less than five or more than six carbon atoms would possess too much strain to be stable. His theory was based on the assumption that all carbons within the ring lie in the same plane. Although this is true for a 3-membered ring, it is not true for rings of four carbons or larger. It is now recognized that there are two main factors influencing the stability or reactivity of the cycloalkane ring system: (1) bond angle strain due to departure from the tetrahedral angle and (2) eclipsed hydrogen strain due to the tendency of some rings to hold the ring hydrogens fixed in an eclipsed arrangement (recall Fig. 2.2 and the higher energy of ethane in the eclipsed conformation). Cycloalkanes of six or more carbons do not have to have all carbons in the same plane but can "pucker" to retain the tetrahedral angle. Any *angle* strain within the ring is thus easily relieved, although strain caused by the eclipsing of hydrogen atoms is present in

---

[6] Adolph von Baeyer (1835–1917). One of the greatest German chemists of his time. A prolific researcher and outstanding teacher. Professor of chemistry, University of Munich. Winner of the Nobel Prize in chemistry (1905).

rings having seven to fourteen carbon atoms. Even 4- and 5-membered rings show a tendency to pucker in order to provide for greater ring stability, their structures being those affording the greatest relief from both types of strain.

Although both **boat** and **chair** forms are possible cyclohexane conformations that are free of angle strain (Figs. 2.6 and 2.7), the more stable chair form

FIGURE 2.6   *Boat Conformation of Cyclohexane. Compare the Ball and Stick Model with the Structure at Right.*

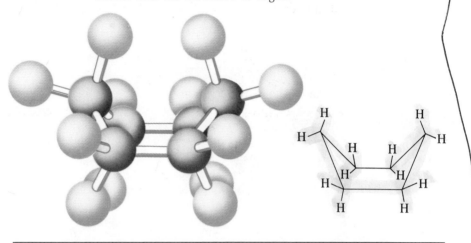

FIGURE 2.7   *Chair Conformation of Cyclohexane. Compare the Ball and Stick Model with the Structure at Right.*

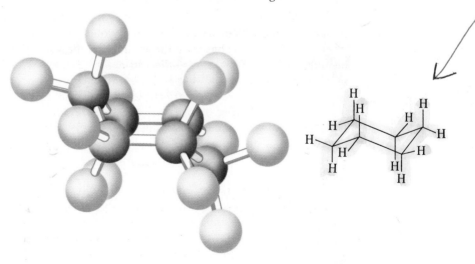

FIGURE 2.8 *Equatorial and Axial Bonds in Cyclohexane*

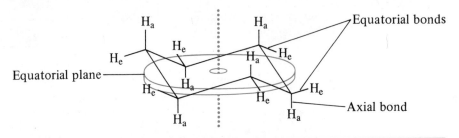

conformation is an essentially strain-free conformation in which all angles are close to tetrahedral and all hydrogen atoms are staggered. The chair conformation has its hydrogen atoms (or other substituents) in two distinct kinds of positions. In one position, six bonds lie roughly in the plane of the ring, or "equatorial belt," and are called **equatorial bonds.** The other six bonds point in a direction parallel to an axis drawn perpendicular to the plane of the ring and are called **axial bonds** (Fig. 2.8). Each carbon atom of cyclohexane has one equatorial and one axial bond. Substituents in axial positions are called axial substituents; those in equatorial positions are called equatorial substituents—as shown by the labels on the hydrogen atoms in Fig. 2.8. Substituents in a 1,3-relationship to each other on the 6-atom ring will show less interference with each other—that is, will have more free space to occupy—if both are in equatorial positions rather than in axial positions.

2.12   cis,trans-*Isomerism in Cycloalkanes*

Cycloalkanes having two or more substituents on the ring may show a form of **stereoisomerism** (Chap. 5) called *cis,trans*-**isomerism.** This form of isomerism in the cycloalkane series is caused by restricted rotation about the single bonds of the ring. The *cis*-isomer has both substituents on the same side of the ring (top or bottom); whereas, the *trans*-isomer has one substituent on each side of the ring. This type of isomerism is illustrated with the *cis* and *trans* isomers of 1,2-dimethylcyclopropane. In naming such isomers, the prefixes, *cis* or *trans,* are placed in front of the cycloalkane name.

cis-1,2-Dimethylcyclopropane   *trans*-1,2-Dimethylcyclopropane

Summary

1. The alkanes, or saturated hydrocarbons, contain only carbon and hydrogen. They are also called paraffins. All carbon atoms in alkanes are singly bonded to four other atoms.

2. The **general formula for an alkane** is $C_n H_{2n+2}$.

3. The names for the alkanes end in "*ane.*" Alkanes may be named unambiguously by the IUPAC system.

4. The **general formula for an alkyl group** is $C_n H_{2n+1}$. Groups are named after the parent hydrocarbon by changing the suffix *ane* to *yl*. The prefixes iso, *sec, tert* are used depending upon which hydrogens have been removed from the parent compound.

5. Saturated hydrocarbons may have straight (that is, in a continuous chain), branched, or cyclic structures.

6. Alkanes may be prepared by:

   (a) Coupling of alkyl groups.

   $$R—Br + Li(R')_2Cu \longrightarrow R—R'$$

   (b) Reduction of an alkyl halide via a Grignard reagent.

   $$R—X + Mg \xrightarrow[\text{ether}]{\text{anhydrous}} RMgX \xrightarrow[\text{(HX)}]{\text{hydrolysis}} R—H + MgX_2$$

7. Chemically, the alkanes are relatively inert. The only reactions which occur without rupture of the carbon-carbon bond are those of substitution. Typical reactions include:

   (a) Oxidation (combustion)

   $$C_n H_{2n+2} + \frac{(3n + 1)}{2} O_2 \longrightarrow n\,CO_2 + (n + 1)\,H_2O$$

   (b) Halogenation

   $$R—H + X_2 \xrightarrow[\text{light}]{\text{uv}} R—X + HX$$

8. Chemical reactions occur when properly oriented molecules collide with sufficient energy to break existing bonds and form new bonds. The energy required to break a chemical bond is called its **dissociation energy.** The same amount of energy is released when the bond is reformed.

9. The heat of reaction, ΔH, is the difference in the heat content of reactants and products. ΔH is negative when the reaction is exothermic; positive when endothermic.

10. The rate of a reaction depends on the height of the energy barrier, or energy of activation.

11. Cycloparaffins are ring compounds of the general formula, $C_n H_{2n}$, which can exist as *cis,trans*-isomeric pairs. Small rings show the effects of angle and eclipsing strain; large rings are free of angle strain but may show eclipsing strain. The chair form of cyclohexane is strain free.

New Terms

activation energy	equatorial
alkyl group	free rotation
anti	gauche
axial	Grignard reagent
bond dissociation energy	heat of reaction
conformation	organometallic compound
cis,trans-isomerism	paraffin
eclipsed	ring strain
energy barrier	

Supplementary Exercises

EXERCISE 2.8   Assign acceptable names (common or IUPAC) to the following structures.

(a) 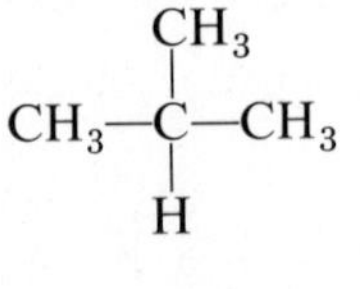
$$CH_3\!-\!\overset{\displaystyle CH_3}{\underset{\displaystyle H}{\overset{|}{\underset{|}{C}}}}\!-\!CH_3$$

(b)

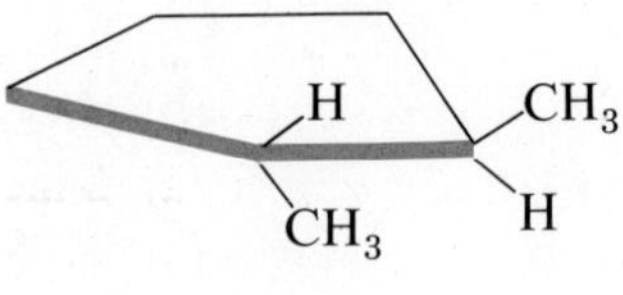

(c) 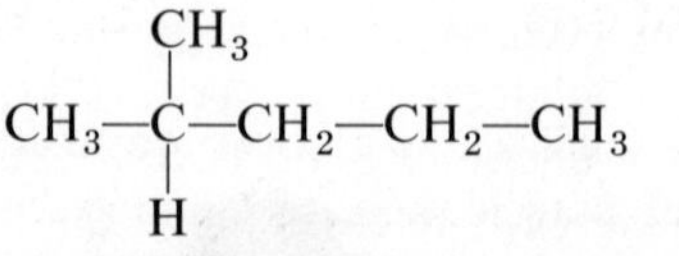
$$CH_3\!-\!\overset{\displaystyle CH_3}{\underset{\displaystyle H}{\overset{|}{\underset{|}{C}}}}\!-\!CH_2\!-\!CH_2\!-\!CH_3$$

(d) 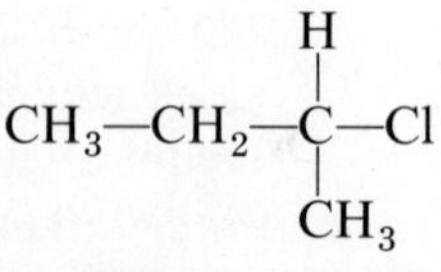
$$CH_3\!-\!CH_2\!-\!\overset{\displaystyle H}{\underset{\displaystyle CH_3}{\overset{|}{\underset{|}{C}}}}\!-\!Cl$$

(e) 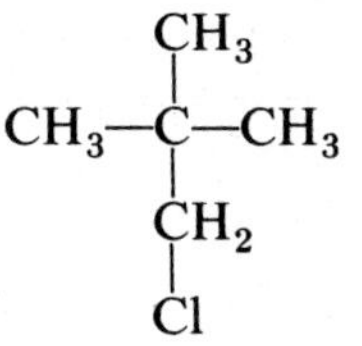 $CH_3-\overset{\overset{\displaystyle CH_3}{|}}{\underset{\underset{\displaystyle Cl}{|}}{\underset{CH_2}{|}}}-CH_3$

(f) 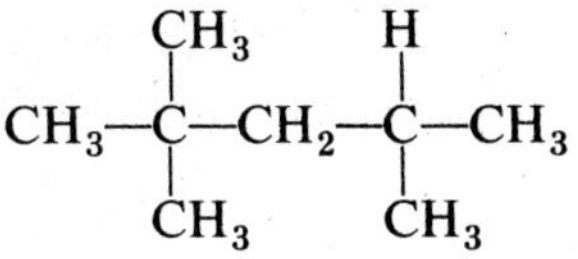 $CH_3-\overset{\overset{\displaystyle CH_3}{|}}{\underset{\underset{\displaystyle CH_3}{|}}{C}}-CH_2-\overset{\overset{\displaystyle H}{|}}{\underset{\underset{\displaystyle CH_3}{|}}{C}}-CH_3$

EXERCISE 2.9   Which of the following structures are the same? What are their IUPAC names?

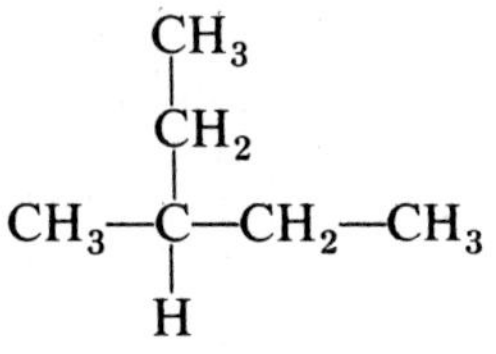

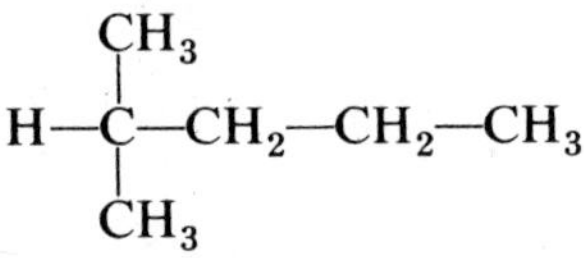

(a) 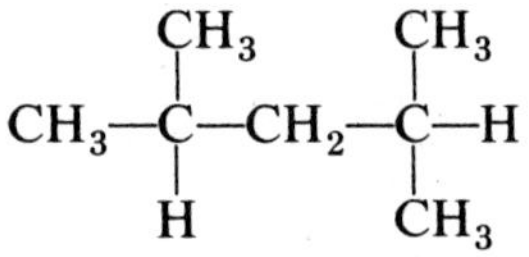 $CH_3-\overset{\overset{\displaystyle CH_3}{|}}{\underset{\underset{\displaystyle H}{|}}{C}}-CH_2-\overset{\overset{\displaystyle CH_3}{|}}{\underset{\underset{\displaystyle CH_3}{|}}{C}}-H$

(b) $CH_3-\overset{\overset{\displaystyle CH_3}{\overset{|}{\underset{|}{CH_2}}}}{\underset{\underset{\displaystyle H}{|}}{C}}-CH_2-CH_3$

(c) $CH_3-\overset{\overset{\displaystyle H}{|}}{\underset{\underset{\displaystyle CH_3}{|}}{C}}-CH_2-\overset{\overset{\displaystyle CH_3}{|}}{\underset{\underset{\displaystyle H}{|}}{C}}-CH_3$

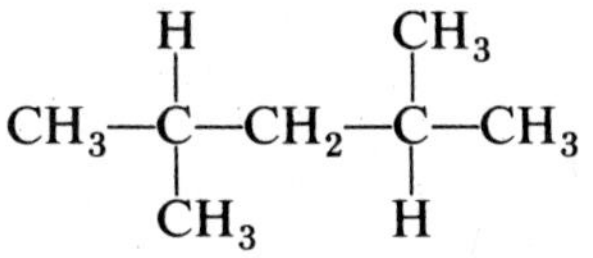

(d) 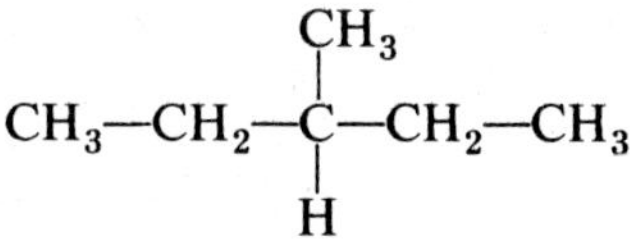 $CH_3-CH_2-\overset{\overset{\displaystyle CH_3}{|}}{\underset{\underset{\displaystyle H}{|}}{C}}-CH_2-CH_3$

(e) 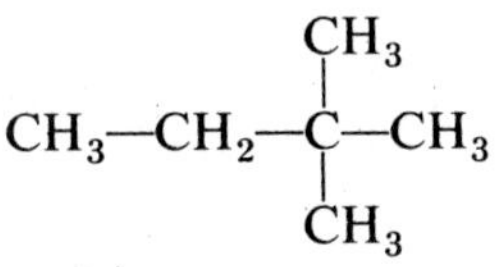 $CH_3-CH_2-\overset{\overset{\displaystyle CH_3}{|}}{\underset{\underset{\displaystyle CH_3}{|}}{C}}-CH_3$

(f) 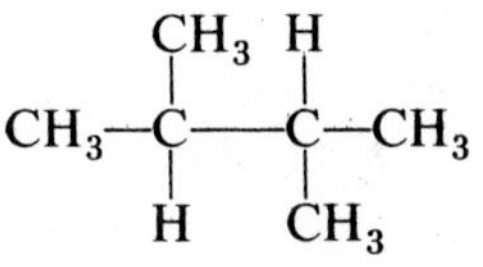 $CH_3-\overset{\overset{\displaystyle CH_3}{|}}{\underset{\underset{\displaystyle H}{|}}{C}}-\overset{\overset{\displaystyle H}{|}}{\underset{\underset{\displaystyle CH_3}{|}}{C}}-CH_3$

(g) $H-\overset{\overset{\displaystyle CH_3}{|}}{\underset{\underset{\displaystyle CH_3}{|}}{C}}-CH_2-CH_2-CH_3$

(h) $CH_3-CH_2-\overset{\overset{\displaystyle CH_3}{|}}{\underset{\underset{\displaystyle CH_3}{|}}{C}}-CH_2-CH_3$

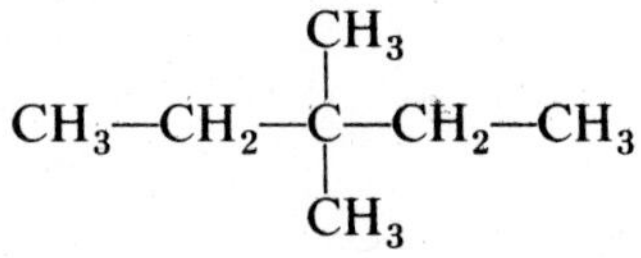

EXERCISE 2.10   The following names are incorrect. In each case tell why the name is wrong and give the correct form.

(a) 2-methyl-3-ethylbutane
(b) 4-methylpentane
(c) 3,5,5-trimethylhexane
(d) 2-isopropyl chloride
(e) 3-methyl-4-chloropentane
(f) 1,3-dimethylcyclopropane

EXERCISE 2.11   Write structural formulas for the following. Also assign systematic names to compounds (c–f).

(a) 2-methylbutane
(b) 2,2,4-trimethylhexane
(c) neopentyl chloride
(d) *tert*-butyl chloride
(e) isobutyl chloride
(f) chloroform
(g) 1,3-dibromopropane
(h) 3-ethyl-2-methylpentane

EXERCISE 2.12   Draw the structures for all of the monochlorinated isomers that could result from the free radical chlorination of 2-methylbutane (isopentane). If the approximate ratios of rates of attack of chlorine on hydrogens located in primary, secondary, and tertiary positions are 1.0 : 3.8 : 5.8, approximately what percent yield of each isomeric mono-substituted product might be expected from isopentane? (Hint: The percent yield of 1-chloropropane (p. 47) was derived from:

$$\frac{6 \times 1}{(6 \times 1) + (2 \times 3.8)} \times 100 = 45\%$$

EXERCISE 2.13   Give all structures possible for $C_3H_6Cl_2$. Name each according to the IUPAC system.

EXERCISE 2.14   Complete the following reactions: (These should be balanced)

    (a) $C_2H_5Br + Mg \xrightarrow{\text{anhydrous ether}}$

    (b) cyclohexane $+ Cl_2 \xrightarrow{\text{light}}$

    (c) $CH_3MgI + H_2O \xrightarrow{H^+}$

    (d) C_3H_8 (bottled gas) $+ O_2 \longrightarrow$

    (e) $CH_4 + Cl_2 \xrightarrow{\text{sunlight}}$

EXERCISE 2.15   Draw and properly identify each part of a reaction coordinate diagram similar to that shown in Fig. 2.5 for the reaction in Exercise 2.5. The $\Delta H^{\ddagger}$ for the reaction of CH_4 and $Br\cdot$ is approximately 18 kcal/mole and for the reaction of $CH_3\cdot$ and Br_2 is approximately 2 kcal/mole. Which step in this reaction would you expect to be the rate determining step?

EXERCISE 2.16   During a certain winter month a home furnace consumed 169 cu. ft. of natural gas. What volume of air (20% oxygen) was required to burn this gas if we assume it to be 80% methane plus noncombustible material? How many liters of water were produced? (1 cu. ft. = approximately 28.2 liters)

EXERCISE 2.17   The orifice in the burner of a water heater using liquefied petroleum, or L. P., gas is much smaller than one found in a similar heater using natural gas. Explain.

EXERCISE 2.18   Calculate ΔH for the following reactions: (*Consult Table 2.4.*)

    (a) $2 H_2 + O_2 \longrightarrow 2 H_2O$
    (b) $C_2H_6 + Cl_2 \longrightarrow C_2H_5Cl + HCl$
    (c) $C_3H_8 + Cl_2 \longrightarrow (CH_3)_2CHCl + HCl$

EXERCISE 2.19   The chair conformation of cyclohexane can be regarded as being in equilibrium with an exactly equivalent conformation (shown without the hydrogen atoms).

A   B

Redraw these structures including the hydrogen atoms, using black balls ● for the equatorial hydrogens and white balls ○ for the axial hydrogens in conformer A. What color balls *must* be used for the equatorial hydrogens in conformer B? For the axial hydrogens?

EXERCISE 2.20 You have available the following alkyl halides: methyl bromide, ethyl bromide, *n*-butyl bromide, and any inorganic reagents of your choosing. Write equations showing how many different alkanes could be synthesized using one or more of these alkyl halides.

EXERCISE 2.21 An alkyl halide (A) forms a Grignard reagent, which on treatment with water yields isopentane. When A is allowed to react with lithiumdimethyl cuprate, 2,3-dimethylbutane is formed. What is the structure of A? Write the equations for the reactions described.

EXERCISE 2.22 The following thermochemical data are for cyclohexane (strain-free, as you will recall):

$$C_6H_{12} + 9\,O_2 \longrightarrow 6\,CO_2 + 6\,H_2O \qquad \Delta H = -944.80\ \text{kcal/mole}$$

When measured under the same conditions, ΔH for the combustion of cyclobutane is -655.86 kcal/mole. Write a balanced equation for the combustion of cyclobutane. Calculate the total strain energy (in kcal/mole) of cyclobutane. (*Hint:* What is the increment of ΔH per $-CH_2-$ in an *unstrained* cycloalkane?)

EXERCISE 2.23 There are three staggered and three eclipsed conformations of *n*-butane; however, two of the staggered forms are equivalent and two of the eclipsed forms are equivalent. Equivalent forms will have the same energy. If the difference in energy between the most stable conformer and the other conformers is 0.9, 3.8, and 4.5 kcal/mole, respectively, draw a diagram like Fig. 2.2 showing the changes in energy as the *n*-butane molecule is rotated around the 2-3 carbon-carbon bond. You will want to use the results of Exercise 2.3 and to extend your diagram to 360°.

EXERCISE 2.24 Organic chemists and biochemists frequently find it necessary to prepare "labeled" compounds in which a hydrogen atom (H) has been replaced with a deuterium atom (D), or a carbon-12 isotope (^{12}C) by a carbon-13 (^{13}C) or carbon-14 (^{14}C) isotope. Assume that you have available D₂O (heavy water) and $^{13}CH_3Br$ and any other unlabeled reagents needed. Show how the following could be synthesized from your starting materials.

(a) $^{13}CH_4$

(b) CH_3D

(c) $^{13}CH_3$—$^{13}CH_3$

(d) CH_3—CH_2—CH_2—CH_2—CH_2—D

3

The Unsaturated Hydrocarbons— Alkenes and Alkynes

Introduction

The unsaturated hydrocarbons include two classes of compounds. In one class are hydrocarbons that contain carbon-carbon double bonds, $-\overset{|}{C}=\overset{|}{C}-$, and are known as **alkenes,** or **olefins.** The second class of unsaturated hydrocarbons include those that contain carbon-carbon triple bonds, $-C\equiv C-$, and are known as **alkynes,** or **acetylenes.** Both classes are described as unsaturated because they do not contain the maximum possible number of hydrogen atoms. They are not as "filled up" or saturated with hydrogen as they could be and potentially are capable of bonding to additional atoms. Unlike the saturated hydrocarbons, more than one pair of electrons is shared between two carbon atoms in these unsaturated compounds. As a result of such unsaturation both the olefins and the acetylenes are much more reactive than the alkanes. They combine readily with other compounds (or with themselves) and yield many useful products.

Part I: The Alkenes (Olefins)

As just pointed out, the alkenes, or olefins, are characterized by the presence of double bonds between adjacent carbon atoms. The general formula C_nH_{2n} corresponds to an open chain olefin if only *one* double bond is present in the molecule. The same general formula also represents a cycloalkane. In the first, or simplest member of the series, $n = 2$, and the formula C_2H_4 is that for ethylene, or ethene.

As a class these compounds are commonly referred to as the olefins (L., *oleum,* oil; *ficare,* to make). The name originated because ethene, in its reaction with chlorine, forms 1,2-dichloroethane, an oily liquid.

$$\underset{\text{Ethene}}{H-\overset{\overset{\displaystyle H}{|}}{C}=\overset{\overset{\displaystyle H}{|}}{C}-H} + Cl_2 \longrightarrow \underset{\text{1,2-Dichloroethane}}{H-\overset{\overset{\displaystyle H}{|}}{\underset{\underset{\displaystyle Cl}{|}}{C}}-\overset{\overset{\displaystyle H}{|}}{\underset{\underset{\displaystyle Cl}{|}}{C}}-H}$$

The name alkene is used in systematic nomenclature when referring to the olefins. Both common and systematic names will be used in the present chapter.

3.1 *Formulas and Nomenclature*

Systematic names for members of the alkene family are formed by replacing the suffix *ane* of the corresponding alkane with *ene*. Common names usually are employed to name the simplest members of the alkenes. Such common names are formed by replacing the suffix *ane* of the corresponding alkane with *ylene*. The following are examples of both systems of nomenclature.

$$\underset{\substack{\text{Ethene}\\\text{(Ethylene)}}}{H-\overset{\overset{\displaystyle H}{|}}{C}=\overset{\overset{\displaystyle H}{|}}{C}-H} \qquad \underset{\substack{\text{Propene}\\\text{(Propylene)}}}{CH_3-\overset{\overset{\displaystyle H}{|}}{C}=\overset{\overset{\displaystyle H}{|}}{C}-H} \qquad \underset{\substack{\text{2-Methylpropene}\\\text{(Isobutylene)}}}{\overset{\displaystyle CH_3}{\underset{\displaystyle CH_3}{C}}=\overset{\overset{\displaystyle H}{|}}{C}-H}$$

The IUPAC rules for naming the alkenes follow those for the alkanes, except for the following modifications:

(1) Instead of simply numbering the longest carbon chain, the longest chain *which includes the functional group* is numbered. In the case of the alkenes

the functional group is the carbon-carbon double bond, $-\overset{\overset{\displaystyle H}{|}}{C}=\overset{\overset{\displaystyle H}{|}}{C}-$. Change the *ane* of the corresponding alkane to *ene*.

(2) Numbering begins from the end of the chain that will confer upon the carbon atom holding the functional group the smaller number. Since the double bond in an alkene appears between two carbons, only the carbon atom with the lower number need be designated. This number precedes and is separated from the name by a hyphen.

(3) The functional group is located by a number each time it appears. The prefixes *di, tri,* etc. appear before *ene* in alkenes with more than one double bond and are incorporated within the name.

(4) In cycloalkenes the double bond is numbered 1,2. Since this is always the case, no number for the double bond is given in the name. If more than one double bond is present in the ring, they are given the lowest numbers possible, and the numbers are given in the name. The ring is numbered in the direction that gives the lower number to the first substituent encountered.

A few examples will help make these points clear.

$$CH_3-\overset{\overset{\displaystyle H}{|}}{C}=\overset{\overset{\displaystyle H}{|}}{C}-CH_3$$
2-Butene

$$\underset{(4)}{CH_3}-\underset{(3)}{CH_2}-\underset{(2)}{\overset{\overset{\displaystyle H}{|}}{C}}=\underset{(1)}{CH_2}$$
1-Butene

$$\underset{(1)}{H}-\underset{(2)}{\overset{\overset{\displaystyle H}{|}}{C}}=\underset{(3)}{\overset{\overset{\displaystyle H}{|}}{C}}-\underset{(4)}{\overset{\overset{\displaystyle H}{|}}{C}}=\underset{}{\overset{\overset{\displaystyle H}{|}}{C}}-H$$
1,3-Butadiene

$$\overset{(5)\ CH_3}{\underset{}{|}} \\ \overset{(4)\ CH_2}{\underset{(3)}{|}} \\ CH_3-\overset{}{\underset{(1)}{C}}=\overset{(2)}{\underset{}{C}}-CH_3 \\ \underset{}{H}$$
3-Methyl-2-pentene

$$\overset{CH_3}{\underset{(1)}{C}} \\ \overset{(2)\ CH}{\ } \quad \overset{(6)}{CH}-CH_3 \\ \overset{(3)\ CH_2}{\ } \quad \overset{(5)}{CH_2} \\ \overset{(4)\ CH_2}{\ }$$
1,6-Dimethylcyclohexene

$$H-\overset{(2)}{C}\underset{(3)}{=\!=\!=}C-H \\ \ \ \ \ (1)\ \ (4) \\ H-\overset{(5)}{C}\underset{}{=\!=}C-H \\ \overset{C}{H_2}$$
1,3-Cyclopentadiene

EXERCISE 3.1

Draw three different structures for C_5H_{10} in which the continuous chain is four carbons long. Name each compound according to IUPAC rules.

3.2 *The Double Bond*

A double bond represents two pairs, or four electrons shared between two carbon atoms. Our simple structural formulas show these as two equivalent covalent bonds, and this formulation can be justified mathematically. However, much of the chemical behavior of alkenes is more conveniently described in terms of a double bond in which the two bonds are not equivalent. You learned in Section 1.4 that, when a hybrid is formed from one $1s$ and three $2p$ orbitals, four sp^3 hybridized orbitals result. However, if only *two* of the three $2p$ orbitals are combined with the $1s$ orbital, then three sp^2 hybrid orbitals are formed and one $2p$ orbital is left. This is the situation for each carbon in ethene. The axes of the three sp^2 orbitals formed lie in the same plane and tend to form angles of 120° with each other. The axis of the remaining $2p$ orbital is perpendicular to the plane (Fig. 3.1). When two carbon atoms with bonding orbitals of this type unite, two sp^2 orbitals (one from each carbon) engage in the usual endwise overlap to form a carbon-carbon sigma orbital; and the two $2p$ orbitals overlap in a *sidewise* manner to form two molecular orbitals of a new type which encompass both carbon nuclei. The *bonding* orbital resembles a cloud lying above and below the plane of the molecule (Fig. 3.2), which has *p*-like symmetry and is called a **π-orbital** (pi orbital).[1] Two of the four shared electrons are assigned to the sigma orbital and two to the π-orbital. A bond formed in this manner is called a **pi** (π) bond to distinguish it from the ordinary covalent or **sigma** (σ) bond. Thus, a double bond can be regarded as made up of one sigma C—C bond and one pi C—C bond.

FIGURE 3.1 *The Three* sp² *Orbitals of Carbon* (*in the Horizontal Plane*) *with the Remaining* p *Orbital Perpendicular to the Plane**

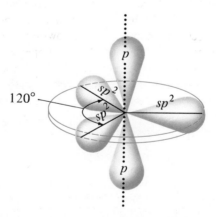

* The shape of an sp^2 orbital is similar to that of an sp^3 orbital (Fig. 1.5) except that the front lobe (the larger lobe) is somewhat larger in the sp^2 orbital and the back lobe is somewhat smaller. The orbital shapes used in this figure are those conventionally employed by chemists to represent the orbitals and were chosen to simplify the drawings.

[1] There is formed also a higher energy *antibonding* π-orbital, which has a nodal plane perpendicular to the plane of the molecule and which bisects the molecule between the carbon atoms.

FIGURE 3.2 *Pi Orbitals: (a) Pi Orbitals, (b) Schematic Representation of the Carbon—Carbon Double Bond of Ethene in Perspective Showing Five σ Bonds and One π Bond*

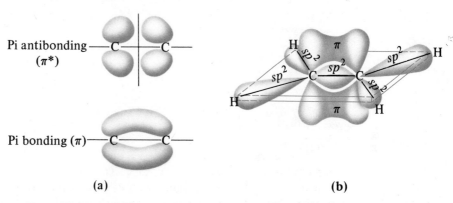

Pi antibonding (π*)

Pi bonding (π)

(a)

(b)

The effect of such double bonding is threefold. One is that two carbon atoms are nearer each other when joined by a double bond than when joined by a single bond; the distance between two doubly-bonded carbon atoms is 1.35 Å compared to 1.54 Å for the C—C single bond. Second, carbon atoms doubly-bonded and the atoms attached to them are now held more or less rigidly *in the same plane*. Free rotation around the bond joining the two carbon atoms is no longer possible, because considerable energy is required to break the *pi* bond. This restriction in a plane bestows upon the double bond a kind of flatness, and the molecule at this site has an upper and lower side. A variation in the space arrangements of groups attached to the carbon atoms makes possible a type of **stereoisomerism** called ***cis,trans*-isomerism** (Sec. 5.7). The *cis* isomer has like groups on the same side of the molecule and the *trans* isomer has like groups on opposite sides of the molecule (Figs. 3.3).[2] Thus, beginning with C_4H_8, a greater number of isomers is possible for any member of the alkene family than was

---

[2] There is some ambiguity in the expressions "same side" and "opposite side" of a double bond. In describing *cis,trans*-isomerism of an alkene the "same side" is defined relative to a line connecting the carbon atoms. In describing the chemical reactions of an alkene, the "same side" is defined relative to the plane of the molecule (upper side or lower side).

cis   *trans*

FIGURE 3.3   *Structural Representations of (a)* trans-*2-Butene and (b)* cis-*2-Butene*

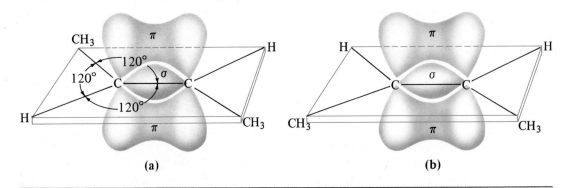

(a)   **(b)**

possible for the corresponding alkane, C_4H_{10}, not simply because the position of the double bond in the carbon chain may vary, but also because the groups attached to it may be *cis* or *trans*. When the two groups (or atoms) attached to a doubly bonded carbon are identical (Fig. 3.4) then, of course, *cis,trans*-isomerism is not possible.

FIGURE 3.4   *Structural Representation of 1-Butene (cis–trans Isomerism Not Possible)*

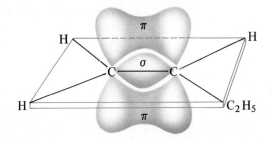

A third effect manifest in doubly-bonded carbon compounds is that the *pi* electrons are not as firmly held as those comprising the sigma bond, but are more readily available to an electrophilic (electron-seeking) reagent. Thus, the region above and below the double bond is electron-rich, and this availability of π-electrons accounts for the greater reactivity of the alkenes compared to the alkanes. Alkenes tend to behave as Lewis bases toward electrophilic reagents.

EXERCISE 3.2

Draw three different structures for C_5H_{10} in which the continuous chain is five carbons long. Name each according to IUPAC rules.

Preparation of the Alkenes

The double bond usually is introduced into a molecule by some type of **elimination reaction** in which groups or atoms are removed from two adjacent carbons.

$$-\underset{\underset{A}{|}}{C}-\underset{\underset{B}{|}}{C}- \longrightarrow AB + -\underset{|}{C}=\underset{|}{C}-$$

In the petroleum industry the elimination of hydrogen atoms from an alkane is accomplished by pyrolysis (Sec. 18.4). Both continuous and branched chain olefins are produced by this cracking process. In a laboratory preparation of an alkene the double bond usually is introduced by eliminating a small molecule such as water, hydrogen halide, or ammonia. Such elimination reactions usually are catalyzed by ions such as H^+, for the removal of water, and OH^-, for the elimination of hydrogen halide, but the double bond also may result from the pyrolytic decomposition of certain ammonium salts (Sec. 13.4-B). Many reactions of both types are known, but we shall consider here only two of the most general procedures for the preparation of alkenes.

3.3 *Dehydration of Alcohols*

Elimination of the elements of water from adjacent carbon atoms of an alcohol will lead to the formation of an alkene. Industrially, water is eliminated from an alcohol by passing the alcohol vapor over heated alumina, Al_2O_3.

$$H-\underset{\underset{H}{|}}{\overset{\overset{H}{|}}{C}}-\underset{\underset{OH}{|}}{\overset{\overset{H}{|}}{C}}-H \xrightarrow[350–360°C]{Al_2O_3} H-\overset{\overset{H}{|}}{C}=\overset{\overset{H}{|}}{C}-H + H_2O$$

Ethyl alcohol   Ethene

In the laboratory, **dehydration** of an alcohol usually is accomplished by heating it either with concentrated sulfuric or phosphoric acid. The acid functions primarily as a catalyst, as will be evident from the discussion of the mechanism of this reaction in Sec. 8.8-B. The reaction is reversible, and the reaction conditions are chosen to favor the forward reaction. Generally, strong acids and high temperatures favor alkene formation. Removal of the usually volatile alkene will force the reaction to the right. Unless removed, some alkenes can react with the sulfuric acid and water to reform the alcohol as illustrated in Sec. 3.8-C. The net effect of heating an alcohol with sulfuric acid may be shown by the following equation for the overall reaction.

$$H-\underset{\underset{H}{|}}{\overset{\overset{H}{|}}{C}}-\underset{\underset{OH}{|}}{\overset{\overset{H}{|}}{C}}-H \underset{170°C}{\overset{H_2SO_4}{\rightleftharpoons}} H_2C=CH_2 + H_2O$$

Ethanol   Ethene

Cyclohexene may be prepared from the cyclic alcohol, cyclohexanol, in a similar manner.

Cyclohexanol   Cyclohexene

3.4   *Dehydrohalogenation of Alkyl Halides*

The elimination of the elements of hydrogen halide (HX) from adjacent carbons is called **dehydrohalogenation.** This reaction is another common method that leads to the preparation of an alkene.

2-Bromopropane   Propene

As this reaction requires a very strong base, the dehydrohalogenation is usually carried out using a solution of potassium hydroxide in alcohol. The elimination of hydrogen halide appears to take place in the following manner.

Hydrogen (as a proton) is removed from the β carbon[3] atom by a strong base (shown in the equation as the hydroxide ion). At the same time, the halogen on the α carbon atom moves off as a halide ion, assisted by its attraction to the solvent (alcohol) molecules. This migration is accompanied by a simultaneous shifting of an electron pair.

Dehydrohalogenation is thus a concerted reaction in which the carbon-hydrogen and the carbon-halogen σ bonds are broken at the same time the new carbon-carbon π bond is formed. Later (Sec. 7.4) a more detailed description will be given of the changes that can occur during the course of a reaction of the foregoing type, which is called an E2 elimination reaction.

---

[3] The carbon atom bonded to the functional group (halogen in the above example) is designated α, adjacent carbons β, next γ, etc.

When either of two β carbon atoms can supply the hydrogen to be eliminated with the halogen atom, obviously two isomeric alkenes are possible. If such is the case the alkene most likely to result will be the more stable one. This appears to be the one with the greater number of alkyl groups attached to the doubly-bonded carbons. Thus, for example, two alkenes are possible when 2-bromobutane is heated with alcoholic potassium hydroxide, but the amount of one isomer produced exceeds that of the other in a ratio of $4:1$.

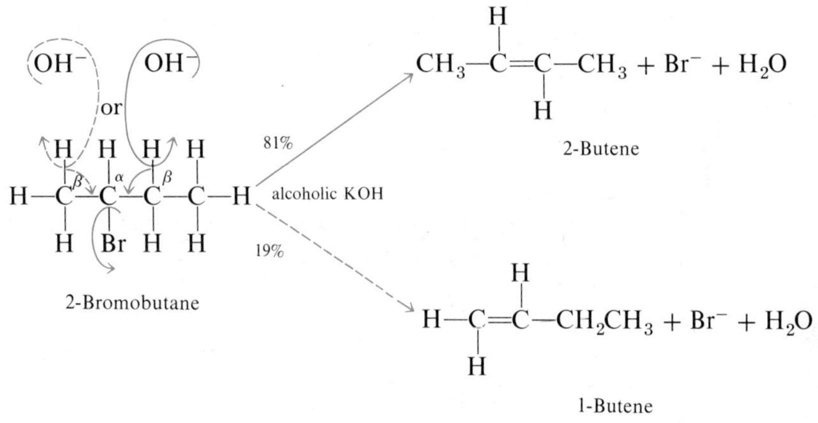

EXERCISE 3.3

Two isomeric alkenes are produced when 2-chloro-2-methylbutane is heated with alcoholic KOH. The yield of one is 86%, the other 14%. Give the name and structure of the isomer produced in greater amount.

3.5 *Properties of the Alkenes*

The physical properties of the alkenes are not too unlike those of the corresponding alkanes. The lower members of the alkene series are gases while the higher members are liquids and solids. The alkenes have slightly higher densities than the alkanes (Table 3.1).

TABLE 3.1   *Physical Constants of the Alkenes*

Name	Formula	M.P., °C	B.P., °C	Density as Liquid
Ethene	$H_2C\!=\!CH_2$	-169.4	-103.8	0.566
Propene	$CH_3CH\!=\!CH_2$	-185.2	-47.7	0.609
1-Butene	$CH_3CH_2CH\!=\!CH_2$	-130	-6.5	0.625

TABLE 3.1   *Continued*

Name	Formula	M.P., °C	B.P., °C	Density as Liquid
2-Butene	$CH_3CH{=}CHCH_3$	−127	1.4	0.630
2-Methylpropene (Isobutylene)	$CH_3{-}\overset{\displaystyle CH_3}{\underset{\displaystyle |}{C}}{=}CH_2$	−140.7	−6.9	0.594
1,3-Butadiene	$CH_2{=}CH{-}CH{=}CH_2$	−113	−4.5	0.650
2-Methyl-1,3-butadiene (Isoprene)	$CH_2{=}\overset{\displaystyle CH_3}{\underset{\displaystyle |}{C}}{-}CH{=}CH_2$	−120	35	0.681
1-Pentene	$CH_2{=}CH(CH_2)_2CH_3$	−138	30.1	0.641
1-Hexene	$CH_2{=}CH(CH_2)_3CH_3$	−141	64.1	0.673

Reactions of the Alkenes

The chemical behavior most characteristic of the alkenes is the ability to form derivatives by addition to the double bond.

$$-\overset{|}{C}{=}\overset{|}{C}- + AB \longrightarrow -\underset{\underset{\displaystyle A}{|}}{\overset{|}{C}}-\underset{\underset{\displaystyle B}{|}}{\overset{|}{C}}-$$

In such reactions the addition of hydrogen, the halogens (Cl_2, Br_2, I_2), the halogen acids (HCl, HBr), the hypohalous acids (HOCl, HOBr), sulfuric acid, and water all result in the formation of saturated compounds. In such reactions the *pi* bond plays the role of an electron pair donor. The principal reactions of the alkenes are discussed in the following sections.

3.6   *Addition of Hydrogen*

Hydrogenation.   Alkenes add hydrogen under pressure and in the presence of a catalyst to produce saturated hydrocarbons.

$$H{-}\overset{\displaystyle \overset{H}{|}}{C}{=}\overset{\displaystyle \overset{H}{|}}{C}{-}H + H_2 \xrightarrow[\text{pressure, heat}]{\text{Ni}} H{-}\underset{\underset{\displaystyle H}{|}}{\overset{\overset{\displaystyle H}{|}}{C}}{-}\underset{\underset{\displaystyle H}{|}}{\overset{\overset{\displaystyle H}{|}}{C}}{-}H$$

Ethene   Ethane

A hydrogenation carried out in this manner is called **catalytic hydrogenation.** Without a catalyst the rate of hydrogenation is negligible even though the addition of hydrogen to the double bond is an exothermic reaction; that is, even

1,2-Dimethylcyclopentene

Surface of catalyst

cis-1,2-Dimethylcyclopentane

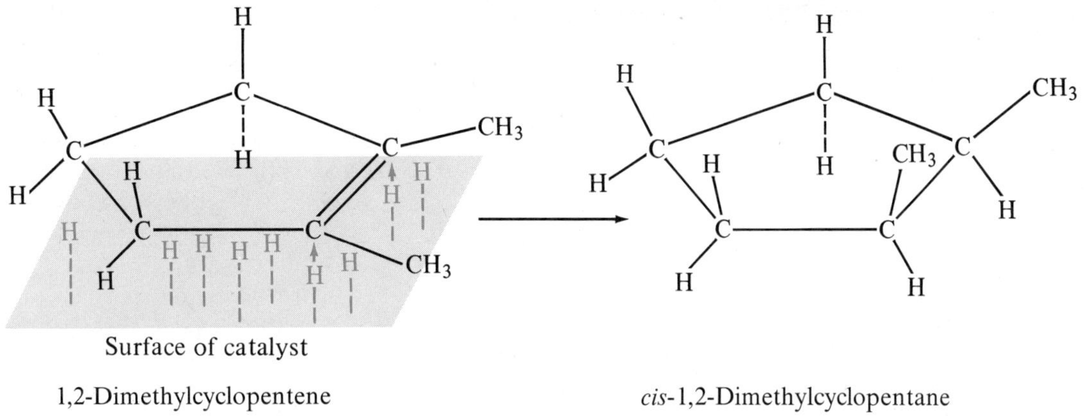

though the products are more stable (have lower energy) than the reactants. The function of the catalyst appears to be to lower the very high energy of activation (Sec. 2.9) of the overall reaction by changing the mechanism of the reaction to one with a lower overall energy of activation. Such a change would increase the rate of hydrogenation. The exact mechanism of hydrogenation is complex and apparently involves carbon-metal bonds. The schematic picture shown in Fig. 3.5 is sufficient for our purposes and shows the tendency of both hydrogen atoms to add to the double bond from the same side—that is, from the side coordinated (complexed) with the catalyst surface by interaction with the π electrons. This mode of addition is called *cis* addition and usually results in the formation of only the *cis*-isomer if isomeric products are possible. For example, the hydrogenation of 1,2-dimethylcyclopentene yields only *cis*-1,2-dimethylcyclopentane.

The best catalysts for hydrogenation of unsaturated compounds are palladium and platinum. Nickel catalysts are also good (and less expensive), but temperatures higher than those needed with palladium and platinum are sometimes necessary.

Hydrogenation of unsaturated compounds is a useful and a commonly employed reaction and one that has a number of important industrial applications. One of these is the hydrogenation of highly unsaturated liquid vegetable oils to solid, butterlike fats called **margarines.**

3.7 Addition of Halogens

Bromination.   An alkene reacts with bromine to add the halogen across the double bond to place one bromine atom on each carbon. The alkene usually is added to a carbon tetrachloride solution of bromine. The reddish color of bromine in carbon tetrachloride disappears as the bromine adds to the alkene. The reaction thus is a test reaction if the presence of the double bond in an unknown compound is suspected. The mechanism by which bromine adds to ethene may be explained as follows.

Ethene, although a nonpolar molecule, may, because of the mobility of the π electrons, cause a polarization of the bromine molecule—that is, a displacement of the pair of electrons joining the bromine atoms toward one of the atoms. One bromine with six electrons is abstracted by the π electrons of ethene to produce a positively charged intermediate which may be either a carbonium ion, (a) or (c), or a bromonium ion, (b).

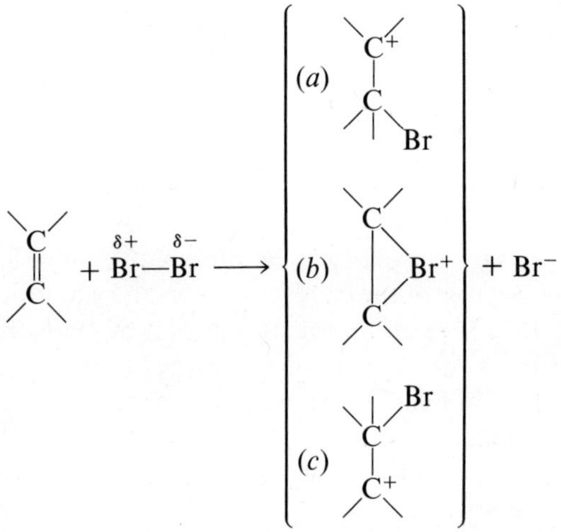

The negative bromide ion then combines with the positively charged intermediate. Experimental studies show that the bromide ion approaches from a direction most remote from the bromine already bonded. The existence of the intermediate bromonium ion or the carbonium ion has been established by the fact that addition of bromine to an alkene in the presence of other negative ions results in a mixture of products.

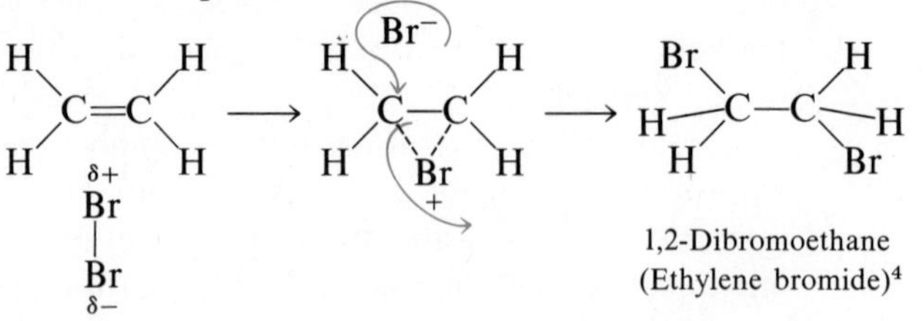

1,2-Dibromoethane
(Ethylene bromide)[4]

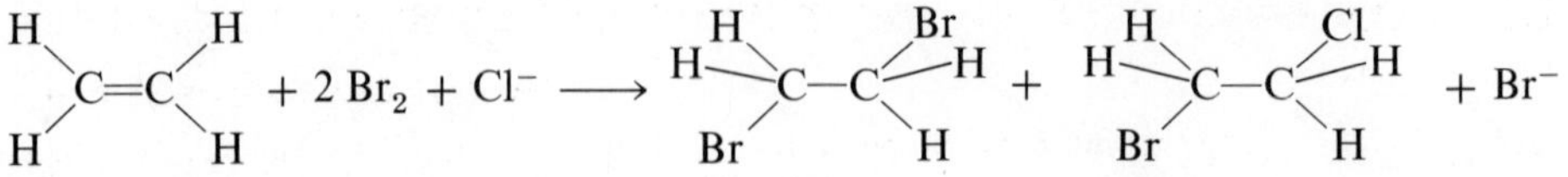

1-Bromo-2-chloroethane

---

[4] The ethylene group is $-\overset{\underset{\displaystyle H}{|}}{\underset{\underset{\displaystyle H}{|}}{C}}-\overset{\underset{\displaystyle H}{|}}{\underset{\underset{\displaystyle H}{|}}{C}}-$. It would be redundant to say ethylene dibromide.

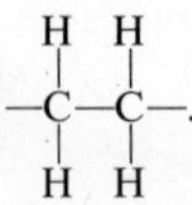

FIGURE 3.6 *Schematic Representation of the Bromination of Cyclopentene*

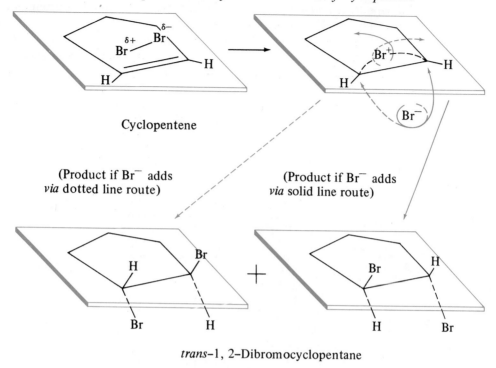

Cyclopentene

(Product if Br⁻ adds
via dotted line route)

(Product if Br⁻ adds
via solid line route)

trans-1, 2-Dibromocyclopentane

This mode of addition is called *trans*-addition and gives the *trans* isomer when isomeric products are possible. The *trans* addition of bromine to the double bond of cyclopentene is illustrated schematically in Fig. 3.6.

3.8 *Addition of Acids to Unsymmetrical Alkenes* (Markovnikov's Rule)

A. Addition of Hydrogen Halide. Hydrogen halides add readily to alkenes to produce alkyl halides, R–X. Thus, hydrogen bromide adds to ethene to produce ethyl bromide according to the following equation.

$$H\!-\!\overset{\overset{\displaystyle H}{|}}{C}\!=\!\overset{\overset{\displaystyle H}{|}}{C}\!-\!H + HBr \longrightarrow H\!-\!\overset{\overset{\displaystyle H}{|}}{\underset{\underset{\displaystyle H}{|}}{C}}\!-\!\overset{\overset{\displaystyle H}{|}}{\underset{\underset{\displaystyle H}{|}}{C}}\!-\!Br$$

Ethene

Ethyl bromide

In this particular reaction there is only one way in which the halogen acid can add across the double bond and therefore only one product is obtainable. This is not the case with an unsymmetrical alkene in which the groups bonded to the unsaturated carbon atoms are different. For example, hydrogen bromide adds to propene to produce *mainly* isopropyl bromide.

$$CH_3-\overset{\overset{\displaystyle H}{|}}{\underset{\delta+}{C}}=\overset{\overset{\displaystyle H}{|}}{\underset{\delta-}{C}}-H \;+\; \boxed{H^+} \;\; \boxed{:\overset{..}{\underset{..}{Br}}:} \;\longrightarrow\; CH_3-\overset{\overset{\displaystyle H}{|}}{\underset{\underset{\displaystyle Br}{|}}{C}}-CH_3$$

Propene

2-Bromopropane
(Isopropyl bromide)
B.P. 59.4°

Note that in this reaction the bromide ion seeks the central rather than the terminal carbon atom. The Russian chemist Markovnikov in 1871 formulated an empirical rule regarding this mode of addition. The rule simply states: *"In the addition of acids to unsymmetrical alkenes, the negative portion of the species added will seek the carbon atom holding the fewer hydrogen atoms."* This empirical rule has few exceptions.

The reason for the success of the Markovnikov rule is easily explained. As was pointed out in Sec. 3.2, the double bond is an electron-rich bond. As such, it tends to behave like a Lewis base and react with electrophilic (electron-seeking) reagents, such as hydrogen bromide.

$$CH_3-\overset{+}{\underset{\underset{\displaystyle H}{|}}{C}}-\overset{\overset{\displaystyle H}{|}}{\underset{\underset{\displaystyle H}{|}}{C}}-H \;+\; :\overset{..}{\underset{..}{Br}}:^- \;\longrightarrow\; CH_3-\overset{\overset{\displaystyle Br}{|}}{CH}-CH_3$$

Major product

$$CH_3-\overset{\overset{\displaystyle H}{|}}{\underset{\underset{\displaystyle H}{|}}{C}}-\overset{+}{\underset{\underset{\displaystyle H}{|}}{C}} \;+\; :\overset{..}{\underset{..}{Br}}:^- \;\longrightarrow\; CH_3-CH_2-CH_2-Br$$

Minor product

In the first step a carbonium ion is formed, which then reacts with bromide ion to form the product.[5] The rate at which a carbonium ion forms in a reaction has been found to depend on its stability, and the stability of a carbonium ion increases with the number of alkyl substituents on the positive carbon atom, or in the order: tertiary > secondary > primary.[6] Therefore, the secondary carbonium ion with two methyl groups is formed as an intermediate more readily than the primary carbonium ion bearing only one ethyl group.

---

[5] The dotted arrow with a cross through it indicates a possible but unfavorable reaction path that is observed to a small extent or not at all in the laboratory.

[6] The *class* of a carbonium ion, carbanion, or radical is determined by the number of substituents on the positive carbon atom, negative carbon atom, or atom bearing the odd electron, respectively: *tertiary,* 3 substituents; *secondary,* 2 substituents; *primary,* 1 substituent.

$$CH_3 \overset{H}{\underset{+}{\mid}} CH_3 \qquad CH_3CH_2 \overset{H}{\underset{H}{\mid}} C^+$$

A secondary carbonium ion	A primary carbonium ion

Although hydrogen bromide adds to unsymmetrical alkenes in accordance with the Markovnikov rule in the dark and in the absence of peroxides, in the presence of light or peroxides hydrogen bromide adds to alkenes in an *anti-Markovnikov* manner. Obviously, the mechanism of the addition under these conditions cannot proceed through a carbonium ion intermediate. The *anti-Markovnikov* mode of addition of HBr can be explained in terms of a free radical mechanism. Indicating the peroxide involved simply as R—O:O—R, we may show the different steps in the "peroxide effect" as follows:

$$R—O:O—R \longrightarrow 2\,R—O\cdot$$
$$R—O\cdot + HBr \longrightarrow R—OH + Br\cdot$$

$$CH_3 \overset{H\ \ H}{\underset{\ \ \ H}{-C-C-}} Br$$
(secondary)

$$CH_3 \overset{H\ \ H}{-C=C-} H + \cdot Br$$
Propene

$$CH_3 \overset{H\ \ H}{\underset{Br}{-C-C-}} H$$
(primary)

$$CH_3 \overset{H\ \ H}{\underset{H}{-C-C-}} Br + HBr \longrightarrow CH_3CH_2CH_2Br + Br\cdot$$
1-Bromopropane
(*n*-Propyl bromide)
B.P. 71°

Since the relative stabilities of free radicals parallel those of carbonium ions, namely, *tertiary* > *secondary* > *primary*, the secondary radical will form almost to the exclusion of the primary radical.

The peroxide-catalyzed addition of hydrogen halide to an alkene, while a useful reaction, unfortunately works only with hydrogen bromide.

EXERCISE 3.5

Two alkyl halides are possible when isobutylene, $(CH_3)_2C=CH_2$, is treated with HBr. Which one will predominate? Why?

B. Addition of Hypohalous Acids. Markovnikov's rule may be extended to cover the addition of other acids. The addition of chlorine or bromine to an alkene, when carried out in the presence of water, has the net result of adding a hypohalous acid, HOX. The reaction results in the formation of *halohydrins*. Halohydrins have a hydroxyl group and a halogen atom on adjacent carbons. The addition of an aqueous solution of chlorine to propene produces propylene chlorohydrin.

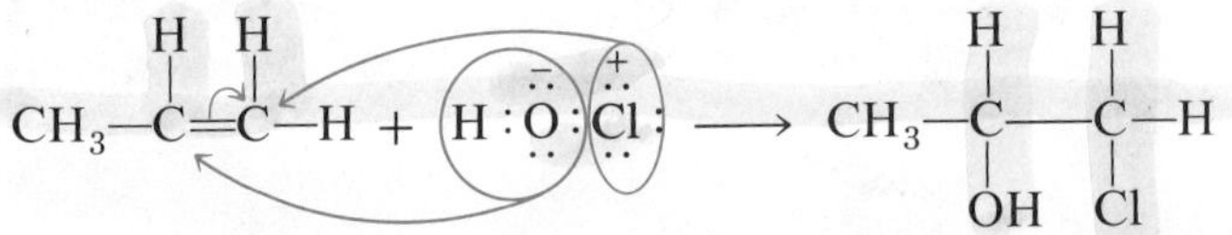

Propene
1-Chloro-2-propanol
(Propylene chlorohydrin)

In this addition, you will note, chlorine is the positive part of the species added.

C. Addition of Sulfuric Acid (Hydration of Alkenes). Alkenes react with sulfuric acid to produce alkyl hydrogen sulfates which, on hydrolysis, yield alcohols. Hydration (addition of water) of an alkene is thus accomplished indirectly in this manner and is an important commercial process for the preparation of alcohols from petroleum products.

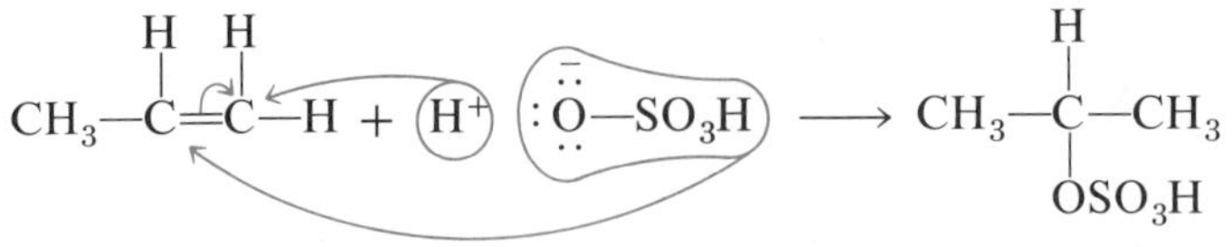

Isopropyl
hydrogen sulfate

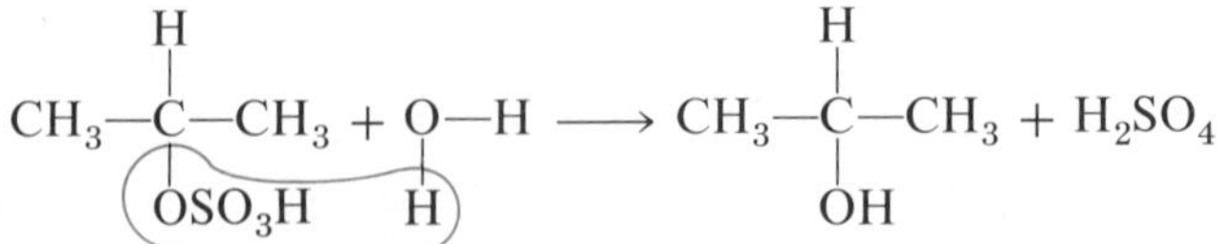

2-Propanol
(Isopropyl alcohol)

3.9 *Addition of Carbenes (Methylenes)*

Carbenes (Sec. 1.10) are electron-deficient, yet have two unshared electrons. Thus, they could function as electrophiles or nucleophiles; however, they usually behave as the former in their reaction with alkenes. The carbene most readily formed is dichlorocarbene, which can be prepared as an unstable intermediate by the α-dehydrohalogenation of chloroform with a very strong base. Dichlorocarbene adds stereospecifically *cis* to alkenes to give 1,1-dichloro-cyclopropanes.

| Chloroform | Potassium *tert*-butoxide | Dichlorocarbene | *tert*-Butyl alcohol |

trans-2-Butene   *trans*-1,1-Dichloro-2,3-dimethylcyclopropane

Another simple synthesis of cyclopropanes involves the addition of the Simmons-Smith reagent, iodomethylzinc iodide, to alkenes. The reagent is easily prepared by the reaction of methylene iodide with zinc dust that has been treated with copper sulfate solution.

$$I—CH_2—I + Zn(Cu) \longrightarrow [I—CH_2—Zn—I]$$

Methylene iodide   Iodomethylzinc iodide

Propene   Methylcyclopropane

Because the reactions of iodomethylzinc iodide resemble those of the carbenes, iodomethylzinc is called a **carbenoid** (carbene-like) and the reaction is called a methylene transfer reaction (a reaction that transfers a methylene group).

3.10 Oxidation

A. The Baeyer Test.
Alkenes are oxidized with cold dilute potassium permanganate to form glycols. In the absence of other easily oxidizable groups (which also react with $KMnO_4$), this reaction serves as an easily recognizable

test for alkenes because as the oxidant is consumed its bright purple color disappears. The reaction is the basis for the Baeyer Test for unsaturation.

$$3 \; H-\overset{\overset{\displaystyle H}{|}}{C}=\overset{\overset{\displaystyle H}{|}}{C}-H + 2 \; KMnO_4 + 4 \; H_2O \longrightarrow 3 \; H-\overset{\overset{\displaystyle H}{|}}{\underset{\underset{\displaystyle OH}{|}}{C}}-\overset{\overset{\displaystyle H}{|}}{\underset{\underset{\displaystyle OH}{|}}{C}}-H + 2 \; MnO_2 + 2 \; KOH$$

1,2-Ethanediol
(Ethylene glycol)

B. Oxidation with Rupture of the Double Bond. Vigorous oxidation of an alkene with potassium permanganate at elevated temperatures results in a cleavage of the double bond. The glycol intermediate formed is destroyed and each of the carbon atoms which comprised the carbon-carbon double bond becomes completely oxidized. A terminal doubly-bonded carbon is oxidized by this treatment to CO_2, and one bearing two alkyl groups is oxidized to the carbonyl group, $\left(\overset{}{\underset{}{}}C=O \right)$.

$$3 \; CH_3-\overset{\overset{\displaystyle}{\underset{\underset{\displaystyle CH_3}{|}}{C}}}{}=CH_2 + 8 \; KMnO_4 \longrightarrow 3 \; CH_3-\overset{}{\underset{\underset{\displaystyle CH_3}{|}}{C}}=O + 3 \; K_2CO_3 + 8 \; MnO_2 + 2 \; KOH + 2 \; H_2O$$

Isobutylene   Acetone

A doubly-bonded carbon atom bearing one alkyl group is oxidized to the carboxyl group $\left(-C\overset{\displaystyle O}{\underset{\displaystyle OH}{<}} \right)$.

$$3 \; CH_3-\overset{\overset{\displaystyle H}{|}}{C}=\overset{\overset{\displaystyle H}{|}}{C}-CH_3 + 8 \; KMnO_4 \longrightarrow 6 \; CH_3-\overset{\overset{\displaystyle O}{||}}{C}-O^- K^+ + 8 \; MnO_2 + 2 \; KOH + 2 \; H_2O$$

2-Butene   Potassium acetate

C. Oxidation with Ozone (O_3). Most alkenes react with ozone to form cyclic peroxide intermediates called **ozonides.** Ozonides may be explosive compounds and thus are not isolated but are directly decomposed with zinc metal and aqueous acetic acid. The **ozonolysis** of an alkene yields two products, each with a doubly-bonded oxygen where originally the doubly-bonded carbon atom had been. Examination of the reaction products thus establishes the location of the double bond. The ozonolysis reaction has its greatest value as a diagnostic procedure, not as a preparative one.

$$CH_3-\overset{\overset{\displaystyle CH_3}{|}}{C}=CH_2 + O_3 \longrightarrow$$

2-Methylpropene
(Isobutylene)

An ozonide

(originally in the alkene)

$$\xrightarrow[\text{CH}_3\text{CO}_2\text{H}]{\text{H}_2\text{O, Zn}} \quad \overset{CH_3}{\underset{CH_3}{\diagup}}C=O + O=\overset{H}{\underset{H}{\diagdown}}C \quad + \tfrac{1}{2}O_2$$

Acetone   Formaldehyde

EXERCISE 3.6

What products would result if propene were treated with a cold, dilute solution of potassium permanganate? With hot concentrated potassium permanganate? With ozone?

3.11 *Polymerization*

Many simple olefinic substances may undergo self-addition to form huge molecules composed of many basic units. Such high molecular weight structures are called **polymers** (Gr., *poly,* many; *meros,* part), and the reaction by which they are formed is called **polymerization.** Polyethylene, polypropylene, Teflon (Sec. 7.3), Orlon (Sec. 3.19), and synthetic rubbers (Sec. 3.22) are examples of familiar polymeric products formed in this manner. One type of polymerization reaction can be initiated by a free radical. The free radical, often produced from peroxide, R—O—O—R, combines in the first step with a **monomer,** that is, with one molecule of the olefin, to produce a new free radical. The chain continues to grow by repeated free radical formation. The following series of reactions illustrate the *initiation, propagation,* and *termination* steps in the free radical polymerization of ethylene.

Initiation: Free radicals are produced by an initiator.

$$R-O:O-R \longrightarrow 2\,R-O\cdot$$

$$R-O\cdot + H-\overset{\overset{\displaystyle H}{|}}{\underset{\underset{\displaystyle H}{|}}{C}}=\overset{\overset{\displaystyle H}{|}}{\underset{\underset{\displaystyle H}{|}}{C}}-H \longrightarrow R-O-\overset{\overset{\displaystyle H}{|}}{\underset{\underset{\displaystyle H}{|}}{C}}-\overset{\overset{\displaystyle H}{|}}{\underset{\underset{\displaystyle H}{|}}{C}}\cdot$$

Ethylene

Propagation: Continued addition of alkene produces a larger free radical.

$$R-O-\underset{\underset{H}{|}}{\overset{\overset{H}{|}}{C}}-\underset{\underset{H}{|}}{\overset{\overset{H}{|}}{C}} \cdot \;+\; n\;H_2C\!=\!CH_2 \;\longrightarrow\; R-O-\left(\underset{\underset{H}{|}}{\overset{\overset{H}{|}}{C}}-\underset{\underset{H}{|}}{\overset{\overset{H}{|}}{C}}\right)_n\!\!-CH_2-\underset{\underset{H}{|}}{\overset{\overset{H}{|}}{C}} \cdot$$

Termination: Two free radicals finally couple or disproportionate.

$$R-O-\left(\underset{\underset{H}{|}}{\overset{\overset{H}{|}}{C}}-\underset{\underset{H}{|}}{\overset{\overset{H}{|}}{C}}\right)_n\!\!-CH_2-\underset{\underset{H}{|}}{\overset{\overset{H}{|}}{C}} \cdot \;+\; \cdot\underset{\underset{H}{|}}{\overset{\overset{H}{|}}{C}}-CH_2-\left(\underset{\underset{H}{|}}{\overset{\overset{H}{|}}{C}}-\underset{\underset{H}{|}}{\overset{\overset{H}{|}}{C}}\right)_m\!\!-O-R$$

Couple   Disproportionate

$$R-O-\left(\underset{\underset{H}{|}}{\overset{\overset{H}{|}}{C}}-\underset{\underset{H}{|}}{\overset{\overset{H}{|}}{C}}\right)_{n+m+2}\!\!-O-R$$

$$R-O-\left(\underset{\underset{H}{|}}{\overset{\overset{H}{|}}{C}}-\underset{\underset{H}{|}}{\overset{\overset{H}{|}}{C}}\right)_{m}\!\!-\underset{\underset{H}{|}}{\overset{\overset{H}{|}}{C}}\!=\!\underset{\underset{H}{|}}{\overset{\overset{H}{|}}{C}}-H \;+\; R-O-\left(\underset{\underset{H}{|}}{\overset{\overset{H}{|}}{C}}-\underset{\underset{H}{|}}{\overset{\overset{H}{|}}{C}}\right)_{n}\!\!-\underset{\underset{H}{|}}{\overset{\overset{H}{|}}{C}}-\underset{\underset{H}{|}}{\overset{\overset{H}{|}}{C}}-H$$

Polyethylene

EXERCISE 3.7

If the polyethylene structure produced in the example just given has a molecular weight of 25,000, what number is represented by n? The weight of the initiator may be neglected.

Many polymeric products are tough, flexible, and unreactive to most chemical reagents. Some have a waxy feel and are insoluble in most solvents. Others, with excellent thermal and electric properties, are useful insulating materials. The simple olefins are used to make several polymeric products, as we have seen, but the double bond in many compounds other than the alkenes is the functional group which makes possible a great number of wonderful polymeric products. Synthetic polymers of the type described not only have revolutionized the packaging industry and the textile industry but also have made possible the creation of new industries. Table 3.2 shows the various polymers and products obtained from substituted ethenes, i.e., vinyls, $H_2C\!=\!CH\!-\!R$.

3.12 *Dienes and Polyenes*

Extremely reactive functional groups often are a part of organic structures. One can be reasonably certain that *simple* compounds containing highly

TABLE 3.2 *Common Addition Polymers*

Structure of R	Name of Monomer	Name of Polymer	Use
—H	Ethylene	Polyethylene	Film, conduit, rubber-like articles.
—CH_3	Propylene	Polypropylene	Molded and extruded plastics, film, fibers for garments and carpeting.
(phenyl ring)	Styrene	Polystyrene	Insulation, breakage-proof packaging, molded articles.
—Cl	Vinyl chloride	Polyvinyl chloride	Electrical insulation, films, rubber-like articles.
—F	Vinyl fluoride	Polyvinyl fluoride	Surface coatings.
—CN	Vinyl cyanide (Acrylonitrile)	Polyacrylonitrile (Orlon)	Fibers for garments, carpeting.
$-O-\overset{\displaystyle O}{\underset{\displaystyle \|}{C}}-CH_3$	Vinyl acetate	Polyvinyl acetate	Films, fibers, molded articles.
$-O-\overset{\displaystyle H}{\underset{\displaystyle C_3H_7}{C}}OCH-CH_2$	Vinyl butyral	Polyvinyl butyral	Laminate in safety glass.

reactive centers do not occur naturally, but are synthetic products. This is the case with the lower members of the alkene series—ethene, propene, and the butenes. Aside from these simple structures, alkenes occur widely in nature in a variety of complex forms. Isoprene, $CH_2=C(CH_3)-CH=CH_2$ (2-methyl-1,3-butadiene), is a **diene** structure which occurs as the building unit in a number of natural products. Isoprene units appear over and over again in a class of natural compounds called the terpenes (Sec. 17.13). Common examples of the terpenes are camphor, rosin, turpentine, natural rubber, and the coloring matter of certain plants, including edible ones. In each of such natural products some multiple of the isoprene unit may be found. An excellent example, illustrating the repeated addition of isoprene units, is β-carotene, the yellow coloring matter of carrots.

β-Carotene

A system of alternate double and single bonds in a carbon chain such as that in β-carotene is called a **conjugated** system. The simplest conjugated system possible is 1,3-butadiene. It also is one of the most important of the dienes; for, with styrene, it is used in the production of one type of synthetic rubber.

A conjugated diene in which two π bonds are separated by one σ bond is more stable than a simple alkene. The stability of a conjugated diene is inherent in its structure. For example, when we show 1,3-butadiene as

$$
\begin{array}{cccc}
\text{H} & \text{H} & \text{H} & \text{H} \\
| & | & | & | \\
\text{H}-\text{C}={}\text{C}-\text{C}={}\text{C}-\text{H} &&&
\end{array}
$$

we are showing its structure as if the two π bonds were independent of each other. Actually, they are not entirely independent. If we redraw butadiene as shown below, we can see that the $2p$ orbitals comprising the basis of the π system can overlap between carbons 2 and 3 as well as between 1 and 2 or 3 and 4. Thus, the π electrons can move to a limited extent over the whole system and are said to be delocalized. The delocalization of the π electrons not only adds somewhat to the stability of the butadiene molecule but also alters its chemical behavior. However, the extent of π bond delocalization in butadiene is sufficiently small that the simple Lewis structure is a satisfactory representation.

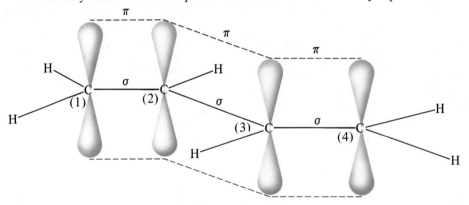

To illustrate the reactivity of butadiene we will examine possible reactions of the compound with bromine.

$$
\begin{array}{cccccc}
{}^{(4)} & {}^{(3)} & {}^{(2)} & {}^{(1)} \\
\text{H}-\text{C}={}\text{C}-\text{C}={}\text{C}-\text{H} & + & :\!\ddot{\text{Br}}-\ddot{\text{Br}}: & \longrightarrow \\
| & | & | & | \\
\text{H} & \text{H} & \text{H} & \text{H}
\end{array}
$$

$$
\left[
\begin{array}{cccc}
&&& \text{Br} \\
&&& | \\
\text{H}-\text{C}={}\text{C}-\overset{+}{\text{C}}-\text{C}-\text{H} \\
| & | & | & | \\
\text{H} & \text{H} & \text{H} & \text{H}
\end{array}
\longleftrightarrow
\begin{array}{cccc}
&&& \text{Br} \\
&&& | \\
\text{H}-\overset{+}{\text{C}}-\text{C}={}\text{C}-\text{C}-\text{H} \\
| & | & | & | \\
\text{H} & \text{H} & \text{H} & \text{H}
\end{array}
\right] + \text{Br}^-
$$

$$
\quad\quad\quad\quad\text{I}\quad\quad\quad\quad\quad\quad\quad\quad\quad\quad\quad\text{II}
$$

$$
\downarrow \text{Br}^- \quad\quad\quad\quad\quad\quad\quad\quad\quad\quad\quad \downarrow \text{Br}^-
$$

$$
\begin{array}{cccc}
& \text{Br} & \text{Br} \\
& | & | \\
\text{H}-\text{C}={}\text{C}-\text{C}-\text{C}-\text{H} \\
| & | & | & | \\
\text{H} & \text{H} & \text{H} & \text{H}
\end{array}
\quad\quad\quad
\begin{array}{cccc}
\text{Br} & & \text{H} & \text{Br} \\
| & & | & | \\
\text{H}-\text{C}-\text{C}={}\text{C}-\text{C}-\text{H} \\
| & | & & | \\
\text{H} & \text{H} & & \text{H}
\end{array}
$$

3,4-Dibromo-1-butene   *trans*-1,4-Dibromo-2-butene
III   IV

The reaction of butadiene with bromine should yield the cation I. However, cation I differs from cation II only in the arrangement of electrons (the shift of a pair of electrons converts I into II, as shown). Thus, I and II are a pair of resonance structures contributing to a hybrid structure for the cationic intermediate (Sec. 1.12). The true structure of the cation is neither that of I nor II, but a hybrid of the two. Resonance hybrids may react as though they had the structure of only one of the contributing forms, or they may react as if they were a mixture of several of the forms. In the present case, the positive charge is almost equally distributed between the two carbon atoms shown with positive charges in I and II; therefore, bromide ion adds to the cationic intermediate to give products derived from both forms I and II to yield 3,4-dibromo-1-butene (III) and *trans*-1,4-dibromo-2-butene (IV). The mode of addition leading to structure III is called **1,2-addition;** that leading to structure IV is called **1,4-addition.**

EXERCISE 3.8

The polymerization of butadiene with free radical catalysts usually produces a type of polymer in which 80% of the recurring units are 1,4-units, $-CH_2-CH=CH-CH_2-$, and 20% are 1,2-units. $\left(CH_2-CH \atop \qquad | \atop CH=CH_2\right)$.

Write a mechanism that will explain the formation of both types of recurring unit.

3.13   *The Diels-Alder Reaction*

One of the most important reactions of conjugated dienes is the Diels-Alder[7] reaction:

1,3-Butadiene	Dimethyl maleate	Dimethyl *cis*-4-cyclohexene-1,2-dicarboxylate
(Diene)	(Dienophile)	(Adduct)

---

[7] Otto Diels (1876–1954), University of Kiel, Nobel prize in 1950. Kurt Alder (1902–1958), a student of Diels before going to the University of Cologne, Nobel prize in 1950.

The reaction involves a **diene** and a **dienophile** to give an **adduct** by a concerted [4 + 2]-**cycloaddition** reaction that requires no catalyst. The designation [4 + 2] indicates that a 4-π-electron system combines with a 2-π-electron system during the course of the reaction. Although the dienophile can be a simple alkene (at the expense of higher reaction temperatures and elevated pressure), electron-attracting groups, such as —COR (aldehyde or ketone), —CO$_2$R (carboxylic acid or its ester), and —C≡N, facilitate the reaction. The reaction is **stereospecific** in that the *cis* or *trans* configuration of the diene and of the dienophile is preserved in the product. Thus, in the example shown, the two ester groups are *cis* relative to the double bond in the dienophile and *cis* relative to the cyclohexene ring in the adduct. This reaction is widely used in the synthesis of cyclohexene derivatives.

Part II: The Alkynes (Acetylenes)

The alkynes, or acetylenes, are unsaturated hydrocarbons which contain a triple bond or three pairs of shared electrons between adjacent carbon atoms. The general formula for an alkyne is C_nH_{2n-2} if only one triple bond is present in the structure. The same general formula fits that of an open chain diene. The first member of the alkyne family is acetylene itself, C_2H_2, the structure of which may be written

$$H—C≡C—H \quad \text{or} \quad H^xC::C^xH$$

EXERCISE 3.9

Draw electronic representations for (a) CS$_2$, (b) HCN, and (c) BeCl$_2$. In what respect are these structures like that of acetylene?

3.14   Formulas and Nomenclature

The alkynes are named systematically by the same rules of nomenclature that were used to name the alkenes. The triple bond is indicated by the suffix *yne* and is located by number in the longest chain which contains it. The alkyl derivatives of acetylene, in which one or both of the singly-bonded hydrogens of acetylene have been replaced by alkyl groups, are named simply as alkyl acetylenes. A few examples will help make this clear.

$$H—C≡C—H \qquad\qquad CH_3C≡C—H$$

Ethyne   Propyne
(Acetylene)   (Methylacetylene)

$$CH_3-C\equiv C-CH_3$$

2-Butyne
(Dimethylacetylene)

(1)   (2)   (3)   (4)   (5)
$$CH_3-C\equiv C-CH_2-CH_3$$

2-Pentyne
(Methylethylacetylene)

(5)   (4)   (3)   (2)   (1)
$$CH_3-CH_2-CH-C\equiv C-H$$
                    |
                  CH_3

3-Methyl-1-pentyne
(*sec*-Butylacetylene)

(1)   (2)   (3)   (4)
$$CH_2=CH-CH\equiv CH$$

1-Buten-3-yne
(Vinylacetylene)

3.15 *The Carbon-Carbon Triple Bond*

When carbon is attached to only two other atoms, hybridization of one $1s$ and one $2p$ orbital will give two sp orbitals. Overlap of two hydrogen $1s$ and four carbon sp orbitals will form three sigma bonds and leave two $2p$ orbitals at right angles to each other on each carbon. Overlap of the $2p$ orbitals will form two π bonds. Now, instead of a planar structure, as in the case of the alkenes, with a cloud of electrons above and below the plane, we have a linear structure with a cloud not only above and below, but also in front and behind the axis as well. Here the electrons of both π bonds overlap and blend into a cylindrical sleeve about the axis of the molecule (Fig. 3.7).

FIGURE 3.7 *Structural Representation of the Carbon—Carbon Triple Bond of Ethyne in Perspective (a) Showing Three σ Bonds and Two π Bonds and (b) As Seen If Viewed along the Carbon—Carbon Axis*

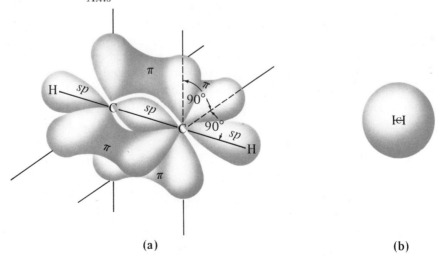

(a)   (b)

EXERCISE 3.10

In the following formula for vinyl acetylene, name the hybrid orbitals representing the labeled sigma bonds (a), (b), (c), and (d). Also, indicate the size of bond angles θ and ϕ.

$$\underset{\text{(a)}}{H_2C} = \underset{\text{(b)}}{CH} - \underset{\text{(c)}}{C} \equiv \underset{\text{(d)}}{C} - H$$

3.16 *Occurrence and Properties of the Alkynes*

The alkynes are somewhat more reactive than the alkenes, and the lower molecular weight members are not found in nature. However, the triple bond does occur as a structural feature in some natural substances. An example is the antibiotic *mycomycin,* which has the remarkable structure

$$\underset{\text{(13)}}{H-C} \equiv \underset{\text{(12)}}{C} - \underset{\text{(11)}}{C} \equiv \underset{\text{(10)}}{C} - \underset{\text{(9)}}{\overset{H}{C}} = \underset{\text{(8)}}{C} = \underset{\text{(7)}}{\overset{H}{C}} - \underset{\text{(6)}}{\overset{H}{C}} = \underset{\text{(5)}}{\overset{H}{C}} - \underset{\text{(4)}}{\overset{H}{C}} = \underset{\text{(3)}}{\overset{H}{C}} - \underset{\text{(2)}}{CH} - \underset{\text{(1)}}{C} \overset{O}{\underset{OH}{}}$$

It has the rather impressive name **3,5,7,8-tridecatetraene-10,12-diynoic acid.** An unusual group of similar highly unsaturated, linear compounds are secreted by various species of fungi. These compounds are straight chain polyacetylenes of eight to fourteen carbon atoms in length with one or more oxygen-containing functional groups.

The chemical properties of the alkynes, with some exceptions, are very similar to those of the corresponding alkenes. The reactions are largely those of addition. One exception is notable, however. *Hydrogen bonded to an acetylenic carbon atom is acidic in character and can be replaced by certain metals to form salts.* Why should a hydrogen attached to a triple-bonded carbon be acidic? If you will consider the orbital hybridization of acetylene (Sec. 3.15) you will see that three pairs of electrons are confined between two carbon nuclei in the linear structure of an acetylenic bond. The repulsion against the single C—H electron pair is less than it would be in the tetrahedral arrangement of alkanes. The pair of electrons between the carbon and hydrogen thus can move a little closer to the carbon atom than it would be able to in a saturated hydrocarbon. As a result, the acetylenic hydrogen is left slightly more positive and easily removed by a strongly basic reagent.

Preparation of the Alkynes

The alkynes are prepared in the laboratory by reactions very much like those employed for the preparation of alkenes. Obviously, *two simple molecules must be eliminated from adjacent carbons in a saturated compound* in order to introduce the triple bond into the molecule.

3.17 *Preparation of Acetylene*

Acetylene (ethyne), C_2H_2, is one of the most important of industrial organic raw materials. It may be prepared by reaction between calcium carbide and water. Calcium carbide is prepared by heating a mixture of coke and limestone in an electric furnace.

$$CaCO_3 \xrightarrow{heat} CaO + CO_2$$

$$3\,C + CaO \xrightarrow{2000°C} CaC_2 + CO$$
$$\text{Calcium}$$
$$\text{carbide}$$

$$CaC_2 + 2\,H_2O \longrightarrow H—C≡C—H + \quad Ca(OH)_2$$
$$\text{Acetylene} \qquad \text{Calcium hydroxide}$$

3.18 *Preparation of Higher Alkynes*

Higher members in the series may be prepared by one of two general methods: (1) elimination of two molecules of hydrogen halide from adjacent carbon atoms in a saturated compound, or (2) replacement of acetylenic hydrogen on an existing carbon atom by an alkyl group.

A. Elimination of Halogen Acid from *Vicinal* Dihalides.[8]   Notice that the product of the first HBr elimination is a substituted vinyl bromide. The vinyl halides are unreactive compounds and a stronger base (NH_2^-) is used to remove the second HBr molecule.

$$CH_3—\overset{H}{\underset{Br}{C}}—\overset{Br}{\underset{H}{C}}—H + KOH \xrightarrow[\text{solution}]{\text{alcoholic}} \quad CH_3—\overset{H}{C}=\overset{Br}{C}—H \quad + KBr + H_2O$$

1,2-Dibromopropane   1-Bromo-1-propene (a vinyl halide)

$$CH_3—\overset{H}{C}=\overset{Br}{C}—H + NaNH_2 \longrightarrow \quad CH_3—C≡C—H \quad + NaBr + NH_3$$
$$\text{Propyne (Methylacetylene)}$$

B. Replacement of Acetylenic Hydrogen by an Alkyl Group.   The first step in this synthesis is the preparation of the sodium salt of acetylene by reaction with sodium metal. The sodium acetylide then is coupled with an alkyl halide.

$$2\,H—C≡C—H + 2\,Na \xrightarrow{\text{liquid } NH_3} 2\,H—C≡C^{:-}\,Na^+ + H_2$$
$$\text{Sodium acetylide}$$

$$H—C≡C^{:-}\,Na^+ + \quad C_2H_5I \longrightarrow H—C≡C—CH_2CH_3 + Na^+I^-$$
$$\text{Ethyl iodide} \qquad \text{1-Butyne}$$

---

[8] Vicinal (L., *vicinalis,* neighboring). Vicinal dihalides have halogen atoms on adjacent carbon atoms.

The second acidic hydrogen, if available, also may be replaced by repeating the process. An alkyne of known structure thus can be synthesized.

$$H-C\equiv C-CH_2CH_3 + Na \xrightarrow[NH_3]{liquid} Na^+:\bar{C}\equiv C-CH_2CH_3 + \tfrac{1}{2}H_2$$

$$C_2H_5Br + Na^+:\bar{C}\equiv C-CH_2CH_3 \longrightarrow CH_3CH_2-C\equiv C-CH_2CH_3 + NaBr$$

3-Hexyne

3.19 *Reactions of the Alkynes*

The chemical reactions of the alkynes, like those of the alkenes, are largely reactions of addition. The mode of addition of hydrogen halides to unsymmetrical alkynes, you will note, also follows Markovnikov's rule.

A. Addition of Br_2, HBr, and H_2 to Acetylenes.

$$CH_3-C\equiv C-H + 2\,Br_2 \longrightarrow CH_3-\overset{\overset{\displaystyle Br}{|}}{\underset{\underset{\displaystyle Br}{|}}{C}}-\overset{\overset{\displaystyle Br}{|}}{\underset{\underset{\displaystyle Br}{|}}{C}}-H$$

Propyne   1,1,2,2-Tetra-
bromopropane

$$CH_3-C\equiv C-H + 2\,HBr \longrightarrow CH_3-\overset{\overset{\displaystyle Br}{|}}{\underset{\underset{\displaystyle Br}{|}}{C}}-\overset{\overset{\displaystyle H}{|}}{\underset{\underset{\displaystyle H}{|}}{C}}-H$$

2,2-Dibromopropane

$$CH_3-C\equiv C-H + 2\,H_2 \xrightarrow{catalyst} CH_3CH_2CH_3$$

Propane

B. Addition Reactions NOT Given by Olefins. Acetylene, unlike the alkenes, is sufficiently reactive to add hydrogen cyanide when the addition is carried out under the influence of acid catalysts. The product, acrylonitrile, is the important industrial chemical from which the synthetic fiber "Orlon" is made.

$$H-C\equiv C-H + (H^+):CN^- \xrightarrow[NH_4Cl\text{-}HCl]{CuCl,} H-\overset{\overset{\displaystyle H}{|}}{C}=\overset{\overset{\displaystyle H}{|}}{C}-CN$$

Acrylonitrile
(Vinyl cyanide)

Acetylene also will add water or acetic acid directly. These useful reactions are catalyzed by mercuric sulfate, the addition of water being effected in an acidic medium. Presumably, reaction with water proceeds through the intermediate formation of vinyl alcohol, $H_2C\!=\!CHOH$, a product which has never been isolated. If formed, vinyl alcohol immediately rearranges to acetaldehyde.

$$H-C\equiv C-H + H_2O \xrightarrow[H_2SO_4]{HgSO_4,} \left[H-\underset{H}{\overset{H}{C}}=\underset{H}{\overset{H}{C}}-O\,\textcircled{H} \right] \longrightarrow CH_3-\underset{H}{\overset{}{C}}{=}O$$

Acetaldehyde

$$H-C\equiv C-H + CH_3-\overset{O}{\overset{\|}{C}}-OH \xrightarrow[75°C]{Hg^{2+},} CH_2{=}CH-O-\overset{O}{\overset{\|}{C}}-CH_3$$

Vinyl acetate

Higher alkynes that still possess an acidic hydrogen may be hydrated to yield methyl ketones, $CH_3-\overset{O}{\overset{\|}{C}}-R$.

$$H-C\equiv C-CH_2CH_3 + H_2O \xrightarrow[H_2SO_4]{Hg^{2+},} \left[H-\underset{\underset{\textcircled{H}}{O}}{\overset{H}{C}}{=}C-CH_2CH_3 \right] \longrightarrow H-\underset{\underset{H}{\overset{|}{C}}}{\overset{H}{\overset{|}{C}}}-\overset{}{\underset{O}{\overset{\|}{C}}}-CH_2CH_3$$

1-Butyne   2-Butanone
(Methyl ethyl ketone)

3.20 *Replacement of Acetylenic Hydrogen by a Metal* (*Salt Formation*)

The principal difference between the chemical behavior of the alkynes and that of the alkenes, as was pointed out previously, is due to the acidic character of the hydrogen atom bonded to an acetylenic carbon. Acetylenic hydrogen atoms can be replaced by metals to produce salts called acetylides. (See Sec. 3.18(b) for reaction of acetylene with sodium in liquid ammonia.) Acetylene, when passed through ammoniacal solutions of cuprous chloride and silver nitrate forms the acetylides of these metals.

$$H-C\equiv C-H + 2\,Cu(NH_3)_2Cl \longrightarrow Cu-C\equiv C-Cu + 2\,NH_4Cl + 2\,NH_3$$

Copper acetylide

$$H-C\equiv C-H + 2\,Ag(NH_3)_2NO_3 \longrightarrow Ag-C\equiv C-Ag + 2\,NH_4NO_3 + 2\,NH_3$$

Silver acetylide

Heavy metal salts of acetylene such as silver acetylide are extremely sensitive to shock when dry and may explode violently.

3.21 *Oxidation of Acetylene* (*Combustion*)

Acetylene burns in an atmosphere of pure oxygen to produce extremely high temperatures (*ca.* 3000°C). This reaction makes possible the common oxyacetylene torch so useful for the welding and cutting of metals.

$$2\,H-C\equiv C-H + 5\,O_2 \longrightarrow 4\,CO_2 + 2\,H_2O + 619.7 \text{ kcal}$$

3.22 *Polymerization*

Acetylene can add to itself to yield a *dimer* when treated with cuprous chloride in aqueous ammonium chloride solution.

$$H-C\equiv C-H + H-C\equiv C-H \xrightarrow{\text{CuCl, NH}_4\text{Cl}} H-C\equiv C-\overset{\overset{\displaystyle H}{|}}{C}=CH_2$$

1-Buten-3-yne
(Vinylacetylene)

The vinylacetylene thus produced will add hydrogen chloride to give 2-chloro-1,3-butadiene or "chloroprene."

$$H-C\equiv C-\overset{\overset{\displaystyle H}{|}}{C}=CH_2 + HCl \longrightarrow H_2C=\overset{\overset{\displaystyle Cl}{|}}{C}-\overset{\overset{\displaystyle H}{|}}{C}=CH_2$$

2-Chloro-1,3-butadiene
(Chloroprene)

Chloroprene can be polymerized through a 1,4-addition of diene units into a synthetic rubber called "neoprene." Neoprene, unlike natural rubber, is resistant to attack by greases and oils and finds numerous uses in the automotive industry. It is used primarily as insulation and in sheathing for wires and cables, in hose and hydraulic lines, and in many other rubber components that are likely to come into contact with lubricants.

Summary Part I:

1. Unsaturated hydrocarbons with double bonds are called alkenes or olefins. The general formula for a continuous chain alkene with one double bond is C_nH_{2n}.

2. The double bond makes possible *cis,trans*-isomerism.

3. The availability of the π electrons causes alkenes to be more reactive than alkanes.

4. Alkenes are prepared by removing the elements of a simple molecule from adjacent carbon atoms by: (a) the dehydration of an alcohol; (b) the dehydrohalogenation of an alkyl halide.

5. Reactions of the alkenes are principally those of addition. Markovnikov's rule is followed in the addition of hydrogen halides with the exception of HBr addition in the presence of peroxides.

   (a) Addition takes place with: halogen, hydrogen, hydrogen halides, sulfuric acid, hypochlorous acid, and carbenes.

(b) Mild oxidation yields glycols; vigorous oxidation yields cleavage products—carboxylic acids, ketones, and CO_2.

(c) Oxidation with ozone results in cleavage of the double bond to yield two carbonyl $\left(\diagup_{\diagdown} C{=}O \right)$ compounds.

(d) Olefins may be polymerized to produce macromolecules called polymers.

6. A system of alternate single and double bonds is a conjugated system. Butadiene, the simplest example of a conjugated system, may add reagents by 1,2- or 1,4-addition.

7. The Diels-Alder reaction yields cyclohexenes by a concerted cycloaddition.

Summary Part II:

1. Unsaturated hydrocarbons with triple bonds are called acetylenes or alkynes. The general formula for a continuous chain alkyne with one triple bond is $C_n H_{2n-2}$.

2. The principal feature that distinguishes the alkynes from the alkenes is the acidic hydrogen on the triple-bonded carbon atom.

3. Alkynes are prepared by: (a) elimination of two simple molecules from adjacent carbon atoms; (b) replacement of the hydrogen on an existing alkyne by an alkyl group. Acetylene itself is prepared from calcium carbide.

4. The reactions of the alkynes are the following.
   (a) Addition (halogens, hydrogen, HCN, H_2O, acetic acid).
   (b) Oxidation (combustion); a highly exothermic reaction.
   (c) Salt formation. The sodium salt is an important intermediate. The heavy metal salts are explosive when dry.
   (d) Dimerization to produce vinylacetylene, chloroprene, and synthetic rubber.

New Terms

1,2- and 1,4-addition	dehydrohalogenation
adduct	Diels-Alder reaction
catalytic hydrogenation	diene
conjugated system	dienophile
cycloaddition	dimer

hydration   pi electrons
hydrolysis   polymerization
Markovnikov's rule   unsaturation
monomer   vicinal dihalide
ozonolysis

Supplementary Exercises

EXERCISE 3.11   Write a structural formula and give another acceptable name for each of the following.
  (a) propylene
  (b) ethylene bromide
  (c) isoprene
  (d) isobutylene
  (e) 3-heptyne

EXERCISE 3.12   Draw the structural formulas for the compounds having the following IUPAC names.
  (a) 2,5-dimethyl-3-hexene   (d) 1,3-cyclohexadiene
  (b) 2-ethyl-3-methyl-1-pentene   (e) 1,7-octadien-4-yne
  (c) *trans*-3,4-dichlorocyclohexene (f) 4-methyl-*cis*-2-pentene

EXERCISE 3.13   Name the following structures according to IUPAC rules.
  (a) $CH_3CH{=}CH{-}CH_2{-}CH_2{-}CH_3$
  (b) $CH_3CH(Cl)CH{=}CHCH_3$
  (c) $CH_3{-}CH_2{-}C{\equiv}C{-}CH_3$
  (d) $CH_3{-}C(Br){=}C(Br){-}CH_3$
  (e) $(CH_3)_2C{=}C(Br){-}CH_3$
  (f) $(CH_3)_2CH{-}CH{=}CH_2$

(g)

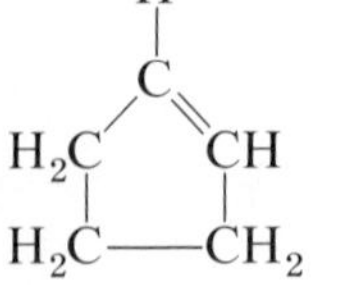

(h)

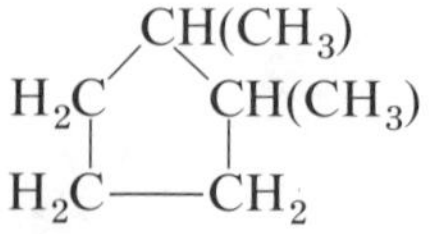

EXERCISE 3.14   Which structures in Exercise 3.13 may show *cis,trans*-isomerism?

EXERCISE 3.15   Beginning with ethyl alcohol, CH_3CH_2OH, as your only starting material and using any other reagent you may require, show how you would prepare:
  (a) ethene   (d) ethylmagnesium bromide
  (b) bromoethane   (e) ethane
  (c) 1-butyne   (f) acetylene

(g) 1,1,2-trichloroethane

(h) 1,2-dibromoethane

(i) acrylonitrile

(j) vinylacetylene

(k) vinyl chloride

(l) cyclopropane

EXERCISE 3.16 Complete each of the following reactions by supplying the necessary but missing component. This could be a reactant, a catalyst, or a product.

(a) $(?) + Br_2$ in $CCl_4 \longrightarrow CH_3CH(Br)CH_2Br$

(b) CH_3—CH=$CH_2 + (?) \longrightarrow CH_3CH(Br)CH_3$

(c) CH_3—CH=$CH_2 + H_2SO_4 \xrightarrow{(?)} CH_3CH(OH)CH_3$

(d) CH_3—CH=$CH_2 + HBr \xrightarrow{(?)} CH_3CH_2CH_2Br$

(e) $CH_3CH(Br)CH_2CH_3 + (?) \longrightarrow CH_3CH$=$CHCH_3$

(f) CH_3CH=$CHCH_3 \xrightarrow{(?)} CH_3COOH$

(g) CH_3CH=$CHCH_3 + O_3 \xrightarrow{\text{then with } Zn/H_2O} (?)$

(h) $(?) + $ dilute $KMnO_4 \xrightarrow{5\% \ Na_2CO_3} HOCH_2CH_2OH$

(i) $(?) \xrightarrow[H_2SO_4]{H_2O, \ Hg^{2+}} \left[H_2C\text{=}CH(OH) \right] \rightarrow CH_3CHO$

(j) 1,2-Dimethylcyclopentene $\xrightarrow[\text{pressure}]{H_2, \ Ni,} (?)$

(k) Cyclopentene $+ Br_2 \longrightarrow (?)$

(l) 1,3-Butadiene $+ HOCl \longrightarrow (?)$

EXERCISE 3.17 For each of the structures below draw two isomeric alkyl halides which, on dehydrohalogenation, would produce the olefin shown. In each case predict which of the two halogen compounds would react more readily to produce the unsaturated compound. Give reasons for your choice.

(a) CH_3CH=CH_2

(b) $(CH_3)_2C$=$CHCH_3$

(c) CH_3CH=$CHCH_2CH_3$

EXERCISE 3.18 A hydrocarbon, C_4H_8, neither decolorized a bromine-carbon tetrachloride solution nor reacted with HBr. When heated to 200° with hydrogen in the presence of a nickel catalyst a new hydrocarbon, C_4H_{10}, was formed. What was the original unknown?

EXERCISE 3.19 Recalling that the alkynes and dienes are isomeric with the type formula C_nH_{2n-2}, draw structures and assign names to two compounds that have the formula C_5H_8, but each of which shows only one of the following sets of properties:

Compound **A** adds two moles of bromine to yield a new substance, $C_5H_8Br_4$. Compound **A** forms a precipitate with ammoniacal $AgNO_3$. It may be hydrated in the presence of mercuric salts to yield a methyl ketone,

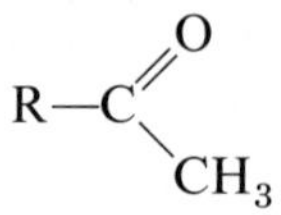

$$R-C\overset{\displaystyle O}{\underset{\displaystyle CH_3}{\diagup}}$$

Compound **B** also adds two moles of bromine to yield a substance of formula $C_5H_8Br_4$. Compound **B** is conjugated. It forms an ozonide which on reductive hydrolysis gives *three different* carbonyl-containing $\left(\diagdown C{=}O\right)$ structures. One of the latter has two carbonyl groups.

EXERCISE 3.20　Predict the products of the following reactions.

(a)

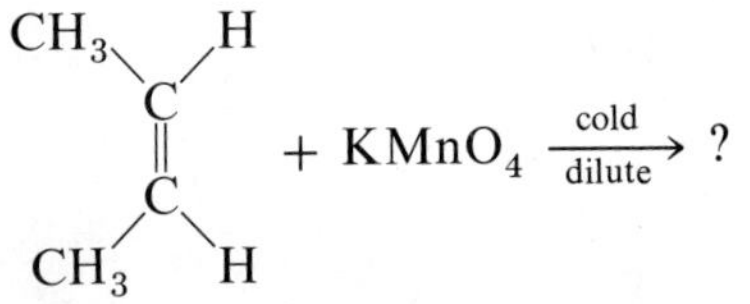

$$\underset{CH_3}{\overset{CH_3}{\diagdown}}\overset{H}{\underset{H}{\diagup}}C{\parallel}C + KMnO_4 \xrightarrow[\text{dilute}]{\text{cold}} ?$$

(b)

$$\underset{CH_3}{\overset{CH_3}{\diagdown}}C{=}CH_2 + H_2O \xrightarrow{H_2SO_4} ?$$

(c)

$$\begin{array}{c} CH_3 \\ | \\ C \\ CH_2 \quad CH \\ | \quad\quad | \\ CH_2 \quad CH_2 \\ \diagdown \quad \diagup \\ CH_2 \end{array} + HBr \xrightarrow{\text{peroxides}} ?$$

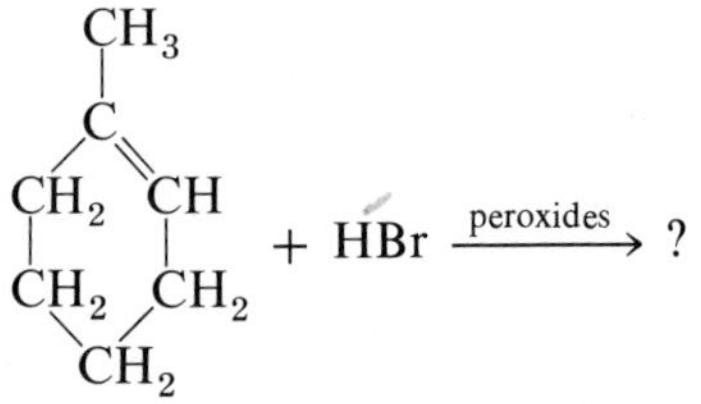

(d)

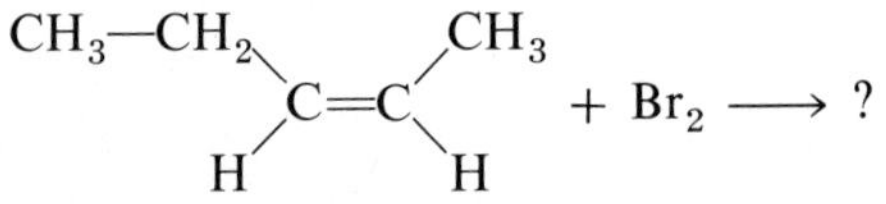

$$CH_3{-}CH_2 \overset{\diagdown}{\underset{H}{\diagup}}C{=}C\overset{CH_3}{\underset{H}{\diagdown}} + Br_2 \longrightarrow ?$$

(e)

$$\begin{array}{c} CH \\ CH_2 \quad CH \\ | \quad\quad | \\ CH_2 \quad CH_2 \\ \diagdown \quad \diagup \\ CH_2 \end{array} + CHCl_3 \xrightarrow[\text{base}]{\text{strong}} ?$$

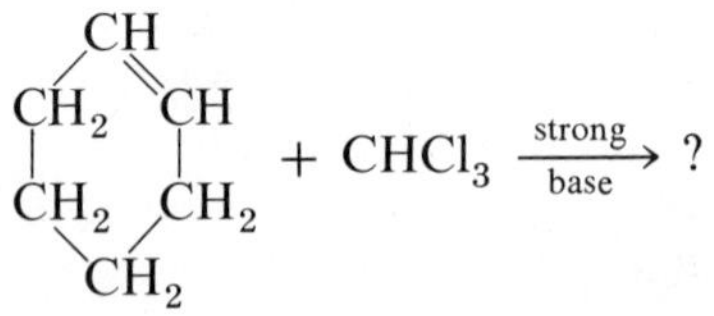

EXERCISE 3.21　Predict the products of the following Diels-Alder reactions.

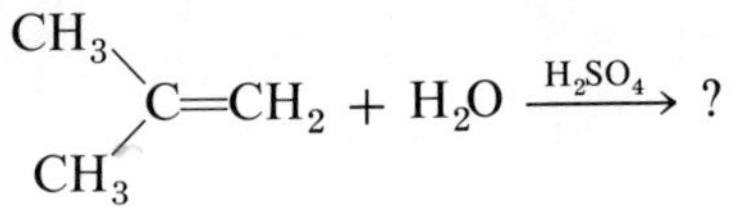

(a)

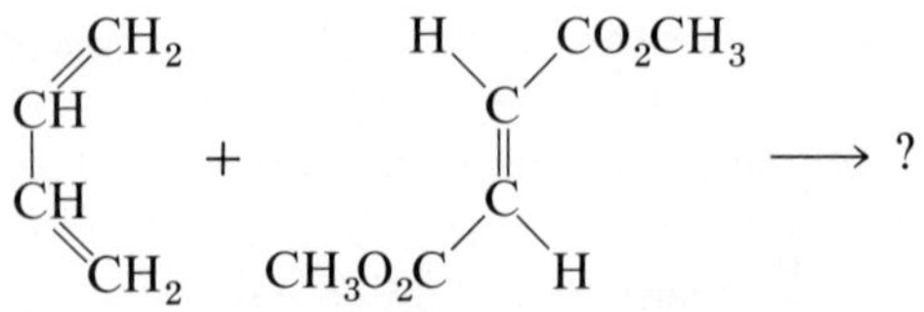

$$\begin{array}{c} CH_2 \\ \diagup \\ CH \\ | \\ CH \\ \diagdown \\ CH_2 \end{array} + \begin{array}{c} H \quad\quad CO_2CH_3 \\ \diagdown \quad\quad \diagup \\ C \\ \parallel \\ C \\ \diagup \quad\quad \diagdown \\ CH_3O_2C \quad\quad H \end{array} \longrightarrow ?$$

(b)

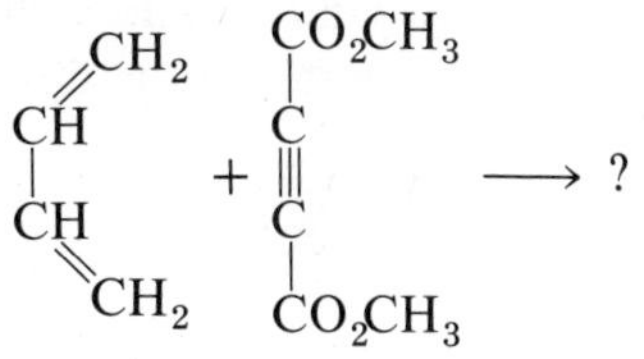

(c)

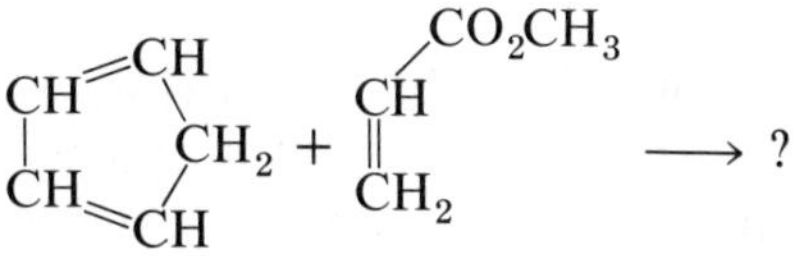

EXERCISE 3.22 Polyethylene prepared as described in Sec. 3.12 is actually a highly branched structure such as

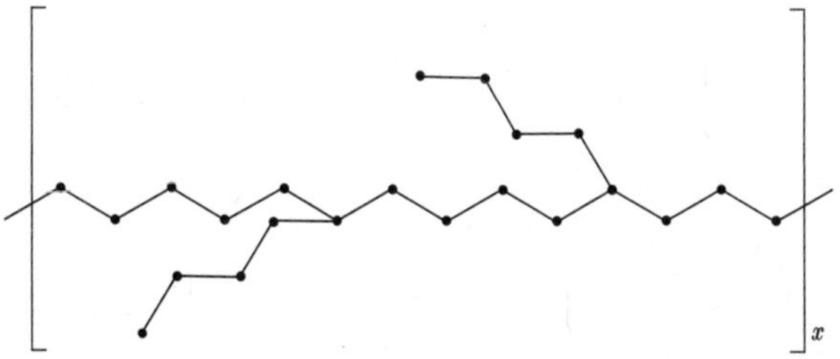

(Linear polyethylene can be made by a more complex *anionic* polymerization process using organometallic catalysts.) Suggest a modification of the mechanism given in Sec. 3.11 that would explain the branching. (*Hint:* Note the length of the branches and study the right-hand drawing in Fig. 2.4.)

EXERCISE 3.23 Draw a reaction coordinate diagram for the addition of hydrogen bromide (HBr) to propene, assuming that the first step of the mechanism for the reaction is the slower step (that having the highest energy of activation). Since there are two possible products (1-bromopropane and 2-bromopropane), you will have to draw a separate curve for the reactions leading to each product, taking into account that 2-bromopropane is formed preferentially.

EXERCISE 3.24 The addition of hydrogen bromide to 1-butene or to 2-butene gives the same product, 2-bromobutane, because both reactions go through the same intermediate, the 2-butyl cation. However, 1-butene reacts faster than 2-butene with hydrogen bromide. Draw a reaction coordinate diagram with curves for each reaction that will explain the experimental results. (*Hint:* Your answer should tell you something about the relative stabilities of the two alkenes.)

EXERCISE 3.25 On the basis of the abbreviated mechanism shown for the reaction of iodomethylzinc iodide with propene (Sec. 3.9), propose a mechanism for the synthesis of cyclopropane given in Sec. 2.10.

EXERCISE 3.26   Write complete equations for the following transformations. Include all conditions and reagents. If an intermediate step is missing, supply it.

      (a) bromocyclohexane → cyclohexene → 1,2-cyclohexanediol
      (b) acetylene → 2-butyne → butane
      (c) 2-butene → 2-butyne → methyl ethyl ketone
      (d) 1-butene → 2-butene

EXERCISE 3.27   The chlorination of propene under free radical conditions (Sec. 2.8-B) gives allyl chloride (3-chloro-1-propene) as the major product. Explain this result, which must have something to do with the stability of the intermediate free radical. (*Hint:* Review Sec. 1.12.)

4

The Aromatic Hydrocarbons, or Arenes

Introduction

The aromatic hydrocarbons are ring hydrocarbons structurally related to benzene, C_6H_6. Although the formula of benzene indicates unsaturation, neither it nor its related compounds are olefinic in behavior. The aromatic hydrocarbons are found in the black, viscous, pitchlike material called coal tar, which is formed as a by-product when bituminous coal is converted to coke. Although coal tar itself is a vile smelling substance, it is a treasure-trove of aromatic hydrocarbons some of which are of rather pleasant odor—hence the name "aromatic." While the term is not descriptive insofar as many of these compounds are concerned, it has come into common usage to classify organic compounds which have as a common structural feature one or more 6-carbon benzene structures. Compounds in this category generally are named **arenes.** Some of these valuable chemicals are now obtainable in large volume from petroleum. From both sources, coal and petroleum, are harvested the basic materials for many pharmaceuticals, dyes, plastics, pesticides, explosives, and a host of other useful, everyday commodities.

4.1 The Structure of Benzene

The discovery of benzene and the elucidation of its structure comprises an interesting and significant chapter in organic chemistry. This remarkable substance first was isolated in 1825 by Michael Faraday. He discovered benzene in the residual, oily condensate which collected in the illuminating gas lines of London. He established its empirical formula as (CH) and called it "carburetted hydrogen." The molecular formula of benzene subsequently was established as C_6H_6 by Mitscherlich, who in 1834 prepared it by heating gum benzoin with lime. For the next thirty years, determination of the structure of benzene posed a problem which occupied the serious attention of the best scientific minds of the time.

Since the molecular formula for benzene provided only one hydrogen atom per carbon, the molecule was expected to be unsaturated and, therefore, very reactive. On the contrary, it appeared to be remarkably stable. Benzene neither decolorized dilute potassium permanganate (the Baeyer test for unsaturation), nor did it add bromine readily, as one would expect an olefin to do. When treated with bromine in the presence of bright sunlight, one gram molecular weight of benzene *added* three gram molecular weights of bromine to yield benzene hexabromide. Obviously, each carbon in the benzene molecule has the potential for bonding to one more atom.

$$C_6H_6 + 3\,Br_2 \xrightarrow{\text{bright sunlight}} C_6H_6Br_6$$

When treated with bromine in the presence of iron, benzene formed a mono-bromo *substitution* product, a behavior remarkably different from that noted above.

$$C_6H_6 + Br_2 \xrightarrow{\text{Fe}} C_6H_5Br + HBr$$

The fact that only one bromobenzene and no isomeric products were obtained indicated to the German chemist Kekulé[1] that each hydrogen atom of benzene must be equivalent to every other hydrogen. In 1865 he proposed a structure for benzene that was in accord with its chemical behavior. The structure Kekulé proposed was a cyclic hexagonal, planar structure of six

---

[1] Friedrich August Kekulé (1829–1896), professor of chemistry successively at Heidelberg, Ghent, and finally at Bonn. Kekulé was an active researcher and a noted teacher but is best remembered for his formulation of the structure of benzene.

carbon atoms with alternate double and single bonds. Each carbon atom was bonded to only one hydrogen atom. The Kekulé structure for benzene (shown below) was widely accepted by his contemporaries and still is used today.

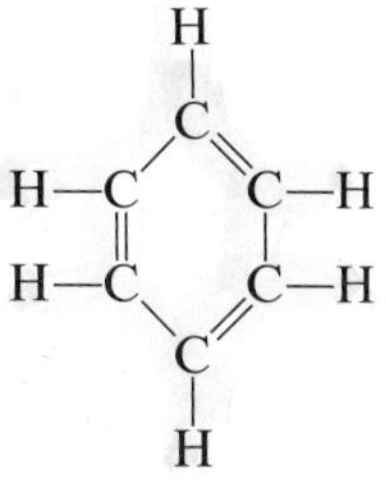

Benzene

The proposed structure for benzene made possible **three** isomeric disubstituted derivatives. In one isomer, adjacent or **ortho** positions could be occupied by the two substituents. In the second isomer, alternate or **meta** positions could be occupied, and in the third isomer, hydrogens bonded to the carbon atoms opposite or **para** to each other in the ring could be substituted.

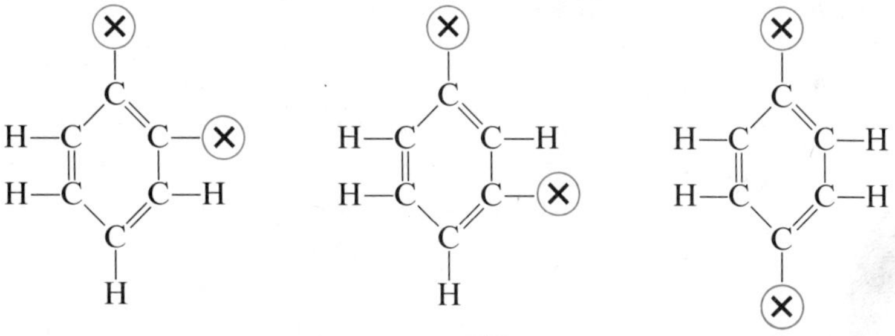

ortho positions   *meta* positions   *para* positions

If Kekulé's formula for benzene were correct, his critics argued, it would appear that **two** *ortho* isomers would be possible. In one *ortho* isomer, the carbon atoms holding the substituents would be separated by a double bond as shown below in structure (a). In the other *ortho* isomer, substituents would be separated by a single bond as shown in structure (b).

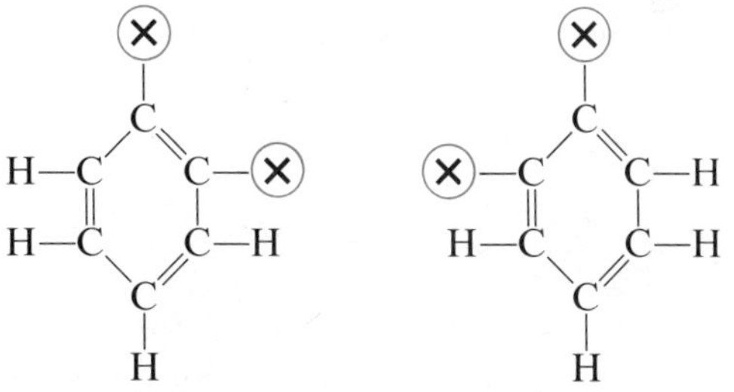

(a)   (b)

Kekulé reconciled the fact that two *ortho* isomers had not been found by stating that the system of alternate double and single bonds in the benzene ring was not a fixed system, but a dynamic one. He suggested that the structure of benzene could be either that of (a) or (b) and that the two structures were in equilibrium with each other.

(a)   (b)

Today we consider the structure of benzene to be that of a resonance hybrid intermediate between structures (a) and (b). Although both structures indicated above are contributing forms, the correct structure of benzene must lie somewhere in between these two extremes.

In formulating a molecular orbital representation of the benzene molecule we form the planar, hexagonal framework by the overlap of sp^2 orbitals on the six carbon atoms. The six $2p$ orbitals, which appear to be the basis of the three double bonds between alternate carbon atoms in the Kekulé structures, combine instead to form six molecular orbitals which encompass the entire carbon framework. There are three bonding molecular orbitals, each occupied by two electrons (still called π electrons), and three empty anti-bonding orbitals. The lowest energy orbital resembles that shown in Fig. 4.1. The next *two* lowest orbitals are of equal energy and also resemble Fig. 4.1, but each has a nodal plane perpendicular to the plane of the ring. Thus, the π electrons are delocalized over the whole ring system. They appear as a cloud above and below the plane of the ring much like two hexagonal doughnuts with all six carbon and hydrogen atoms sandwiched in between. The bonding molecular orbitals and their relative energies can be represented schematically with the conventional symbols for $2p$ orbitals interconnected by lines to show overlap (page 101, top).

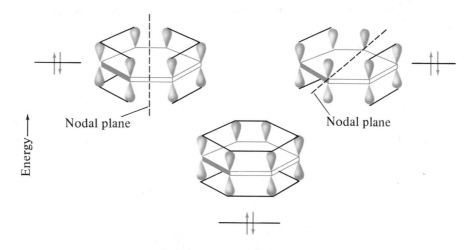

Nodal plane Nodal plane

Energy →

FIGURE 4.1 *Structural Representation of Benzene Showing Overlap of Six π Electrons*

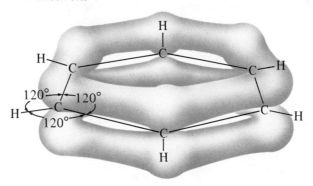

FIGURE 4.2 *Molecular Models of Benzene: (a) Scale Model, (b) Ball and Stick Model*

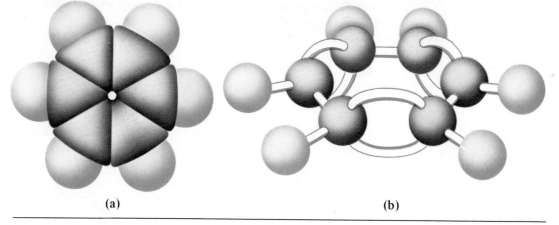

(a)   (b)

4.2   The Aromatic Sextet and Aromatic Character

To be classified as aromatic a compound must meet the following requirements. It must be a cyclic structure in order for the delocalized π electrons to be in an orbital ring. It must be a planar structure in order to permit a maximum overlap of p orbitals. Finally, the number of p electrons must be equal to $4n + 2$, where n is an integer. The latter requirement is referred to as the Hückel Rule. If $n = 1$, we have the sextet of π electrons possessed by benzene. For other values of n the $4n + 2$ rule gives the necessary number of π electrons required for stability, and any cyclic, planar structure with this number of π electrons may exhibit aromaticity. Aromaticity implies a high degree of stability. The π electrons above and below the plane of the molecule are able to participate in the formation of more than one bond. The results are strong bonds and a stable molecule. The additional stability possessed by benzene is called **stabilization** or **resonance energy** and the benzene structure is said to be **stabilized by resonance.** The high degree of stability conferred upon the benzene structure by resonance may be illustrated by comparing its heat of hydrogenation with that of an unsaturated 6-carbon ring structure in which little or no resonance occurs. For example, we might reasonably expect the heat of hydrogenation for a ring structure with three double bonds to be three times that of a compound such as cyclohexene which has only one double bond. The heat evolved when one mole of cyclohexene is hydrogenated is 28.6 kcal.

Cyclohexene   Cyclohexane

If benzene were simply a cyclohexatriene, its heat of hydrogenation should be three times 28.6 kcal or 85.8 kcal. Actually, the heat of hydrogenation of benzene is 49.8 kcal per mole.

Benzene

85.8 kcal (expected heat) − 49.8 kcal (experimental heat) = 36 kcal
(stabilization energy)

Benzene, therefore, gives up 36 kcal less energy than the expected amount because it must contain less energy. This means that benzene is more stable by 36 kcal than the hypothetical cyclohexatriene would be.

A frequently used graphic formula for benzene which embodies all the concepts of π-bonding, resonance, and aromaticity is a regular hexagon with an inscribed circle.

The hexagon represents the six σ bonds between adjacent carbon atoms and the inner circle represents the π-electron cloud. Each corner of the hexagon is understood to be occupied by one carbon with an attached hydrogen atom. All carbon and hydrogen atoms lie in the same plane and the bonds between them describe angles of 120°. Although the hexagon-inscribed-circle notation is a simple one to draw and use in writing the reactions of benzene, for pedagogical reasons we shall more often use the Kekulé representation in our discussion, with the same features being implied. Further, when any hydrogen atoms have been replaced, we shall show only the substituents with the understanding that the other hydrogens remain.

EXERCISE 4.1

Which of the following structures may be classified as *aromatic*? Which may be classified as *non-aromatic*? (a) cyclopropenyl cation; (b) cyclopentadienyl anion; (c) cyclobutadiene; (d) cycloheptatrienyl cation; (e) cyclohexene; (f) cyclooctatetraene.

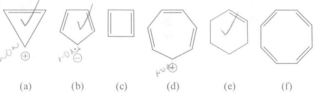

(a)   (b)   (c)   (d)   (e)   (f)

4.3 Formulas and Nomenclature

The naming of aromatic compounds may be done in several different ways. As in the aliphatic[2] or open chain systems, aromatic compounds also have both trivial and systematic names. Compounds with one or more hydrogen atoms of

---

[2] Gr., (*aleiphatos,* fat). Many of the first open chain structures studied were derived from fatty acids. In present usage, aliphatic pertains to noncyclic carbon compounds, or to carbon compounds other than aromatic.

benzene replaced by other atoms or groups may be named as substituted benzenes. Examples of such substitutive names are the following:

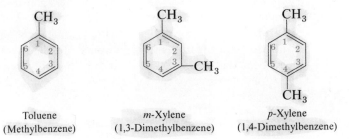

Ethylbenzene   Bromobenzene   Nitrobenzene

Trivial, or common names, usually of early origin, offer no clue to the nature of the substituent. However, such common names are so frequently used that they must be learned. When two substituents appear on the benzene ring, their relative positions must be designated. This is done either by using the prefixes (*ortho* (*o-*), *meta* (*m-*), and *para* (*p-*) as previously described (Sec. 4.1))[3], or by locating substituents on the ring by numbers. A few examples will illustrate these rules.

Toluene
(Methylbenzene)

m-Xylene
(1,3-Dimethylbenzene)

p-Xylene
(1,4-Dimethylbenzene)

When a common or trivial name is used, the group which is responsible for the name, i.e., methyl in the case of toluene, or hydroxyl in the case of phenol, is considered to be on carbon 1. When a trivial name is not used, IUPAC rules require that the lowest number be given to the principal functional group, if a functional group is present. If not, the lowest number (1) is given to the group cited first in the name.

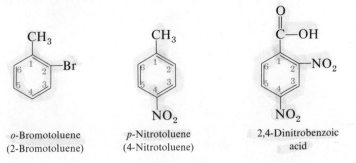

o-Bromotoluene
(2-Bromotoluene)

p-Nitrotoluene
(4-Nitrotoluene)

2,4-Dinitrobenzoic
acid

---

[3] In 1972 the use of *o-*, *m-*, and *p-* was discontinued by *Chemical Abstracts,* the principal abstracting and indexing journal in the field of chemistry. The IUPAC rules still permit such use.

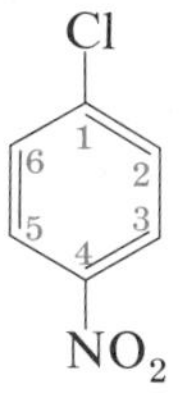

p-Chloronitrobenzene
(1-Chloro-4-nitrobenzene)

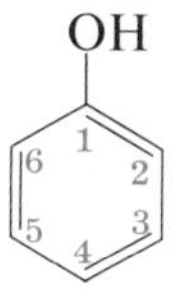

Phenol
(Hydroxybenzene)

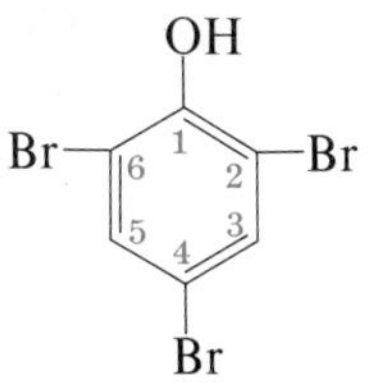

2,4,6-Tribromophenol

The order of naming usually is alphabetical when three different substituents occupy positions upon the ring.

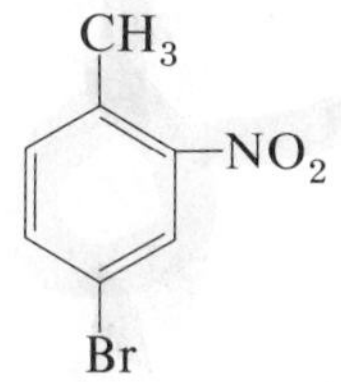

4-Bromo-2-nitrotoluene

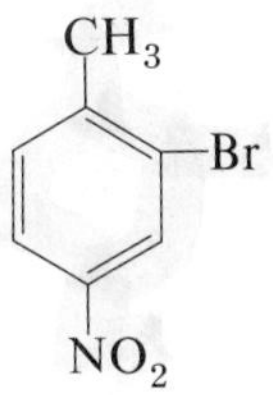

2-Bromo-4-nitrotoluene

4.4   *Position Isomerism*

A disubstituted benzene, as you have already learned, may be one of three isomeric forms—*ortho, meta,* and *para.* If both substituents in each of these three isomeric forms are identical, a third similar group, if substituted in the ring, also makes possible three isomers—**vicinal** (*vic*), **asymmetric** (*asym*), and **symmetric** (*sym*). Thus, the vicinal, asymmetric, and symmetric trimethyl benzenes are produced when a third methyl group is substituted into each of the three isomeric xylenes. Ring positions available in each xylene molecule, relative to the methyl groups already on the ring, are designated by *o, m,* and *p.* For example, the four positions on *p*-xylene open to an incoming methyl group are all the same. Each open position is *ortho* to one methyl group and *meta* to the other, and is labeled (*o, m*).

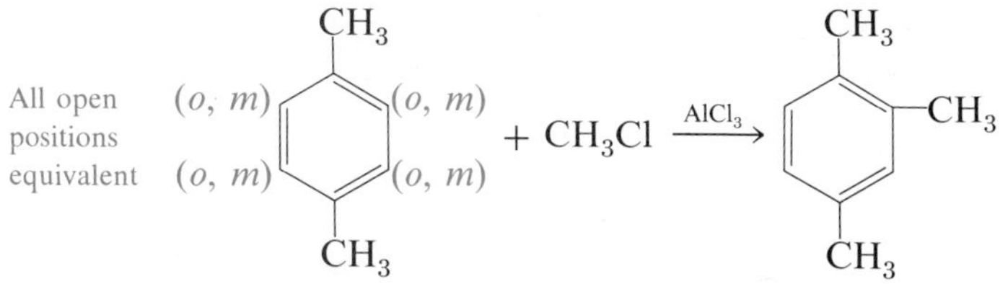

All open positions equivalent

p-Xylene

1,2,4-Tri-methylbenzene

Isomer (A)
(asymmetric)

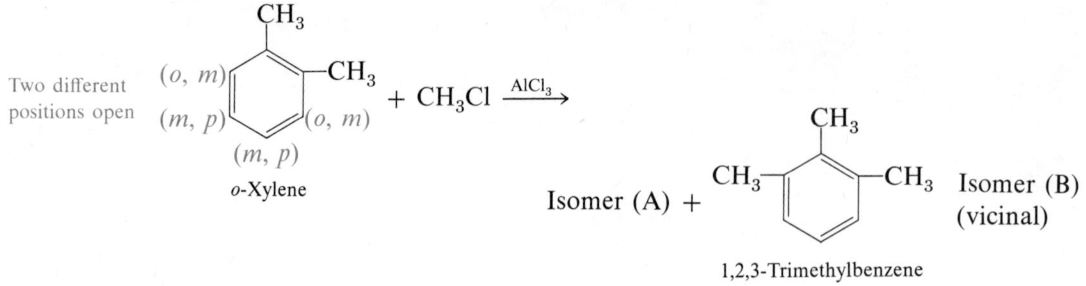
Two different positions open
CH3
CH3
(o, m)
(m, p)
(o, m)
(m, p)
+ CH3Cl
AlCl3
o-Xylene
Isomer (A) +
CH3
CH3
CH3
Isomer (B)
(vicinal)
1,2,3-Trimethylbenzene

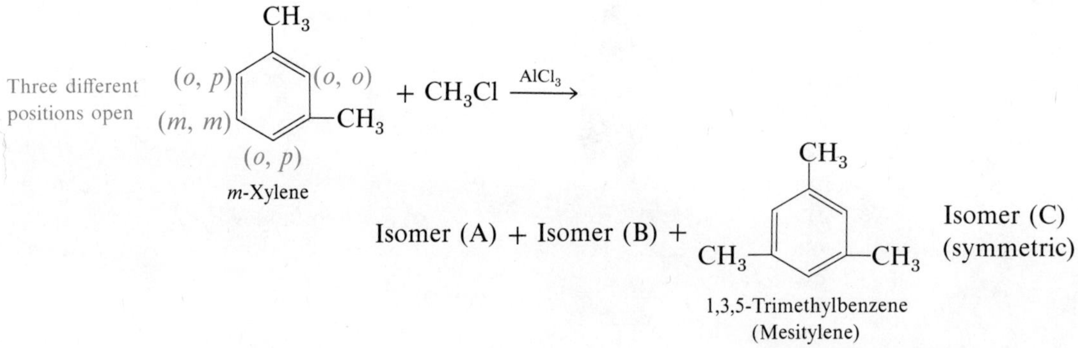
Three different positions open
CH3
(o, p)
(o, o)
(m, m)
CH3
(o, p)
+ CH3Cl
AlCl3
m-Xylene
Isomer (A) + Isomer (B) +
CH3
CH3
CH3
Isomer (C)
(symmetric)
1,3,5-Trimethylbenzene
(Mesitylene)

EXERCISE 4.2

Draw Kekulé structures for the six isomeric products possible if a bromine atom replaces a hydrogen as a third ring substituent in each of the three xylenes.

4.5 *Other Aromatic Hydrocarbons*

Common aromatic hydrocarbons, other than the alkyl substituted benzenes, include a number of polycyclic systems. In these systems the rings may be either condensed—that is, share a side in common or be separate but joined through carbon-carbon bonds. Many of these hydrocarbons, like benzene, also are coal tar derivatives. The following structures, along with their numbering systems, are those which appear most frequently as part of polycyclic organic molecules.

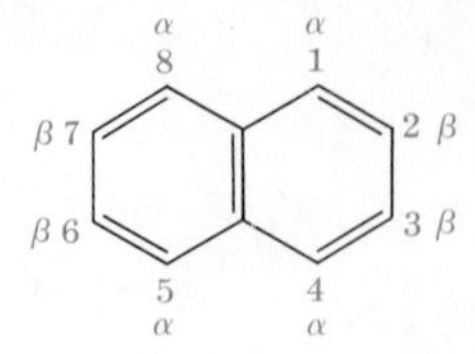
α
8
α
1
β 7
2 β
β 6
3 β
5
α
4
α

Napthalene

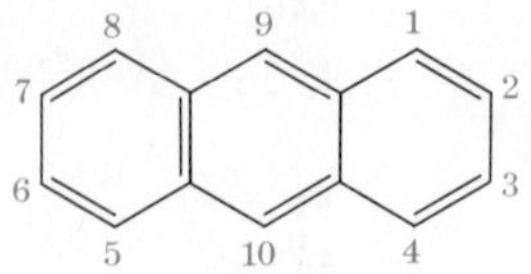
8
9
1
7
2
6
3
5
10
4

Anthracene

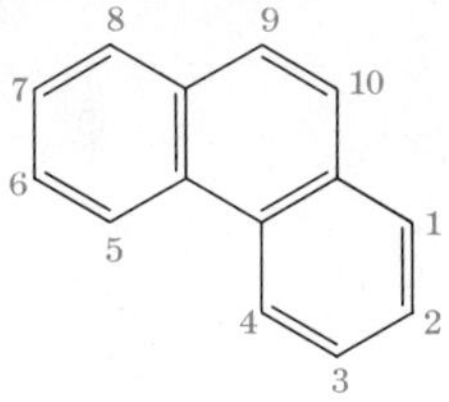

Phenanthrene

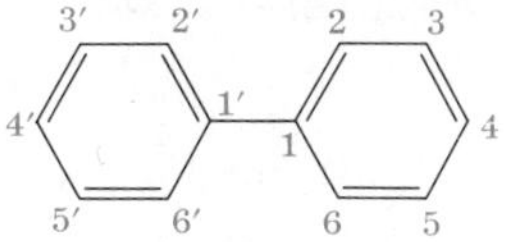

Biphenyl

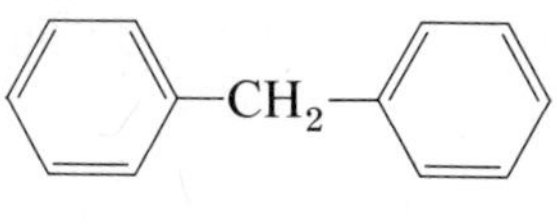

Diphenylmethane

EXERCISE 4.3

Draw two additional structures for naphthalene other than the one shown above. Which structure is the most important contributor to the resonance hybrid?

4.6 Aromatic Groups

The removal of a hydrogen atom from an aromatic hydrocarbon, or arene, produces an **aryl group** (Ar). The use of such group names is limited largely to benzene and naphthalene derivatives. The *phenyl* group, C_6H_5—, stems from *phene,* an early name for benzene. Table 4.1 gives the structures and names of the most commonly used aryl groups.

TABLE 4.1 *Common Aryl Groups*

Structure	Name of Group	Example of Usage	Name of Example
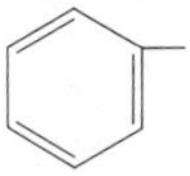	Phenyl	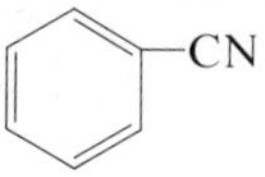 —CN	Phenyl cyanide
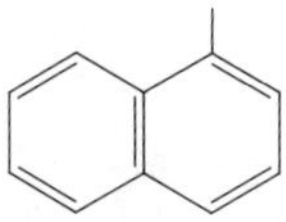	α-Naphthyl (1-Naphthyl)	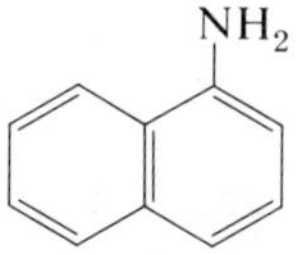 NH$_2$	α-Naphthylamine
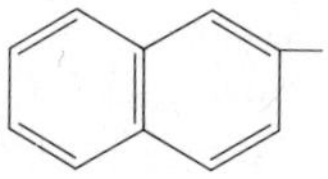	β-Naphthyl (2-Naphthyl)	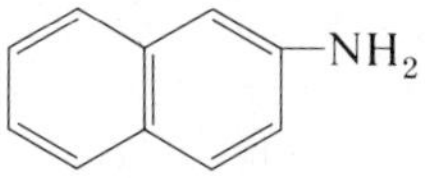 —NH$_2$	β-Naphthylamine
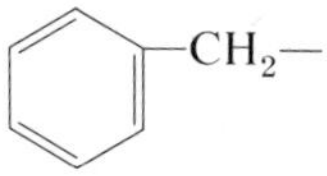 —CH$_2$—	Benzyl	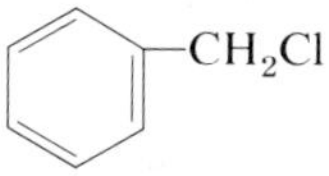 —CH$_2$Cl	Benzyl chloride
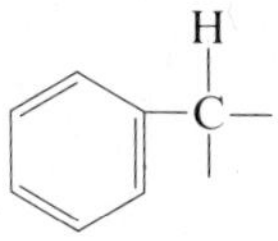	Benzal	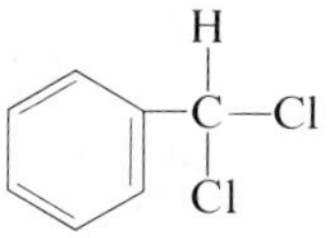	Benzal chloride

EXERCISE 4.4

How many mono-substituted products are possible for naphthalene? How many di-substituted products?

4.7   *Reactions of the Aromatic Hydrocarbons*

The reactions of benzene and other aromatic hydrocarbons are mainly reactions of substitution in the aromatic ring. Substitution reactions involving the aromatic ring, unlike those of the alkanes, are controllable and much more useful. The benzene ring is attacked in nearly every case by an electrophilic reagent which may be either a cation or a neutral but polarized molecule. The electron-deficient group, by sharing in an electron pair supplied at one of the ring carbons, first forms with the benzene ring a cationic intermediate, called a **sigma complex,** in which the positive charge is distributed among the remaining carbon atoms. In the second step a proton is lost to regenerate the more stable aromatic ring.

Catalysts often are necessary to generate the electrophile. Such catalysts, as you will see in the following sections, often are Lewis acids, each of which lacks an electron pair. The result of attracting a pair of electrons from the substituting

FIGURE 4.3   *Reaction Profile for Electrophilic Substitution of Benzene*

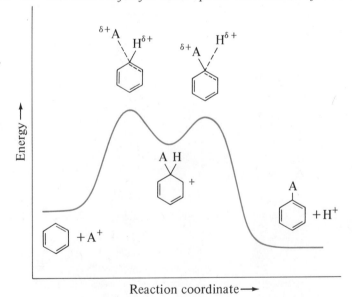

reagent by the catalyst is the formation of either a positive ion or a polarized molecule. In either case, the positive ion or the positive part of the molecule is the ring-attacking group. The mechanism of electrophilic aromatic substitution may be illustrated by the following.

sigma complex

$$H^+ + B—\bar{C} \longrightarrow H—B + C$$

A reaction profile for the reaction would resemble Fig. 4.3.

The principal reactions of benzene are described in the following sections.

A. Halogenation.   Halogenation of the aromatic ring may be illustrated by bromination. Bromine attacks the aromatic ring in a manner analogous to its attack upon the double bond of an olefin, except that a catalyst is necessary to assist in the polarization of the bromine molecule. Iron, usually in finely divided form, is used for this purpose. Presumably, ferric bromide, $FeBr_3$, is the actual catalyst and this could form as shown in the first of the equations representing the bromination of benzene.[4]

$$3 \, Br_2 + 2 \, Fe \longrightarrow 2 \, FeBr_3$$

Ferric
bromide

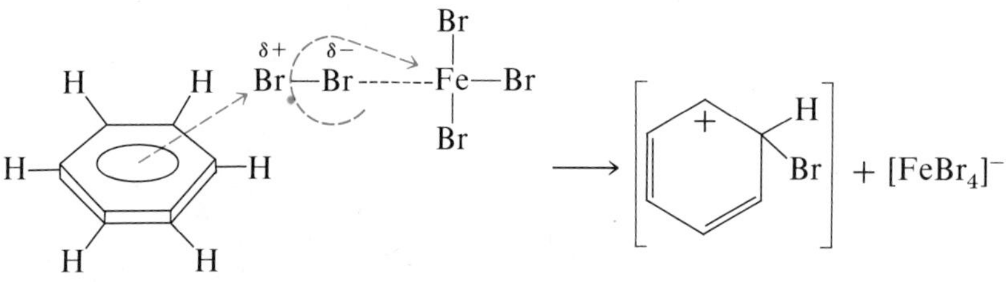

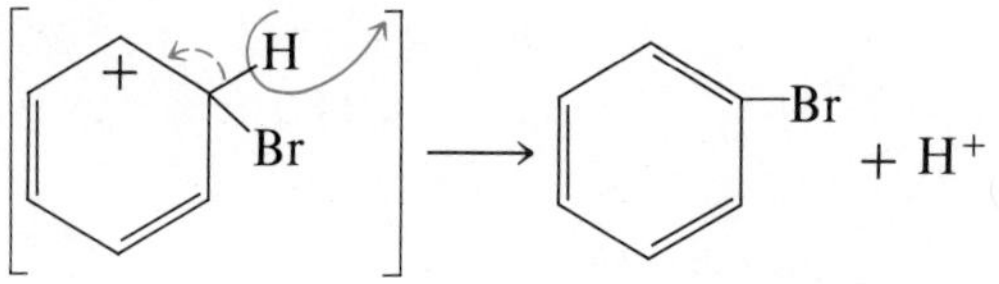

$$H^+ + FeBr_4^- \longrightarrow FeBr_3 + HBr$$

Although we have indicated the bromination of benzene as a series of discrete steps, the net equation for the over-all reaction may be written simply as

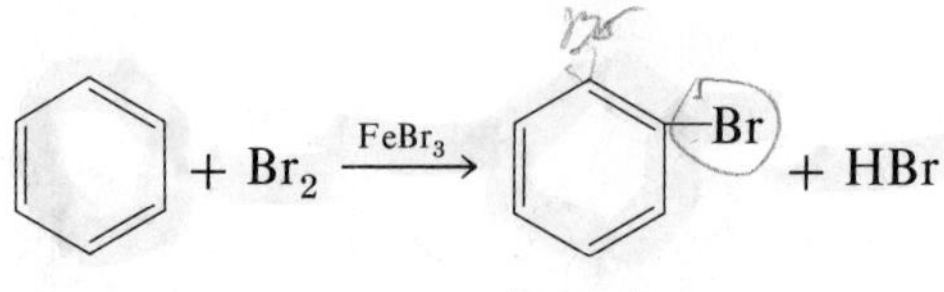

Bromobenzene

Substitution of benzene by chlorine also may be accomplished in a manner similar to that shown for the bromination reaction. As before, a catalyst is necessary. Anhydrous aluminum chloride, $AlCl_3$, is an excellent catalyst for the chlorination reaction. If iron is used, the catalyst presumably is $FeCl_3$.

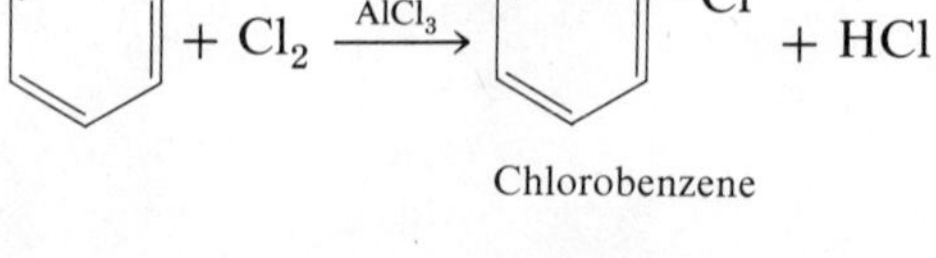

Chlorobenzene

------

[4] In this and subsequent mechanisms for electrophilic aromatic substitution only one resonance form, in brackets, may be given for the cationic intermediate. The student should practice writing all three (or more) forms, referring to those on p. 109 as necessary.

B. Nitration of Benzene. Nitration of the aromatic ring is a very important reaction and produces a number of useful compounds unobtainable in any other way. Nitration usually is accomplished by treating benzene with a mixture of concentrated nitric and sulfuric acids. A mixture of these two acids is referred to as a nitrating mixture. The electrophilic species in the nitration of benzene appears to be the nitronium ion, $^+NO_2$, which may be formed from nitric acid by the action of sulfuric acid. The mechanism usually given for the nitration of benzene is illustrated by the following sequence of reactions.

$$HO-N\overset{O}{\underset{O}{}} + 2\ H_2SO_4 \longrightarrow 2\ HSO_4^- + H_3O^+ + \quad \oplus N\overset{O}{\underset{O}{}}$$

Nitronium ion

$$+\ ^{\oplus}N\overset{O}{\underset{O}{}} \longrightarrow \left[\right] \longrightarrow \quad -NO_2 \quad +\ H^+$$

Nitrobenzene

The net equation for the over-all reaction may be written

$$+\ HNO_3 \xrightarrow{H_2SO_4} \quad NO_2 \quad +\ H_2O$$

Nitrobenzene

C. Alkylation of Benzene. The simple alkyl benzenes such as toluene (methylbenzene) and the xylenes (dimethylbenzenes) are readily available from petroleum. On occasion, however, it may be necessary or desirable to introduce an alkyl group into the benzene ring. One method for accomplishing this substitution is a reaction known as the **Friedel-Crafts** reaction.[5] In the Friedel-Crafts reaction an alkyl group is attached to the ring by treating benzene with an alkyl halide in the presence of anhydrous aluminum chloride, $AlCl_3$. The electrophilic reagent which attacks the ring in the alkylation reaction appears to be a carbonium ion or incipient carbonium ion. Inasmuch as alkenes (equation 2) and alcohols (equation 3) also are capable of forming carbonium ions (Sec. 8.8-B), they, too, can be used to alkylate the benzene ring. The mechanism of the Friedel-Crafts reaction is illustrated in the following equations for the preparation of isopropylbenzene.

$$(1)\quad CH_3 \overset{CH_3}{\underset{H}{\overset{|}{C}}} Cl + Al\overset{Cl}{\underset{Cl}{-}}Cl \longrightarrow {}^{\oplus}\overset{CH_3}{\underset{CH_3}{C}}\cdots\cdots Cl\overset{(-)}{\underset{Cl}{-}}Al\overset{Cl}{\underset{Cl}{-}}Cl$$

Isopropyl chloride

---

[5] Charles Friedel (1832–1899), a French chemist, known mainly for his discovery, in collaboration with an American chemist, James M. Crafts (1838–1917), of a convenient method of synthesis of aromatic ketones.

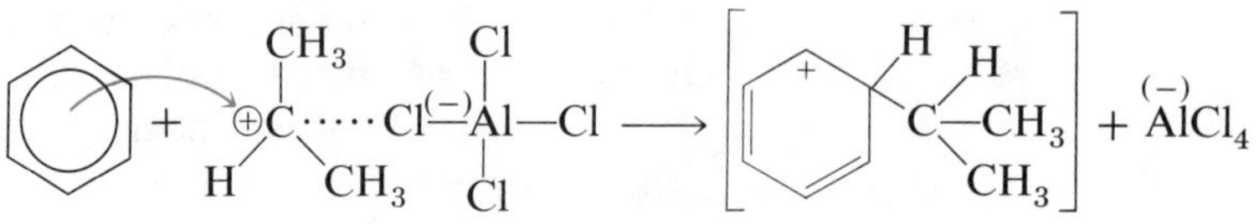

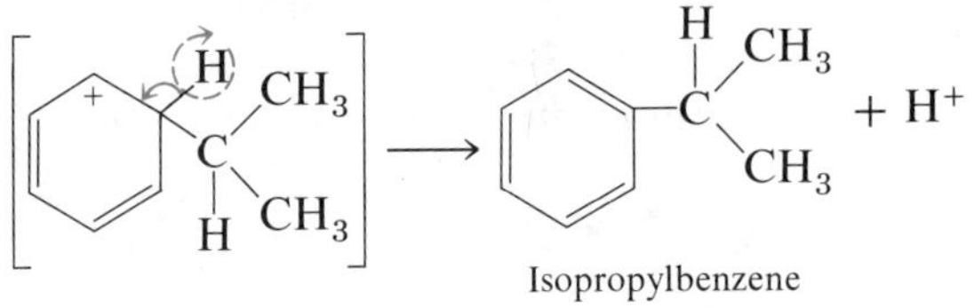

$$
\text{Isopropylbenzene (Cumene)}
$$

$$
H^+ + \overset{(-)}{AlCl_4} \longrightarrow AlCl_3 + HCl
$$

$$
(2) \quad CH_3\overset{H}{\underset{}{-}}C \overset{\frown}{=} CH_2 + HCl + AlCl_3 \longrightarrow CH_3\overset{H}{\underset{\oplus}{-}}C-CH_3 + \overset{(-)}{AlCl_4}
$$

$$
(3) \quad CH_3\overset{H}{\underset{OH}{-}}C-CH_3 + 2\,H_2SO_4 \longrightarrow CH_3\overset{H}{\underset{\oplus}{-}}C-CH_3 + H_3\overset{+}{O} + 2\,HSO_4^-
$$

Isopropyl alcohol

The isopropyl cation then combines with the aromatic ring in the same manner as indicated in the first equation above.

The Friedel-Crafts reaction, although useful, has the disadvantage of giving not only polysubstituted products, but also rearranged products if the group to be substituted into the benzene ring is unbranched and larger than ethyl. For example, if benzene is treated with *n*-propyl chloride and aluminum chloride, the product obtained is largely isopropylbenzene, or cumene. Only small amounts of the normal isomer are obtained. Formation of the more highly branched secondary carbonium ion, which may be accomplished by a rearrangement of a hydrogen atom and a pair of electrons, appears to be favored because it represents a more stable intermediate than the primary carbonium ion.

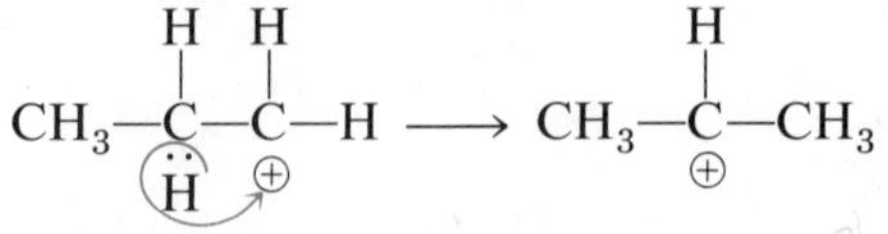

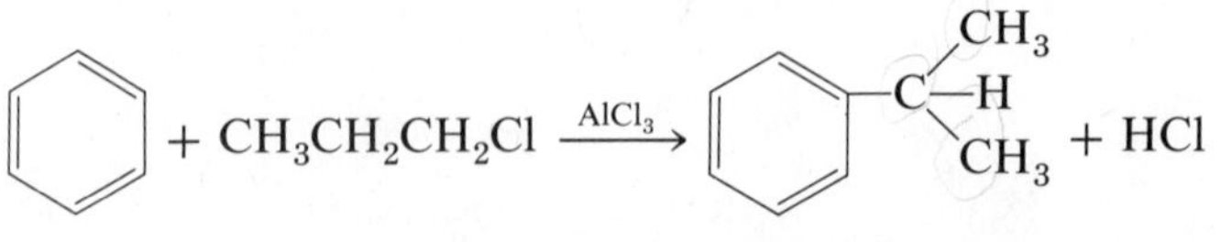

$$
CH_3\overset{H}{\underset{\overset{\cdot\cdot}{H}}{\overset{|}{-}}}C\overset{H}{\underset{\oplus}{-}}C-H \longrightarrow CH_3\overset{H}{\underset{\oplus}{-}}C-CH_3
$$

Isopropylbenzene
(Cumene)

D. Sulfonation of Benzene.   Benzenesulfonic acid is produced when benzene is heated with concentrated sulfuric acid. The ring-attacking group in the sulfonation reaction is not a positive ion as was the case in the previous reactions. Instead, it appears to be the electron-deficient sulfur trioxide molecule produced by the reaction of two molecules of sulfuric acid.

$$2 \; H_2SO_4 \rightleftharpoons H_3O^+ + \overset{\cdot\cdot\cdot}{\underset{\cdot\cdot\cdot}{\overset{\cdot\cdot}{O}}} \; \overset{\cdot\cdot}{S} \overset{\cdot\cdot}{O} \cdot\cdot + HSO_4^-$$

The sulfonation of aromatic compounds is a very important reaction, especially in the preparation of dyestuffs. Sulfonation of the aromatic ring is a convenient method for making aromatic compounds water soluble, because many of the sulfonic acids form water-soluble metal salts, and many aromatic sulfonic acids are themselves quite soluble in water.

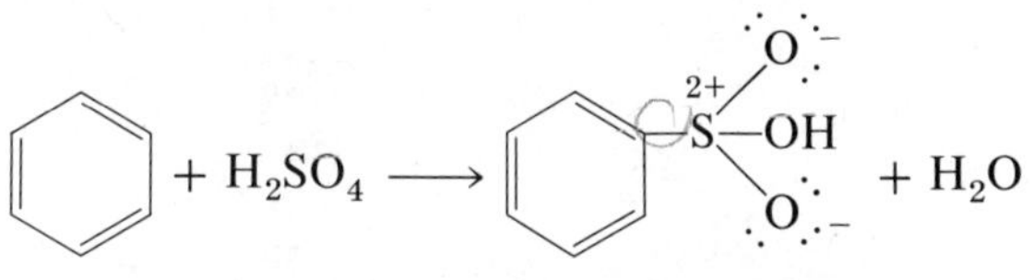

Benzenesulfonic acid

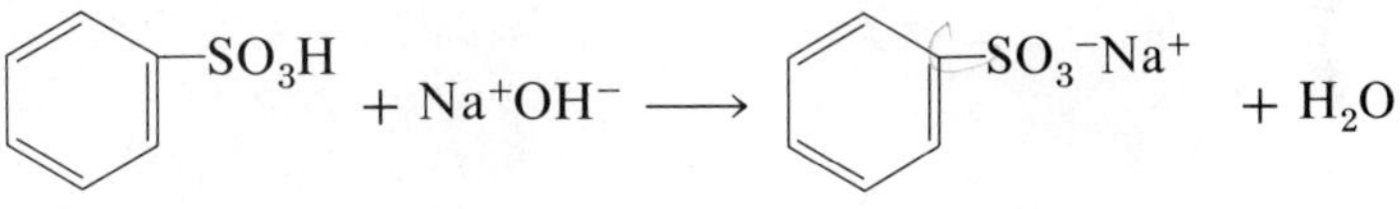

Sodium benzenesulfonate

E. Oxidation of Aromatic Hydrocarbons.   Neither benzene nor polycyclic aromatic compounds are reactive to the usual oxidizing reagents such as potassium permanganate ($KMnO_4$) or potassium dichromate ($K_2Cr_2O_7$). However, the benzene ring can be ruptured and oxidized when treated with oxygen in the presence of vanadium pentoxide at high temperatures. The anhydride of maleic acid, a dicarboxylic acid (Sec. 12.1), may be produced in this manner.

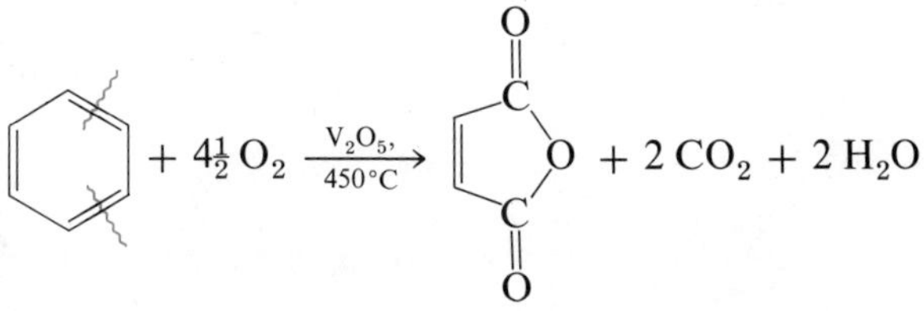

Maleic
anhydride

Naphthalene also is oxidized under these same conditions. One benzene ring appears to facilitate the oxidation of the other. The anhydride of phthalic acid, an important industrial chemical used in the preparation of the glyptal resins (Sec. 12.4), is produced from naphthalene in this manner.

$$\text{Naphthalene} + 4\tfrac{1}{2}\,O_2 \xrightarrow[350^\circ C]{V_2O_5,} \text{Phthalic anhydride} + 2\,CO_2 + 2\,H_2O$$

| Naphthalene | Phthalic anhydride |

Other reactions of naphthalene are similar to those shown for benzene. In nearly every instance the substituting group exhibits a predilection for the **alpha** positions (Sec. 4.5).

4.8 *Oxidation of Alkyl Benzenes*

An alkyl group attached to the benzene ring undergoes oxidation quite readily. Regardless of its length, the carbon *side chain* is degraded to the last ring-attached carbon atom. The ring-attached carbon is converted to a carboxyl group, **—COOH.** All other carbon atoms in the chain are oxidized to carbon dioxide. Hot potassium permanganate or potassium dichromate in sulfuric acid usually is used for the oxidation of side chains.

$$\text{Toluene}\;(-CH_3) + K_2Cr_2O_7 + 4\,H_2SO_4 \xrightarrow{\text{heat}} \text{Benzoic acid}\;(-C(=O)-OH) + 5\,H_2O + Cr_2(SO_4)_3 + K_2SO_4$$

| Toluene | Benzoic acid |

EXERCISE 4.5

Write a balanced equation for the oxidation of isopropylbenzene (cumene) with potassium dichromate and sulfuric acid.

4.9 *Directive Influence of Ring Substituents*

A substituent already on the ring not only influences the facility with which a second group enters but also determines the position on the ring that the incoming group will occupy, that is, determines the **orientation** of further substitution. If a substituent has one or more pairs of nonbonded electrons, the group may serve as an electron-pair donor and tend to stabilize preferentially the cationic intermediates formed by addition of the reagent to the *ortho* and *para* positions of the ring. Substitution at these sites is enhanced, and little, if any, substitution will occur at the *meta* position. Groups capable of orienting new substituents to the *ortho* and *para* positions in this manner are called **ortho-para directors.** Consider, for example, the electrophilic substitution of phenol and the cationic intermediates formed by the addition of the electrophile, A⁺, to the *ortho, meta,* and *para* positions. Note that there are four contributing forms, including one with a positive charge on oxygen, for the *ortho* and *para* intermediates, but only three forms for the *meta* intermediate.

The stability of a charged species depends in part on the dispersal of the charge over as many atoms as possible, and the stability of a resonance hybrid depends upon the number of reasonable resonance forms that can be written. Both of these factors tell us, not only that the *ortho* and *para* intermediates should be more stable than the *meta* intermediate, but also that they should be more stable than the intermediate (p. 109) derived from (unsubstituted) benzene (three forms). Thus, the hydroxyl group is not only an *o-p* director but also has an activating effect on the ring causing it to be more reactive than benzene.

In contrast to the foregoing, an electron-attracting group tends to *deactivate* the ring and make it less reactive than benzene. Such substituents usually are groups in which the atom attached to the ring is multiply-bonded to other atoms of greater electronegativity. Substitution at the *ortho* and *para* positions will be inhibited because the cationic intermediates will be destabilized somewhat more than the intermediate for *meta* substitution. The nitro group is one of the most powerful electron-withdrawing groups and serves as a good model for other groups of this type.

ortho

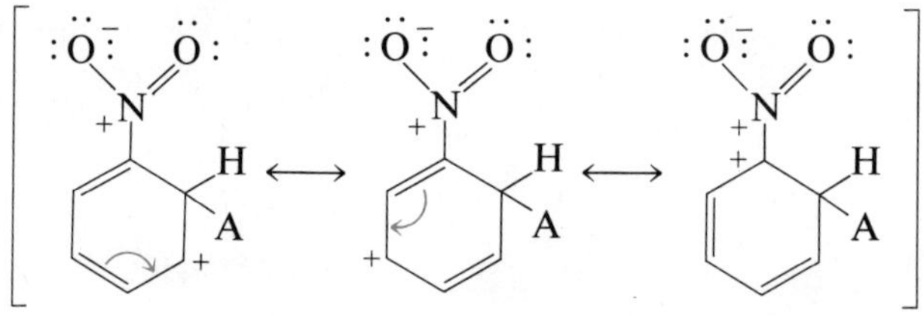

poor form

meta

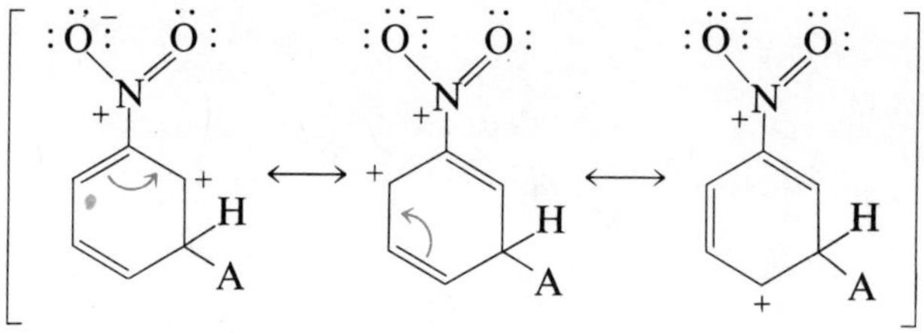

para

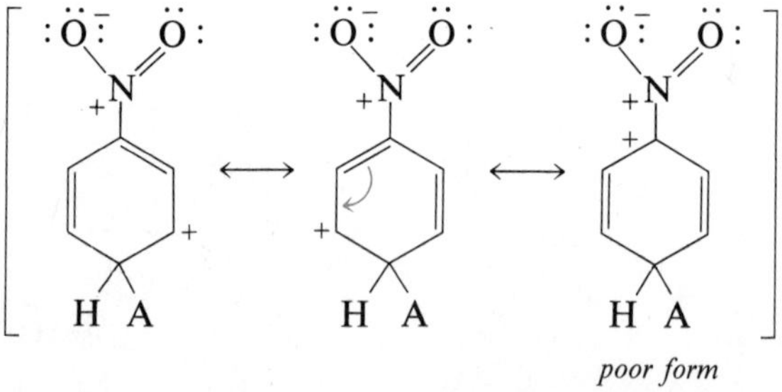

poor form

All of the contributing forms are destabilized by the electrostatic repulsion of the cationic positive charge and the strong dipole of the nitro group (Sec. 1.5).

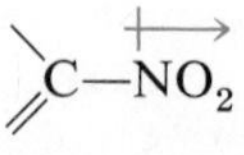

$$\diagdown C - NO_2$$

The third contributing form for both the *ortho* and *para* cationic intermediates will be affected the most because the cation's positive charge is on the carbon atom attached to the strongly electron-withdrawing nitro group, that is, at the positive end of the nitro group dipole. Although all three of the cationic intermediates are destabilized (relative to the intermediates for benzene), the *ortho* and *para* intermediates are destabilized the most. Thus, an incoming substituent, if it bonds to the ring at all, must take next best and affix itself to one of the *meta* carbons. Groups capable of orienting other substituents to *meta* positions are called **meta directors.** The nitro group is a very strong electron attractor, deactivates the ring, and is a *meta* director. The deactivating power of the nitro group is sufficiently strong to prevent nitrobenzene from entering into a Friedel-Crafts reaction. Nitrobenzene, because of its unreactivity, frequently is used as the solvent for the Friedel-Crafts reaction.

The orientation and activation effects in aromatic electrophilic substitution are shown diagrammatically in the reaction profiles for the first step of the substitution reaction in Fig. 4.4. As always, the *rates* of the three competitive reactions, *ortho-, meta-,* and *para*-substitution, depend on the energy of activation (the height of the "hump"), that is, they depend on the relative stabilities of the *transition states* whose geometries we cannot know. Apparently, the curves do not cross each other; therefore, we are reasonably safe in explaining the observed effects, as we have done, on the basis of the stabilities of the cationic *intermediates*.

FIGURE 4.4   *Reaction Profile for the First Step in the Electrophilic Substitution of (a) Phenol and (b) Nitrobenzene*

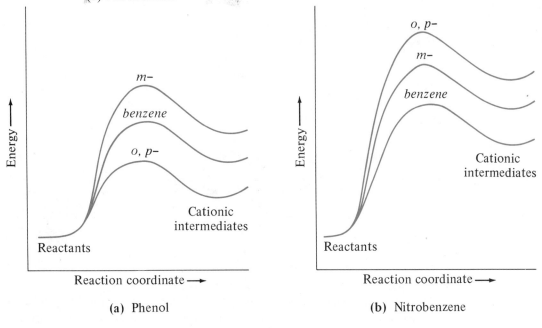

(a) Phenol

(b) Nitrobenzene

TABLE 4.2 *Table of* ortho-para *and* meta *directing Groups*

Ortho-Para Directors (Activating)	Representative Compounds	Meta Directors (Deactivating)	Representative Compounds
—OH	Phenol,	$-\overset{+}{N}H_3$	Anilinium chloride,
—NH₂	Aniline,	—N→O with O	Nitrobenzene,
—OCH₃	Anisole,	—C≡N	Benzonitrile,
—N(H)—C(O)—CH₃	Acetanilide,	—S(O)(O)—OH	Benzenesulfonic acid,
—O—C(O)—CH₃	Phenyl acetate,	—C(O)—CH₃	Acetophenone,
—CH₃	Toluene,	—C(O)—H	Benzaldehyde,
—Cl, —Br, —I*	Bromobenzene,	—C(O)—OH	Benzoic acid,

*The halogens are *ortho-para* directing by virtue of their non-bonded electron pairs, but they are also electron-attracting because of their high electronegativity. The second influence destabilizes all three intermediates and results in ring deactivation.

Table 4.2 lists the common *ortho-para* and *meta* directors in approximate order of diminishing directive power.

The directive power of substituents must be considered in any synthetic route which leads to a polysubstituted aromatic ring. For example, if *m*-bromo-nitrobenzene is to be prepared, one should not attempt to nitrate bromobenzene, but instead, should brominate nitrobenzene.

Bromobenzene + HNO₃ →(H₂SO₄) *o*-Bromo-nitrobenzene (38%) + *p*-Bromo-nitrobenzene (62%)

Nitrobenzene + Br₂ →(Fe) *m*-Bromo-nitrobenzene (nearly 100%)

Summary

1. Aromatic hydrocarbons are derived from coal tar and petrolum. They are cyclic, relatively stable systems containing one or more six-carbon benzenoid structures.

2. Aromatic hydrocarbons may be homologs of benzene or polycyclic systems. The latter may be either fused or isolated rings.

3. The benzene structure is a resonance hybrid.

4. Aromatic compounds may be named by trivial names or as derivatives of benzene. Relative positions of substituents must be indicated.

5. Reactions of benzene include
   (a) halogenation
   (b) nitration
   (c) alkylation (Friedel-Crafts)
   (d) sulfonation
   (e) oxidation of the side chain.

6. Substituents already on the ring govern the ease of further substitution and possess a directive influence upon the ring position which entering groups assume.

7. *Ortho-para* directors (except the halogens) activate the ring and enhance further substitution; *meta* directors deactivate the ring and inhibit further substitution.

New Terms

alkylation

aromaticity

Friedel-Crafts reaction

meta director

nitrating mixture

ortho-para director

ring-activating group

ring-deactivating group

side chain

stabilization (resonance) energy

Supplementary Exercises

EXERCISE 4.6   Assign an acceptable name to each of the following:

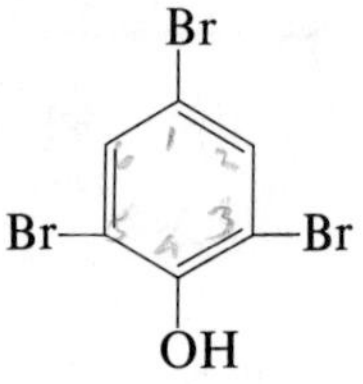

(a) 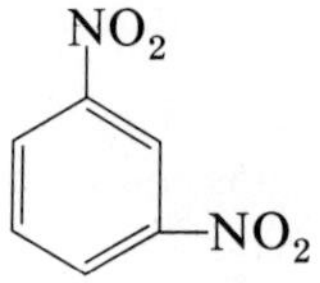   (b) 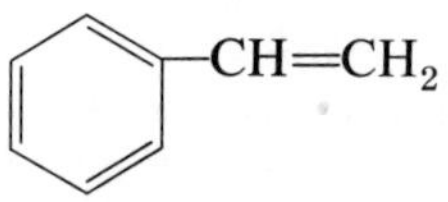   (c) 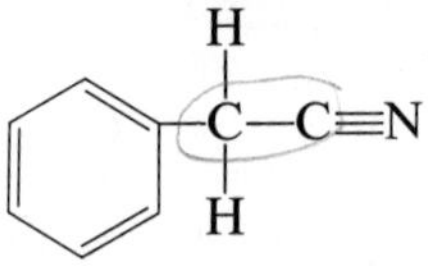   (d)

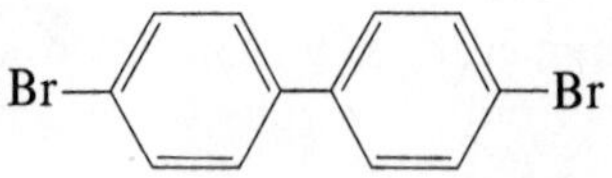

(e) 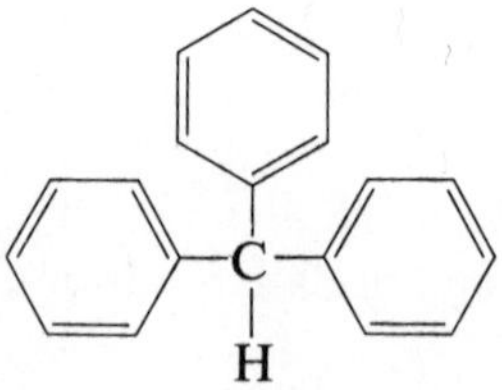   (f) 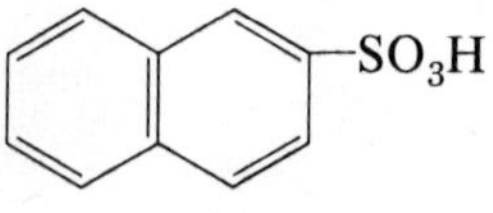   (g)

EXERCISE 4.7   Draw structural formulas for the following useful substances:
(a) T.N.T. (2,4,6-trinitrotoluene, a high explosive)
(b) Salicylic acid (*o*-hydroxybenzoic acid, a pharmaceutical)
(c) Pentachlorophenol (a wood preservative and fungicide)
(d) *p*-Dichlorobenzene (a moth repellent and larvicide)

EXERCISE 4.8   Draw structures for all isomeric homologs of benzene with the formula C_9H_{12}. Assign an acceptable name to each.

EXERCISE 4.9 In the bromination of benzene iron filings were used as a "carrier." Explain by what mechanism iron could promote the formation of an electrophilic ring-seeking bromine atom, when halogens are by nature such strong electronegative elements.

Do you think that $AlCl_3$ could serve as a catalyst in the bromination reaction?

EXERCISE 4.10 Write structural formulas for:

(a) A compound, C_8H_{10}, which can give only one theoretically possible monobromo ring substitution product.

(b) A compound, C_9H_{12}, which can give only one theoretically possible mononitro ring substitution product.

EXERCISE 4.11 An unidentified liquid is thought to be either benzene, cyclohexene, or cyclohexane. What simple chemical test will identify it?

EXERCISE 4.12 Using toluene or benzene as your only aromatic organic starting material and any other reagents you may require, devise synthetic routes which will lead to the following products.

(a) *o*-nitrobenzoic acid
(b) *m*-bromonitrobenzene
(c) *p*-bromobenzoic acid
(d) *p*-toluenesulfonic acid
(e) cumene (isopropylbenzene)
(f) benzyl chloride
(g) 1-bromo-4-nitrobenzene

EXERCISE 4.13 Show the probable mono-nitrated products when each of the following is treated with a nitrating mixture.

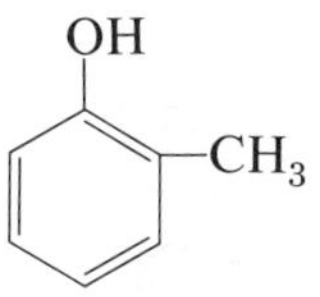

(a) 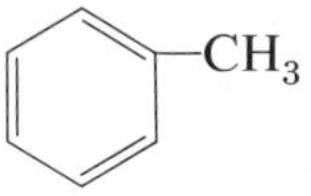 (b) 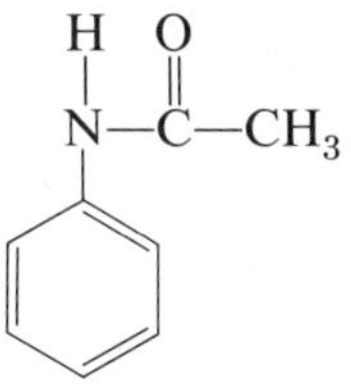 (c)

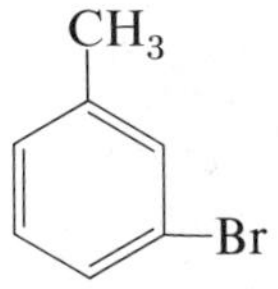

(d) 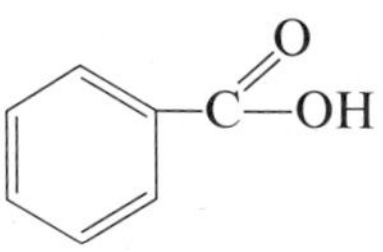 (e) 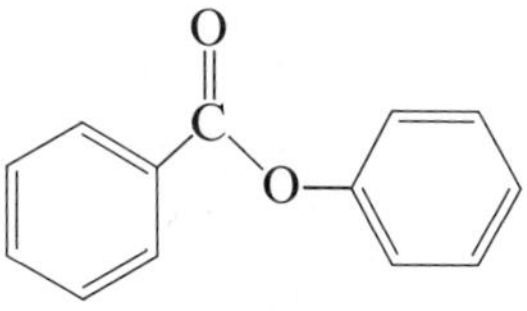 (f)

EXERCISE 4.14   Complete the following reactions showing the principal products, if any, expected in each.

(a)

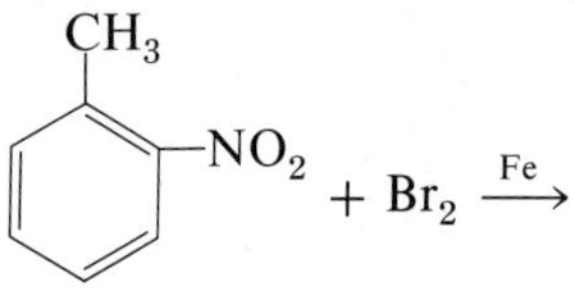

$$\underset{\substack{\text{CH}_3 \\ \text{-NO}_2}}{\bigcirc} + \text{Br}_2 \xrightarrow{\text{Fe}}$$

(b)

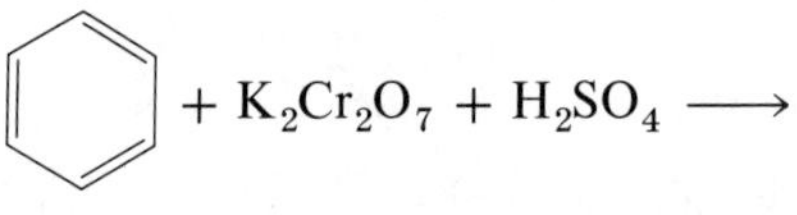

$$\bigcirc + \text{K}_2\text{Cr}_2\text{O}_7 + \text{H}_2\text{SO}_4 \longrightarrow$$

(c)

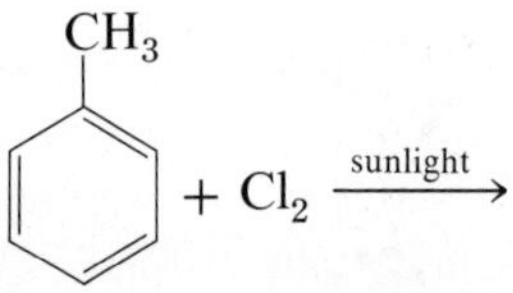

$$\underset{\text{CH}_3}{\bigcirc} + \text{Cl}_2 \xrightarrow{\text{sunlight}}$$

(d)

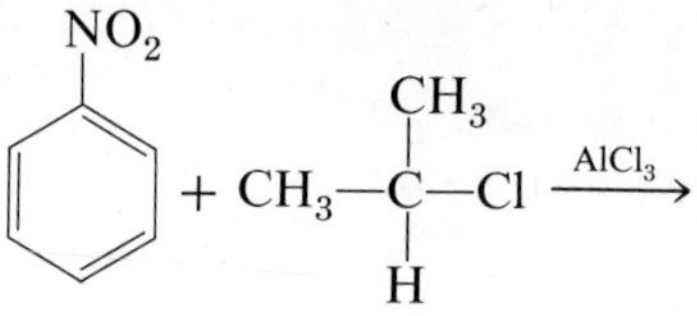

$$\underset{\text{NO}_2}{\bigcirc} + \text{CH}_3 - \underset{\underset{\text{H}}{|}}{\overset{\overset{\text{CH}_3}{|}}{\text{C}}} - \text{Cl} \xrightarrow{\text{AlCl}_3}$$

(e)

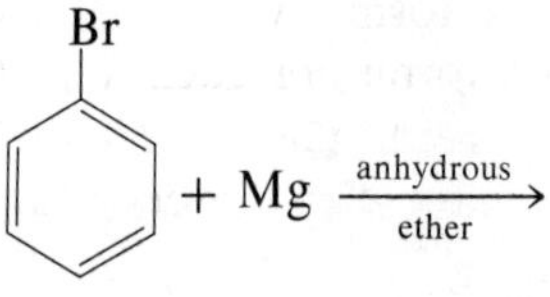

$$\underset{\text{Br}}{\bigcirc} + \text{Mg} \xrightarrow[\text{ether}]{\text{anhydrous}}$$

(f)

$$\bigcirc + \text{H}_2\text{SO}_4 \longrightarrow$$

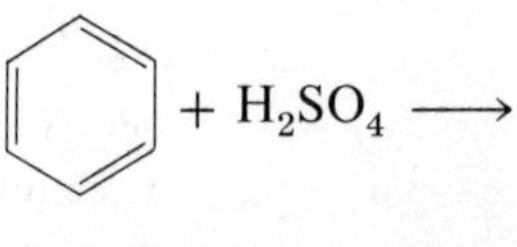

EXERCISE 4.15   A hydrocarbon of the formula C_8H_{10} yields two monobromo derivatives. On strong oxidation it gave an acid identical with the oxidation product of naphthalene. What was the original hydrocarbon?

EXERCISE 4.16   Albert Ladenburg[6] in 1869 proposed the prism structure shown to account for the one monosubstitution product and the three disubstituted products found

---

[6] Albert Landenburg (1842–1911), professor of chemistry at Kiel (1873–1889); at Breslau (1889–1911).

for benzene. Where would the *ortho, meta,* and *para* positions be on the Ladenburg structure?

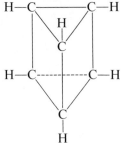

EXERCISE 4.17 Draw the more important resonance structures for the following compounds. In each example you should have five forms (including the two Kekulé forms) because π electrons from the ring may be delocalized into the substituent or from the substituent into the ring. In the former case the ring will have a net positive charge, and in the latter, a net negative charge. If done properly, there will be a consistent pattern for the location of such charges around the ring.

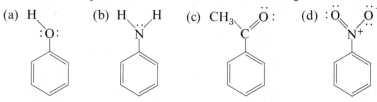

(a) Phenol   (b) Aniline   (c) Acetophenone   (d) Nitrobenzene

EXERCISE 4.18 Draw the principal resonance forms for benzyl radical and benzyl cation. You should have a total of five forms for each, including the Kekulé forms (see Exercise 4.17). Would you expect either or both of these reactive intermediates to be more stable than primary alkyl radicals and cations?

$$\text{Benzyl radical} \qquad \text{Benzyl cation}$$

EXERCISE 4.19 The methyl group of toluene is an *o,p*-director and activates the ring toward electrophilic substitution. Draw the resonance forms for the cationic intermediates in the electrophilic substitution (by the general reagent, A^+) at all three positions. Suggest an explanation for the *o,p*-direction and ring activation. (*Hint:* Look at each contributing form carefully to see whether or not it might be more stable than the others, recalling what you have learned about the stability of *simple* carbonium ions (Sec. 3.8-A).)

EXERCISE 4.20 Draw a reaction coordinate diagram similar to those in Fig. 4.4 for the electrophilic substitution of chlorobenzene (*o,p*-director but makes ring less reactive than benzene).

EXERCISE 4.21   Starting with benzene, toluene, ethylbenzene, and any alkyl halides of your choice, show how the following compounds could be synthesized. The review of earlier chapters may be helpful. You may use as many steps as are necessary.

(a) CH_2-D (b) CH_3 (c) $CH_2CH_2CH_2CH_3$ (d) $CH=CH_2$

D

(e) $-CH_2-CH_2-$ (f) $-C\equiv CH$

(g) CH_2-CH_2 / CH_2 $CH-CH_3$ / CH_2-CH_2 (h) $-CH\langle\begin{matrix}CH_2\\CH_2\end{matrix}$

EXERCISE 4.22   The bromination of ethylbenzene under free radical conditions gives exclusively 1-bromo-1-phenylethane. Explain this result. (*Hint:* Review Sec. 2.8-B and Exercise 4.18).

CH_2-CH_3 $+ Br_2 \xrightarrow{light}$ $\overset{Br}{\underset{}{CH}}-CH_3$ $+ HBr$

EXERCISE 4.23   As you saw in Exercise 4.1 aromaticity is not limited to compounds with six-membered rings. The important factor in the stabilization of cyclic structures is the delocalization of π electrons over the entire ring. Using this concept and the drawing of resonance structures as a means of visualizing delocalization, explain the aromaticity of the following compounds.

(a) CH $\parallel$ $C=\ddot{O}:$ CH

(b) $HC-CH$ HC CH $\ddot{O}$

(c) H N $H-B$ $\ddot{N}$ $B-H$ $H-\ddot{N}$ $:\ddot{N}-H$ B H

(d) H H $C=C=C=C$ $H-C$ $C-H$ $C-C\equiv C-C$ H H

(e) H C $H-C$ $C-H$ $H-C$ $C-H$ $\ddot{N}$

(f) O C $H-\ddot{N}:$ $C-H$ C $\ddot{N}$ $C-H$ O N H

5

Stereoisomerism

Introduction

The possibility of isomerism was first discussed in Section 1.8 where it was pointed out that two different structures could be drawn to represent the molecular formula C_4H_{10}. These two structural variations represented two *different* molecular species—namely, *n*-butane and isobutane. Because these two hydrocarbons are different compounds, each has physical and chemical properties different from those of the other. The substitution reactions of the alkanes and the arenes, as well as the addition reactions of the alkenes, we observed, were reactions that could result in the formation of isomeric products. As we continue our study of organic chemistry we realize that isomerism is an important aspect of it, and that if we understand *why* isomeric products are formed then we also will understand *how* organic molecules react.

The kind of isomers we shall consider in the following sections are molecular structures in which the same atoms and groups that are joined together in one isomer are similarly joined in the other. The atomic linkages are the same in both isomers but the spatial arrangement of the atoms and groups in each are different. Such isomers are called **stereoisomers** (Gr., *stereos,* solid). Our definition of stereoisomers does not include such spatial variations as those allowed 1,2-dichloroethane in its staggered and eclipsed forms (Sec. 2.5) nor those for cyclohexane in its boat and chair forms (Sec. 2.11). Such spatial arrangements are interconverted by the expenditure of relatively small amounts of energy and may be accomplished by the simple expedient of rotation about single bonds. Such different arrangements, you will recall, are referred to as conformations. Stereoisomers, on the other hand, have different **configurations** and can not be interconverted without breaking bonds and rearranging groups.

5.1   *Optical Isomerism (Mirror-Image Isomerism). Polarized Light*

Optical isomerism (or **mirror-image isomerism**) gets its name from the unique properties of certain isomeric *pairs* of compounds called **enantiomers.** Enantiomers have the same chemical properties and the same physical properties—with one notable exception. They differ in the direction in which each is able to rotate a beam of **plane-polarized light.**[1] The degree of rotation is the same for each enantiomer, but the direction of rotation is opposite.

In order to understand the phenomenon of "optical activity," it would be helpful to review the nature of plane-polarized light. Ordinary white light exhibits an electromagnetic wave motion in which waves of varying lengths are vibrating in all possible planes at right angles to the path of the ray. Monochromatic light, used in the measurement of optical activity, is light of only one wavelength. It usually is produced in one of two ways: (1) all unwanted wavelengths of ordinary light are removed by means of a colored filter, or (2) light of one wavelength is generated from a special source such as a sodium or a mercury lamp. Monochromatic light, like ordinary light, also consists of waves vibrating in all possible planes at right angles to the path of propagation.

Certain substances such as tourmaline crystals, polaroid, or specially prepared prisms, called Nicol prisms, act as screens when light is passed through them. Waves vibrating in only one plane pass through such special screens and all those vibrating in other planes either are rejected or absorbed. Such specially filtered light is said to be *plane-polarized.*

---

[1] There are several types of isomers, which may or may not show optical activity, included under the broad terms, optical isomerism or mirror-image isomerism. Only enantiomers show equal but opposite optical rotation.

Enantiomers may vary in their rates of reaction with other optically active substances. This is especially true in biological systems. Thus, one enantiomer may be attacked by a bacterium, the other not. Again, one of a pair of enantiomers may have a hormonal activity far greater than that of the other. These differences are due to the spatial requirements of each reactant in combining with or attacking the other.

FIGURE 5.1 *Polarization: (a) A Beam of Monochromatic Light Vibrating in All Planes, (b) A Beam of Plane-Polarized Light*

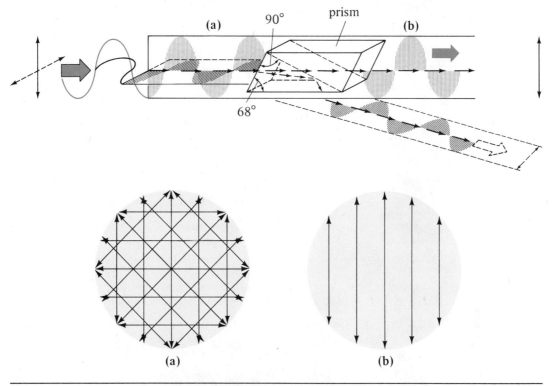

We can illustrate the action of such a **polarizer** best with the Nicol prism. The Nicol prism is an ingenious device made of calcite, $CaCO_3$. Calcite is a clear crystal with the shape of a rectangular rhombohedron and having an unusual optical property called *birefringence,* or double refraction. A ray of light entering the crystal is refracted or bent in two slightly different directions to produce two rays. To form the Nicol prism, the crystal is cut in a plane diagonally through the obtuse angles and perpendicular to the two end faces. The cut surfaces then are polished and recemented with Canada balsam. After entering the crystal and striking the cemented surface, one of the refracted rays vibrating in one plane is transmitted and others are reflected out of the crystal. Figure 5.1 illustrates graphically how a beam of monochromatic light vibrating in more than one plane might appear before (a) and after (b) it had passed through a Nicol prism polarizer.

A maximum transmission of plane-polarized light may be observed through a second prism, called the **analyzer,** only when the latter is oriented in the same optical plane as the polarizer. On the other hand, when the analyzer is rotated, the intensity of the emergent beam, as seen by the viewer, is gradually dimin-

FIGURE 5.2 *Transmission: (a) Maximum Transmission of Plane-Polarized Light, (b) Plane-Polarized Beam Blocked*

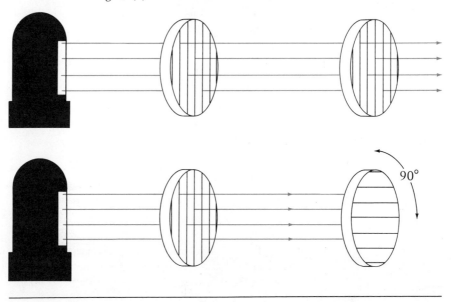

FIGURE 5.3 *Schematic Diagram of the Polarimeter, Showing Components*

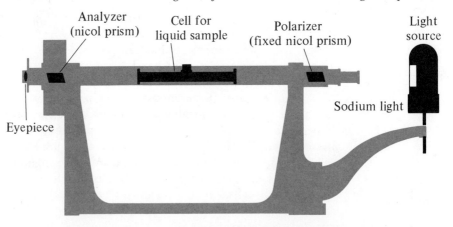

Analyzer (nicol prism) Cell for liquid sample Polarizer (fixed nicol prism) Light source

Sodium light

Eyepiece

ished. A point of minimum transmission is reached when the second prism has been rotated through an angle of 90° from the point of maximum transmission. The phenomenon can be illustrated graphically by using two polaroid lenses (Fig. 5.2).

An **optically active** compound is one which is capable of rotating the plane of polarized light. The compound is said to be **dextrorotatory** (L., *dexter,* right) when the plane of polarized light is rotated to the right, or clockwise. A dextrorotatory substance is always indicated by the positive (+) sign. A compound is described as **levorotatory** (L., *laevus*, left) when the direction of rotation is to the left, or counterclockwise. In this case the negative (−) sign is used to describe its rotation. The angle of rotation, α, is measured in degrees.

Optical activity in organic compounds is observed and measured by means of an instrument called a **polarimeter** (Fig. 5.3). The value of the specific rotation, $[\alpha]$, depends upon the length of the tube containing the sample solution, measured in decimeters, and upon the concentration of the solution measured in grams per milliliter. If the sample is a pure liquid, its density is used in place of the concentration.

$$[\alpha] = \frac{\text{observed rotation in degrees}}{\text{length of sample tube in dm} \times \text{concentration (g/ml)}}$$

In order to standardize the specific rotation values of optically active substances, both the temperature at which observations are made and the light source employed are indicated. For example, the term $[\alpha]_D^{25°} = +54°$ indicates that the specific rotation of an optically active compound is 54° to the right when the measurement is made at 25°C and the D line of the spectrum (sodium light source, 5893 Å) is used.

5.2   *Chirality and Optical Isomerism*

The characteristic structural feature of optically active compounds is **chirality**—or "handedness". This feature enables these compounds to be optically active. The relationship that pairs of such **chiral molecules** bear to each other is that which the right hand bears to the left. The enantiomers are nonidentical mirror images, or **nonsuperimposable** mirror images. A molecule whose mirror image is superimposable (or identical) with the molecule is said to be **achiral** or to lack chirality and will not show optical activity. The most common feature of chiral molecules is the presence of an **asymmetric carbon atom** (sometimes called a chiral carbon atom).[2] An asymmetric carbon atom is one which has *four different* groups attached to it, and, in a formula, is usually indicated by an asterisk placed near it. For example, in the formula for *sec*-butyl chloride the number two carbon atom is indicated as being asymmetric.

$$\begin{array}{ccccc} & H & H & H & H \\ & | & | & | & | \\ H- & C- & C- & \overset{*}{C}- & C-H \\ & | & | & | & | \\ & H & H & Cl & H \end{array}$$

sec-Butyl chloride

---

[2] Molecules and objects can be chiral—not atoms. Therefore, an asymmetric carbon atom is a chiral center or point of asymmetry. All molecules that contain *only* one asymmetric carbon atom are chiral, but molecules containing more than one asymmetric carbon atom may be chiral or achiral. Not all chiral molecules contain an asymmetric carbon atom: the necessary and sufficient condition for chirality is that the molecule have no plane, center, or four-fold alternating axis of symmetry.

FIGURE 5.4   *Nonsuperimposable Mirror Images*

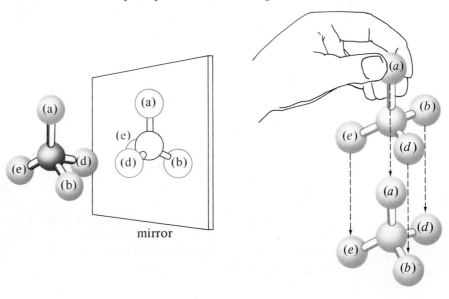

The mirror-image relationship of a pair of stereoisomers with one asymmetric carbon atom is shown in Fig. 5.4. One isomer cannot be positioned above the other with all four of the different substituents of the isomer above coinciding with those of the isomer below. Mirror-image, optically active isomers are called enantiomers (enantiomorphs (Gr., *enantios,* opposite; *morph,* form)).

EXERCISE 5.1

Which of the following objects may exist in enantiomeric forms? (a) glove, (b) shoe, (c) cap, (d) axe, (e) wood screw, (f) a pair of pliers, (g) a pair of scissors.

5.3   *Projection Formulas and Relative Configurations*

Most molecules containing carbon are three-dimensional. The four bonds of a saturated carbon atom, as was pointed out in the first chapter, are directed to the vertices of a tetrahedron. Although tetrahedral structures can, and sometimes must, be drawn as perspective (pseudo-3-dimensional) formulas (Exercise 2.3), or as Newman projections (Fig. 2.2), these are often more difficult to work with than **Fischer**[3] **projection** formulas, which are drawn with all four groups in the same plane as the carbon atom to which they are bonded (Fig. 5.5(b)). The spatial positions of four groups bonded to a central carbon atom is of little importance if two or more of the groups are identical, for in this case there will be no difference in the relative positions of the four bonded groups, regardless of their arrangement about the carbon atom. However, a random bond assignment must not be made when drawing and naming the structures of optical isomers because two configurations are now possible.

Because the **absolute** (or true) structures of optical isomers were unknown to the early chemists working with optically active compounds, Fischer decided to relate as many configurations as possible to that of a standard structure, one whose absolute structure was unknown but was to be arbitrarily defined by Fischer. His choice for the standard structure was that of glyceraldehyde,

$\overset{*}{\text{HOCH}_2\text{CH(OH)CHO}}$ (for reasons that will become apparent in (Sec. 14.4)). In the Fischer projection of glyceraldehyde, the carbon chain is drawn vertically with only the asymmetric carbon atom in the plane of the paper. Both the carbonyl and the hydroxymethyl groups are drawn as if behind the plane—the carbonyl group (—CHO) at the top and the hydroxymethyl group (—CH₂OH)

FIGURE 5.5 (a) Perspective and (b) Fischer Projection Formulas of D(+)-Glyceraldehyde

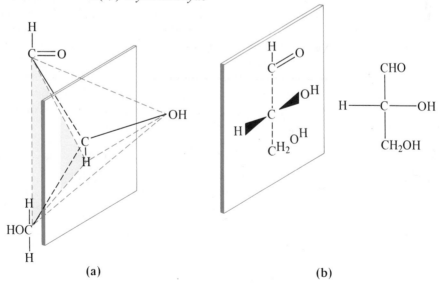

(a)   (b)

[3] Emil Fischer (1852–1919), Professor of Chemistry, University of Berlin. Winner of the Nobel Prize in Chemistry, 1902.

at the bottom. The hydroxyl group and the hydrogen atom attached to the asymmetric carbon atom are drawn as if in *front* of the plane of the paper—the hydroxyl group to the right and the hydrogen to the left (Fig. 5.5(a)). Flattening out the structure produces the projection formulas shown in Fig. 5.5(b). This configuration was arbitrarily designated as the D-configuration of glyceraldehyde by Fischer and is identified by a small capital D. Its mirror-image enantiomer, with the opposite configuration, is identified as belonging to the L series.

The structure of any other optically active compound of the type

$$\text{R—CHX—R}', \text{ is drawn with the carbon chain } \overset{\displaystyle \text{R}}{\underset{\displaystyle \text{R}'}{-\text{C}-}} \text{ in the vertical direction}$$

with the lowest numbered carbon atom at the top. If the X group (—OH, —NH$_2$, or halogen usually) is on the right, the relative configuration is D-(dee); if it's on the left, the configuration is L-(ell).

The designations "D" and "L" refer to configuration only and are not to be interpreted as "dextro" and "levo." The latter terms refer to the direction of rotation and are symbolized by the plus and minus signs respectively. They sometimes are abbreviated; dextro is *d* and levo is *l*. Although the D-form of glyceraldehyde was arbitrarily chosen as the dextrorotatory isomer without a knowledge of its absolute or true configuration, the choice was a fortuitous one; for in 1951, with the use of modern analytical methods, the dextro isomer was definitely established as being the D-form. It is possible, you see, for an optically active substance to have the L-configuration and still be dextrorotatory.

If Fischer projections are being used to assign D- or L-configurations, they *must* be drawn, as shown, with the carbon chain in the vertical direction. For *other* uses, the Fischer projections may be rearranged by carefully following two rules applicable to all molecules containing asymmetric carbon atoms: (1a) the exchange of any two groups on *each and every* asymmetric carbon atom converts the molecule into its enantiomer, and (1b) two exchanges of the type described in (1a) merely convert the molecule into itself (as should be apparent from rule (1a)). Rule 2 will be given later. These rules work with all types of projections and molecular models but are especially valuable with Fischer projections, as shown for D-glyceraldehyde.[4]

$$\begin{array}{ccc}
\text{CHO} & \text{CHO} & \text{H} \\
\text{H}-\!\!\!-\text{OH} \xrightarrow[\text{H and OH}]{exchange} \text{HO}-\!\!\!-\text{H} \xrightarrow[\text{H and CHO}]{exchange} \text{HO}-\!\!\!-\text{CHO} \\
\text{CH}_2\text{OH} & \text{CH}_2\text{OH} & \text{CH}_2\text{OH} \\
\text{D}- & \text{L}- & \text{D}- \\
\textbf{I} & \textbf{II} & \textbf{III}
\end{array}$$

---

[4] The asymmetric carbon atom is often omitted in drawing Fischer projections, and other simplifying abbreviations are sometimes used (Sec. 14.4).

Thus, I and II are enantiomers (rule 1a); I and III are simply different Fischer projections of the same enantiomer (rule 1b).

Unfortunately, the D,L-system is not applicable to many molecules of interest to the organic chemist; therefore, the IUPAC has adopted a system called the $(R),(S)$-system, which is capable of specifying the absolute configuration of any optical isomer. The "right-handed" enantiomer of a pair of optical isomers is designated in this system as the (R)-form, and the "left-handed" enantiomer as the (S)-form. These prefixes are derived from the Latin *rectus,* right, and *sinister,* left. To use this system we must assign a priority to each of the four atoms or groups attached to the asymmetric carbon atom. The group of lowest priority is placed in the position most remote from us as we view the asymmetric carbon atom. The remaining three groups are assigned positions in a *clockwise* order of decreasing priority for the (R)-isomer, and in a *counterclockwise* order of decreasing priority for the (S)-isomer. The detailed rules for assigning priorities to atoms and groups are beyond the scope of this text; however, there are two basic rules:

(1) Higher atomic number is of higher priority than lower atomic number (and higher atomic mass over lower atomic mass, if the atomic numbers are the same).

(2) Priorities of groups are determined by working outward from the point of attachment to the asymmetric carbon atom using rule (1) until a point of difference is reached, treating double and triple bonds as two or three of the same atom single bonds.

The priority sequence for the most common atoms and groups is as follows.

$$(\textit{High priority}) \quad -\text{I}, \ -\text{Br}, \ -\text{Cl}, \ -\text{SO}_2\text{R}, \ -\text{SOR}, \ -\text{SR}, \ -\text{SH}, \ -\text{F}, \ -\text{O}-\overset{\displaystyle\overset{\text{O}}{\|}}{\text{C}}-\text{R}, \ -\text{OR},$$

$$-\text{OH}, \ -\text{NO}_2, \ -\text{NH}-\overset{\displaystyle\overset{\text{O}}{\|}}{\text{C}}-\text{R}, \ -\text{NR}_2, \ -\text{NHR}, \ -\text{NH}_2, \ -\text{CCl}_3, \ -\overset{\displaystyle\overset{\text{O}}{\|}}{\text{C}}-\text{Cl}, \ -\overset{\displaystyle\overset{\text{O}}{\|}}{\text{C}}-\text{OR},$$

$$-\overset{\displaystyle\overset{\text{O}}{\|}}{\text{C}}-\text{OH}, \ -\overset{\displaystyle\overset{\text{O}}{\|}}{\text{C}}-\text{NH}_2, \ -\overset{\displaystyle\overset{\text{O}}{\|}}{\text{C}}-\text{R}, \ -\overset{\displaystyle\overset{\text{O}}{\|}}{\text{C}}-\text{H}, \ -\text{C}\equiv\text{N}, \ -\text{CR}_2-\text{OH}, \ -\overset{\displaystyle\overset{\text{OH}}{|}}{\text{CH}}-\text{R}, \ -\text{CH}_2\text{OH},$$

$$\text{C}_6\text{H}_5-, \ -\text{CR}_3, \ -\text{CHR}_2, \ -\text{CH}_2\text{R}, \ -\text{CH}_3, \ -\text{D}, \ -\text{H} \quad (\textit{Low priority})$$

In molecules with several asymmetric carbon atoms an (R)-substituent has higher priority than an (S)-substituent.

The proper placement of all groups is quite easy if, as has been suggested, one simply visualizes the asymmetric carbon atom as the hub of a three-spoked steering wheel with the atom of lowest priority attached to the steering column. The (R)- and (S)-enantiomers of 2-chlorobutane are shown.

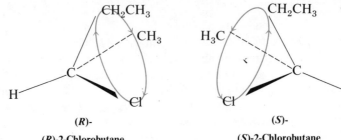

(R)-   (S)-

(R)-2-Chlorobutane   **(S)-2-Chlorobutane**

Fischer projections may be used to assign (R)- and (S)-configurations by the following simple procedure. First, if necessary, following the rules on p. 132, make *two* exchanges of groups to give a Fischer projection in which the *lowest* priority group is at the top *or* bottom. Next, draw the shortest curved arrow from the highest priority group to the second highest and then to the third highest. Where the arrow passes the lowest priority group is not important. If the curved arrow describes a *clockwise* path, the configuration is (R); if it describes a *counterclockwise* path, the configuration is (S). Thus, for D-glyceraldehyde, using Fischer projection III on p. 132, we have:

Priorities:

1. HO
2. CH=O
3. CH₂—OH
4. H

①HO——⌇——CH=O②

H ④ (top)

③ CH₂OH

Group 4 is at the top. The sequence 1 → 2 → 3 is clockwise.

(R)-Glyceraldehyde

Therefore, D-glyceraldehyde is also (R)-glyceraldehyde.

EXERCISE 5.2

Rotate structure I for D-Glyceraldehyde 90° in the plane of the paper and determine whether the rotated structure is (R) or (S). Next, rotate structure I 180° in the plane of the paper and determine whether the rotated structure is (R) or (S). From your results formulate a rule that defines the permissable rotation of Fischer projections without change in configuration.

5.4   *Asymmetric Synthesis*

The structure of 2-chlorobutane, as we have seen, is that of an asymmetric molecule since there are four different groups attached to the number two carbon.

However, the preparation of 2-chlorobutane by the direct chlorination of *n*-butane leads to an optically inactive product, because an equal number of (R)- and (S)-isomers are produced. Such 50-50 mixtures of equal parts of enantiomorphs are called **racemates,** *d,l*-mixtures, or (±)-mixtures. The optical activity of one half of the isomers in the mixture is nullified by that of the other half, and the mixture is said to be optically inactive by what is called **external compensation.**

It is not difficult to understand why equal amounts of enantiomeric forms of 2-chlorobutane result when one considers the planar structure of the hydrocarbon free radical that forms as one of the reaction intermediates (Sec. 2.8). Obviously, there is an equal opportunity for the chlorine molecule to react with the free radical from either side. Since Avogadro's number (the number of molecules in one mole) is about 6×10^{23}, the number of molecules involved is enormous, and the probability of the chlorine molecule approaching from either side is exactly one half. Employing the convention for indicating bond positions as that used in Exercise 2.3, we can show two reaction sequences graphically to illustrate the synthesis of racemic 2-chlorobutane.

(±)-2-Chlorobutane (a racemic mixture)

A laboratory synthesis that leads to the production of a chiral molecule from an achiral molecule nearly always produces a racemic mixture. In nature, however, it appears that a stereospecificity governs both the synthesis and the reactions of optically active substances. For example the catalytic reduction of pyruvic acid, $CH_3—\overset{\overset{\displaystyle O}{\|}}{C}—COOH$, in the laboratory leads to racemic lactic acid, or a *d,l*-mixture. On the other hand, pyruvic acid is reduced by yeast to *levo*-lactic acid. Lactic acid isolated from muscle tissue is *dextro*-lactic acid.

Most natural products are optically active and usually are stereospecific in their reactions. A study of optical isomerism, therefore, must be a prelude to any study of natural substances.

EXERCISE 5.3

Using solid, dotted, and wedge-shaped bonds draw two stereochemical formulas for 2-pentene. (a) Does the addition of HBr to either structure lead to resolvable optical isomers? (b) Could the addition of HBr to either structure lead to an optically inactive molecule?

5.5   *Compounds Containing Two Asymmetric Carbon Atoms*

(*Case A*) *Two Unlike Asymmetric Carbon Atoms.* The total number of optical isomers possible when a molecule contains n unlike asymmetric carbon atoms is 2^n. This is a statement of the **van't Hoff rule.**[5] The rule may be illustrated by considering the stereoisomerism possible for a molecule with a configuration like that exhibited by the ball and stick model shown.

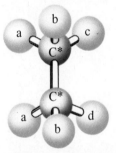

Designating the upper asymmetric carbon α and the lower one β in structures IV to VII, it can be shown that two configurations are possible for each asymmetric carbon atom. Structures IV and V represent a pair of enantiomers, as do VI and VII. Although structures V and VI (also structures V and VII, IV and VI, and IV and VII) are optically active stereoisomers, they are not mirror images. Stereoisomers that are not mirror images are called **diastereoisomers.**

If we let a = H, b = CH_3, c = Br, and d = Cl, we can draw the Fischer projections of all the optical isomers of 2-bromo-3-chlorobutane. We must be

---

[5] J. H. van't Hoff (1852–1911), Dutch physical chemist, was one of the first to recognize asymmetry in a compound in which four different groups were attached to a carbon atom. He postulated that the four different groups in such an asymmetric structure could have two possible spatial arrangements.

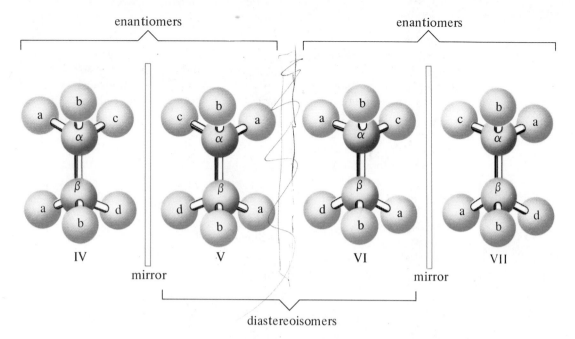

very careful in doing this, for we will be drawing the Fischer projections from the point of view of an observer looking at them from the back side of the page toward us. This is what we draw.

We can now complete our set of two rules for working with optical isomers: rule (2), in a molecule with two or more asymmetric carbon atoms, the exchange of any two groups on any *one* (and only one) asymmetric carbon atom converts the molecule into one of its diastereomers (exchange of H and Cl in IV gives VI).

EXERCISE 5.4

Show that for 2-bromo-3-chlorobutane the following configurational assignments are correct: IV, (*R*)-2-bromo-(*S*)-3-chlorobutane; V, (*S*)-2-bromo-(*R*)-3-chlorobutane; VI, (*R*)-2-bromo-(*R*)-3-chlorobutane; VII, (*S*)-2-bromo-(*S*)-3-chlorobutane. (*Hint:* Make exchanges carefully until the lowest priority group on one asymmetric carbon atom is at the top and the lowest priority group on the other is at the bottom. You should be able to do this by making two exchanges on *each* asymmetric carbon atom.)

(*Case B*) *Two Similar Asymmetric Carbon Atoms.* The structure of 2,3-dibromobutane is an example of a compound having two asymmetric carbon atoms that are similar—that is, both carbons hold identical groups. A compound with two similar asymmetric carbon atoms can have three isomeric forms. Two forms will be optically active mirror images and the third optically inactive.

The order of substituents in the upper part of structure VIII (below), when viewed from above and clockwise, is opposite to that of its lower half. Structure

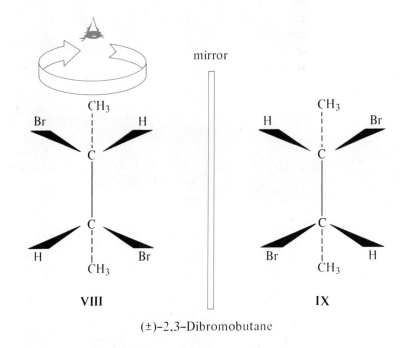

(±)–2,3–Dibromobutane

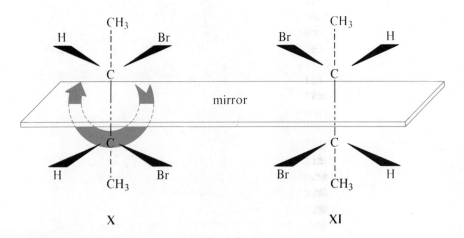

VIII has its mirror image in structure IX. In each structure the configuration of one half of the molecule, when viewed from above, is exactly opposite to that of the other half. Structures VIII and IX are nonsuperimposable mirror images. On the other hand, the order of substituents in the upper half of either structure X or XI is the same as that of each corresponding lower half when viewed from above and clockwise. In fact, structure XI is but structure X rotated (in the plane of the paper) 180°. Structures X and XI, therefore, are one and the same. In each, the upper half is a mirror image of the lower half. Such an arrangement of substituents makes for an optically inactive, or a *meso,* form. A meso form is thus optically inactive by internal compensation. Any effect the upper half of such a structure might have upon plane-polarized light is exactly opposite to that produced by the lower half. The meso form of 2,3-dibromobutane, as might be expected, has physical properties unlike those of the optically active (+) and (−)-forms.

EXERCISE 5.5

In Section 3.6 we learned that the catalytic hydrogenation of 1,2-dimethyl-cyclopentene gave *cis*-1,2-dimethylcyclopentane. Are any of the carbon atoms in the disubstituted cycloalkane bonded to four different groups? Is *cis*-1,2-dimethylcyclopentane optically active?

5.6 *Resolution of Optical Isomers*

A. Mechanical Separation. The separation of a racemic mixture into its optically active (*R*)- and (*S*)-forms is called **resolution.** The first resolution of a racemic mixture was the historic accomplishment of Pasteur in 1848. He observed that the crystalline tartaric acid salts (Sec. 12.5), when precipitated from solution, appeared to consist of "right- and left-handed" crystals. He carefully selected a number of each kind by using a pair of tweezers and a magnifying glass. Separate solutions of the same concentration were prepared of each form and their rotations observed. The rotation values of the two solutions were found to be alike *but of opposite direction.* Mechanical resolutions such as the one done by Pasteur are seldom possible. Other methods of resolution are usually necessary.

B. Preparation and Separation of Diastereoisomers. Reaction of a racemic mixture with an optically active reagent results in the preparation of diastereoisomers. The physical properties of diastereoisomers are different, and they may be separated by crystallization, distillation, or by some other technique. An example of such chemical resolution is the preparation of salts from a racemic acid and a *levo*-base.

Enantiomers　　　　　　　　　　　　　　*Diastereoisomers*

$$\begin{Bmatrix} \textit{dextro-}\text{acid} \\ \textit{levo-}\text{acid} \end{Bmatrix} + \textit{levo-}\text{base} \longrightarrow \begin{Bmatrix} \textit{dextro-}\text{acid} \cdot \textit{levo-}\text{base salt} \\ \textit{levo-}\text{acid} \cdot \textit{levo-}\text{base salt} \end{Bmatrix}$$

Each diastereoisomer, having physical properties different from those of the other, may be separated by fractional crystallization. Subsequent treatment with a mineral acid liberates the optically active organic acid.

$$\textit{dextro-}\text{acid} \cdot \textit{levo-}\text{base salt} + HCl \longrightarrow \textit{dextro-}\text{acid} + \textit{levo-}\text{base} \cdot HCl$$
$$\textit{levo-}\text{acid} \cdot \textit{levo-}\text{base salt} + HCl \longrightarrow \textit{levo-}\text{acid} + \textit{levo-}\text{base} \cdot HCl$$

C. Biochemical Resolution.　Certain microorganisms preferentially attack one isomer to the exclusion of the other when allowed to "feed" upon a racemic mixture. One optical form is thus destroyed. An example of this kind of microbiological separation is the resolution of racemic tartaric acid by *Penicillium glaucum*. This mold uses up the *dextro*-tartaric acid and leaves the *levo* form unattacked. Biochemical resolutions as preparative procedures, however, are limited in their application.

Frequently the resolution of a *d,l*-mixture becomes the most difficult step in a preparation.

5.7　cis,trans-*Isomerism*

As was shown in Secs. 2.12 and 3.2, *cis,trans*-isomerism is a form of stereoisomerism that results when the rotation of one carbon atom with respect to another is restricted by the presence of a double bond, or when such rotation is prevented by the incorporation of the atoms in a ring. Several examples of this type of isomerism have been given in the previous chapters. Although the terms *cis* and *trans* will continue to be used for simple alkenes, they are in the process of being replaced by the IUPAC for more complex structures with the more general and less ambiguous designations: (*E*)- (from the German *entgegen,* across) and (*Z*)- (from the German *zusammen,* together). In the (*E*), (*Z*)-system (no pun intended), the two groups attached to each end of the double bond are assigned priorities (exactly as in the (*R*),(*S*)-system). If the two higher priority groups are on the same side of the double bond, the configuration is (*Z*); if on the opposite sides, the configuration is (*E*).

Relative priorities are encircled.

(*E*)-2-Bromo-3-methyl-2-pentene　　　　　　(*Z*)-2-Bromo-3-methyl-2-pentene

For the cycloalkanes, the *cis* and *trans* designations of configuration have been retained. For flexible ring systems, such as cyclohexane, the ring may be treated as if it were planar for purposes of assigning configuration.

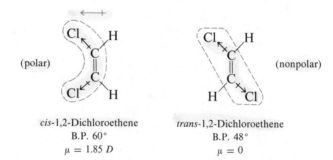

cis-1,3-Dimethylcyclohexane

cis,trans-Isomers, unlike optical isomers, have widely different physical properties and often vary in their chemical behavior. Measurable differences in these properties allow us to distinguish between *cis* and *trans* forms rather easily. For example, *cis*-1,2-dichloroethene has a measurable **dipole moment,**[6] while *trans*-1,2-dichloroethene has a zero dipole moment. The dipole moment of

a molecule with polar bonds such as the $\overrightarrow{C-Cl}$ bond, will be the vector sum of the individual bond (dipole) moments. These bond moments cancel each other in the *trans*-isomer but add to give a net dipole moment in the *cis*-isomer.

(polar)   (nonpolar)

cis-1,2-Dichloroethene   *trans*-1,2-Dichloroethene
B.P. 60°   B.P. 48°
$\mu = 1.85\ D$ $\mu = 0$

The chemical behavior of *cis* and *trans* isomers may also vary sufficiently to enable us to distinguish one from the other. For example, the two carboxyl groups of maleic acid are on the same side of the double bond and thus properly oriented in space to permit the easy removal of a water molecule, when heated, to give maleic anhydride. Its *trans* isomer, fumaric acid, cannot form the anhydride.

---

[6] The dipole moment of a compound measures the concentration of positive and negative charges in different parts of the molecule. It is defined by the expression, $\mu = e \times d$, where e is the charge (in electrostatic units) and d is the distance between the centers of positive and negative charge in Angstroms (Å). Dipole moments are measured with a device called a dipolimeter and are expressed in Debye units, D.

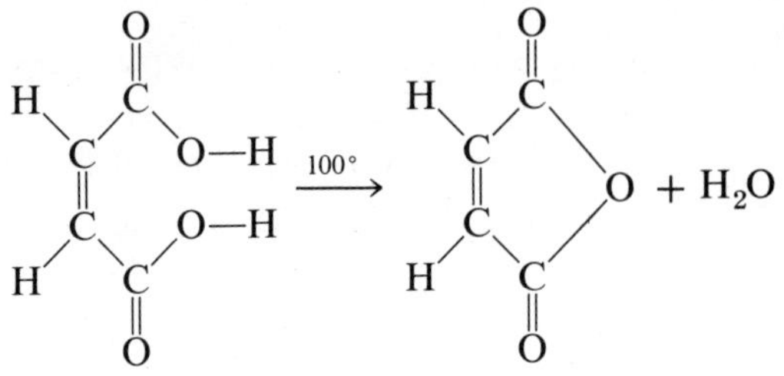

Maleic acid
(*cis*-Butenedioic
acid)
M.P. 130°

Maleic anhydride

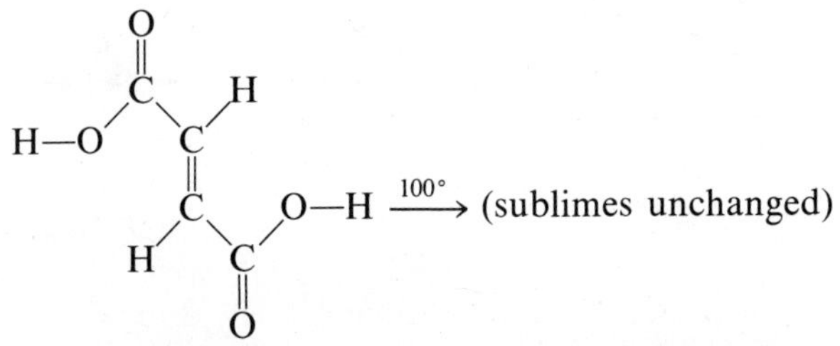

Fumaric acid
(*trans*-Butenedioic acid)
M.P. 270°

EXERCISE 5.6

Natural rubber and its isomer, gutta-percha, are polymers of isoprene. Natural rubber has a *cis*-configuration; gutta-percha has a *trans*-configuration. Draw partial structures of each natural product.

Summary

1. Isomers that have similar structural formulas but different configurations (spatial arrangements) are called **stereoisomers.**

2. Stereoisomers may be either **optical** or **cis,trans**-isomers.

3. An **optically active** compound is one capable of rotating the plane of polarized light. Compounds that are **chiral** are optically active. Chirality is the property of "handedness."

   An asymmetric carbon atom is one with four different atoms or groups bonded to it.

4. Plane-polarized light is light in which all rays are vibrating in a single plane.

5. Optical isomers that are mirror images, or enantiomers, rotate plane-polarized light to the same degree, but in opposite directions.

6. The direction of rotation, if to the right, is designated **dextro** ($+$); if to the left, **levo** ($-$). *Dextro* and *levo* forms also may be referred to as *d* and *l* forms.

7. The configurations of many optically active compounds may be related to that of glyceraldehyde and designated as D- or L-forms.

8. Absolute configurations of optically active compounds may be designated as (R)- or (S)-forms.

9. Optical activity can be measured by means of an instrument called a **polarimeter.**

10. A synthesis which produces a compound containing an asymmetric carbon atom usually results in a 50-50 mixture of (R)- and (S)-forms. Such a mixture is called a **racemic mixture.** Racemic mixtures are optically inactive.

11. A molecule with *n different* asymmetric carbon atoms can have 2^n optical isomers.

12. A molecule with *similar* asymmetric carbon atoms can have, in addition to optically active forms, one or more **meso** forms that are optically inactive.

13. **Diastereoisomers** are optical isomers that are not mirror images.

14. Separation of a mixture of enantiomers is called **resolution.** Resolution can be accomplished mechanically, chemically, or biologically.

New Terms

absolute configuration	diastereomers
achiral	D- and L-configurations
chiral	*E*- and *Z*-configurations
configuration	enantiomers
dextrorotatory	Fischer projections

levorotatory

meso form

polarimeter

polarized light

polarizer

(*R*)- and (*S*)-configurations

racemic mixture

relative configuration

resolution

Supplementary Exercises

EXERCISE 5.7   Predict the number and kind of stereoisomers possible for each of the following.
(a) 2-Butenoic acid, $CH_3CH=CH—COOH$
(b) 1,2-Dimethylcyclopentane
(c) 2,3-Butanediol, $CH_3CH(OH)CH(OH)CH_3$
(d) 3-Bromocyclohexene
(e) Glucose, $HOCH_2(CHOH)_4CHO$

EXERCISE 5.8   Indicate the asymmetric carbon atoms in each of the following natural substances.

(a)

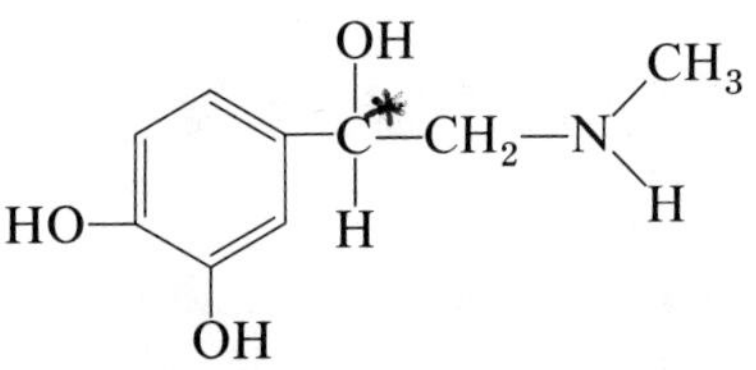

Epinephrine

(b)

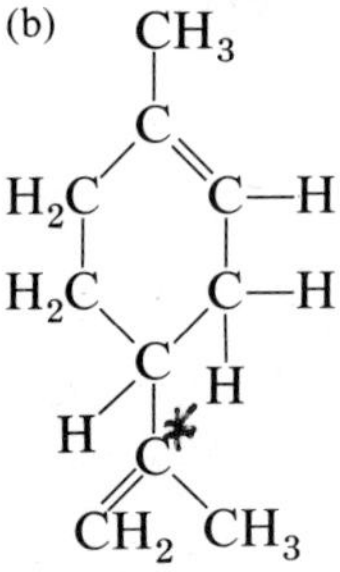

Limonene

(c)

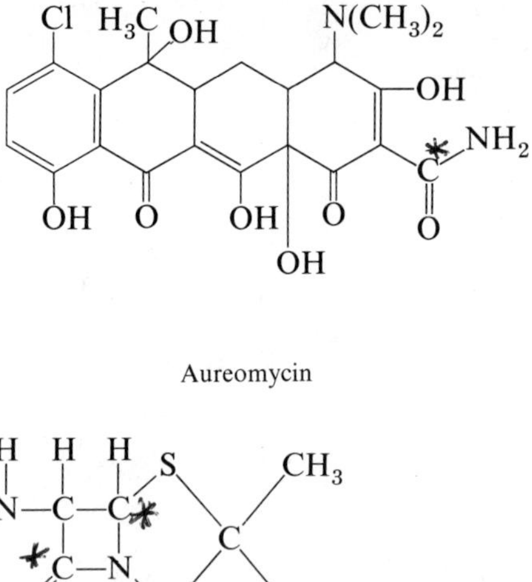

Aureomycin

(d)

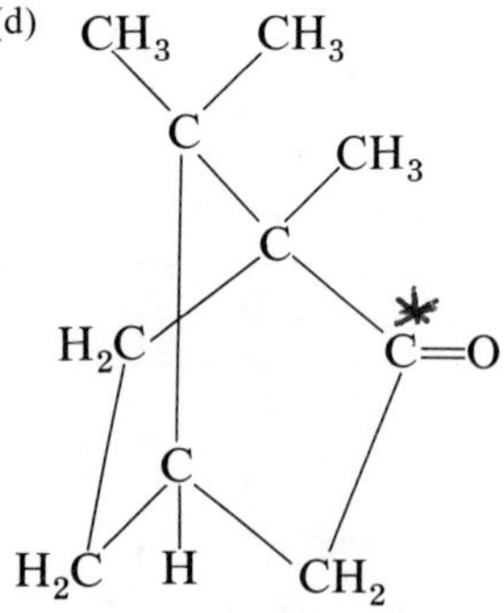

Camphor

(e)

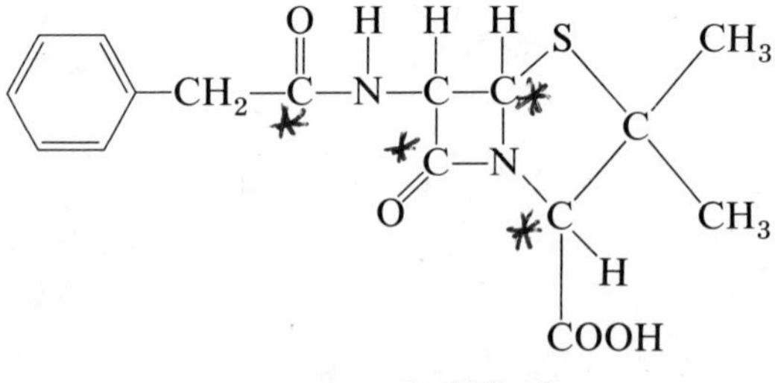

Penicillin G

(f) $CH_3—(CH_2)_5—CH(OH)—CH_2—CH=CH—(CH_2)_7COOH$
Ricinoleic acid

EXERCISE 5.9   Could the following compounds be resolved into (+) and (−) forms?

(a) $CH_3CH_2CH(OH)CH_3$

(b) $CH_3CH(Br)CH_3$

(c) 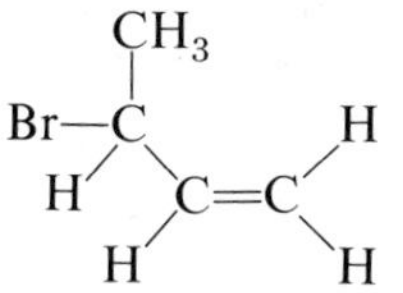
$$Br-\overset{\displaystyle CH_3}{\underset{\displaystyle H}{C}}\overset{\displaystyle}{\underset{\displaystyle C=C}{}}\overset{H}{\underset{H}{}}$$

(d)

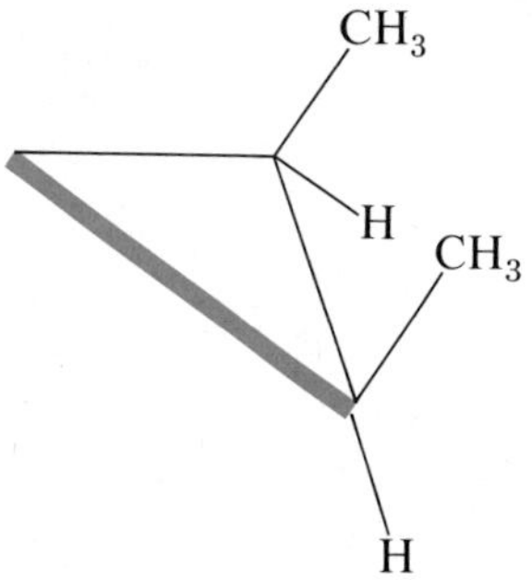

(e)

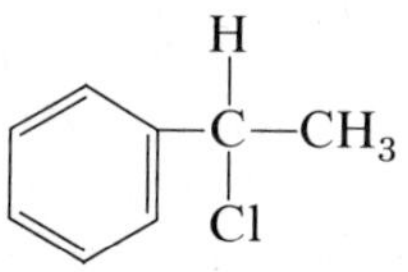

EXERCISE 5.10   Draw *R* and *S* configurations for Compound (c), in Exercise 5.9. (*Note: Treat the vinyl group,* $-CH\!=\!CH_2$ *as though its structure were* $-\overset{CH_2-}{\underset{CH_2-}{C}}-H$ *in determining its priority.*)

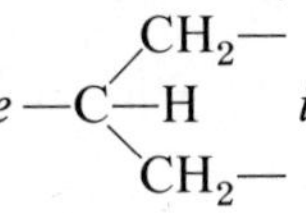

EXERCISE 5.11   Draw projection formulas for

(a)

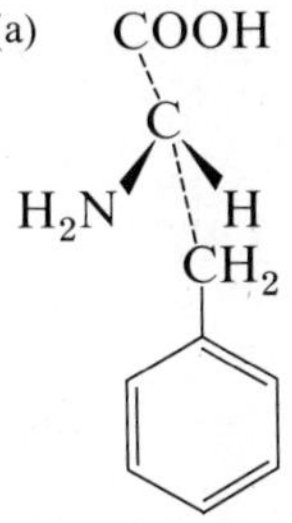

Phenylalanine

(b)

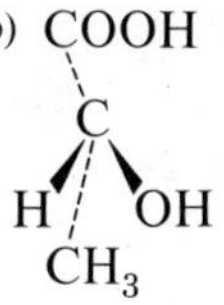

Lactic acid

Identify each as D or L; as *R* or *S*.

EXERCISE 5.12   Draw stereochemical formulas for three unsaturated compounds of formula C_5H_9Br that would show optical activity. Do any of these also have *cis* and *trans* forms?

EXERCISE 5.13   Complete the following reactions. Are the principal products resolvable into optically active forms?

(a) 1-Butene + HBr $\longrightarrow$

(b) 2-Butene + HBr $\longrightarrow$

(c) Cyclopentene + HBr $\longrightarrow$

(d) Cyclopentene + Br$_2$ $\longrightarrow$
(e) 2-Butene + HOCl $\longrightarrow$
(f) Propylene + HOCl $\longrightarrow$
(g) Propylene + cold, dilute KMnO$_4$ $\longrightarrow$

EXERCISE 5.14   If ethyl alcohol had a pyramidal structure such as that shown in (I) below, what kind of isomerism would be possible? What kind of isomerism would be possible if ethyl alcohol had a planar structure such as that shown in (II)?

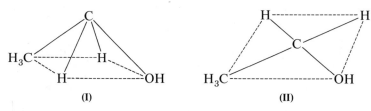

Why is there but one structure possible for ethyl alcohol?

EXERCISE 5.15   What stereoisomers are possible when bromine is added to *trans*-2-butene? What stereoisomers are possible when bromine is added to *cis*-2-butene? (*Hint: See Section 3.7. Also draw three-dimensional structures or Fischer projections for reactants and products.*)

EXERCISE 5.16   An optically active liquid (A), C$_4$H$_9$Cl, when heated with alcoholic KOH lost its optical activity. The product, an optically inactive gas (B), C$_4$H$_8$, was bubbled through bromine-carbon tetrachloride solution to give a new substance (C), C$_4$H$_8$Br$_2$, that could not be resolved into (+) and (−) forms. Write the reactions that lead to products (B) and (C) and show the stereochemistry involved by drawing three-dimensional formulas.

EXERCISE 5.17   Draw the Fischer projection of the compound whose Newman projection is given below.

Properly done, the answer will provide a useful device for interconverting Fischer and Newman projections.

EXERCISE 5.18   Label each of the double bonds in the following naturally occurring substances as (*E*) or (*Z*).

(a)

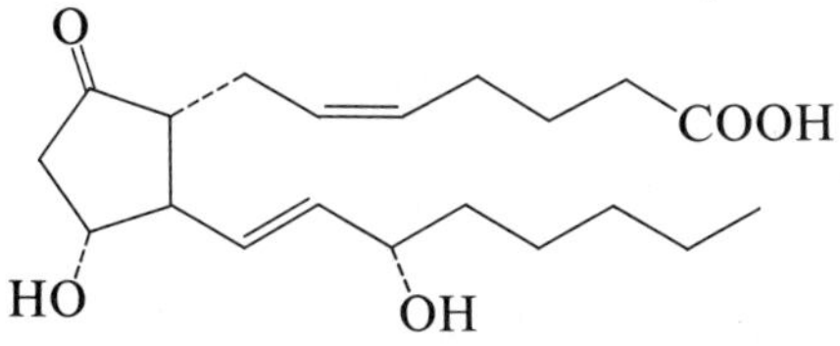

Prostaglandin PGE$_2$

(b) $CH_3(CH_2)_3$

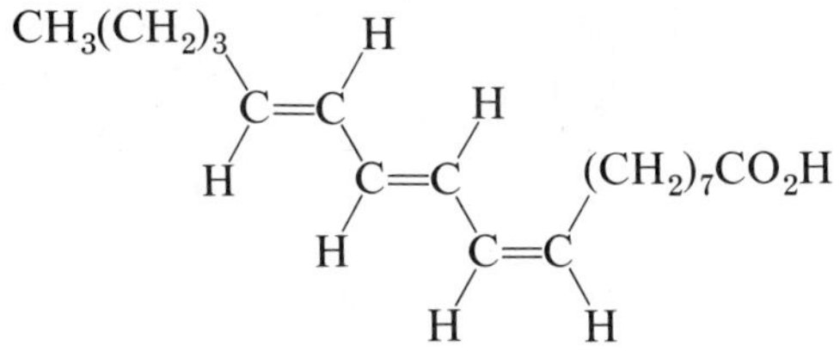

Eleostearic acid

EXERCISE 5.19 Determine which, if any, of the following is a *meso*-isomer.

(a)
```
        CH3
  HO ——— H
   H ——— CH3
        OH
```

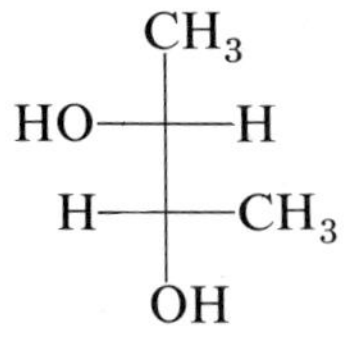

(b)
```
        Cl
   H ——— CH3
   H ——— Cl
        CH3
```

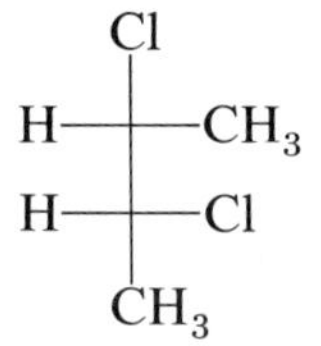

(c)
```
        Cl
   H ——— Cl
   H ——— Cl
        Cl
```

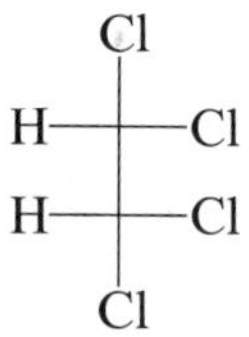

(d)
```
        H
  HO ——— CH3
  HO ——— H
  CH3 ——— H
        OH
```

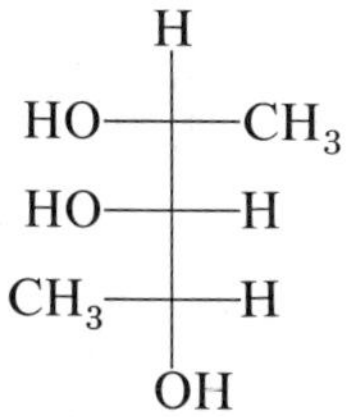

EXERCISE 5.20 Label each of the asymmetric carbon atoms in the following compounds as (*R*) or (*S*).

(a)
```
        CH3
   H ——— OH
   H ——— Cl
        CH3
```

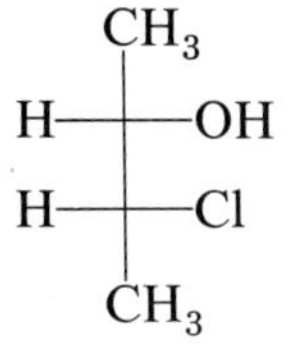

(b)
```
        CH3
   H ——— Cl
  Cl ——— H
        CH3
```

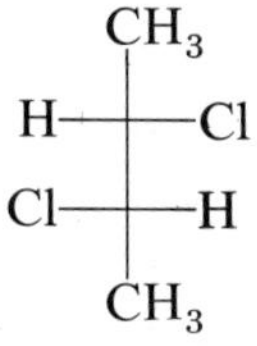

(c)
```
        CH3
   H ——— OH
   H ——— OH
        CH3
```

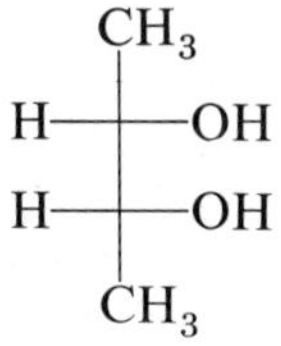

(d)
```
        CH3
   H ——— OH
   H ——— Cl
   H ——— OH
        CH3
```

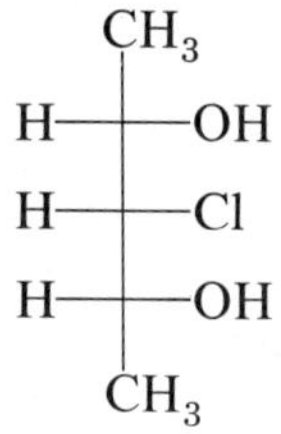

(e)
```
        CH=O
   H ——— OH
   H ——— OH
   H ——— OH
   H ——— OH
        CH2OH
```

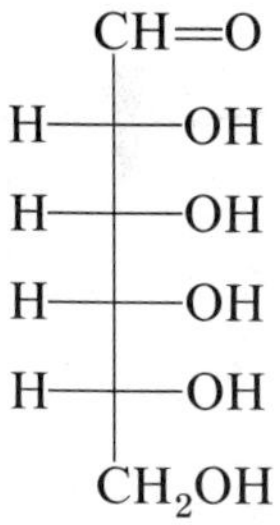

EXERCISE 5.21 Draw the Fischer projection of (*R*)-2-chlorobutane. Which of the following structures are also representations of (*R*)-2-chlorobutane? Which are representations of (*S*)-2-chlorobutane?

(a)
```
          H
  CH3 ——— CH2CH3
          Cl
```

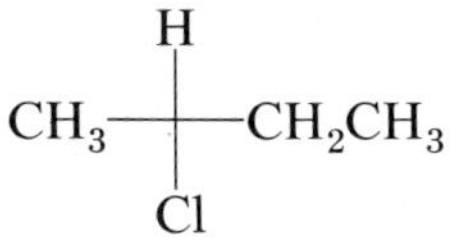

(b)
```
          CH2CH3
   H ———— Cl
          CH3
```

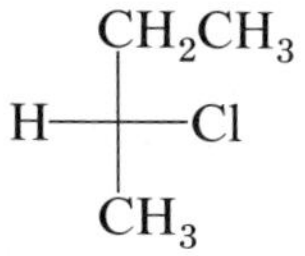

(c)

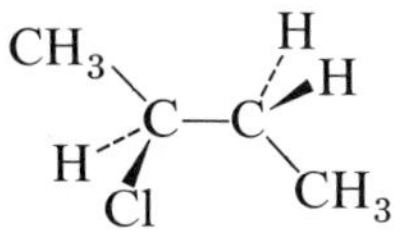

(d)

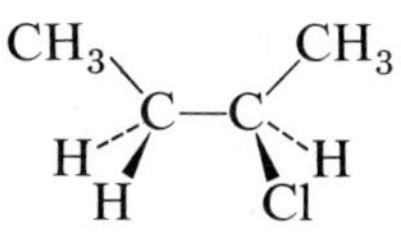

(e)

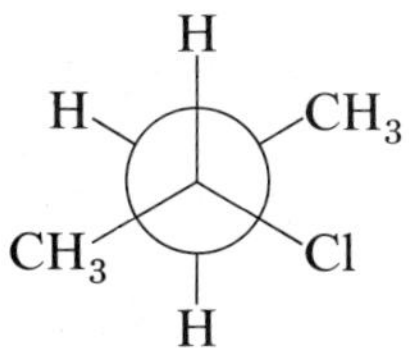

(f)

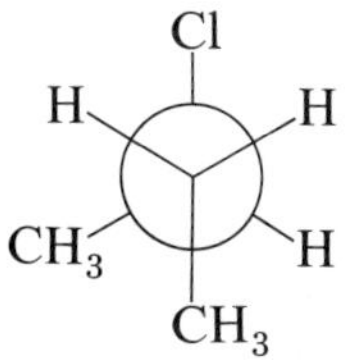

EXERCISE 5.22 A solution of 0.90 g of limonene (Exercise 5.8.-(b)) in sufficient methanol (CH_3OH) to make 50 ml of solution gave an observed rotation, α, of $+1.91°$ in a 10-cm polarimeter tube at 20°C. Calculate $[\alpha]_D^{20}$.

EXERCISE 5.23 The following molecules can be treated as having fairly rigid structures for the reasons given with each structure. Test each of these compounds to see which, if any, could exist in enantiomeric forms. For those that can, draw the structures of both isomers. The use of models may be helpful.

(a)

2,2'-Dicarboxy-6,6'-dinitro-biphenyl

Rings perpendicular to each other; rotation around central bond restricted.

(b) trans-Cyclooctene

Conformation stable below 120° because interconversion goes through highly strained form.

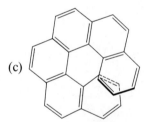

(c)

Hexahelicene

Ring system overlaps with one end locked below the other.

(d)

2,3-Pentadiene

Planes defined by the $CH_3—C—H$ at each end of the molecule are perpendicular.

Determination of Molecular Structure; Spectroscopy

Introduction

Let us assume that we have in hand a small amount of a substance that we believe to be a pure compound. It may be the product of a reaction that we have carried out in the laboratory, or it may be a naturally occurring material that we have isolated from a plant or animal source after the expenditure of considerable time, labor, and ingenuity. It may be one of the more than two million organic compounds that have already been prepared or isolated and whose structures are known, or it may be a new substance with exciting chemical or biological properties whose structure has yet to be determined. How do we go about the task of identifying our compound if it is one of those already known or of establishing its structure if the compound is an unknown substance? During the first century of the development of modern organic chemistry, the identification of known compounds or the determination of the structure of new substances was based on the use of a few physical and biological properties (mp, bp, refractive index, color, odor, taste, etc.) and a host of chemical tests and reactions. It was not uncommon for a chemist to spend months or years on the structural characterization of a single important compound of complex structure. However, during the past thirty years the determination of the structure of an organic compound has been greatly simplified by the development of spectroscopic methods of analysis and structure determination. The combination of **spectroscopy,** other physical properties, and, if need be, chemical tests or reactions has made possible the rapid determination of most organic compounds, often with no more than a few milligrams of sample. Spectroscopy is a very broad term which encompasses all methods based on the separation of beams of electromagnetic radiation or elementary particles or ions. Our discussion will be concentrated on **absorption spectroscopy,** which may be defined as a procedure based on the selective absorption of electromagnetic radiation by a system and the experimental determination of the frequencies and amounts of radiation absorbed.

In a typical spectroscopic analysis using an absorption **spectrometer,** a sample is placed in the path of a beam of electromagnetic radiation (such as light). The frequency of the radiation is varied, and the amount of radiation that is absorbed by the sample at each frequency is determined by the spectrometer, which plots the results in the form of a **spectrum.** The spectrum usually consists of one or more **peaks** or **bands** (broad or overlapping peaks), whose frequency and size (intensity) can be correlated with some structural feature of the molecule.

Amount of
radiation
absorbed

Frequency of radiation ⟶

6.1   *General Characteristics*

Of the many types of spectroscopy known, those most frequently used at present in the analysis and determination of structure of organic molecules are infrared (ir), ultraviolet (uv), and nuclear magnetic resonance (nmr) spectroscopy. Although these three spectroscopic techniques differ in the details of their theory, instrumentation, and application, they are based on the same fundamental principle: the selective absorption of energy from an electromagnetic wave by the molecule being studied, accompanied by a change in the energy state of the molecule. Since the molecule can exist only in discrete energy states (characterized by its quantum numbers), the absorption of energy is said to be quantized and is found to be related to the frequency (ν) of the electromagnetic wave by the following equation.

$$\Delta E = h\nu$$

Here ΔE is the energy absorbed and is equal to the difference in energy between the initial (lower) and final (upper) energy states of the molecule. The proportionality constant, h, is Planck's constant (6.6262×10^{-34} joules/Hz).

This relationship is illustrated in Figure 6.1 for a hypothetical molecule which is allowed only three energy states, E_1, E_2, and E_3. The energy of the molecule cannot take on values intermediate between E_1 and E_2, nor between E_2 and E_3. Therefore only those frequencies of electromagnetic radiation meeting the criteria

$$\nu_1 = \frac{\Delta E_1}{h} = \frac{(E_2 - E_1)}{h}$$

$$\nu_2 = \frac{\Delta E_2}{h} = \frac{(E_3 - E_2)}{h}$$

$$\nu_3 = \frac{\Delta E_3}{h} = \frac{(E_3 - E_1)}{h}$$

may be absorbed by the molecule. Whether or not these frequencies *will* be absorbed depends on rules, called **selection rules,** which tell us which changes in structure (or energy state) are **allowed** for a given type of spectroscopy (nmr, ir, uv-vis, etc.). If a change in energy state, called a **transition,** is **forbidden** by the

FIGURE 6.1 *Energy Levels of a Hypothetical Molecule and Transitions Between Them**

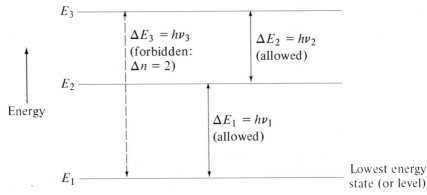

* Transitions between the levels E_n are shown for a typical selection rule: $\Delta n = \pm 1$ for allowed transitions.

selection rules, the probability of that transition occurring is likely to be very low. An allowed transition will tend to give rise to a strong or intense absorption peak, and a forbidden transition to a weak absorption peak, or to no peak. In summary we conclude that the absorption of energy from an electromagnetic wave of frequency v is likely to take place only when $\Delta E = hv$ and the selection rules for the particular type of spectroscopy involved allow the transition corresponding to ΔE.

For a given organic molecule there are many energy states and allowed transitions; thus, electromagnetic energy of many frequencies may be absorbed by the molecule. Our problem is to relate the frequencies and energy absorption in a qualitative and quantitative fashion to the structure of the molecule. Fortunately, the absorption of light in the uv portion of the electromagnetic spectrum (Fig. 6.2) can be related to the different **electronic energy states** associated with the movement of an electron from one orbital to another, and the absorption of light in the ir region of the electromagnetic spectrum can be

FIGURE 6.2 *Electromagnetic Spectrum*

λ = Wavelength (meters)

10^8 10^6 10^4 10^2 1 10^{-2} 10^{-4} 10^{-6} 10^{-8} 10^{-10} 10^{-12} 10^{-14} 10^{-16}

10^{19} ———————— γ-rays

3×10^{16}–10^{19} ———— X-rays

7×10^{14}–3×10^{16} ——— ultraviolet

3.95×10^{14}–7.90×10^{14} – visible

3×10^{11}–4×10^{14} ———— infrared

3×10^9–3×10^{11} ——— microwave

2×10^7–4×10^8 ——— nuclear magnetic resonance

6×10^5–1.6×10^6 – broadcast radio (am)

1–10^9 ———————————— radio and radar

1 10^2 10^4 10^6 10^8 10^{10} 10^{12} 10^{14} 10^{16} 10^{18} 10^{20} 10^{22} 10^{24}

v = Frequency (Hertz)

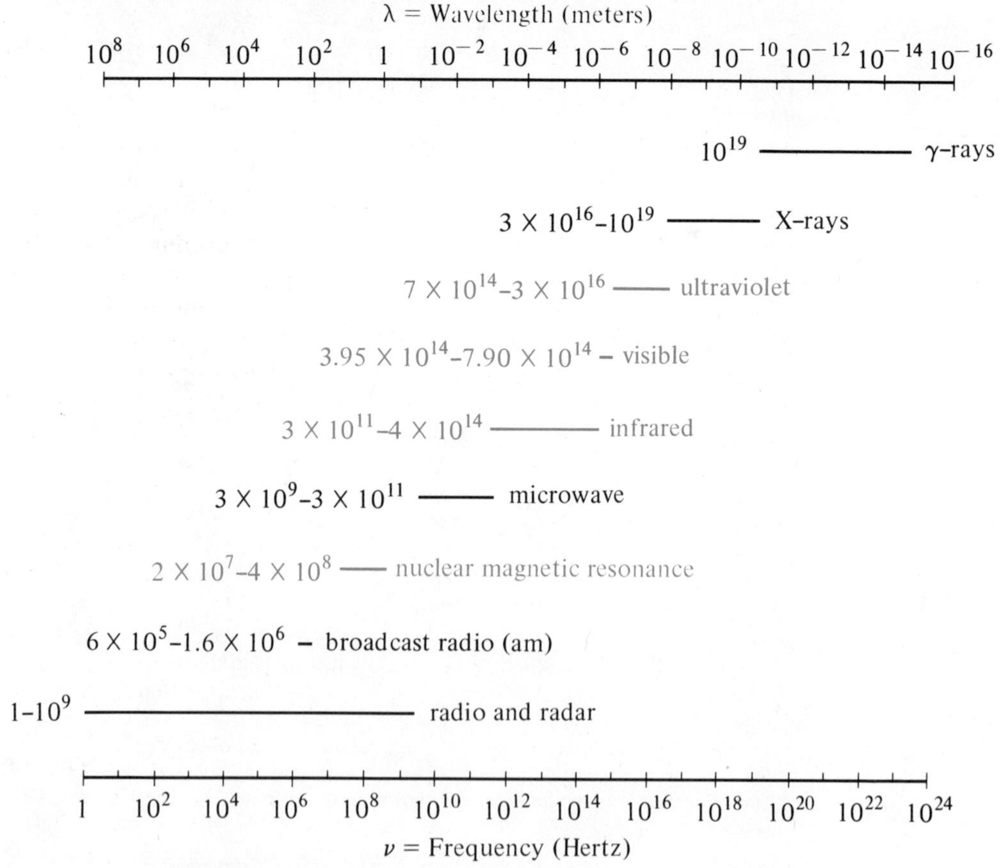

correlated with the different **vibrational** and **rotational energy states** associated with the motions of atoms in space. Absorption of electromagnetic energy in the high radio frequency portion of the spectrum is dependent on the **magnetic environment** of certain atoms in the molecule and makes possible one of the most versatile of all analytical techniques, nmr spectroscopy.

6.2 *Nuclear Magnetic Resonance*

An atomic nucleus which contains either an odd number of protons (odd atomic number) or an odd number of neutrons (odd mass number), or both, is said to have a non-zero nuclear magnetic moment that arises from nuclear "spin." The nature of nuclear spin is thought to be complex, but for our purposes the nucleus can be regarded as a charged body spinning on an axis and generating a magnetic field pointing along the direction of the spin axis. If the **spin quantum number,** I, of the nucleus is $\frac{1}{2}$ (as it is for $^{1}_{1}H$, $^{13}_{6}C$, $^{15}_{7}N$, $^{17}_{8}O$), the "spinning" nucleus can be treated as a tiny bar magnet. If the spin quantum number is zero (as it is for $^{12}_{6}C$ and $^{16}_{8}O$, the common isotopes of these elements), the nucleus will not display magnetic properties and will not give an nmr spectrum. Let us focus our attention on the hydrogen nucleus, which we will call the proton ($^{1}_{1}H$), and on proton nmr (sometimes called pmr) spectroscopy. We will represent the nuclear magnetic moment of the spinning proton as an arrow (or vector) whose length is proportional to the strength of the proton's magnetic field and whose direction is that of the field. The arrow will point along the direction of the spin axis with its head at the north (N) pole of the tiny nuclear magnet (Fig. 6.3(a)). In the absence of a strong, external magnetic field a collection of nuclear magnets will be oriented in a random fashion relative to each other (Fig. 6.3(b)). However, in a strong, static magnetic field, H_0, the nuclear magnets are allowed only two orientations relative to the H_0 field direction.[1] In the more stable orientation (the lower energy state), the nuclear magnet is aligned with the field; and in the less stable orientation (the higher energy state), it is aligned against the field (Fig. 6.3(c)).[2] In this text we will

---

[1] For nuclei with nuclear spin quantum numbers greater than $\frac{1}{2}$, the number of orientations permitted is $(2I + 1)$. For example, if $I = 1$ (as it does for deuterium, $^{2}_{1}H$ or D, or for $^{14}_{7}N$), the nuclear magnets may be oriented in three directions relative to the H_0 field: with the field, against the field, and perpendicular to the field.

[2] The magnetic field between the poles of a magnet is represented by an arrow (vector) pointing from the north pole to the south pole (leaving the north pole and entering the south pole). The magnetic field *within* the magnet can be represented by an arrow pointing from the south to the north pole. In the earth's magnetic field and in the absence of local magnetic effects, the north pole of a compass needle (a tiny magnet) would point directly toward the earth's North Pole. This is no contradiction of Fig. 6.3(c) for the earth's North Pole is a south *magnetic* pole. At the nuclear level in the more stable orientation the tiny nuclear magnets do not "point" directly toward the south pole of the magnet generating the field H_0, but do align themselves at some constant angle θ relative to the field direction. However, it can be shown that, for our purposes, we may treat the system of nuclear magnets as though they do line up with the field or against it.

FIGURE 6.3   *Nuclear Magnetic Resonance: (a) Equivalence of Spinning Nucleus and Small Bar Magnet, (b) Orientation of Nuclear Magnetic Moments in the Absence of an External Magnetic Field, (c) Orientation of Nuclear Magnetic Moments in a Strong External Magnetic Field,* H_0, *(d) Energy Level Diagram for a Nucleus for which* $I = \frac{1}{2}$, *(e) Effects of Neighboring Protons*

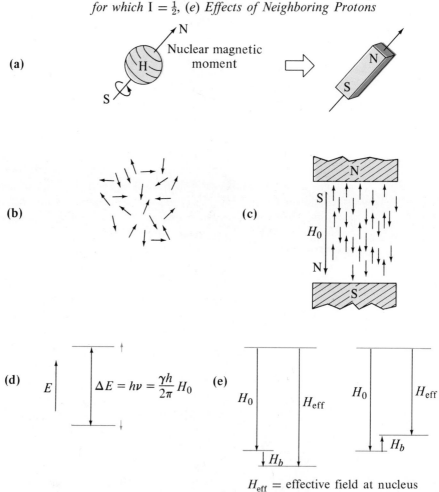

H_{eff} = effective field at nucleus

assume that the field H_0 "points" in the downward or $-z$ direction (from the top of the page toward the bottom), and we will represent the nucleus which is aligned with H_0 by the symbol $\downarrow$ and the nucleus aligned against H_0 by the symbol $\uparrow$. As is suggested in Figure 6.3(c), the number of nuclear magnets aligned with the field H_0 is only very slightly larger than the number aligned against the field (an excess of only about six nuclei per million!). Thus, *slightly* more than half of the nuclei are in the lower energy state, and slightly less than half in the upper energy state.

   If the assembly of nuclear magnets in the field H_0 is irradiated with electromagnetic energy of the proper frequency (and of proper orientation relative to H_0), some of the nuclei aligned with the field can absorb a quantum of energy and reorient themselves against the field. The proper frequency is defined by the following simple relationship (Fig. 6.3(d)):

$$\nu = \frac{\gamma H_0}{2\pi}$$

where γ, the gyromagnetic ratio, is a constant characteristic of each isotope. This equation defines the nmr resonance condition. For the hydrogen nucleus (proton) $\gamma/2\pi = 42.57602$ MHz/T.[3] Thus, the equation tells us that if $H_0 = 14{,}092.44$ gauss (1.409244 tesla), the *only* frequency of electromagnetic radiation that will be absorbed by protons is

$$\nu = 42.57602 \times 1.409244 = 60.00000 \text{ MHz}$$

Most of the proton nmr spectrometers in use today have 14,092 gauss electromagnets or permanent magnets and 60 MHz radiofrequency generators (oscillators). However, proton spectrometers operating at 100 MHz (23,487 gauss) are becoming increasingly common, and operation at 220 MHz or 360 MHz is possible using super-conducting magnets capable of generating the powerful magnetic fields required at these higher frequencies.

EXERCISE 6.1

Calculate the magnetic field required for a proton magnetic resonance spectrometer operating at 220 MHz.

   Thus far, we have considered only bare hydrogen nuclei (protons), which in a 14,092-gauss magnetic field would absorb electromagnetic radiation of only the one frequency, 60 MHz. However, in an organic molecule each hydrogen nucleus is surrounded by an electron cloud of valence electrons which binds that nucleus to another atom such as C, N, O, etc. The effect of the valence electrons is to **shield** the nucleus from the magnetic field, H_0. Thus, the effective field *at the nucleus* is less than the applied field, H_0. Therefore, the **shielded** proton will not be in resonance under the same experimental conditions that apply to the bare proton.

---

[3] The unit of frequency is hertz (Hz) (formerly cycles per second or cps), and 1 MHz = 1 megahertz = 1×10^6 Hz. The unit of magnetic flux density, whose symbol is B, is tesla (T), where 1 T = 10^4 gauss (G). In nmr spectroscopy the symbol H, magnetic field strength, has been used instead of B by most authors. In some very recent advanced texts on nmr spectroscopy you will find B rather than H.

In addition to the shielding by the valence electrons, the field *at the nucleus* may be altered by the effects of the magnetic fields of nearby atoms, bonds, or functional groups. These effects may be **shielding** (field-reducing) or **deshielding** (field-reinforcing). Thus, in general, protons in different **chemical** environments in a molecule will be in different **magnetic** environments when the molecule is placed in a strong magnetic field. These protons will experience different net magnetic fields *at the nucleus* (that is, different shielding) for a given value of H_0. The effects of shielding can be accommodated by rewriting the equation for the resonance condition:

$$\nu = \frac{\gamma}{2\pi}H_0(1 - \sigma)$$

where σ is a (very small) constant which has a different value for each **chemically non-equivalent proton** in the molecule. We will define **chemically equivalent** protons as protons in the same molecule that are chemically indistinguishable. Chemically equivalent protons will share the same value of σ. For example, in 2-chloro-2,4,4-trimethylpentane there are three sets of non-equivalent protons, as shown by the circle, rectangle and triangle surrounding them.

2-Chloro-2,4,4-trimethylpentane

Within each set the protons are equivalent, but no proton in one set is chemically (or magnetically) equivalent to any of the protons in the other sets. Therefore, this molecule should absorb 60-MHz electromagnetic radiation at three slightly different values of H_0 near 14,092 gauss.

EXERCISE 6.2

Label the sets of chemically non-equivalent protons in (1) isopentane, (2) methylcyclohexane, (3) toluene, and (4) 1-butene.

There are two ways of bringing each of the shielded but chemically non-equivalent protons to the resonance condition. We may hold the frequency ν constant and *increase* H_0 until the modified equation defining the resonance condition is satisfied, or we can hold the field H_0 constant and *decrease* the frequency ν until the equation is satisfied. The difference in the two values of H_0 required to bring two nuclei in different molecular environments into resonance at constant frequency is called the **chemical shift.** Alternatively, the chemical shift can be defined as the difference in resonance frequencies of two nuclei at constant field H_0. The two definitions are exactly equivalent. For practical reasons many nmr spectrometers are designed so that we determine the chemical shift by holding ν constant and varying H_0, but we will use units that might make it *appear* that we had held H_0 constant and varied ν.

A highly simplified schematic of an nmr spectrometer is shown in Figure 6.4. The sample, as a liquid or a solution (in CCl_4, $CDCl_3$, or D_2O, or some other *nonprotonic* solvent), is contained in a slender glass tube inserted into a wire coil situated in a strong, highly homogeneous (uniform) magnetic field. A radio frequency oscillator feeds a high frequency alternating current to the coil causing it to generate a high frequency electromagnetic field. In the simpler instruments the frequency (ν) of the oscillator is held constant, and the strength of the magnetic field (H_0) is slowly increased. As each chemically (and hence magnetically) nonequivalent proton in the sample comes into resonance, energy is absorbed from the electromagnetic field, causing a change in the flow of current through the coil. The change in current is monitored by a sensitive

FIGURE 6.4 *Schematic Diagram of a NMR Spectrometer*

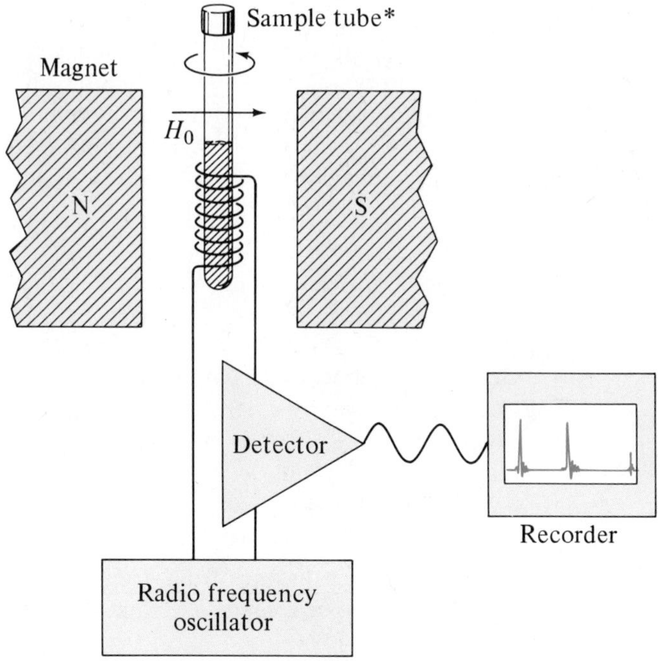

* The sample tube is spun by a small air turbine to average out inhomogeneities in the magnetic field.

detector and is recorded as a plot of energy absorbed (ordinate) *versus* the chemical shift (abscissa). In more complex or more recent instruments either H_0 or ν may be held constant and the other varied.

EXERCISE 6.3

Assume that you have a molecule with two non-equivalent protons present and that the values of σ for the protons are 5×10^{-6} and 7×10^{-6}, respectively. Your spectrometer frequency is fixed at 60,000,000 Hz. Calculate the two values of H_0 required to bring the two protons into resonance. Subtract the smaller value of H_0 from the larger to calculate the chemical shift in gauss. Multiply the chemical shift *in gauss* by 4257.602 to calculate the chemical shift *in Hz* (since $\Delta\nu = 4257.602\,\Delta H_0$ for ν in Hz and H_0 in gauss). Because the differences are small, you will have to carry your calculation out to at least eight significant figures.

It is the practice in nmr spectroscopy to make chemical shift measurements relative to a standard. The most widely used standard is tetramethylsilane, $(CH_3)_4Si$, which is called TMS and which was chosen because its twelve protons are equivalent and are highly shielded.[4] Therefore, TMS gives a strong, single peak which is found at a higher field than, and is well separated from, the protons in most organic compounds. TMS has the further desirable properties of being relatively inexpensive, inert, easily removed from the sample after use (its bp is 26.5°C), and soluble in most nmr solvents (except D_2O, in which water-soluble derivatives of TMS can be used).

$$CH_3-\underset{\underset{CH_3}{|}}{\overset{\overset{CH_3}{|}}{Si}}-CH_3$$

Generally, chemical shifts are reported in frequency units ($\Delta\nu$) relative to TMS *for a specific spectrometer frequency* (ν_0), which *must* be specified, or in δ units that are independent of the magnitude of the spectrometer's nominal operating frequency. Thus,

$$\Delta\nu = \text{distance from TMS in Hz (at a specified } \nu_0)$$

$$\delta = \frac{\text{distance from TMS in Hz}}{\text{spectrometer frequency in MHz}} = \frac{\Delta\nu}{\nu_0}$$

---

[4] Silicon has $+I$ effect (footnote 5, p. 290) relative to carbon, $\overset{\longrightarrow}{Si-C}$, which tends to increase the electron density at the carbon atom. The increased electron density at carbon in turn increases the electron density in the C—H bond, resulting in increased shielding of the TMS protons.

FIGURE 6.5 *NMR Spectrum of Methyl* tert-*Butyl Ketone*

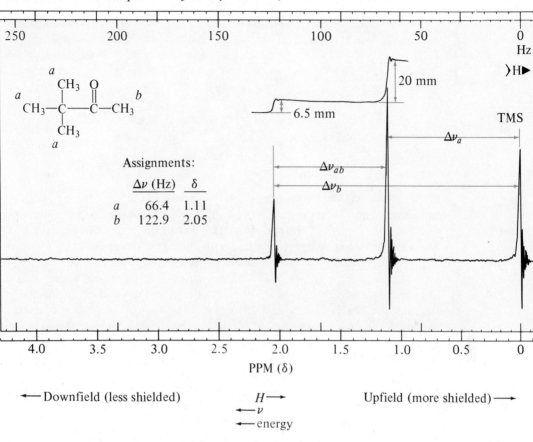

The unit of δ is parts per million (ppm); however, the unit is usually not specified in present practice. Positive values of δ correspond to chemical shifts **downfield** (lower field, higher frequency) from TMS, and negative values (seldom seen) correspond to chemical shifts **upfield** (higher field, lower frequency) from TMS. Since δ is independent of the spectrometer frequency, it is the preferred unit for the chemical shift. Occasionally, chemical shifts are given in tau (τ) units, where $\delta = 10 - \tau$. Proton chemical shifts cover a range of about 1000 Hz at 60 MHz.

Figure 6.5 shows the nmr spectrum of methyl *tert*-butyl ketone run at

$$CH_3-\underset{\underset{CH_3}{|}}{\overset{\overset{CH_3}{|}}{C}}-\overset{\overset{O}{\|}}{C}-CH_3$$

Methyl *tert*-butyl ketone

EXERCISE 6.4

An nmr spectrum shows two peaks (in addition to that of TMS) at δ 1.45 and δ 1.55. Calculate the chemical shifts of these two peaks relative to TMS and relative to each other in frequency units ($\Delta \nu$ in Hz) for spectrometer frequencies of 60 MHz and 220 MHz. Plot the 60-MHz and 220-MHz spectra, one above the other, on a sheet of graph paper using 1-inch vertical lines to represent the peaks and plotting frequency units (Hz) along the abscissa (x-axis). Based on your plot, how does increasing the spectrometer frequency affect the separation of the nmr peaks?

60 MHz with TMS as the internal standard. Note that there are three peaks, two for the non-equivalent protons in the sample and one for the TMS protons, and that the sizes (areas) of the sample peaks vary in the approximate ratio, 9:3. Most nmr spectrometers are supplied with electronic integrators that measure the area under each peak and record it as the **integral,** usually in the form of a separate trace on the graph.[5] Since the amount of electromagnetic energy that is absorbed by a molecule is roughly proportional to the number of protons in resonance at that frequency, and the recorder responds to the amount of energy absorbed, the area under each peak will be *approximately* proportional to the number of protons responsible for the peak. However, all that we can determine in this way is the *ratio* of the number of protons in each set of non-equivalent protons, not the exact number in each set. Thus, we cannot distinguish between sets of 9:3 and 3:1, and in the example given in Figure 6.5 we would need information from some other source to decide whether the proton ratio was 3:1, 6:2, or 9:3, etc. Also, from Figure 6.5 it should be clear that the term chemical shift ($\Delta \nu$) is sometimes used in two different ways: (1) to indicate the difference in resonance frequencies between the peaks of the sample and that of TMS, and (2) to indicate the difference in resonance frequencies between peaks in the sample molecule itself.

The chemical shift of a proton will depend largely on its location relative to heteroatoms (atoms other than carbon or hydrogen), double or triple bonds, aromatic rings, or functional groups. Protons two or more carbon atoms away from such structural features usually will show relatively small chemical shifts (less than δ 1.7); whereas, protons that are attached to or are a part of these structural units may show rather large chemical shifts. Some typical proton chemical shifts are summarized in Table 6.1 and detailed in Figure 6.6.

---

[5] To obtain maximum accuracy usually the integral is made as large as possible, overlapping the spectrum. For clarity in Figure 6.5, the integral was reduced below normal size.

FIGURE 6.6 *Chemical Shifts of Protons in Different Environments**

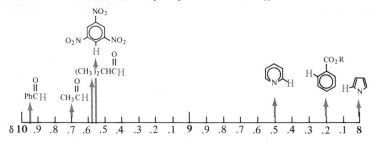

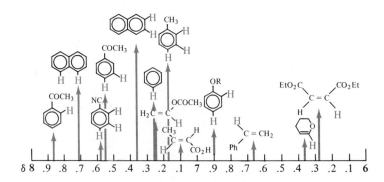

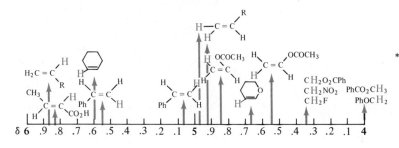

*The chemical shifts shown are for the protons drawn with the large H's. Some of the formulas are only partial and show only the essential parts of the molecule. Assume that any missing parts are R (alkyl) groups. For example, at δ 1.1 the structure CH₃C— CO—R appears. This structure should be read as

$$CH_3-\underset{\underset{R}{|}}{\overset{\overset{R}{|}}{C}}-\overset{\overset{O}{\|}}{C}-R$$

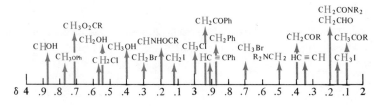

The symbols Ph— and ⬡— are common representations of the phenyl group. Chemical shift values are probably accurate to within 0.1–0.3 ppm. (Adapted from John A. Landgrebe, *Theory and Practice in the Organic Laboratory,* 2nd ed., D.C. Heath and Company, Lexington, Massachusetts, 1977, p. 170.)

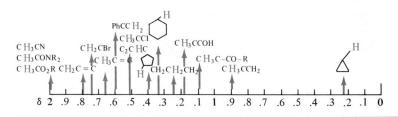

TABLE 6.1 *Proton Chemical Shift Values**

Type of Proton	δ (ppm)
 $\triangleright\!CH_2$	0.3
$CH_3-R,\ CH_2R_2,\ CHR_3$	0.8–1.7
$CH_3-CR\!=\!CR,\ CH_3-C\!\equiv\!CR,\ CH_3-\overset{\overset{\displaystyle O}{\|}}{C}-R,\ CH_3-C\!\equiv\!N,$ $CH_3-NR_2,\ CH_3-SR,\ CH_3-Ar,\ H-C\!\equiv\!CR$	1.7–2.6
$CH_3Br,\ CH_3Cl,\ CH_3-NRAr,\ CH_3OR,\ CH_3-\overset{\overset{\displaystyle R}{\|}}{N}-\overset{\overset{\displaystyle O}{\|}}{C}-R,$ $CH_3-O-\overset{\overset{\displaystyle O}{\|}}{C}-R$	2.7–3.9
$H-CR\!=\!CR$	4.4–7.0
$Ar-H$	7.0–8.5
$R-\overset{\overset{\displaystyle O}{\|}}{C}-H$	9.3–10.2
$R-\overset{\overset{\displaystyle O}{\|}}{C}-OH$	10.0–13.00
$R-OH,\ RNH_2$	variable (1–6)
$Ar-OH$	variable (4–8)

* For CH_3, CH_2, and CH groups attached to the same functional group the order of chemical shifts is usually $CH_3 < CH_2 < CH$ with a difference of 0.3–1.0 ppm between CH_3 and CH.

EXERCISE 6.5

Based on Fig. 6.6 predict the approximate chemical shifts of the protons in the following compounds:

(a) $C_6H_5-CH_3$

(b) $CH_3CH_2-\overset{\overset{\displaystyle O}{\|}}{C}-H$

(c) 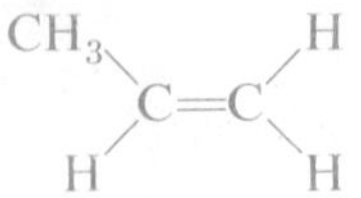 $\overset{CH_3}{\underset{H}{}}\!\!C\!=\!C\!\overset{H}{\underset{H}{}}$

(d) CH_3CH_2OH

(e) $(CH_3)_4C$

FIGURE 6.7 *NMR Spectrum of Uracil in Basic D₂O*

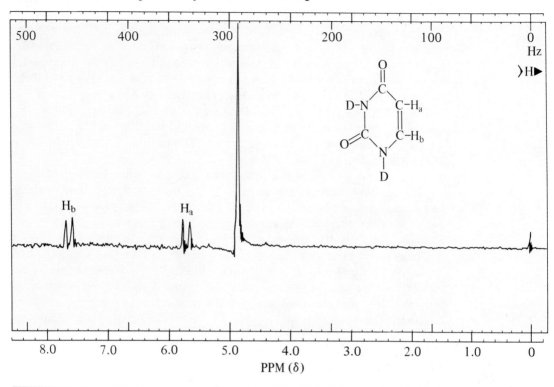

Many nmr spectra, even of relatively simple molecules, are much more complex than that shown in Figure 6.5. Consider the spectrum of the **hetero-aromatic**[6] compound, uracil, shown in Figure 6.7. Uracil has the structure shown below and is one of several heteroaromatic compounds playing vital roles in the biosynthesis of protein (Sec. 16.9).

Uracil

---

[6] A heteroaromatic compound is a cyclic compound containing a heteroatom (usually, but not exclusively, oxygen, nitrogen, or sulfur) in a ring that obeys Hückel's rule (Sec. 4.2) and, therefore, is aromatic. (See Exercise 4.23(f).)

The spectrum of uracil illustrates two important aspects of nmr spectroscopy. First, the spectrum was run in slightly basic deuterium oxide (D_2O or "heavy water"). In this solvent hydrogen attached to oxygen, nitrogen, or sulfur is often *exchanged* for deuterium. Although deuterium compounds do give nmr spectra, their deuterium resonance peaks are found about 32 miles to the right (upfield) of TMS (at an H_0 field of 14.1 kilogauss). Thus, treating a compound containing —OH, —NH, or —SH groups with D_2O is a useful way of either (1) eliminating the proton nmr peaks of such groups from the spectrum or (2) establishing which nmr peaks are due to these groups. The peak at δ 4.68 is that of the D-O-H formed in the exchange reaction.

The second feature of the uracil nmr spectrum is the number of peaks remaining after deuterium exchange. Although there are only two non-equivalent protons remaining in the molecule, there are two *pairs* of peaks, or two **doublets,** rather than two **singlets** in the spectrum. This **splitting** of singlets into **multiplets** is the result of **spin-spin coupling,** a phenomenon in which the spin of one nucleus interacts with the spin of another and which may cause the nmr peaks of each nucleus to be split. Although the coupling of spins of nuclei can make the interpretation of some nmr spectra much more difficult, this phenomenon does provide much valuable information as to the structure of the molecule. The spin-spin splitting shown in Figure 6.7 is called **first-order** splitting and is relatively easily explained and interpreted. More complex splitting patterns are frequently encountered in nmr spectroscopy but will not be discussed in detail in this text.

First-order splitting is caused by the effect of the magnetic moments of neighboring nuclei on the chemical shift of the nucleus being observed. Consider the following simple system of magnetically non-equivalent protons, H_a and H_b, in which the chemical shift of H_a is downfield of that of H_b. What will

$$
\begin{array}{ccc}
\text{Y} & \text{M} & \\
| & | & \\
\text{X}-\text{C}-\text{C}-\text{N} \\
| & | & \\
\text{H}_a & \text{H}_b &
\end{array}
$$

be the effect of H_b's magnetic moment on H_a's resonance frequency? There are two possibilities, depending on the orientation of H_b's magnetic moment relative to that of H_a: $H_a(\downarrow)H_b(\downarrow)$ and $H_a(\downarrow)H_b(\uparrow)$. Recall that we have taken H_0 as pointing downward on the page. Therefore, H_a must be aligned with H_0 if we are to observe H_a absorb energy and realign itself against H_0. Thus, we will ignore combinations involving $H_a(\uparrow)$ as long as we are observing H_a. However, H_b could be aligned with or against the field as shown. If H_b is aligned with the field, its magnetic moment will increase slightly the effective field at the nucleus of H_a as shown in Figure 6.3(d). Therefore, H_0 will not have to be made quite as large to bring H_a into resonance (as it would in the absence of H_b), and the H_a peak will come at a *lower* value of H_0. However, if H_b is aligned against the field, its magnetic moment will decrease slightly the effective field at the nucleus of H_a; and H_0 will have to be made larger (than it would in the absence of H_b) to bring H_a into resonance, causing the H_a peak to come at a *higher* value of H_0.

Since both orientations of H_b are almost equally probable, the effects of both will be observed, and the H_a peak will be split into two peaks (a doublet) of almost equal intensity. Since proton H_a will affect the chemical shift of H_b in a similar fashion, the H_b peak will also be split into a doublet. The spectrum will look something like the following:

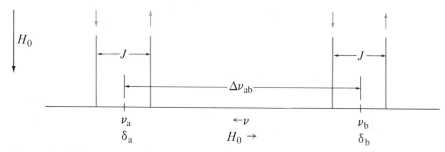

No peak will appear at the resonance frequencies (chemical shifts) of H_a and H_b; however, two peaks will appear on either side of the true chemical shifts. The separation of each doublet will be identical and equal to J Hz where J (or J_{ab}) is called the **coupling constant.** As its name implies, J is a constant for a given coupling and is independent of the spectrometer frequency. Typically, J is of the order of 0–15 Hz for protons (and may have either a $+$ or $-$ sign). The small arrows ($\downarrow$ and $\uparrow$) over each line or peak in the doublets indicate the orientation of the nucleus *not* being observed (that is, H_b for the doublet centered on v_a).

Next we examine the system

$$X-\underset{\underset{H_a}{|}}{\overset{\overset{X}{|}}{C}}-\underset{\underset{H_b}{|}}{\overset{\overset{Y}{|}}{C}}-H_b$$

where the resonance frequency of H_a is greater than that of H_b (that is, H_a is downfield of H_b). As H_0 increases toward the resonance condition for H_a, $H_a(\downarrow)$ senses the H_b protons in the alignments: (1) $\downarrow\downarrow$, (2) $\downarrow\uparrow$, (3) $\uparrow\downarrow$, and (4) $\uparrow\uparrow$. Alignment (1) increases the magnetic field at the H_a nucleus, and alignment (4) decreases the field at the nucleus. In both (2) and (3) the two H_b spins are opposed; therefore, they will cancel each other and have no effect on the magnetic field at the H_a nucleus. Therefore, as H_a comes into resonance, we will expect three peaks: the first corresponding to alignment (1) of the H_b proton spins, the second to alignments (2) and (3), and the third to alignment (4). Since each combination of H_b spins is about equally probable, the second (or middle) peak should be about twice as large as the first and third because the second peak is the result of *two* different, but equivalent, alignments, (2) and (3). We will also expect that the second peak will come at the H_a resonance frequency (chemical shift) since the effects of the H_b spins have been cancelled for this peak. As we continue to increase H_0 and approach the resonance condition for the H_b protons, these protons sense H_a in two alignments: (1) $\downarrow$ and (2) $\uparrow$. Therefore, the H_b protons will appear as a doublet symmetrically disposed about the H_b chemical shift. Our complete nmr spectrum will consist of a $1:2:1$ triplet and a $1:1$ doublet in which the ratio of

TABLE 6.2 *NMR Coupling Constants (J)**

Structural Unit	J (Hz)	Structural Unit	J (Hz)
One-bond coupling, ^1J		*Long-range coupling (^4J, ^5J, etc.)*	
^{13}C—H	100–260	H···H structure (—C—C—C—)	~0
Two-bond (geminal) coupling, ^2J		H···H structure (—C—C=C—)	0.5–3
C with two H	6–15	H structure (—C—C≡C—H)	2–3
C=C with two H	0–4	H···H structure (—C—C=C—C—)	2–3
Three-bond (vicinal) coupling, ^3J		*Aromatic ring couplings*	
—C—C— (H, H) (freely rotating)	6–8	H_o J_o	6–10
Newman projection HH (rigid)	8–14	H_m J_m	1–3
Newman projection H, H (rigid)	2–3	H_p J_p	0–1.5
Newman projection H / H (rigid)	8–14		
H, H cis C=C	6–14		
H / H trans C=C	11–18		
—C—C=C— (H H)	4–10		

* Coupling constants may be positive or negative. The sign is important but not needed for most structural work and is omitted from this table. The superscript number before the *J* indicates the number of bonds between the coupled nuclei (for example, ^3J = three bonds between the two protons).

the *total* areas of the triplet:doublet will be 1:2. The spectrum will be approximately as follows:

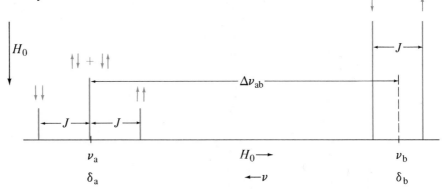

In simple first-order systems the **multiplicity** (number of lines or peaks) of an nmr signal will be $n + 1$ where n is the number of equivalent hydrogens on the *adjacent* carbon atoms.[7] These are often called *neighboring* hydrogens. The relative intensities of the lines in a multiplet are given approximately by Pascal's triangle (whose construction will be obvious after careful inspection).

$n + 1$	Area ratios						
1				1			
2			1		1		
3		1		2		1	
4	1		3		3		1
5	1	4		6		4	1
6	1	5	10		10	5	1

EXERCISE 6.6

Add another line to Pascal's triangle. This addition may prove useful in the solution of problems in this text.

In Table 6.2 are given some selected, typical values of coupling constants (J) for protons in common structural units. Although J values are sensitive to structural changes, these values will be suitable for relatively uncomplicated molecules or preliminary analysis.

With the foregoing rather lengthy introduction you should be able to decipher many *first-order* nmr spectra; however, most nmr spectra either are not first-order or contain other complicating features. Even with such complex spectra you will be able to deduce the structure, or part of the structure, of the

---

[7] For nuclei with $I > \frac{1}{2}$, the multiplicity is $2nI + 1$.

FIGURE 6.8 *NMR Spectrum of 1,2-Dimethoxyethane in CDCl₃*

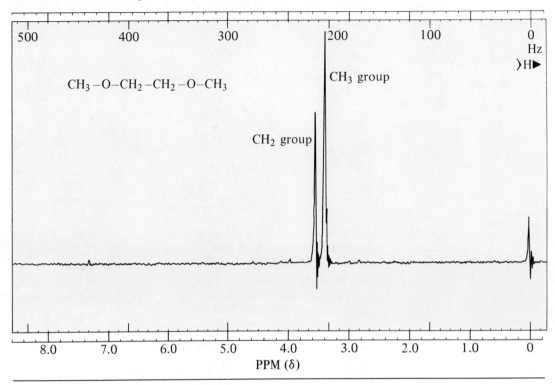

molecule. The following rules or guidelines will be of value in dealing with more complex spectra.

1. Nuclei having the same chemical shift do not split each other. They are *coupled* but not *split;* thus, these two terms are not synonyms. In the spectrum of 1,2-dimethoxyethane (Fig. 6.8) the four protons in the two —CH₂— groups have the same chemical shift; therefore, they do not split each other and appear as one peak, as do the six protons in the two —CH₃ groups.

2. First-order splitting is not observed unless the chemical shift ($\Delta\nu$) is much larger than *J*. Generally, for first-order splitting the following should be true

$$\Delta\nu > 10 \times J$$

For example, in the nmr spectrum of 1-bromo-2-chloroethane (Fig. 6.9) the chemical shift is about 12.6 Hz (at 60 MHz) and the coupling constants between the protons on C-1 and C-2 are about 6–9 Hz. The spectrum is clearly not first order and contains at least 20 lines. Such spectra can be analyzed only with the aid of a computer.

3. *Chemically* equivalent nuclei are not necessarily *magnetically* equivalent. To be magnetically equivalent two nuclei must be not only

FIGURE 6.9 *NMR Spectrum of 1-Bromo-2-chloroethane in CDCl₃*

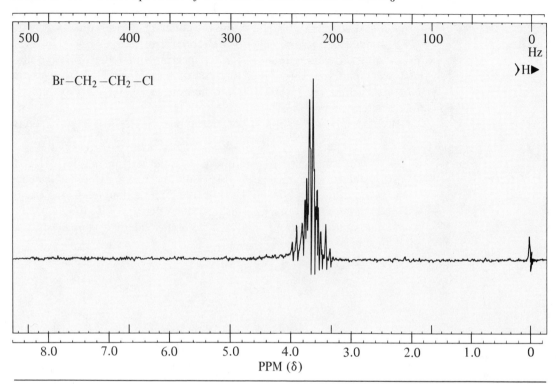

indistinguishable chemically but also must be coupled in the same way to all other nuclei in the molecule. Consider the aromatic compound shown below and the coupling of the two ortho protons to H_m. One proton, H_o, is ortho to H_m, and the other proton, $H_{o'}$, is para to H_m. Therefore, the two ortho protons are not equally coupled to H_m (as is shown by the coupling constants given with the structure). Similarly, the two meta protons are not equally coupled to either of the ortho protons. Therefore, the two ortho protons are not *magnetically* equivalent although they are clearly chemically equivalent. The two meta protons are likewise not magnetically equivalent. Many simple aromatic compounds give complex nmr spectra for the ring protons. Others give simple spectra because the chemical shifts of the ring protons just happen to be the same. For example, toluene gives a single, sharp peak at δ 7.17 $\pm$ 0.03 (in CDCl₃ solution) for its five aromatic protons.

$$J_{om} = 6\text{--}10 \text{ Hz}$$

$$J_{o'm} = 0\text{--}1.5 \text{ Hz}$$

Very few compounds of the structure $X-CH_2-CH_2-Y$ give truly first-order spectra because the protons are not magnetically equivalent in the CH_2 groups (for reasons beyond the scope of our discussion). However, many of these compounds give nmr spectra with patterns resembling first-order patterns. First-order analysis of these spectra often gives the correct *structure,* but no reliable information on J values. An example of this type is given in the Exercises.

4. In a *first-order* system of the type, $-\overset{\displaystyle |}{\underset{\displaystyle H_a}{C}}-\overset{\displaystyle |}{\underset{\displaystyle H_b}{C}}-\overset{\displaystyle |}{\underset{\displaystyle H_c}{C}}-$,   unless $J_{ab} = J_{bc}$, the simple $(n+1)$ multiplicity rule will not apply to H_b. If $J_{ab} \neq J_{bc}$, the H_b resonance peak will be split by H_a into a doublet *and* each of the peaks of the doublet will be split by H_c into a doublet, giving rise to a **doublet** of **doublets** (not a quartet!). Assume that $J_{ab} > J_{bc}$ (and $J_{ac} = 0$). Then, the splitting can be analyzed with the aid of a simple splitting diagram.

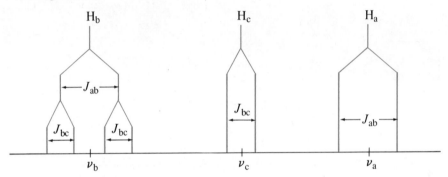

5. In order to observe a proton in a given location relative to the rest of the molecule, the proton must remain in that location for a rather long time (0.1 to 1 sec or longer). If the proton changes position or location, by any means, rather rapidly, the nmr spectrometer will report only its average location. If the proton changes position or location very slowly, the nmr spectrometer may report its several locations as separate peaks (because in the sample there will be protons in each of the locations slowly changing to other locations). Generally, fast exchange (of position) is more common. If it were not for the fast interconversion of conformers (Sec. 2.5) by rotation around single bonds, we would see nmr peaks for each of the more stable conformers and nmr spectra would be even more complex. Indeed, the use of nmr techniques (particularly at low temperatures where rotations about bonds are slowed) makes possible the study of conformer populations. Fast exchange has considerable influence on the spectra of compounds having OH and NH groups. Very often the protons on such groups are moving from one molecule to another very rapidly (especially in the presence of traces of acids or bases) on the nmr time scale. There are two effects of this chemical exchange. First, the chemical shifts of OH and NH protons tend to

FIGURE 6.10 *NMR Spectrum of Ethyl Alcohol in CDCl₃*

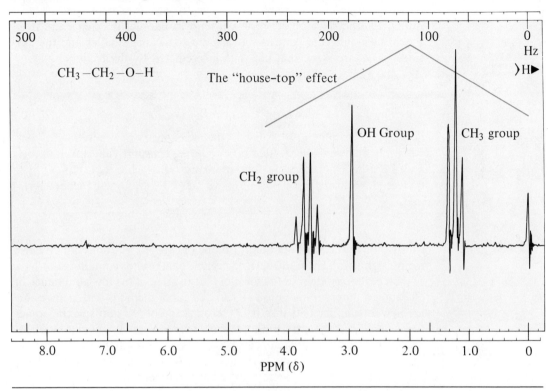

vary more widely than those of protons on carbon atoms and are more sensitive to experimental conditions. Second, the rapid exchange of protons between OH groups and between NH groups tends to eliminate the effects of the coupling of OH and NH protons with protons of adjacent CH groups. Thus, in the spectrum of ethyl alcohol, CH_3CH_2OH, as run under ordinary conditions (Fig. 6.10), the CH_3 and CH_2 groups split each other in a first-order fashion, but the OH group does not split the CH_2 group, nor is it split by the CH_2 group.

The use of nmr spectroscopy in the study of time-dependent phenomena (rate of rotation about bonds, equilibria, chemical exchange reactions, etc.) is one of its more important applications.

6. Some practical suggestions for working with nmr spectra include the following. If there appears to be only one multiplet in an nmr spectrum, it is not truly a multiplet but a group of uncoupled singlets, or there is another multiplet somewhere off the chart, perhaps way off the chart, in the nmr spectrum of another nucleus such as ^{19}F, D, or ^{13}C, which can and does couple to protons. Beware of parts of multiplets buried under strong singlets

or lost in the "noise" at the baseline. The itensities of multiplets follow Pascal's triangle only approximately. If you draw a "house-top" over a set of *two* multiplets (for example, see Fig. 6.10) with its peak between the two multiplets, the slant of the roof will tend to parallel the departure from Pascal's triangle: the "outside" peaks are somewhat smaller and the "inside" peaks are somewhat larger than predicted by the triangle.

EXERCISE 6.7

Predict the approximate appearance of the nmr spectra of the following compounds using chemical shift values and coupling constant values from Tables 6.1 and 6.2 and Figure 6.6: (a) chloroethane (ethyl chloride); (b) propyne (methylacetylene); (c) 1,3-dichloropropane; (d) 1,1-dibromoethane; (e) cyclohexane.

Although the most abundant isotope of carbon, ^{12}C, does not give rise to nmr signals, all naturally derived organic compounds contain about 1.1% of their carbon content in the form of ^{13}C. In spite of the low abundance of ^{13}C and an inherent low sensitivity to detection, methods and instruments are now available for the routine determination of ^{13}C nmr spectra (sometimes called cmr spectra). Within a very short time the use of cmr spectroscopy is likely to be as common as pmr is today. Although the theory for cmr spectroscopy is essentially the same as that for pmr spectroscopy, the procedures used tend to be quite different and involve the use of computers. We will not discuss this rapidly evolving method of structural analysis; however, a typical cmr spectrum run on a sample of ordinary table sugar, sucrose ($C_{12}H_{22}O_{12}$) (Sec. 14.12), is given in Figure 6.11. In this spectrum the ^{13}C atoms are **decoupled** from the protons in the compound; thus, each peak corresponds to one carbon atom, although peak areas do not tell us the number of carbon atoms per peak in this type of nmr. Two of the peaks are poorly separated, but each of twelve carbon atoms does give rise to a separate peak.

6.3 *Infrared Spectroscopy*

Although we tend to describe organic molecules as rather rigid structures with fixed bond lengths and precise bond angles, the bond lengths and bond angles in drawings or models are only average values of interatomic distances and angles because the atoms within the actual molecules are in a constant state of motion. Bonds are stretching and contracting in a periodic, quantized fashion. Bond angles are opening and closing, and groups of atoms are rotating relative to each other about their common bond axes. We will separate the

FIGURE 6.11 *The ^{13}C NMR Spectrum of Sucrose in D_2O (25.2 MHz)*

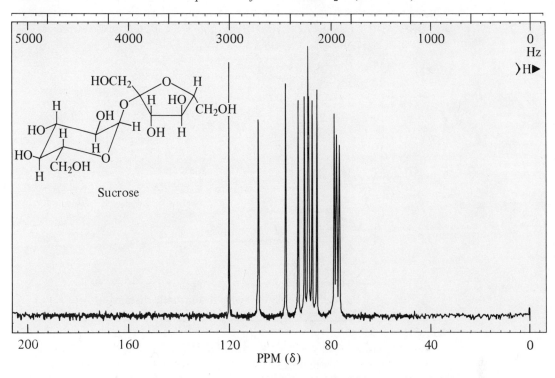

various types of intramolecular motion into two classes: (1) vibration, which includes stretching and bending motions of bonds, and (2) rotation. We will be concerned principally with molecular vibrations because it is from such motions and their dependence on the absorption of infrared light that the organic chemist can obtain the most useful structural information. Molecular vibrations are quantized; that is, for each molecule there are discrete vibrational energy levels, and the frequencies ν (and amplitudes) of molecular vibration are related to these energy levels and to the absorption of infrared light by the equation

$$\Delta E = h\nu$$

where the frequency (ν) now refers to both the frequency of radiation absorbed and the frequency of the vibrational motion.

The familiar diagram for the energy levels of a diatomic molecule is given in Fig. 6.12. Although the vibrations of a typical organic molecule are much more complex than those of a simple diatomic molecule, we can often treat complex molecules as though the bonds between two or three atoms were stretching or bending independently of the motions of the other atoms in the molecule. Therefore, we will use the energy level diagram of the diatomic molecule to represent the vibrational levels associated with, for example, the stretching motion of a carbon-carbon double bond. At room temperature most

FIGURE 6.12   *Energy Levels for Bond Vibration Based on Diatomic Model*

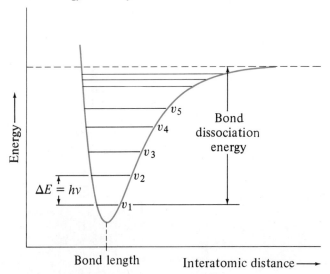

of the molecules of an organic compound will be in the lowest vibrational energy level (v_1), and the amplitude of molecular vibration will be at a minimum. When the organic molecule is irradiated with electromagnetic radiation of the appropriate frequency near 10^{14} Hz (that is, with light in the infrared region), the molecule can absorb some of the radiation and pass from the lowest vibrational energy level (v_1) to the second vibrational level (v_2) in which the frequency of the vibration will be equal to that of the infrared light and the amplitude of the vibration will be increased (as shown by the length of the horizontal line for v_2 in Fig. 6.12).

A graphical plot of the infrared energy absorbed (as the ordinate) versus the frequency or wavelength of the infrared light (as the abscissa) is called an infrared (ir) spectrum. Several systems of units are used for both the intensity of absorption and the position of the ir absorption bands. The best current practice requires that band positions be reported in frequency units; however, the use of the fundamental unit of frequency v (vibrations per second, or Hz) gives very large numbers (near 10^{14} Hz). Therefore, spectroscopists have converted these large numbers into more convenient numbers by means of the following equation

$$\tilde{v} = \frac{v}{c}$$

where $\tilde{v}$ is the wavenumber (or number of waves per cm) and c is the velocity of light (3×10^{10} cm/sec). For example, if $v = 10^{14}$ Hz (or 10^{14} sec^{-1})

$$\tilde{v} = \frac{10^{14} \text{ sec}^{-1}}{3 \times 10^{10} \text{ cm sec}^{-1}} = 3333 \text{ cm}^{-1}$$

Thus, the unit of the wavenumber is *centimeters to the minus one* or *reciprocal centimeters* or *wavenumbers*. It is current and acceptable practice to describe an ir band at $\tilde{\nu} = 3333$ cm^{-1} as occurring at a *frequency* of 3333 cm^{-1}. Although technically incorrect, this practice is based on the assumption that division by c is understood. Indeed, spectroscopists often use $\tilde{\nu}$ as a unit for energy because of the relationship

$$E = hc\tilde{\nu}$$

It is because of the direct relationship between E and $\tilde{\nu}$ that $\tilde{\nu}$ is the preferred unit for ir band position. However, with some of the older infrared instruments and in many older texts, band positions were given in wavelength (λ) units based on the relationship between ν, λ, and c *in a vacuum.*

$$\lambda = \frac{c}{\nu} \quad \text{or} \quad \lambda = \frac{1}{\tilde{\nu}}$$

The common unit of wavelength is the *micron* (μ) where

$$1 \text{ micron } (\mu) = 10^{-4} \text{ cm} = 10^{-6} \text{ m}$$

The intensity of absorption is most commonly reported as percent transmission or percent absorption, although absorbance (A) and transmittance (T) units are sometimes encountered (see Sec. 6.4).

$$A = \log\frac{1}{T} = \log\frac{100}{\% \text{ transmission}}$$

The infrared region is divided for convenience into three regions: the near infrared (from the red end of the visible region, 12,500 cm^{-1} to 4000 cm^{-1}), the middle infrared (4000–200 cm^{-1}), and the far infrared (200–10 cm^{-1}). In organic chemistry most instruments operate over the range of about 4000–650 cm^{-1}. You may have noted that these numbers seem to run backwards, from 4000 down to 650. This is because of the usual practice of plotting ir spectra with "frequency" units *decreasing* from left to right on the chart. Other special features of ir spectra can be seen by examination of Figure 6.13. First, it is apparent that the peaks or bands have been plotted upside down relative to nmr (or uv-vis) spectra. This is simply a matter of custom and one with which you will rapidly become familiar. Second, many of the absorption bands are relatively broad compared with nmr peaks. This broadening is the result of transitions of very similar energies between the various rotational energy levels associated with each vibrational energy level (Fig. 6.14), leading to many closely spaced or overlapping lines that cannot be resolved by the optical system of the **infrared spectrometer** (or **spectrophotometer**).

Block diagrams of typical infrared and ultraviolet-visible spectrometers are given in Figure 6.15. Both types of instrument consist of three basic units: (1) a **source** of continuous or **polychromatic** radiation that is a mixture of light of all the frequencies required; (2) a **monochromator** which selects from the continuous radiation **monochromatic** light of the desired single frequency, or, more realistically, of a very narrow range of frequencies centered on the desired single

FIGURE 6.13   *Infrared Spectrum of Polystyrene**

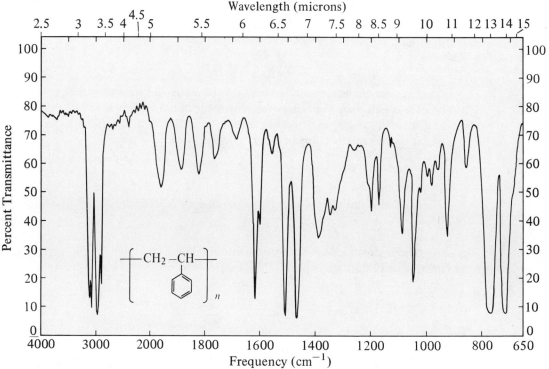

* Run on a thin film of polystyrene supplied with the instrument for checking the frequency accuracy and resolution (ability to separate closely spaced peaks). Polystyrene is a good model for the spectra of monosubstituted benzene derivatives.

frequency; and (3) a detector, usually called a **photometer.** Polychromatic light from the source is passed through both a **sample** and **reference** compartment, although the reference compartment will contain a reference cell filled with pure solvent only if the sample is being studied in solution. As the two polychromatic light beams pass through the monochromator, the monochromator selects one-by-one the monochromatic frequencies to be transmitted to the photometer. The photometer determines the intensity of the transmitted light in the sample beam and compares it with that of the reference beam, thereby compensating for variations in source intensity and the absorption of ir light by the air and by the solvent, if any. Electronic signals from the monochromator (frequency) and photometer (intensity) are fed to a recorder, which plots the spectrum.

Infrared spectra may be obtained with samples in the solid, liquid, gaseous, or solution form. Because the most common infrared sample cells are constructed with sodium chloride "windows", solvents such as methylene chloride, chloroform, carbon tetrachloride, and carbon disulfide are commonly used. Unfortunately, most solvents absorb strongly in some regions of the spectrum, and in these regions the spectrometer ceases to provide useful information because no light reaches the photometer. Solid samples are ground with potassium bromide powder and pressed into a clear pellet or wafer.

Since there are many bonds and bond angles in all but the simplest organic molecules, there are many possible vibrational motions: $3n - 6$ fundamental vibrational modes for a nonlinear molecule containing n atoms. Furthermore, there may be interaction between the various vibrational modes, and sometimes "forbidden" transitions do occur to the third vibrational level (v_3). Thus, ir spectra tend to be rather complex. Therefore, we do not usually try to interpret an entire ir spectrum (as we would an nmr spectrum) but rather take advantage of the empirical observation that many molecules containing a certain functional group or structural unit will absorb ir light in a rather narrow frequency range characteristic of that group or structural unit. Such absorptions are called **characteristic group frequencies** or **vibrations.** For example, any compound that contains a carbonyl group, C=O, will have a strong ir absorption band in the region between 1870 and 1630 cm^{-1}. In effect, we may treat *some* molecular vibrations as though they were caused by the stretching or bending of an

FIGURE 6.14 *Transitions Between Vibrational and Rotational Energy Levels**

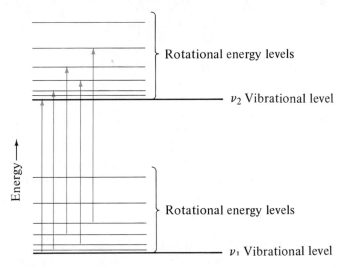

* In the gas phase it is possible to observe fine structure due to transitions between rotational levels. In the liquid or solid phase the rotational levels are less well defined and only broad bands can be observed.

FIGURE 6.15 *Block Diagrams of (a) Infrared and (b) Ultraviolet–Visible Spectrometers*

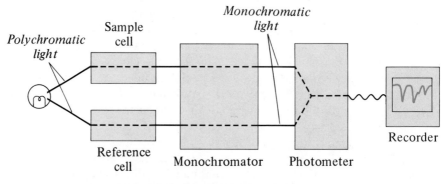

(*a*) Infrared spectrometer

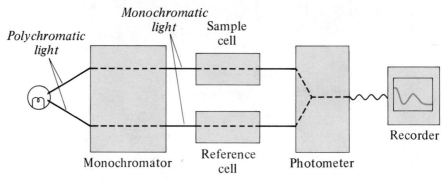

(*b*) Ultraviolet–visible spectrometer

individual bond, as was mentioned previously. Some of the typical stretching and bending vibrations of organic molecules are illustrated in Figure 6.16. In some of the drawings the bonds are portrayed as small coil springs, an analogy that is sometimes useful in ir spectroscopy.

For about 15 years before the development of nmr spectroscopy as a common structural tool, ir spectroscopy was the most powerful probe of structure available to the organic chemist. Through the detailed study of characteristic group frequencies, the structures of many organic molecules were deduced. However, such study requires considerable skill, knowledge, and experience. At present most organic chemists tend to use ir spectroscopy as an adjunct to nmr spectroscopy for the detection of specific functional groups and structural units

FIGURE 6.16 *Typical Vibrations of Organic Molecules*

Stretching Vibrations

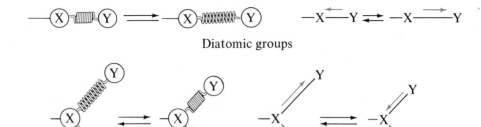

Diatomic groups

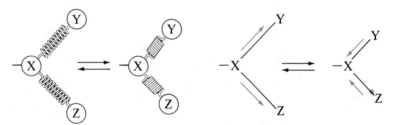

Triatomic groups—antisymmetric stretch

Triatomic groups—symmetric stretch

Bending Vibrations

Scissoring (in–plane bending)

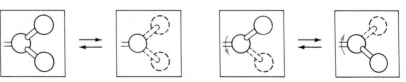

Wagging (in and out of plane)   Twisting (in and out of plane)

TABLE 6.3 *A Simplified Correlation Table of the IR Absorption Frequencies and Functional Groups*

$\bar{\nu}$ (cm^{-1})	Compounds	Characteristic Vibrations
3700–3100	Alcohols, amines, phenols, carboxylic acids, amides	—O—H, —N—H
3100–3000	Aromatic hydrocarbons, unsaturated hydrocarbons	—C=C—H
3000–2800	Saturated hydrocarbons	—C—H
2400–2000	Acetylenes, nitriles	—C≡C—, —C≡N
1870–1630	Aldehydes, ketones, carboxylic acids, carboxylic acid derivatives	C=O
1680–1620	Alkenes	C=C
1615–1515	Aromatic rings	
1550–1200	Nitro compounds, methylene groups (—CH$_2$—), methyl groups (—CH$_3$)	—N$^+$(=O)O$^-$, —CH$_2$—, —CH$_3$
1300–1000	Ethers, alcohols, phenols, esters, acid anhydrides	—C—O—
1000–650	All hydrocarbons (C—H bending)	—C—H

rather than for a complete structural analysis. Table 6.3 gives a brief survey of the principal regions of the ir spectrum used by the organic chemist. Thus, a typical approach to an ir spectral problem would involve a quick scan of the regions above 1600 cm^{-1} to determine whether or not any of the stretching vibrations of the functional groups listed are present. If any of these groups are present, a more detailed examination is made using a chart or table of characteristic group frequencies. A rather brief table of this type is given in Table 6.4. More detailed tables are available in some chemical handbooks and in numerous books on organic spectroscopy. If there is no absorption due to OH, NH,

C≡C, C≡N, C=O, or C=C, then the rest of the spectrum is examined for other structural clues, such as the strong absorption characteristic of ethers in the 1000–1300 cm^{-1} region, or of nitro groups (NO_2) at 1550–1330 cm^{-1}, or of C—Br and C—Cl at 850–650 cm^{-1}. Analysis of the region below 1500 cm^{-1} is more difficult because of the large number of absorptions of various types that are found in this region, sometimes called the "fingerprint" region; that is, each molecule tends to have its own "fingerprint" in this region.

TABLE 6.4 *Characteristic Group Infrared Frequencies**

(a) Stretching Vibrations

Group	$\tilde{\nu}$ (cm^{-1})	Group	$\tilde{\nu}$ (cm^{-1})
C—H Groups		*C—O Groups*	
—C—H	2980–2850 (s)	C—O	1340–1000 (m-s)
=C—H	3100–3000 (s)	C=O	1870–1630 (s)
≡C—H	3300–3250 (m-s)	RCHO	1740–1720 (vs)†
		ArCHO	1720–1680 (vs)†
O—H Groups		R_2C=O	1740–1700 (vs)
R—OH	3670–3600 (w, free OH)	ArCOR	1700–1650 (vs)
	3400–3200 (s, b, H-bonded)	RCOOH	1800–1740 (s, monomer)
RCOOH	3550 (w, monomeric)		1720–1680 (s, dimer)
	3000–2500 (vb, dimeric)	ArCOOH	1700–1680 (s, dimer)
		RCO_2^-	1650–1550
N—H Groups			1440–1350 (vs, b, two bands)
RNH_2	3500–3330	RCO_2R	1750–1730 (vs)
	3420–3250 (m-s, two bands)	$ArCO_2R$	1800–1760 (vs)
R_2NH	3350–3300 (m)	$RCONH_2$	1700–1670 (s, dilute solution)
⟩N⁺—H	3000–2200 (vb)		1680–1630 (vs, solid)
		$(RCO)_2O$	1825–1815
			1755–1745 (vs, two bands)
$RCONH_2$	3530–3390 (s, two bands, dilute solution)	RCOCl	1810–1760 (vs)
	3360–3170 (s, two bands, solid)	*N—O Groups*	
RCONHR′	3450–3400 (m, dilute solution)	R—NO_2	1560–1530 (vs)
	3350–3000 (s, solid)		1390–1370 (m-s) (two bands)
		Ar—NO_2	1540–1500 (vs)
C—C Groups			1370–1330 (s) (two bands)
C—C	1250–1150 (m)		
C=C	1680–1620 (m-s)	*C—Cl, C—Br*	850–650 (m)
C≡C	2250–2100 (w)		
C—N Groups			
C—N	1350–1070 (m-s)		
C=N	1690–1620 (m-s)		
C≡N	2260–2220 (var)		*(continued next page)*

TABLE 6.4 *Continued*

(b) Bending Vibrations

Compound or Group	$\tilde{\nu}$ (cm^{-1})
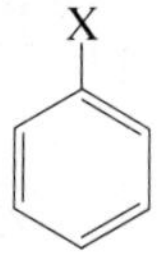	770–730 (vs); 710–690 (s); four bands at 2000–1720 (w)§
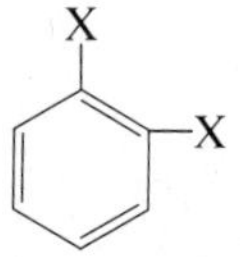	770–730 (vs); four bands near 1940, 1900, 1780, 1680 (w)§
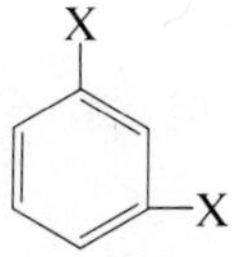	810–750 (vs); 725–680 (m-s); three bands at 1930–1740 (w)§
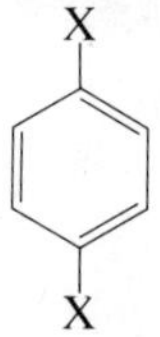	860–800 (vs); two bands at 1900–1750 (w)§
R—CH=CH$_2$	1000–980 (s); 920–900 (s); overtone near 1800
R$_2$C=CH$_2$	900–880 (s); overtone near 1800
trans-R—CH=CH—R	980–950 (vs)
cis-R—CH=CH—R	750–650 (m-s)
R$_2$C=CHR	830–780 (m-s)
—CH$_2$—	1470–1440 (vs)
CH$_3$—	1470–1440 (vs); 1460–1360 (s); 1250–800 (m-s)
(CH$_3$)$_2$C⟨	1470–1400 (vs); two bands near 1380–1360; 1250–800 (m-s)

* Abbreviations: s = strong; w = weak; b = broad; v = very; var = variable.
† Also has a doublet near 2800–2700 (w-m) cm^{-1}.
§ Bands between 2000 and 1700 cm^{-1} are overtones (almost forbidden transitions between the first and third vibration levels). Aromatic compounds also have stretching vibrations at 1630–1590 (m) and at 1520–1480 (m) cm^{-1}.

In order for a molecule to absorb infrared light and undergo a specific vibration, the dipole moment (Sec. 5.7) of the molecule must change during the vibration. If the dipole moment does not change, the vibration is said to be **infrared inactive** and no absorption peak corresponding to the vibration will appear in the infrared spectrum. Thus, when the triple bond of methylacetylene (propyne) stretches and contracts, the dipole moment increases and decreases; and a strong infrared band is found at 2150 cm^{-1}. However, when the triple bond of dimethylacetylene (2-butyne) stretches and contracts, the dipole moment remains the same (essentially zero); therefore, there is no triple bond absorption near 2150 cm^{-1}. Symmetrical alkenes and alkynes may show weak or no absorption in the double and triple bond regions, respectively, because of this requirement.

$$CH_3-C\equiv C-H \rightleftharpoons \overset{\longleftarrow \; +}{CH_3-C\equiv C-H} \qquad CH_3-C\equiv C-CH_3 \rightleftharpoons \overset{\longleftarrow \; +}{CH_3-C\equiv C-CH_3}$$

Changing dipole moment   Unchanging dipole moment

Apart from its use in structural determination ir spectroscopy has been used to study electronic effects in molecules by examination of changes in the frequencies of stretching vibrations when electron-withdrawing or electron-releasing groups are introduced on or near a functional group, to study hydrogen bonding (Sec. 8.3) in a variety of compounds, and to explore the effects of the various types of strain in organic molecules (Sec. 2.11).

EXERCISE 6.8

If an organic compound has the following properties, which functional group(s) may be present in the molecule: (a) a hydrocarbon absorbs strongly at 1660 cm^{-1}; (b) a hydrocarbon shows weak absorption at 1800 cm^{-1}, strong absorption at 900 cm^{-1}, but no absorption near 1000 cm^{-1}; (c) a hydrocarbon shows weak absorption near 2150 cm^{-1} and reacts with water and sulfuric acid in the presence of mercury salts to give a compound absorbing strongly at 1730 cm^{-1}.

6.4   *Electronic Spectroscopy (Ultraviolet-Visible Spectroscopy)*

The absorption of electromagnetic radiation (light) in the ultraviolet (uv) and visible (vis) region of the electromagnetic spectrum is a property of all organic molecules and is the result of an **electronic excitation** in which an electron is promoted from its lowest energy or ground state to a higher electronic state. For our purposes we can employ the usual approximation that such an excitation corresponds to the transition of an electron from a bonding (σ or π) or non-bonding (n) molecular orbital to an anti-bonding (σ^* or π^*) orbital. The relationships between the energies of the various types of orbitals and the possible transitions between energy levels are shown in Fig. 6.17. As in the other types of spectroscopy the energy absorbed is related to the frequency of absorbed light by the familiar equation

$$\Delta E = h\nu$$

where ΔE is (approximately) the difference in energies between the two orbitals linked by the transition.

In electronic or uv-vis spectroscopy the present practice in this country is to use wavelength (λ) rather than frequency (ν) or wavenumber ($\tilde{\nu}$) to define the nature of the electromagnetic radiation. The relationships between the various units that could be used are the same as for ir spectroscopy,

$$c = \lambda\nu \qquad \text{and} \qquad \tilde{\nu} = \frac{1}{\lambda} \qquad \text{where } c = \text{velocity of light}$$

The common units of wavelength in uv-vis spectroscopy are either nanometers (nm)[8] or Angstroms (Å), where

$$1 \text{ nm} = 10^{-9} \text{ m} = 10 \text{ Å}$$
$$1 \text{ Å} = 10^{-10} \text{ m}$$

The magnitude of the energy absorbed in a transition in the uv-vis region can be calculated by the following equations.

$$\Delta E = \frac{28{,}635}{\lambda} \text{ kcal/mole} \qquad \text{for } \lambda \text{ in nm}$$

$$\Delta E = 28.635 \times 10^{-4} \times \tilde{\nu} \text{ kcal/mole} \qquad \text{for } \tilde{\nu} \text{ in cm}^{-1}$$

FIGURE 6.17 *(a) Electronic Energy Levels and Transitions, (b) $\pi \rightarrow \pi^*$ Transition in 1,3-Butadiene*

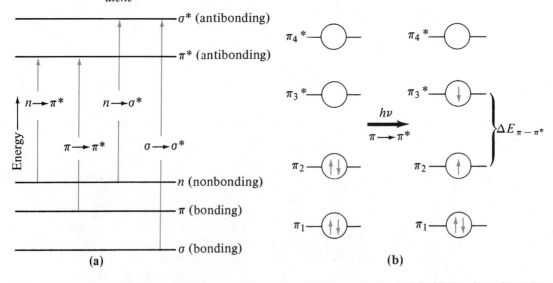

(a) (b)

---

[8] Until recently the common unit for wavelength was the millimicron (mμ); however, this unit has been replaced by its exact equivalent, the nanometer, that is, 1 mμ = 1 nm.

FIGURE 6.18 *Electronic and Infrared Spectral Regions and Associated Energies*

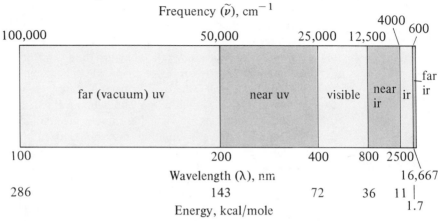

The range of energies involved in uv-vis spectroscopy is shown in Fig. 6.17 from which you can see that the energy absorbed may be greater than that required for the dissociation of some bonds (Table 2.4). It is for this reason that we sometimes use uv light to initiate free radical reactions.[9] Indeed, a whole branch of organic chemistry called **photochemistry** is based on the reactions of organic molecules that take place when these molecules are irradiated with uv-vis light. Similarly, the current concern about the possible depletion of the ozone layer in the upper atmosphere due to excessive use of aerosol sprays (Sec. 6.3) or nitrogen-supplying fertilizers is based on the supposition that the strong absorption of uv radiation from the sun by ozone helps to protect us from certain types of skin cancer caused by uv-light–induced chemical changes (Sec. 16.10).

The uv-vis region includes wavelengths of about 50–800 nm. Because the human eye can see light only in the 400–750 nm region and because below about 190 nm we must employ vacuum (or other) techniques to avoid the strong absorption by atmospheric oxygen, electronic spectroscopy is divided for convenience into three regions: vacuum or far uv (50–200 nm), near uv (200–400 nm), and visible (400–800 nm) (Fig. 6.18). However, except for these practical considerations there is no fundamental difference in the three regions.

In uv-vis spectroscopy, as in ir spectrocopy, sharp absorption peaks or lines are rarely observed. As we have seen, organic molecules are subject to the continual stretching and bending of bonds (vibrations) and the rotation of atoms or groups about their bonds. Therefore, transitions between two electronic energy levels can involve several vibrational or rotational levels (Fig.

---

[9] As an example we can rewrite the first equation of Sec. 2.8-B as a $\sigma \rightarrow \sigma^*$ transition followed by bond cleavage. $\quad\quad\quad Cl{-}Cl \xrightarrow{uv} Cl^*{\cdot}Cl \rightarrow 2\,Cl\cdot$

6.19). Because the spacing between vibrational and rotational levels is very small compared to that between electronic energy levels, the result is a large number of closely spaced or overlapping lines that cannot be resolved. Thus, most uv-vis spectra consist of a small number of rather broad absorption bands.

Electronic excited states have very short lifetimes (10^{-7} sec or less). As the excited molecule returns to its ground state, the absorbed energy may be (1) reemitted as light of a longer wavelength than the original exciting radiation (a phenomenon called fluorescence or phosphoresence depending on the electronic pathway followed to the ground state), (2) converted to thermal energy (heat), or (3) used to initiate a photochemical reaction. In part because of these various possibilities, uv-vis spectrometers are generally constructed so that the light from the source is separated into monochromatic light before passage through the sample (Fig. 6.15) rather than after passing through the sample as in ir spectroscopy.

FIGURE 6.19 *Transitions from the Ground State to the Vibrational Levels of the First and Second Electronic States of a Molecule**

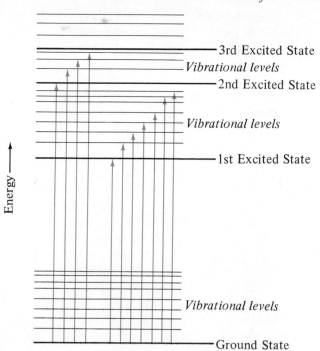

* Between each pair of vibrational levels there is an array of rotational levels, which have been omitted for the sake of clarity.

The intensity of absorption in the uv-vis region is proportional to the number of absorbing molecules in the light path (Beer-Lambert law) and is generally described in terms of the following equation

$$\log \frac{I_0}{I} = \varepsilon c l = A$$

where I_0 = the intensity of incident light, I = the intensity of transmitted light, ε = the **molar absorption coefficient** (molar extinction coefficient), c = the concentration in moles/liter, l = the pathlength of the cell in cm, and A = the **absorbance** (which is generally recorded by the spectrometer). Although the proper procedure would be to use the area under an absorption band as the measure of its intensity, common practice is to *represent* the intensity by the value of molar absorption coefficient (ε_{max}) at the wavelength (λ_{max}) at which ε reaches a maximum. A spectrum may contain several ε_{max} and their corresponding λ_{max}. The intensity of uv-vis bands (as measured by ε_{max}) varies widely (over a range of 1 to 10^5) because of the selection rules controlling "allowed" and "forbidden" transitions. Allowed transitions result in intense bands and forbidden transitions in weak bands.

Of the electronic transitions represented in Fig. 6.17 the $\sigma \rightarrow \sigma^*$ transition involves excitation of an electron in a σ-bond and requires such high energy that absorption occurs in the relatively inaccessible far (vacuum) uv region (below 200 nm) and is generally of limited use. The $n \rightarrow \sigma^*$ transitions involve excitation of an electron from a nonbonding orbital containing an unshared pair (as in O, N, S, or halogen compounds) to an antibonding σ^* orbital and give rise to relatively weak absorption bands (ε_{max} from 100 to 4000). Because of their low absorption in the 200–800-nm region hydrocarbons (λ_{max} below 190 nm) and alcohols (ε_{max} about 500 at λ_{max} 180–185 nm) are commonly used as solvents in uv-vis spectroscopy.

The most useful transitions are the $n \rightarrow \pi^*$ and $\pi \rightarrow \pi^*$ transitions. The $n \rightarrow \pi^*$ transition involves the excitation of an electron from a nonbonding orbital to an antibonding π^* orbital. These low intensity (ε about 10–100) bands occur at relatively long wavelengths and are important factors in determining the color of certain dyes and pigments (Chapter 19). Compounds containing the

$$C\!=\!\ddot{O}, \; C\!=\!\ddot{S}, \; \ddot{N}\!=\!\ddot{N}, \; \text{and} \; \overset{+}{N}\!=\!\ddot{O} \quad \text{structural units will show } n \rightarrow \pi^* \text{ transition}$$
$$\underset{\ddot{O}:^-}{|}$$

bands.

The most important transitions from the standpoint of structure determination are the $\pi \rightarrow \pi^*$ transitions, which involve the excitation of an electron from a π orbital to a π^* orbital. Compounds having double or triple bonds or aromatic rings undergo such transitions; however, the absorption bands may fall in the far uv region unless the π system is a conjugated system. As the length of the conjugated system increases, the wavelength and usually the intensity of absorption increase. The effect of increased conjugation is illustrated with several polyenes with three to six conjugated double bonds in Table 6.5. The yellow to orange color of carotene (Sec. 3.11) (λ_{max} 452 nm) from carrots and the red color of lycopene (λ_{max} 469 nm) from tomatoes and paprika are due to the extended conjugation of the π system and the corresponding low energy (high wavelength) $\pi \rightarrow \pi^*$ transitions in these compounds. The explanation for this phenomenon is that, as the π system becomes more extended, the energy difference between the highest occupied bonding and lowest unoccupied anti-bonding orbitals descreases, as shown in Fig. 6.20.

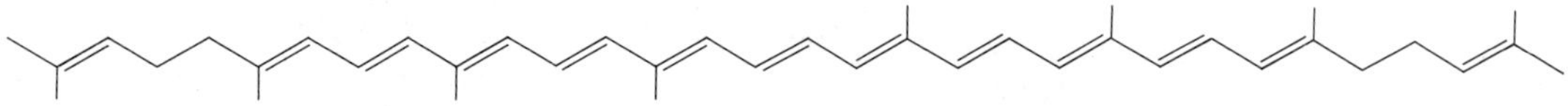

Lycopene

FIGURE 6.20   *π and π^* Energy Levels in Conjugated Polyenes*

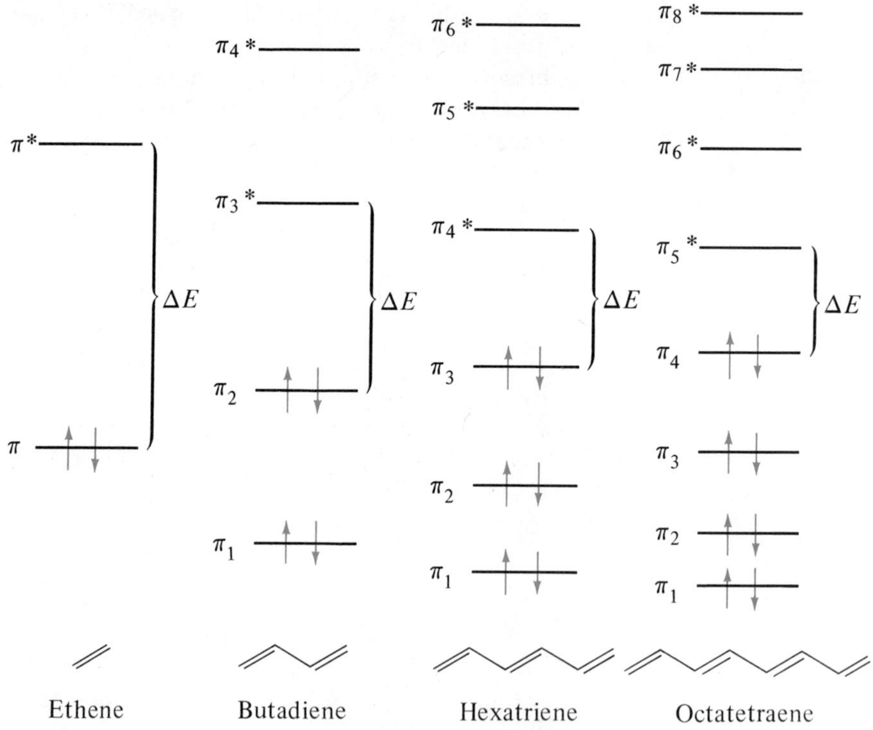

A structural unit in a molecule that can undergo an electronic transition is called a **chromophore.** A nonchromophoric group, usually containing unshared electron pairs, which causes a shift in the wavelength of an absorption peak and

TABLE 6.5 *Electronic Absorption Bands*

Compound	Transition	λ_{max} (nm)	ε_{max}	Solvent
CH_3-CH_3	$\sigma \longrightarrow \sigma^*$	135	7000	vapor
CH_3CH_2OH	$n \longrightarrow \sigma^*$	181	325	vapor
CH_3I	$n \longrightarrow \sigma^*$	259	400	hexane
$(CH_3CH_2)_2NH$	$n \longrightarrow \sigma^*$	193	193	hexane
$CH_3CH{=}CHCH_3$ (*trans*)	$\pi \longrightarrow \pi^*$	178	13,000	vapor
$CH_3(CH{=}CH)_3CH_3$	$\pi \longrightarrow \pi^*$	275	30,000	hexane
$CH_3(CH{=}CH)_4CH_3$	$\pi \longrightarrow \pi^*$	310	76,500	hexane
$CH_3(CH{=}CH)_5CH_3$	$\pi \longrightarrow \pi^*$	342	122,000	hexane
$CH_3(CH{=}CH)_6CH_3$	$\pi \longrightarrow \pi^*$	380	146,500	chloroform
$HC{\equiv}CH$	$\pi \longrightarrow \pi^*$	173	600	vapor
$CH_3-\overset{\overset{O}{\|}}{C}-CH_3$	$\pi \longrightarrow \pi^*$ / $n \longrightarrow \pi^*$	190 / 275	1000 / 22	cyclohexane
$CH_3-\overset{\overset{O}{\|}}{C}-\overset{\overset{O}{\|}}{C}-CH_3$	$\pi \longrightarrow \pi^*$ / $n \longrightarrow \pi^*$	282 / 420	19 / 10	ethanol
$CH_3-C\overset{O}{\underset{OH}{}}$	$n \longrightarrow \pi^*$	204	41	ethanol

usually an increase in its intensity is called an **auxochrome.** The absorption bands of the chromophores in a number of simple molecules are given in Table 6.5. The π system in benzene is an important chromophore, giving rise to three bands at 184, 204, and 254 nm. The band at 254 nm is of greatest use in structural assignments; however, it is a rather weak band because the electronic selection rules do not favor this particular transition. The effects of several auxochromes on the wavelength and intensity of the 254-nm band of benzene can be compared by examination of the entries in Table 6.6. The importance of chromophores and auxochromes in determining the colors of organic molecules and in the chemistry of dyes is discussed in Sec. 19.1.

EXERCISE 6.9

Proteins are high molecular weight polymers made up largely of amino acids linked together. The structures of the amino acids from which proteins are derived are given in Table 15.1. Only three of the amino acids absorb in the uv-vis region above 240 nm, at 257, 275, and 280 nm. Can you identify these three amino acids, which are the basis of protein assay by uv spectroscopy, and assign to them the appropriate λ_{max}? (*Hint:* Use Table 6.5 as a guide.)

TABLE 6.6   *Electronic Absorption Bands of Benzene Derivatives*

Compound	λ_{max} (nm)	ε_{max}	λ_{max} (nm)	ε_{max}	λ_{max} (nm)	ε_{max}	Solvent
Benzene	184	68,000	204	8,800	254	250	hexane
Chlorobenzene			210	7,500	257	170	ethanol
Toluene	189	55,000	208	7,900	262	260	hexane
Aniline			230	8,600	280	1,400	water
			203	7,500	254	160	water + HCl
Phenol			211	6,200	270	1,450	water
			236	9,400	287	2,600	water + NaOH
Nitrobenzene*			252	10,000	280	1,000	hexane
4-Nitrophenol			226	6,900	318	10,000	water
Acetophenone†			243	13,000	279	1,200	ethanol

* Also an $n \rightarrow \pi^*$ band at 330 nm (ε 140).

† Also an $n \rightarrow \pi^*$ band at 315 nm (ε 55).

The most common structural application of uv-vis spectroscopy is the detection of conjugation and the determination of its nature and extent. However, at present the analytical applications far outnumber structural applications, and uv-vis spectrometry is widely used in the assay of vitamins, antibiotics, enzyme activity, etc. and in the determination of the rates of organic reactions. Measurement of the absorption at 280 nm is a rapid and convenient method of estimating the protein content of a solution, and absorption at 260 nm has been used to assay nucleotide (Sec. 16.4) content and monitor the denaturation of DNA (Sec. 16.5). The uv spectrum of uracil, whose nmr spectrum was given in Figure 6.7, is given in Figure 6.21 as determined in both neutral and basic solution. The shift in λ_{max} in basic solution toward the red end of the spectrum (a **bathochromic shift**) is similar to that observed with phenol in basic solution (Table 6.6) and is caused by the removal of a proton from the oxygen atom in phenol and from one of the nitrogen atoms in uracil, which increases the number of nonbonding (n) electrons available to the conjugated system. By determining uv-vis spectra in solutions having a range of pH values, it is possible to study the interaction of certain functional groups with the aromatic ring and to explore keto-enol tautomerism (Sec. 9.5).

6.5   *Mass Spectroscopy; X-Ray Crystallography*

Two other methods for structural investigation are of such importance to the organic chemist that they require at least brief mention: mass spectroscopy and X-ray crystallography. In mass spectroscopy a very small sample of an

FIGURE 6.21   *Ultraviolet Spectrum of Uracil (8.44 × 10⁻⁵M Solution in Water) at pH 7.0 and 10.0*

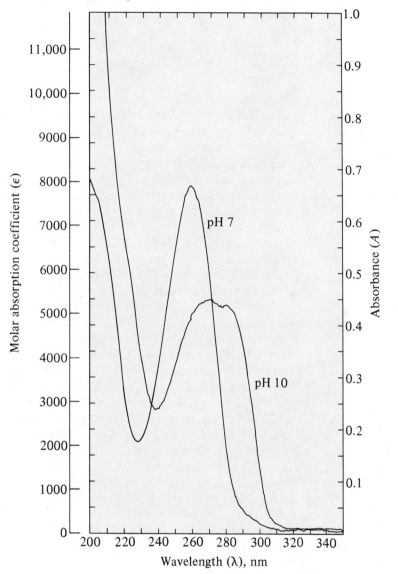

organic compound is bombarded by a high-energy beam of electrons in a positively charged, evacuated chamber. The electron beam removes an electron from the molecule forming a radical cation M^+, called the **molecular ion.** This ion generally has so much excess energy that one or more of its bonds break yielding a large number of fragment ions (and neutral radicals or smaller molecules).

$$
\begin{array}{c}
\text{R}' \\
| \\
\text{R}''\!-\!\overset{\displaystyle |}{\text{C}}\!:\!\text{R} \;-\; e \\
| \\
\text{R}'''
\end{array}
\longrightarrow
\left[
\begin{array}{c}
\text{R}' \\
| \\
\text{R}''\!-\!\overset{\displaystyle |}{\text{C}}\!\cdot\!\text{R} \\
| \\
\text{R}'''
\end{array}
\right]^{+\cdot}
\longrightarrow
\begin{array}{c}
\text{R}' \\
| \\
\text{R}''\!-\!\overset{\displaystyle |}{\text{C}}\!\cdot \\
| \\
\text{R}'''
\end{array}
\;+\;
\text{R}^{+}
$$

| Molecule | Molecular ion | Radical | Fragment ion |

The positively charged ions are repelled by the high positive charge on the chamber and are ejected out of the chamber and propelled at high velocity down a curved tube into the field of an electromagnet. In a magnetic field a charged particle follows a curved path dependent on the *intensity* of the field and the *mass* of the particle. By varying the field intensity in a regular fashion each positive ion's path can be controlled to cause that ion to strike an ion collector, which counts the ions electronically. A recorder in the system plots the intensity (ion count) of each ion against its mass number (that is, the magnetic field intensity required to cause the particle to strike the collector). The result is a mass spectrum, which consists of a series of many peaks at the mass numbers of the fragment ions. Molecules tend to fragment in predictable ways; thus, the fragmentation pattern shown by the mass spectrum can often be used to deduce the structure of the molecule. Measurement of the exact mass of the molecular ion (to four decimal places) gives not only the molecular weight but also the molecular formula (or a short list of formulas, from which the correct formula can be chosen on the basis of other information). For example, the mass number of 112.0273 corresponds to the molecular formula of uracil, $C_4H_4N_2O_2$, and to no other molecular formula.

When a beam of X rays is directed at a *single crystal* of an organic compound at the proper angle, the X rays are scattered by the atoms of the molecule, which are arranged in a regular pattern throughout the crystal. The scattered X rays tend to be arranged in a pattern of alternating regions of high and low intensity, depending on the arrangement of the atoms in the crystal. If the scattered X rays are allowed to fall on a piece of photographic film, an array of spots of varying darkness (density) in a definite pattern, called a diffraction pattern, appears on the developed film. If the angle of the X-ray beam is varied in a regular way, the nature of the pattern changes. These patterns may also be collected electronically and fed directly to a small computer. Analysis of thousands of the spots by the computer gives an unequivocal structure for the molecule in many instances. The analysis takes from several days for a small molecule to many months for a large molecule such as a protein. However, for small molecules, not only does the computer give accurate bond lengths and bond angles, but also from a recorder connected to the computer we can obtain a perspective drawing of the molecule. An example of such a drawing is given on the next page. X-ray crystallography is the ultimate technique for the determination of the precise geometry of molecules in the crystalline state.

3-Carbomethoxybicyclo[1.1.0]butane-
1-carboxylic acid

6.6   *Solving a Structural Problem*

Let us now apply some of the techniques described in the previous sections to the characterization of an "unknown" compound, that is, to the determination of its structure. This compound is a colorless liquid (bp 221°), containing only carbon, hydrogen, and oxygen. The mass spectroscopy laboratory reports that the exact mass of the molecular ion is 148.0884 ± 0.0015 and that, based on this exact mass, the molecular formula is probably $C_{10}H_{12}O$. The uv, ir, and nmr spectra of the compound are given in Figure 6.22.

Before analyzing the spectra, it would be wise for us to use the molecular formula to calculate the number of rings and/or double bonds in the molecule, which we will represent by the symbol θ (a ring with a double bond through it). For a molecule of the formula, $C_cH_hN_nO_oX_x$ (where X = halogen)

$$\theta = \text{number of rings and/or double bonds} = c - \frac{h}{2} + \frac{n}{2} - \frac{x}{2} + 1$$

The formula gives the total number of rings and/or double bonds (C=C, C=O, N=N, etc.), where a triple bond (C≡C or C≡N) is equivalent to two double bonds and a benzene ring is equivalent to a ring plus three double bonds. For our "unknown"

$$\theta = 10 - \frac{12}{2} + 1 = 5$$

Generally, a molecule with such a high value of θ is likely to be aromatic; however, the possibility of an unsaturated acyclic compound cannot be ignored.

Next we turn to the ir spectrum to determine, if we can, what functional groups may be present. The strong band at 1690 cm^{-1} is characteristic of the

FIGURE 6.22 (a) *Infrared Spectrum of an Unknown* (*Neat Liquid*)

Wavelength (microns)

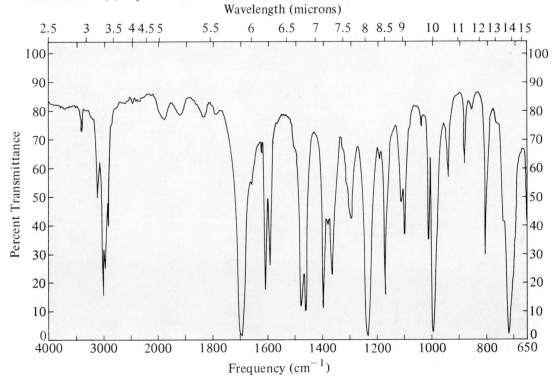

(b) *NMR Spectrum of an Unknown* (*CDCl$_3$ Solution*)

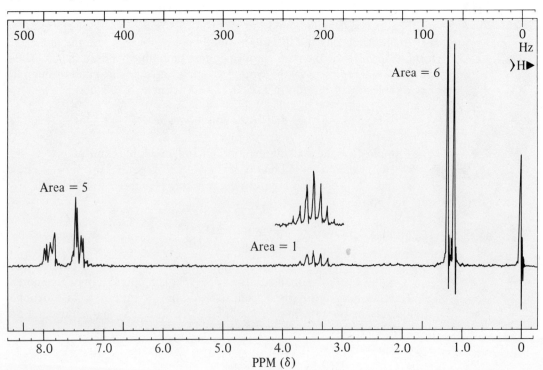

(c) *Ultraviolet Spectrum of an Unknown (Ethyl Alcohol Solution)*

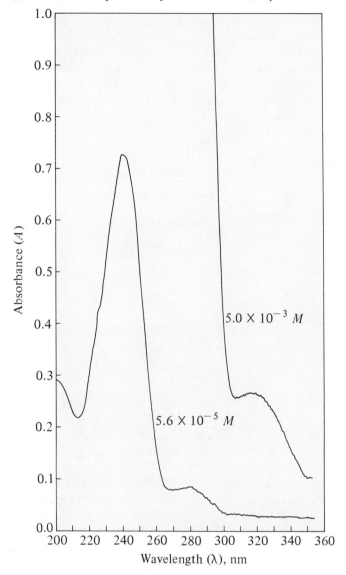

C=O group according to Tables 6.3 and 6.4. From Table 6.4 we find that the aromatic ketones (ArCOR, 1700–1650 cm⁻¹) have absorption bands in the same region as our unknown. The band at 3100 cm⁻¹, the bands near 1600 and 1480 cm⁻¹, the four weak bands between 1980 and 1780 cm⁻¹, and the strong band at 720 cm⁻¹ (which could be two overlapping bands) are all suggestive of a monosubstituted aromatic compound (Table 6.4(b)).

Rather than try to analyze the ir spectrum in greater detail (which could be done in this instance), we next examine the nmr spectrum. The one-proton septet at δ 3.47 requires six neighboring protons (two CH₃ groups?), and the six-proton doublet at δ 1.17 requires a single neighboring proton. Putting this information together we conclude that our compound contains an isopropyl group, $(CH_3)_2CH$—. The complex multiplet between 440–480 Hz is probably that of a benzene ring, but we will not try to analyze this higher order multiplet in detail. With a benzene ring ($\theta = 4$) and an isopropyl group and a carbonyl (C=O) group ($\theta = 1$), we have accounted for all of our carbon atoms, hydrogen atoms, and oxygen atom as well as our θ value of 5. Thus, at this stage we have the following structural units

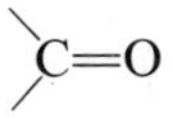

$$\text{C=O} \qquad \text{possibly} \quad Ar{-}\overset{\overset{\displaystyle O}{\|}}{C}{-}R$$

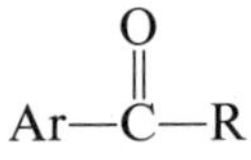

probably monosubstituted

$(CH_3)_2CH$—

Putting the units together in a logical fashion we arrive at the structure

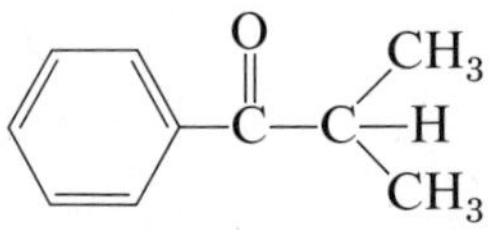

Isopropyl phenyl ketone

If our structure is to be acceptable, it must fit all of the data, not just that which we have used to this point. Therefore, we next check to be sure that the chemical shift data are reasonable. From Figure 6.6 we find that the ketone with the highest chemical shift (shown in the figure) is CH_2COPh (RCH_2COPh) at δ 2.95, substantially lower than δ 3.47. However, no ketone of the structure $CHCOPh$ ($R_2CHCOPh$) is shown, and we infer that such a ketone would have a chemical shift above δ 2.95 (on the basis of the usual trend in chemical shifts for $(CH_3)_2CH > CH_3CH_2 > CH_3$). The doublet at δ 1.17 is very close to the value on the chart (δ 1.1) for the structural unit, $CH_3C{-}CO{-}R$. Thus, the chemical shift values are reasonable ones for isopropyl phenyl ketone.

The uv spectrum serves to confirm our assignment of structure. The bands at 240 and 280 nm are characteristic of substituted benzenes, and the very weak band at 318–320 nm could be the $n \to \pi^*$ band of a ketone. Indeed, the spectrum appears to resemble that of acetophenone (methyl phenyl ketone) (Table 6.6).

All of the data seem to be consistent with the structural assignment that we have made. Naturally, not all structural problems are so readily solved, but many are no more difficult than this one, and all structural problems look simpler as we learn through practice and experience.

Summary

1. Organic absorption spectroscopy is based on the correlation of structure with the selective absorption of electromagnetic radiation by organic molecules.

2. The principal types of absorption spectroscopy used in organic chemistry are
   (a) Nuclear magnetic resonance (nmr) spectroscopy, in which absorption is related to the magnetic environment of protons or other "magnetic" nuclei in the molecule.
   (b) Infrared (ir) spectroscopy, in which absorption is related to the vibrations of bonds in the molecule.
   (c) Electronic (ultraviolet-visible, uv-vis) spectroscopy, in which absorption is related to the promotion of electrons from the lowest (ground) electronic energy state to a higher electronic state.

3. Nuclear magnetic resonance spectroscopy:
   (a) The **chemical shift** is the difference in the **resonance frequencies** of two nuclei at a constant value of the magnetic field.
   (b) Proton chemical shifts are usually measured relative to a standard, tetramethylsilane (TMS).
   (c) Proton chemical shifts are characteristic of the chemical environment of the protons. Protons in the same chemical environment are said to be chemically equivalent (chemically indistinguishable).
   (d) The area under an nmr peak is roughly proportional to the number of protons responsible for the peak.
   (e) Protons on adjacent carbon atoms may **couple** (or exchange spin information) with each other, resulting in a **splitting** of singlets into doublets, triplets, etc.
   (f) In simple, first-order systems the number of peaks (multiplicity) in a multiplet will be $n + 1$, where n is the number of protons on the adjacent carbon atom(s).
   (g) If a proton changes location or position in a molecule rapidly, the nmr spectrometer reports an average position; if it changes slowly, the spectrometer may report both (or all) positions as separate peaks.

4. Infrared spectroscopy:
   (a) The frequency of absorbed infrared light can be correlated with the **stretching** and **bending** motions (vibrations) of bonds in functional groups and small, specific structural units.
   (b) Many molecules containing a given functional group or structural unit will absorb infrared light of a **characteristic group frequency;** therefore, characteristic group frequencies can be used to establish the presence or absence of specific functional groups.
   (c) Only those vibrations that involve a changing dipole moment can be caused by the absorption of infrared light.

5. Electronic spectroscopy:
    (a) The absorption of uv-vis light depends on the presence in the molecule of **chromophores,** structural units that can undergo an electronic transition.
    (b) Transitions occur from occupied bonding or nonbonding orbitals to unoccupied antibonding orbitals.
    (c) The most useful transitions are the $n \rightarrow \pi^*$ and $\pi \rightarrow \pi^*$ transitions.
    (d) The $\pi \rightarrow \pi^*$ transitions may be used in the detection of conjugation and the determination of its nature and extent.
    (e) **Auxochromes** are nonchromophoric groups which cause a shift in the wavelength of an absorption peak.

6. Mass spectroscopy may be used to determine molecular weight and molecular formula and to provide structural information based on the fragmentation patterns of organic molecules.

7. X-ray crystallography is used to determine the precise geometry of molecules in the crystalline state.

New Terms

absorbance

absorption band

auxochrome

bending vibration

characteristic group frequency

chemical equivalence

chemical shift

chromophore

coupling constant

deshielded nucleus

integral

magnetic equivalence

molar absorption coefficient

molecular ion

monochromator

peak

photochemistry

photometer

radical ion

shielded nucleus

spectrometer

spectrum

spin-spin coupling

spin-spin splitting

stretching vibration

Supplementary Exercises

EXERCISE 6.10   The nmr pattern of an ethyl group which is not coupled to any other protons is one of the most common patterns in nmr spectroscopy. A typical example is the ethyl group in ethyl iodide (CH_3CH_2I). The nmr spectrum of ethyl iodide consists of two groups of peaks at δ 1.83 and δ 3.20, respectively. Answer the following questions about this spectrum:

(a) What is the multiplicity of the group of peaks at δ 1.83?
(b) What is the multiplicity of the group of peaks at δ 3.20?
(c) Which group of peaks corresponds to the CH_3 group?
(d) Which group of peaks corresponds to the CH_2 group?
(e) What is the chemical shift of the CH_2 group in Hz at 60 MHz?
(f) What is the chemical shift of the CH_2 group in Hz at 100 MHz?
(g) The coupling constant (J) in the 60 MHz spectrum is 7 Hz. What is the coupling constant in the 100 MHz spectrum?

EXERCISE 6.11   Using lines of appropriate height and spacing (δ and J), sketch the nmr spectra expected for the following compounds.

(a) $CH_3-\underset{\underset{Br}{|}}{CH}-CH_3$

(b) $CH_3-O-CH_2CH_3$

(c) $CH_3-\underset{\underset{Br}{|}}{CH}-\underset{\underset{Br}{|}}{CH}-CH_3$

(d) $H-C\equiv C-CH_2-Br$

(e) $Cl-CH_2-CH_2-CH_2-Cl$

(f) $CH_3-\underset{\underset{Br}{|}}{CH}-CH_2-\underset{\underset{Br}{|}}{CH_2}$ (all ^3J's equal)

EXERCISE 6.12   Estimate the chemical shifts (in Hz and in δ values) and coupling constants (in Hz) for the peaks in the 60-MHz spectra in:

(a) Figure 6.7
(b) Figure 6.8
(c) Figure 6.10

(d) Figure 6.11
(e) Figure 6.22(b)

EXERCISE 6.13   Assign as many absorption bands as you can in Figure 6.14 to the corresponding characteristic group frequencies in Table 6.4.

EXERCISE 6.14   Assign as many absorption bands as you can in Figure 6.22(a) to the corresponding characteristic group frequencies in Table 6.4.

EXERCISE 6.15   Calculate ε_{max} for uracil (Figure 6.21) at pH 7.0 and 10.0 (cell pathlength = 1.0 cm). Check your results against the ε scale provided at the left side of the spectrum.

EXERCISE 6.16   Calculate ε_{max} for each band in the uv spectrum of isopropyl phenyl ketone (Figure 6.22(c)) (cell pathlength = 1.0 cm).

EXERCISE 6.17   Hooke's law for relating the frequency of vibration of two masses, m_1 and m_2, connected by a spring of force constant (strength), k, is sometimes used to illustrate the validity of the ball and spring model for molecular stretching vibrations. Reduced to molecular dimensions and units, Hooke's law tells us that

$$\tilde{\nu} = \frac{1}{2\pi c}\sqrt{\frac{k(m_1 + m_2)}{m_1 \times m_2}} = 1303 \sqrt{\frac{k(m_1 + m_2)}{m_1 \times m_2}} \ \text{cm}^{-1}$$

where k is the bond force constant (strength of the bond). For single, double, and triple bonds k = 5, 10, and 15 mdyne/Å (the proper units for the equation), respectively. Using this approximation, calculate the stretching frequencies of the following bonds. Compare your results with the values in Table 6.4.

(a) C—C

(b) C=C

(c) C≡C

(d) C≡N

(e) C—H

(f) O—H

(g) C=O

(h) C—Cl

EXERCISE 6.18   Using Hooke's law, estimate the stretching frequency for the C—D bond based on a C—H stretching frequency of 2900 cm^{-1} and the assumption that both bonds have the same force constant (k). (*Hint:* Use the C—H stretching frequency to calculate the force constant or devise a procedure for calculation that does not require the use of k as long as the k's are equal.)

EXERCISE 6.19   Compound A contains nitrogen and chlorine and has a molecular weight of 89.5. The compound gave the ir and nmr spectra shown in Figure 6.23. From these data deduce the structure of compound A.

EXERCISE 6.20   Compound B, $C_9H_8O_2$, gave the ir and nmr spectra shown in Figure 6.24. Suggest a reasonable structure for compound B, including the proper stereochemistry, if necessary.

EXERCISE 6.21   Compound C, $C_5H_{11}Cl$, gave the nmr spectrum shown in Figure 6.25. Suggest a reasonable structure for compound C.

EXERCISE 6.22   Compound D, $C_4H_9O_2Cl$, gave the nmr spectrum shown in Figure 6.26. Suggest a reasonable structure for compound D.

FIGURE 6.23 *(a) Infrared Spectrum of Compound A (Neat Liquid)*

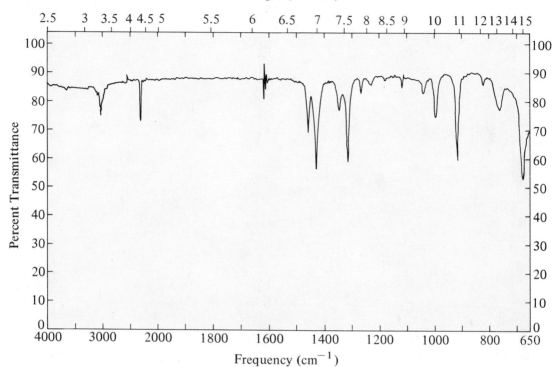

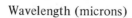
Wavelength (microns)

(b) NMR Spectrum of Compound A (CCl₄ Solution)

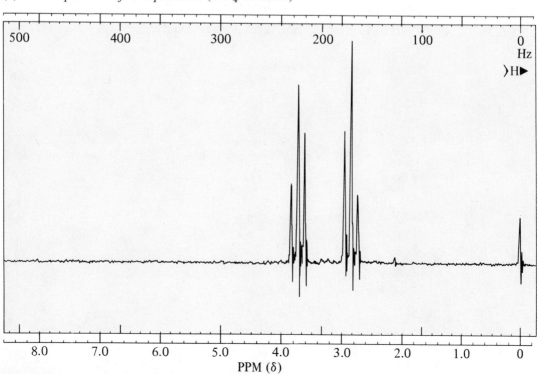

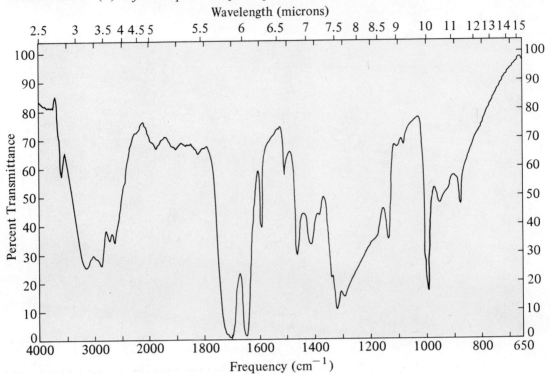

(b) *NMR Spectrum of Compound B* (*CDCl₃ Solution*)

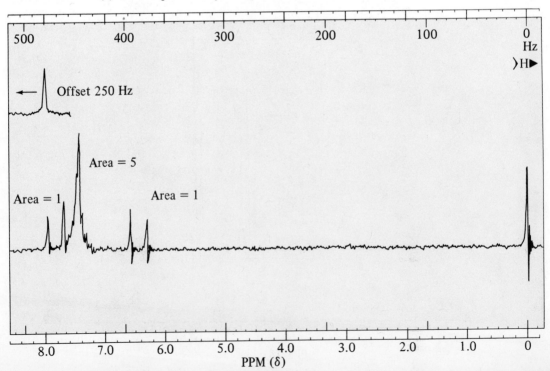

FIGURE 6.25 *NMR Spectrum of Compound C (CCl₄ Solution)*

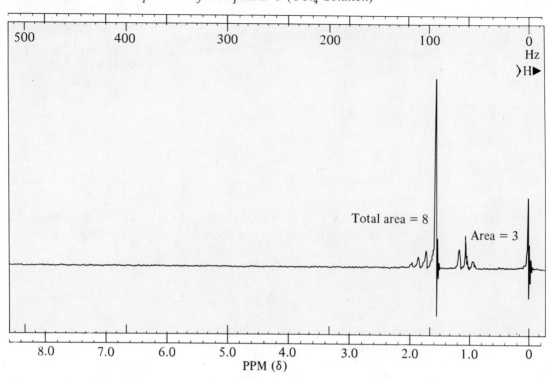

FIGURE 6.26 *NMR Spectrum of Unknown D (CDCl₃ Solution)*

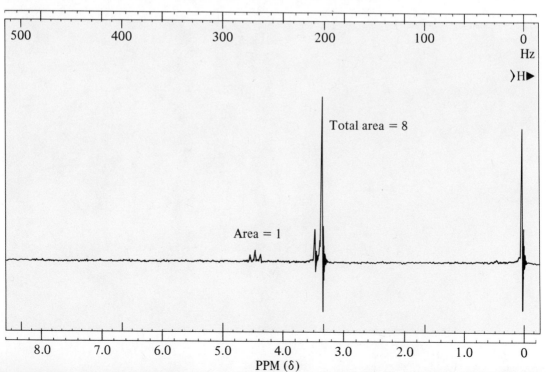

EXERCISE 6.23   The nmr spectrum of an organic compound with three non-equivalent protons, A, B, and C, shows the following pattern of twelve lines (three doublets of doublets). Draw a splitting diagram that will explain this pattern. (*Hint:* The coupling constants J_{AB}, J_{AC}, and J_{BC}, have different values.)

EXERCISE 6.24   In molecules with a CH_2 (methylene) group near an asymmetric center the methylene protons, H_a and H_b, are likely to have different chemical shifts.

$$Z-\underset{\underset{H_b}{|}}{\overset{\overset{H_a}{|}}{C}}-\underset{\underset{X}{|}}{\overset{\overset{H_c}{|}}{C}}-Y \qquad X \neq Y; Z \neq H$$

Although they are chemically indistinguishable with *achiral* reagents, H_a and H_b may not react identically with *chiral* reagents (such as an enzyme); thus, H_a and H_b are not chemically equivalent. Draw Newman projections of the more stable conformations of this molecule, and use these projections to show that H_a and H_b are never in exactly the same chemical environment.

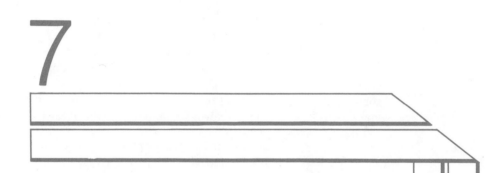

7

Organic Halogen Compounds

Introduction

The earth's crust (including the oceans) contains inorganic halides of every kind in great abundance. In contrast to this, there are relatively few naturally occurring organic compounds in which halogen atoms are covalently bonded to carbon. Yet, the importance of this family of substances is so great that many synthetic routes lead to end products only by way of some halogen intermediate. Many organic halogen compounds are synthesized, therefore, simply to be used as chemical reagents. Others are prepared as useful commodities in their own right and are employed as propellants in aerosol sprays, and as refrigerants, pesticides, dry cleaners, and plastics.

We found double and triple bonds in the alkenes and alkynes to be the structural features which bestowed a characteristic reactivity upon these compounds. We shall now study for the first time a functional group composed of a single atomic species—the covalently bound halogen.

205

7.1 *Classification and Nomenclature*

All organic halogen compounds may be classified either as aliphatic or aromatic. If aliphatic, they are alkyl halides that may be classified further according to structure—that is, they are either primary, secondary, or tertiary alkyl halides. This classification is based upon the nature of the carbon atom to which the halogen atom is bonded.

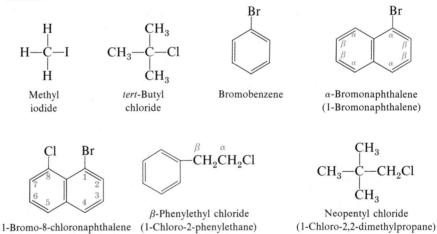

| A primary | A secondary | A tertiary |
| alkyl halide | alkyl halide | alkyl halide |

The nomenclature of the halogen compounds is not difficult to learn. Common names frequently are used for the simpler members of this family. In such names the alkyl group to which the halogen is attached is given first, and this is then followed by the name of the halogen. The halogen (as halide) is a separate word. Alkyl halides of five carbons or greater and the aromatic halogen compounds are named more conveniently as substituted hydrocarbons according to IUPAC rules. The following examples illustrate each kind of nomenclature.

Methyl iodide

tert-Butyl chloride

Bromobenzene

α-Bromonaphthalene (1-Bromonaphthalene)

1-Bromo-8-chloronaphthalene

β-Phenylethyl chloride (1-Chloro-2-phenylethane)

Neopentyl chloride (1-Chloro-2,2-dimethylpropane)

EXERCISE 7.1

Draw all structures possible that have the molecular formula $C_5H_{11}Cl$. Classify each as a primary, secondary, or tertiary chloride. Name each according to the IUPAC system.

7.2 *Preparation of Organic Halogen Compounds*

A. Preparation by Direct Halogenation of Hydrocarbons. The aromatic halogen compounds in which the halogen is attached to the ring may be prepared by electrophilic (ionic) halogenation (Sec. 4.7-A). The halogenation of alkanes by a free radical chain process (Sec. 2.8) is difficult to control and produces a mixture of isomers and polyhalogenated products. The chlorination of methane, for example, produces all of the possible mono- and polysubstituted methanes. For this reason direct halogenation of alkanes is seldom used as a means of preparing alkyl halides in the laboratory, although it is used in some industrial processes.

Chlorination of the methyl group of toluene under free radical conditions (light catalysis) also results in the production of varying amounts of polysub- stituted derivatives and is an industrial, not a laboratory, process.

| Toluene | Benzyl chloride | Benzal chloride | Benzotrichloride |

With side chains longer than methyl, chlorine reacts with alkylbenzenes to give a mixture of isomeric halides. However, bromine, which reacts relatively slowly with alkanes, reacts readily with alkylbenzenes under free radical con- ditions to give only the product in which the bromine atom is attached to the carbon atom next to the ring (called the benzylic carbon or **benzylic position**). The reason for this highly selective attack is that an odd electron on the benzylic carbon may be delocalized into the ring, as may be seen from the resonance structures for the benzyl radical. This resonance interaction stabilizes the free radical intermediate and favors its formation. We shall see this same phenom- enon manifest throughout chemical behavior—namely, the reaction that leads to a stabilized intermediate, or product, is often (but not always) the favored reaction.

| Ethylbenzene | α-Phenylethyl bromide (1-Bromo-1-phenylethane) |

EXERCISE 7.2

If you have not already done so (Exercise 4.18), draw the resonance forms for the benzyl radical.

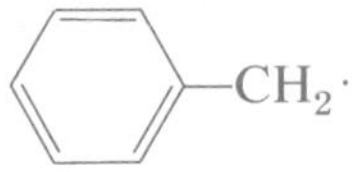

$$\langle \bigcirc \rangle - CH_2 \cdot$$

(*Hint:* Use one-electron shifts rather than two-electron shifts to derive the forms, such as is shown below with fish-hooks (single-barbed arrows) for the allyl radical.

$$[CH_2 = CH - CH_2 \cdot \longleftrightarrow \cdot CH_2 - CH = CH_2]$$

B. Preparation from Alcohols.

A laboratory preparation of an alkyl halide may begin with an alcohol as starting material. Treatment of the appropriate alcohol with any of the following reagents results in replacement of the hydroxyl group, —OH:

(a) concentrated hydrobromic acid, HBr.
(b) concentrated hydriodic acid, HI.
(c) concentrated hydrochloric acid, HCl + $ZnCl_2$.
(d) phosphorus tribromide, phosphorus triiodide, or phosphorus pentachloride.
(e) thionyl chloride, $SOCl_2$.

Two examples of the preparation of organic halogen compounds from alcohols by the use of these reagents are given below.

$$3 CH_3CH_2OH + PBr_3 \longrightarrow 3 CH_3CH_2Br + H_3PO_3$$

Ethyl alcohol   Ethyl bromide

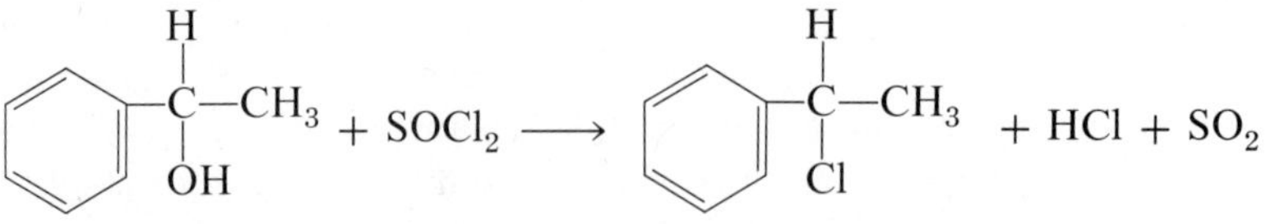

$$\langle \bigcirc \rangle - \overset{\overset{\displaystyle H}{|}}{\underset{\underset{\displaystyle OH}{|}}{C}} - CH_3 + SOCl_2 \longrightarrow \langle \bigcirc \rangle - \overset{\overset{\displaystyle H}{|}}{\underset{\underset{\displaystyle Cl}{|}}{C}} - CH_3 + HCl + SO_2$$

1-Phenylethanol   1-Chloro-1-phenylethane

Thionyl chloride, $SOCl_2$, is a superior chlorinating reagent because the by-products formed in this case are gases and escape from the reaction mixture.

C. Preparation from Olefins.

Halogen acids (HBr, HCl, HI) will add to olefins to yield alkyl halides (Sec. 3.8-A). The mode of addition follows Markovnikov's rule, except for the addition of hydrogen bromide in the presence of air, light, or peroxides.

$$CH_3—\overset{\displaystyle \overset{H}{|}}{C}—CH_3$$
$$\underset{Br}{|}$$

Isopropyl bromide
(2-Bromopropane)

$$CH_3—\overset{\displaystyle \overset{H}{|}}{C}=CH_2$$

Propylene
(Propene)

$\xrightarrow[\text{no light}]{HBr}$

$\xrightarrow[\text{light}]{HBr}$

$$CH_3—CH_2—CH_2—Br$$

n-Propyl bromide
(1-Bromopropane)

D. Preparation *via* Halogen Exchange. Certain halides, in some instances, are best prepared from other halides by an exchange. This method lends itself especially well to the preparation of iodo and fluoro compounds. Sodium iodide, unlike sodium chloride or sodium bromide, is soluble in anhydrous acetone. If a primary or secondary alkyl chloride is dissolved in a solution of sodium iodide in acetone, the halogens exchange. The equilibrium favors formation of the alkyl iodide because the sodium chloride which forms is not soluble in acetone and precipitates.

$$R—Cl + Na^+I^- \xrightarrow{\text{acetone}} R—I + NaCl$$

Primary or
secondary

Fluorine compounds often are prepared *via* a halogen exchange through the use of inorganic fluorides, rather than by reactions utilizing fluorine or hydrogen fluoride because the latter chemicals are extremely reactive and difficult to handle.

$$2\,CH_3Br + Hg_2F_2 \longrightarrow 2\,CH_3F + Hg_2Br_2$$

Methyl   Mercurous   Fluoro-   Mercurous
bromide   fluoride   methane   bromide

EXERCISE 7.3

If you were given propene and benzene as your only organic starting materials and any other inorganic reagents that you might require, how would you prepare (a) isopropyl chloride, (b) *n*-propyl bromide, (c) *n*-propyl iodide, and (d) 2-chloro-2-phenylpropane?

7.3 *Polyhalogen Compounds*

Halogenated products prepared from methane become increasingly dense, higher boiling, and nonflammable as halogen successively replaces the four hydrogen atoms. Polychlorinated methanes may be prepared by the direct halogenation of methane (Sec. 2.8-B) or, as in the case of carbon tetrachloride, by using replacement methods.

$$C + 2\,S \longrightarrow CS_2$$
$$CS_2 + 3\,Cl_2 \longrightarrow CCl_4 + S_2Cl_2$$
$$2\,S_2Cl_2 + CS_2 \longrightarrow CCl_4 + 6\,S$$

The chlorine atoms of carbon tetrachloride, in turn, can be exchanged for fluorine in the preparation of the fluorinated methanes. For example,

$$CCl_4 + 2\,SbF_3 \longrightarrow CCl_2F_2 + 2\,SbF_2Cl$$

Dichlorodifluoromethane
(Freon-12)

Carbon tetrachloride is an excellent solvent for oils, waxes, fats, and greases and has been used as a dry cleaner. While it is nonflammable, it is very volatile and repeated inhalation of its vapor can cause cumulative liver damage. Carbon tetrachloride as a fire extinguishing agent has been replaced largely by lique-fied carbon dioxide, bromotrifluoromethane ($CBrF_3$), and dibromodifluoro-methane (CBr_2F_2). The fluorinated methanes may be used safely in confined spaces such as submarines, aircraft, boats, and automobiles. Carbon tetrachlo-ride, on the other hand, decomposes when heated in the presence of oxygen to give phosgene, an extremely poisonous gas.

$$\overset{\displaystyle O}{\underset{\displaystyle Cl-C-Cl}{\|}}$$

Phosgene

The trihalomethanes, chloroform, bromoform, and iodoform, may be prepared by treating acetone or ethyl alcohol with alkaline solutions of the halogens, as is described in Sec. 9.8.

$$CH_3-CH_2-OH + 4\,Cl_2 + 6\,NaOH \longrightarrow HCCl_3 + H-\overset{O}{\underset{O^-Na^+}{C}} + 5\,NaCl + 5\,H_2O$$

Chloroform   Sodium formate

At present chloroform is prepared industrially by the reduction of carbon tetrachloride or the chlorination of methane. It is a heavy, sweet-tasting, nonflammable, volatile liquid. Medicinally, it found some use at one time as an inhalation anesthetic and in cough medicines. It is now considered not only too toxic for anesthesia but also possibly carcinogenic. "Halothane", $CF_3-CHClBr$, is an effective, relatively nontoxic inhalation anesthetic. Iodoform, CHI_3, a yellow powder with a characteristic, pungent odor, has found some application in medicine as a topical antiseptic.

The fluorochloro compounds, marketed under the name of **"Freons,"** are extremely inert, nontoxic, noncorrosive, low-boiling liquids and gases. They are widely used as refrigerants and have also been used extensively as propellants in aerosol sprays of all kinds. In the latter application they have been used to dispense from pressurized cans shaving creams, hair sprays, perfumes, paints, toothpaste, insecticides, and even food (whipped cream, cheeses). Unfortunately, it has been suggested recently, based on theoretical studies, that the widespread release of the Freons into the upper atmosphere from such use may result in an eventual depletion of the atmosphere's ozone layer—thought to protect us from some of the harmful effects of the sun's ultraviolet radiation.

Polyfluoro- and polychloroethylenes are used as the structural units in several very useful plastics. One of these, "Teflon," is a polyfluoroethylene.

$$x \ \underset{(x\,=\,5{,}000\text{--}20{,}000)}{F-\overset{\overset{\displaystyle F}{|}}{C}=\overset{\overset{\displaystyle F}{|}}{C}-F} \ \xrightarrow[\text{45–50 atm.}]{\text{catalysts,}} \ \left(\!-\overset{\overset{\displaystyle F}{|}}{\underset{\underset{\displaystyle F}{|}}{C}}-\overset{\overset{\displaystyle F}{|}}{\underset{\underset{\displaystyle F}{|}}{C}}-\!\right)_{\!x}$$

"Teflon"

Teflon is a waxy plastic completely inert to nearly all reagents and ideal for gasket materials and valve packings where corrosive chemicals and high temperatures are encountered. Its great resistance to high temperatures suggested its use as nose cone material for missiles and spacecraft. Teflon is widely used as a liner in frying pans and muffin tins, and on other utensils and tools to provide nonsticking surfaces.

Vinyl chloride (CH_2=CHCl, Sec. 7.5) is the organic halogen compound manufactured in greatest amount in the United States (some 5 billion tons/year), principally for the preparation of the important polymer, polyvinyl chloride, used in the manufacture of leather substitutes, tubes and pipes, packaging films, etc. Although the polymer is safe (unless decomposed by excessive exposure to high temperatures), the monomer, vinyl chloride, poses a serious health problem because repeated inhalation can cause a rare form of cancer. Polymerization of vinyl chloride with vinylidene chloride (1,1-dichloroethene, CH_2=CCl_2) produces a copolymer with properties superior to those of the homopolymer, polyvinyl chloride, for some uses. The popular packaging film, "Saran Wrap," is an example of such a copolymer.

Many polychlorinated hydrocarbons are effective pesticides, but, unfortunately, these substances are "hard" insecticides; that is, they are very resistant to biological degradation. Because of the persistent and damaging effects they have had on wildlife, especially birds and fish, the indiscriminate use of polychlorinated insecticides in agriculture has been severely criticized in recent years. The use of DDT, which is estimated to have saved 75,000,000 lives from the ravages of malaria and other mosquito-borne diseases but which is also a hard insecticide, was banned in 1972 for agricultural use in the United States. The "hard" pesticide, "Aldrin," (and a related derivative, "Dieldrin") has been banned for all but limited use in termite control and a few other applications because of the high persistence of the substance in the environment and a presumptive link with cancer in certain rodents.

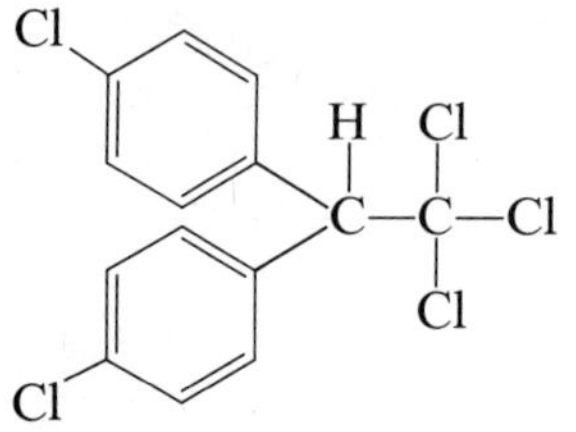

*Di*chloro*di*phenyl*tri*chloroethane

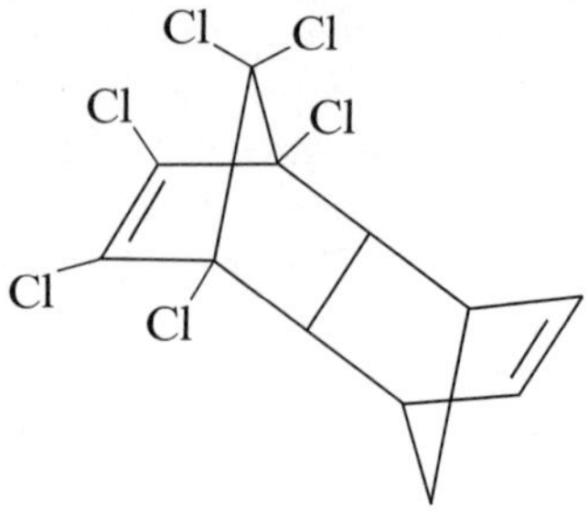

Aldrin

Ethylene bromide, 1,3-dichloropropene, and 1,2-dibromo-3-chloropropane are used as soil fumigants for the control of nematodes. Another hard insecticide, benzene hexachloride, abbreviated as BHC or 666 is prepared by the addition of chlorine to benzene. Nine different isomers are possible for this structure, but, strangely enough, only one of the seven forms thus far isolated (designated the gamma, γ, form) shows the insecticidal property. This insecticide is sometimes referred to as "Gammexane" or "Lindane."

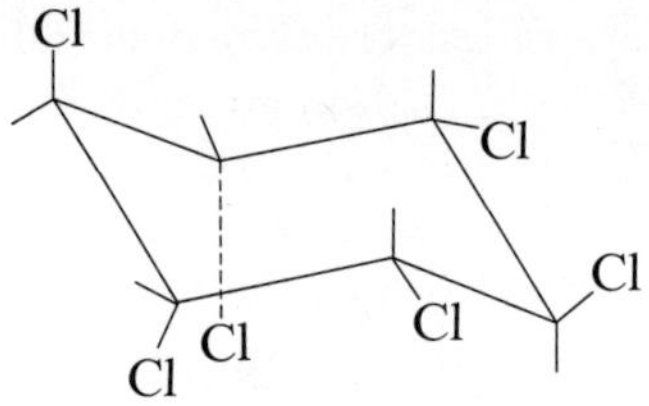

γ-Benzene hexachloride (Lindane)

For many years homemakers protected woolen blankets and clothing in storage by use of "moth balls," originally composed of naphthalene. Later naphthalene was replaced for this purpose by *p*-dichlorobenzene. Now there is some evidence that this compound, and its ortho isomer, *o*-dichlorobenzene, may be carcinogenic.

p-Dichlorobenzene

Although it may appear to you at this stage that most organic halogen compounds are toxic, the student of chemistry must maintain the proper perspective. Thorough toxicological studies have been carried out on relatively few organic compounds; and, in many instances, only qualitative rather than quantitative evaluations of the hazards associated with the use of chemicals are available. On the basis of the principle, "better safe than sorry," many chemicals are being labeled "toxic" or "cancer suspect" on the basis of preliminary evidence, pending thorough investigation. Thus, the following halides are just a sample of those on the list[1] of cancer suspect substances: methyl iodide, *sec*-butyl chloride, *tert*-butyl chloride, trichloroethylene ($ClCH=CCl_2$), chloromethyl methyl ether ($Cl-CH_2-O-CH_3$), and 2-chloroethyl ethyl ether ($Cl-CH_2-CH_2-O-CH_2CH_3$). The prudent student will treat *all* organic chemicals as potentially hazardous and will minimize exposure to them by avoiding breathing or ingesting them or spilling them on the skin and by the use of good laboratory practices: cleanliness, use of available safety equipment such as fume hoods, proper disposal techniques that avoid contamination of the environment, etc. With continued attention to the results of ongoing toxicological research on chemicals and superior laboratory technique and facilities, even relatively hazardous chemicals can be and are being used with safety in the laboratory.

7.4 *Reactions of the Alkyl Halides*

The reactions of the alkyl halides are principally of two types—**substitution** and **elimination.** The halogen atom in a substitution reaction is displaced by an electron-rich *nucleophilic* reagent (Sec. 1.11). Such nucleophilic reagents are Lewis bases and often are negative ions. Substitution reactions may take place by two different mechanisms. In one of these, the rate of reaction depends upon the concentration of *one* reacting species and is called a *substitution, nucleophilic, unimolecular,* or simply S_N1. In the other substitution mechanism the rate of reaction *usually* depends upon the concentration of *two* reacting species and is called a *substitution, nucleophilic, bimolecular,* or S_N2. The rate at which a reaction proceeds is never greater than that of the slowest step, and this phase of a reaction is called the **rate-determining step.**[2] In S_N1 reactions, the rate-determining step is the formation of a carbonium ion. Any factors such as steric effects (Sec. 8.4-B), resonance stabilization (Sec. 1.12), or inductive effects (Sec. 10.2) which help to promote the formation of the carbonium ion will increase the rate of reaction. The reaction rate may be increased further through the assistance of a polar solvent, which can solvate both cation and anion. The ionic charges are then dispersed over solvent molecules as well as the ions, thereby assisting in ion formation. An example of a reaction that proceeds largely by way of the S_N1

---

[1] *Registry of Toxic Effects of Chemical Substances,* Superintendent of Documents, U.S. Government Printing Office, Washington, D. C. 20402. Most chemistry libraries have one or more reference volumes on the hazards of chemicals, and some chemical companies attempt to label the more hazardous chemicals as such in their catalogs, *e.g.,* the *Aldrich Catalog/Handbook of Organic Chemicals and Biochemicals,* Aldrich Chemical Company.

[2] The terms unimolecular and bimolecular refer to the *molecularity* of the reaction, which is determined by the number of molecules (or other species) that are involved in the *transition state* for the rate-determining step. Often the *rate* of a *unimolecular* reaction will depend on the concentration of only *one* reactant and that of a *bimolecular* reaction will depend on the concentration of *two* reactants, but this is not invariably true.

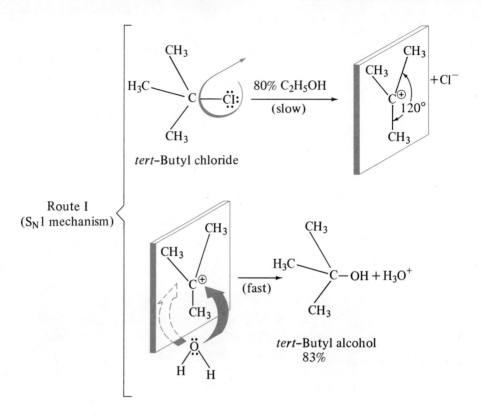

Route I
(S_N1 mechanism)

tert-Butyl chloride

80% C₂H₅OH
(slow)

+Cl⁻

120°

(fast)

tert-Butyl alcohol
83%

mechanism is the hydrolysis of *tert*-butyl chloride in an 80% aqueous alcoholic solution (above).

Recall that the intermediate carbonium ion is likely to be in a planar configuration (Sec. 1.10). This being the case, you will note that the attacking nucleophile (water) may approach the positive carbon from either front or back. The end result in the hydrolysis of *tert*-butyl chloride is the same, but a choice in the direction of combination may lead to racemization if optically active halides are hydrolyzed in this manner. Let us illustrate this point using an optically active form of α-phenylethyl chloride (see facing page).

Regardless of which optical isomer of α-phenylethyl chloride is hydrolyzed, the product is always a partially racemized mixture. Racemization is seldom complete to the extent that an equal number of (+) and (−) forms are produced because the departing anion (chloride, in this case) may still be near enough to one face of the carbonium intermediate to interfere with the approach of the nucleophile from that side. The isomeric product resulting from the backside approach thus usually predominates. However, racemization is characteristic of all S_N1 reactions.

Let us now return to the *tert*-butyl chloride hydrolysis reaction and see what has happened to the remainder of our starting material. In order to account for the other 17% we must consider another possible reaction route. If, instead of combining with the nucleophilic reagent, the carbonium ion loses a proton

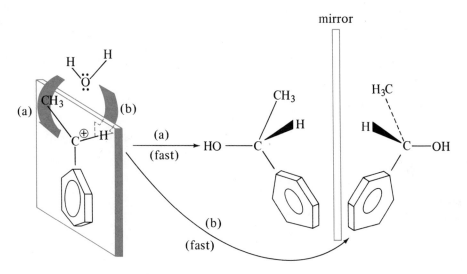

$(+)$ or $(-) - \alpha - $ Phenylethyl chloride

$(\pm) - \alpha - $ Phenylethyl alcohol

(Route II) in the second (fast) step, then the product of the reaction is an alkene, and the reaction mechanism is referred to as E1. As in the S_N1 mechanism, the number in the abbreviation refers to unimolecular. The E, in such cases, indicates elimination. Because of the competition between substitution and elimination, more than one product often results.

Route II
(El mechanism)

The S_N2 reaction mechanism may be illustrated by the alkaline hydrolysis of *sec*-butyl chloride to *sec*-butyl alcohol. In this case in the slow step of the reaction sequence both *sec*-butyl chloride and hydroxide ion are involved. The hydroxide ion displaces the chloride ion, not by a head-on approach to the face of the carbon atom bonded to the chlorine atom, but by an attack from the rear. In making this backside approach a new bond forms between the carbon atom and the hydroxyl group as the bond between the carbon atom and chlorine atom is broken. Thus, the reaction proceeds through a transition state in which the carbon atom is partially bonded to both the hydroxyl group and the chlorine atom. Transition states are often shown in brackets to emphasize that they are a transient, high energy species.

As the transition state collapses to the products, the groups bonded to the carbon atom change their relative positions as shown and the configuration becomes exactly opposite that of the starting material. Such a reversal of configuration is referred to as a **Walden inversion,** and invariably occurs in an S_N2 reaction. Thus, if we were to begin with "left-handed" *sec*-butyl chloride, we would obtain as our product, "right-handed" *sec*-butyl alcohol. Indeed, it was through such stereochemical studies that the S_N2 mechanism was established.

Route I (S_N2 mechanism)

sec-Butyl chloride   (transition state)   sec-Butyl alcohol

It should be pointed out that there is also an E2 reaction path which may enter into competition with the S_N2 mechanism through the elimination of a β-hydrogen atom. You will remember the dehydrohalogenation reactions used in the preparation of the alkenes (Sec. 3.4) as examples of such β-eliminations.

sec-Butyl chloride   2-Butene

The stereochemistry of E2 elimination is such that in the transition state the four atoms of the alkyl halide that are involved (hydrogen, two carbons, and halogen) must lie in the same plane and the hydrogen and halogen must be *anti* to each other (Sec. 2.5). This arrangement is called anticoplanar. As shown, *trans*-2-butene is the principal product, although smaller amounts of both 1-butene and *cis*-2-butene are formed. The relative amounts of the three isomers will depend on the nature of the base and of the halogen (iodide, bromide, or chloride); but, as pointed out in Sec. 3.4, the most stable alkene usually predominates, and the most stable alkene is generally that with the most substituents on the double bond. These observations suggest that the transition state for E2 elimination is stabilized by the same structural factors that stabilize alkenes.

EXERCISE 7.4

Based on the stereochemistry shown for the bimolecular E2 elimination mechanism, draw a reasonable picture of the E2 transition state.

The characteristic differences between the two-step S_N1 and one-step S_N2 reactions are further illustrated by comparison of their reaction coordinate diagrams in Figure 7.1.

Which of the four reaction mechanisms predominates depends upon the concentration of reagents, the structures of the alkyl halides involved, the solvent used, and the nature of the nucleophile. Unbranched, primary halides react with good nucleophiles principally by the S_N2 route (and *not* by S_N1). Only with strong, bulky bases do they give E2 reactions. Tertiary halides react, in the absence of strong base, by S_N1 and E1 routes, but elimination usually predominates, and tertiary halides are normally not used for substitution reactions. In the presence of strong bases, tertiary halides give good yields of elimination products by the E2 route. Secondary halides are intermediate in behavior and are less predicable; however, with good nucleophiles the S_N2

FIGURE 7.1 *Reaction Profiles for the (a) S_N1 and (b) S_N2 Nucleophilic Substitution Reactions of an Alkyl Halide*

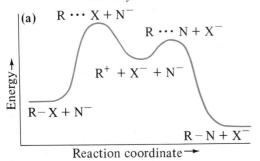
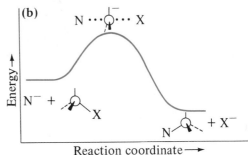

mechanism is favored, and with strong bases good yields of elimination products are formed by the E2 mechanism. The order of reactivity in elimination reactions is *tertiary > secondary > primary*.

The more important nucleophilic substitutions of alkyl halides are illustrated by the following general equations. The nucleophile in each case is underlined in color.

$$R—X + Na^+\underline{OH}^- \longrightarrow R—OH + Na^+X^-$$

Sodium   An
hydroxide   alcohol

$$R—X + Na^+\underline{OR}^- \longrightarrow R—OR + Na^+X^-$$

Sodium   An
alkoxide   ether

$$R—X + Na^+\underline{SH}^- \longrightarrow R—SH + Na^+X^-$$

Sodium   A thio
hydrosulfide   alcohol

$$R—X + Na^+\underline{CN}^- \longrightarrow R—CN + Na^+X^-$$

Sodium   An alkyl
cyanide   cyanide
             (nitrile)

$$R—X + Na^+\underline{NH_2}^- \longrightarrow R—NH_2 + Na^+X^-$$

Sodium   An
amide   amine

$$R—X + Na^+:\underline{\bar{C}{\equiv}C—H} \longrightarrow R—C{\equiv}C—H + Na^+X^-$$

Sodium
acetylide

$$R—X + :\underline{NH_3} \longrightarrow R\overset{+}{N}H_3 + X^-$$

Ammonia   An ammonium
                ion

A generalization helpful in selecting the proper halogen compound to employ in a reaction is that the iodides are more reactive than the bromides, and the bromides are more reactive than the chlorides. Unfortunately, their cost parallels this same order.

EXERCISE 7.5

What would be the stereochemical result if (*R*)-2-Chloro-3-methylbutane were heated with alcoholic KOH and the resulting product treated with bromine?

7.5 *Reactions of Aryl, Vinyl, and Allyl Halides*

The displacement reactions of the alkyl halides discussed in the previous section are not applicable to the aryl halides unless, of course, halogen is in the side chain. Halogen held to a doubly-bonded carbon such as that found in chlorobenzene is unreactive. The ring halogen is not sufficiently labile to engage in displacement reactions similar to those illustrated for the alkyl halides. Only when the *ortho* and/or the *para* positions of the aromatic ring are occupied by strong electron-attracting groups can the ring halogen be displaced by a nucleophile. Electron-attracting groups in the *ortho* or *para* positions diminish the electron density at the halogen-attached ring carbon and make it susceptible to attack. For example, 2,4-dinitrochlorobenzene may be hydrolyzed in a potassium hydroxide solution to produce 2,4-dinitrophenol. Under the same conditions chlorobenzene fails to react.

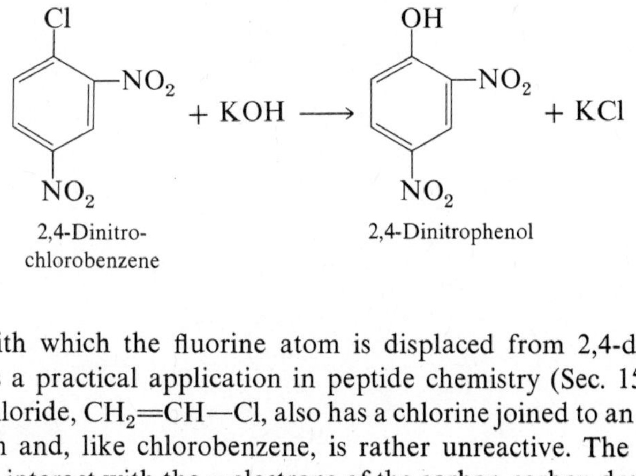

2,4-Dinitro-
chlorobenzene

2,4-Dinitrophenol

The ease with which the fluorine atom is displaced from 2,4-dinitrofluorobenzene has a practical application in peptide chemistry (Sec. 15.8).

Vinyl chloride, $CH_2\!\!=\!\!CH\!-\!Cl$, also has a chlorine joined to an unsaturated carbon atom and, like chlorobenzene, is rather unreactive. The electrons of chlorine can interact with the π electrons of the carbon-carbon double bond to shift *away from* the chlorine as shown in the resonance structure below. As a result, the chlorine atom is held somewhat closer to the carbon, the carbon-halogen bond is strengthened, and displacement of chlorine is made more difficult. Chlorine on the benzene ring represents a similar system, except that in this case the electrons of chlorine can interact with the π electrons of the benzene ring.

$$\left[\begin{array}{c} \underset{\delta-}{\overset{H}{\underset{|}{C}}} = \underset{\delta+}{\overset{H}{\underset{|}{C}}} - \ddot{\underset{..}{\text{Cl}}}\colon \end{array} \longleftrightarrow \begin{array}{c} H - \overset{H}{\underset{|}{C}} - \overset{H}{\underset{|}{C}} = \overset{..}{\text{Cl}} \\ - \qquad + \end{array} \right]$$

Resonance forms of vinyl chloride

Contributing forms to the chlorobenzene structure

In contrast to vinyl halides, **allyl halides,** $CH_2\!=\!CH\!-\!CH_2X$, are extremely reactive. In allylic halides the halogen is on the carbon next to the double bond. A different factor now comes into play. The allyl cation which remains, once the halide ion has separated, is stabilized by resonance (as shown in Sec. 1.12).

$$CH_2\!=\!CH\!-\!CH_2\!-\!Cl \longrightarrow [CH_2\!=\!CH\!-\!\overset{\oplus}{C}H_2 \longleftrightarrow \overset{\oplus}{C}H_2\!-\!CH\!=\!CH_2] + Cl^-$$

Benzyl chloride, which has a similar allylic structure, reacts in a manner similar to that of allyl chloride.

Benzyl chloride

Contributing forms to the benzyl cation hybrid

Halogens on carbon atoms other than that next to the ring (the benzylic carbon) behave like simple alkyl halides.

Halogen compounds that react by way of an ionic intermediate (tertiary, allyl, and benzyl) readily precipitate silver halide when warmed with alcoholic silver nitrate. This is a convenient test for highly reactive alkyl halides. Vinyl and aryl halides do not react under these conditions.

We now can generalize further the order of reactivity of halogen compounds in S_N1 reactions as follows:

tertiary > allylic > benzylic > secondary ≫ primary > vinyl

For S_N2 reactions the order is:

benzylic > allylic > primary > secondary ≫ tertiary > vinyl

EXERCISE 7.6

Which type of reaction mechanism, S_N1, S_N2, E1, E2, is illustrated by each of the following reactions?

(a) $CH_2=CH-CH_2Cl + AgNO_3 + C_2H_5OH \longrightarrow$

$$AgCl + CH_2=CH-CH_2OC_2H_5 + HNO_3$$

(b) $(CH_3)_3CCl + KOH \xrightarrow{\text{alcohol}} (CH_3)_2C=CH_2$

(c) $CH_3CH_2CH_2CH_2Cl + NaI \xrightarrow{\text{acetone}} CH_3CH_2CH_2CH_2I + NaCl$

(d) $CH_3-CH=CHCl + \overset{+}{Na}\overset{-}{NH_2} \longrightarrow CH_3-C\equiv C-H + NH_3 + NaCl$

7.6   *Reactions of Halogen Compounds with Metals*

The reaction of an alkyl halide with magnesium to form the important Grignard reagents already has been discussed (Sec. 2.7-A). The Grignard reagent also may be prepared from aryl halides. An example of a frequently prepared aryl Grignard reagent is phenylmagnesium bromide.

$$\text{Br} + \text{Mg} \xrightarrow[\text{ether}]{\text{anhydrous}} \text{MgBr}$$

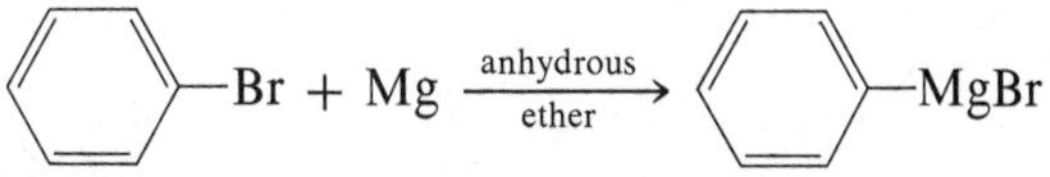

Bromobenzene   Phenylmagnesium
                                bromide

The Grignard reagent is one of the most useful of all organometallic compounds and plays a very important role in numerous synthetic procedures. The various applications of the Grignard reagent are illustrated in later sections.

The **Wurtz reaction,** which involves the reaction of primary alkyl halides with sodium, can be used for the preparation of alkanes.

$$2\,CH_3-(CH_2)_3-CH_2-Br + 2\,Na \longrightarrow CH_3-(CH_2)_8-CH_3 + 2\,NaBr$$

However, this reaction, first reported in 1855, is less attractive than the more general reaction of alkyl halides with lithiumdialkyl cuprates (Sec. 2.7-B).

Another reaction that proceeds through a copper derivative, an unstable arylcopper species, is called the **Ullmann reaction.** For example, iodobenzene, when heated in the presence of finely divided copper, produces biphenyl.

$$2 \bigcirc\!\!-\!\!I + 2\,Cu \xrightarrow{\text{heat}} \bigcirc\!\!-\!\!\bigcirc + 2\,CuI$$

Iodobenzene   Biphenyl

Two reactions especially important to the petroleum industry involve the reaction of alkyl halides and lead. Ethyl and methyl chlorides, when treated with sodium-lead alloy, react to produce the important anti-knock gasoline additives, tetraethyllead (**TEL**), and tetramethyllead (**TML**) (Sec. 18.5).

$$4\,C_2H_5\!-\!Cl + Na_4Pb \longrightarrow (C_2H_5)_4Pb + 4\,NaCl$$

Sodium   Tetraethyl
lead   lead
alloy

Summary

1 Halogen compounds may be classified as:
   (a) Aliphatic (R—X) or aromatic (Ar—X).
   (b) Primary, secondary, or tertiary halogen compounds.
   (c) Allylic, benzylic, and vinylic.
   (d) Iodo, bromo, chloro, or fluoro compounds.

2. Halogen compounds can be prepared by
   (a) direct halogenation (most useful in the preparation of aromatic halogen compounds).
   (b) replacement of the hydroxyl group (—OH) of alcohols.
   (c) addition of HX to alkenes (Markovnikov's Rule is followed).
   (d) exchange for other covalently bound halogens (most useful in the preparation of iodo and fluoro compounds).
   (e) the haloform reaction which yields trihalomethanes.

3. Reactions of the halogen compounds include
   (a) displacement of the halogen atom (as halide) by another negative group—i.e., hydroxy (OH⁻), alkoxy (OR⁻), amino (NH₂⁻), mercapto (SH⁻), cyano (CN⁻), acetylide (H—C≡C⁻).
   (b) elimination (dehydrohalogenation) to form alkenes.
   (c) reactions with metals: Wurtz, Friedel-Crafts, Ullmann, preparation of the Grignard reagent, preparation of TEL and TML.

4. Important polychloro compounds include chloroform, carbon tetrachloride, and polyvinyl chloride.

5. The polyfluoromethanes, ethanes, and cyclobutanes known as "Freons" are useful aerosol spray propellants and refrigerants. "Teflon" (polyfluoroethylene) is a very useful inert, heat-resistant plastic.

6. The polyhalogenated compounds, such as DDT, BHC, and *p*-dichlorobenzene, are powerful pesticides but are not easily removed from the biosphere.

New Terms

E1 reaction

E2 reaction

nucleophilic substitution

S_N1 reaction

S_N2 reaction

Ullmann reaction

Walden inversion

Wurtz reaction

Supplementary Exercises

EXERCISE 7.7   Name each of the following compounds.

(a) $CH_2{=}CH{-}CH_2Br$ (b) CH_3MgI (c) $CH_3{-}\overset{\overset{\displaystyle H}{|}}{\underset{\underset{\displaystyle Br}{|}}{C}}{-}CH_3$ (d) HCl_3

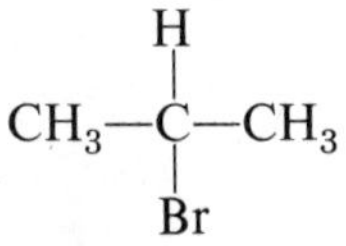

(e)

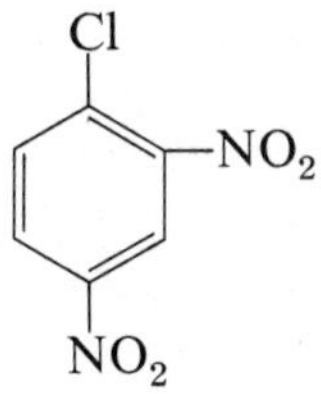

(f)

(g) $Cl{-}$$-CH_2Cl$

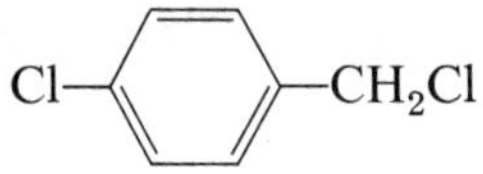

(h)

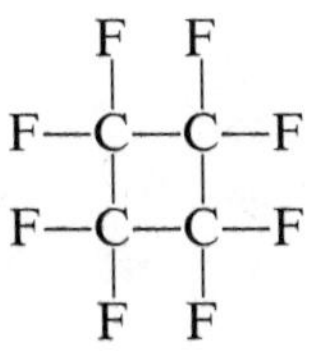

(i)

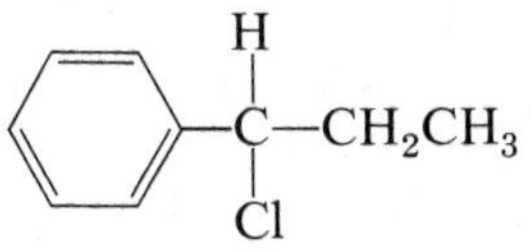

EXERCISE 7.8   Write structures for and assign acceptable names to all isomers with the molecular formula (a) C_4H_9Br; (b) $C_4H_8Cl_2$. Are any of the structures you have drawn those of optically active compounds?

EXERCISE 7.9   Indicate the principal organic products (if any) given by the following reactions:

(a) *tert*-C_4H_9OH + HCl $\longrightarrow$

(b) $CH_3CH_2CH{=}CH_2$ + HBr $\xrightarrow{\text{peroxide}}$

(c) Bromobenzene + Mg $\xrightarrow{\text{anhydrous ether}}$

(d) Product of (a) + alcoholic KOH $\longrightarrow$

(e) Product of (b) + aqueous KOH $\longrightarrow$

(f) Product of (e) + I_2 + NaOH $\longrightarrow$

(g) Chlorobenzene + alcoholic KOH $\longrightarrow$

(h) Benzoic acid + Br_2 $\xrightarrow{\text{Fe}}$

(i) *cis*-2-Butene + Br_2 $\longrightarrow$

(j) 2,4-Dinitrochlorobenzene + NH_3 $\longrightarrow$

(k) *n*-C_4H_9Cl + NaI $\xrightarrow{\text{acetone}}$

(l) Ethylbenzene + Cl_2 $\xrightarrow{\text{sunlight}}$

(m) *n*-Butyl chloride + benzene $\xrightarrow{\text{AlCl}_3}$

(n) Ethyl bromide + sodium acetylide $\longrightarrow$

(o) Ethyl bromide + sodium methoxide $\longrightarrow$

(p) (*R*)-2-Bromobutane + NaOH $\xrightarrow{S_N2}$

(q) Isopropylbenzene + Cl_2 $\xrightarrow{\text{uv light}}$

(r) 2-Methyl-2-butene + HBr $\longrightarrow$

(s) *iso*-C_3H_7 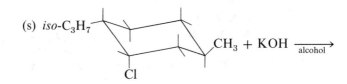 CH_3 + KOH $\xrightarrow{\text{alcohol}}$

EXERCISE 7.10   Show how the following transformations can be effected.

(a) isopropyl alcohol → isopropyl bromide → propane

(b) benzene → ethylbenzene → styrene

(c) isopropyl bromide → *n*-propyl bromide

(d) 1,2-dibromopropane → 2,2-dibromopropane

(e) toluene → *p*-bromotoluene → *p*-bromobenzoic acid

EXERCISE 7.11   Give the structure of the major product expected, if any, when *n*-propyl chloride is treated with each of the following reagents.

(a) NaOH, H_2O (b) Mg, anhydrous ether

(c) cold, concentrated H_2SO_4

(d) benzene, anhydrous $AlCl_3$

(e) KI, acetone

(f) $CH_3-C\equiv C:^- Na^+$

(g) KOH, C_2H_5OH

(h) NH_3

EXERCISE 7.12 Complete the table below by making a comparison between S_N1 and S_N2 reaction mechanisms with respect to each of the characteristics in the left column.

	S_N1	S_N2
(a) Molecularity	____	____
(b) Stereochemistry	____	____
(c) Relative rates of reaction of primary, secondary, tertiary alkyl halides	____	____
(d) Effect of doubling concentration of nucleophile	____	____
(e) Effect of increasing polarity of solvent	____	____

EXERCISE 7.13 What simple chemical test, i.e., test tube reactions, will serve to distinguish between:

(a) cyclohexene and chlorobenzene

(b) α-phenylethyl chloride and β-phenylethyl chloride

(c) chlorobenzene and 2,4-dinitrochlorobenzene

(d) *p*-bromotoluene and benzyl bromide

EXERCISE 7.14 A halogen compound contained 58% bromine. When the compound was heated with alcoholic potassium hydroxide a combustible gas evolved. The gas, when hydrogenated in the presence of a catalyst, absorbed one mole of H_2. On combustion, the gas united with oxygen in a ratio of 1:6 by volume (STP) to give carbon dioxide and water as the only products. When this same gas was subjected to chemical oxidation with hot, concentrated $KMnO_4$, only acetic acid, CH_3COOH, was obtained as the oxidation product. Give the formula for the original compound.

EXERCISE 7.15 An analysis of an organic compound established its formula as C_8H_9Cl. What would be its structure and name if it had shown the following behavior:

(a) It was optically active, precipitated AgCl when warmed with an alcoholic silver nitrate solution, and on oxidation yielded benzoic acid.

(b) It was optically inactive, precipitated AgCl when warmed with alcoholic silver nitrate, and on oxidation yielded phthalic acid, $C_6H_4(COOH)_2$.

(c) It was optically inactive and on oxidation yielded *p*-chlorobenzoic acid.

EXERCISE 7.16   When *cis*-3-methylcyclopentyl chloride is heated with potassium hydroxide only *trans*-3-methylcyclopentanol is formed.

>   (a) Draw stereochemical structures for both starting material and product.
>   (b) Did this displacement proceed by the S_N1 or by the S_N2 reaction pathway? Explain.

EXERCISE 7.17   When *tert*-butyl bromide is hydrolyzed in an ethanolic solution, three products result. What are they and which of the three would you expect to predominate? Why?

EXERCISE 7.18   When ethyl bromide is heated with a solution of potassium hydroxide in ethyl alcohol, the principal product is ethyl ether.

$$CH_3-CH_2-Br + KOH \xrightarrow{\text{CH}_3-\text{CH}_2-\text{OH}} CH_3-CH_2-O-CH_2-CH_3$$

Explain this result. What does this result tell you about the nucleophilic species actually present in alcoholic solutions of potassium hydroxide?

EXERCISE 7.19   The E2 dehydrohalogenation of 2-bromopentane with potassium hydroxide in ethyl alcohol solution gives a mixture of 31% 1-pentene, 17% *cis*-2-pentene, and 51% *trans*-2-pentene. If the stability of the alkenes were the *only* factor governing the direction of elimination, the composition of the mixture should have been 3% 1-pentene, 23% *cis*-2-butene, and 74% *trans*-2-pentene. Draw three Newman projections of 2-bromopentane showing the conformations required to give the products, and suggest a reason for the high yield of 1-pentene.

EXERCISE 7.20   The base-catalyzed E2 elimination of *one* molecule of hydrogen bromide from (*R,R*)-1,2-dibromo-1,2-diphenylethane gives only 1-bromo-*trans*-1,2-diphenylethene, while (*R,S*)-1,2-dibromo-1,2-diphenylethane gives 1-bromo-*cis*-1,2-diphenylethene. Explain these results.

EXERCISE 7.21   One mechanism for the substitution reactions of aryl halides is an *addition-elimination* mechanism (addition of a nucleophile, elimination of halide, in that order). This bimolecular mechanism is thought to involve an intermediate sigma complex, one of whose resonance form is shown. Write structures for the important resonance forms of the sigma complex that would be formed from 2,4-dinitro-1-chlorobenzene and hydroxide ion. Using these structures, explain the greater reactivity of 2,4-dinitro-1-chlorobenzene compared to chlorobenzene.

EXERCISE 7.22   Halides of the type shown below show little tendency to react by either S_N1, S_N2, E1, or E2 reactions. Suggest an explanation for the failure of S_N1 and S_N2 substitutions. (*Hint:* Looking at a model may help.)

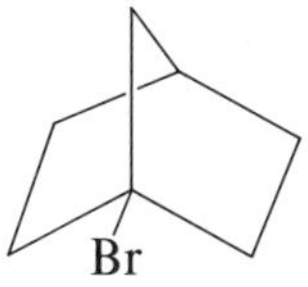

$$+ \text{ good nucleophile} \xrightarrow[S_N2]{S_N1 \text{ or}} \text{ no reaction}$$

EXERCISE 7.23   An organic compound, C_3H_6ClI, gave an nmr spectrum with a triplet at δ 3.6, another triplet at δ 3.3, and a quintet at δ 2.2. All coupling constants were approximately 7 Hz, and the integral indicated that each multiplet represented two hydrogen atoms. Suggest a structure for the compound.

EXERCISE 7.24   An organic compound, $C_5H_8Br_4$, gave an nmr spectrum in which there was only one peak, a singlet, at δ 3.3. Suggest a structure for the compound.

8

Alcohols, Phenols, and Ethers

Introduction

Alcohols, phenols, and ethers represent three classes of oxygen-containing compounds in which the oxygen atom is singly bonded (—O—) to two other atoms. The oxygen atom bridges carbon to hydrogen in both the alcohols and phenols. In ethers, the oxygen is the bridge between two carbon atoms. All three classes of compounds provide us with a great number of useful products. These include germicides, antifreeze agents, pharmaceuticals, explosives, solvents, anesthetics, and plastics.

8.1 Structure of the Alcohols and Phenols

Alcohols and phenols may be considered as hydroxyl substituted hydrocarbons of the general formulas R—OH, and Ar—OH respectively. The **hydroxyl group** (—OH) is the functional group which characterizes both alcohols and phenols. Compounds that have hydroxyl groups joined to carbon atoms of alkyl groups are alcohols. Compounds that have hydroxyl groups joined to carbon atoms like those found in the aromatic ring compounds are phenols. Both types may be illustrated by the following general formulas and specific examples.

$$R—\overset{..}{O}{\overset{\textstyle .}{}}_{\diagdown H}\qquad CH_3—OH \qquad CH_3CH_2—OH$$

An alcohol Methyl alcohol Ethyl alcohol

$$Ar—\overset{..}{O}{\overset{\textstyle .}{}}_{\diagdown H}$$

A phenol   Phenol β-Naphthol

8.2 Classification and Nomenclature of Alcohols and Ethers

The alcohols, like the alkyl halides, may be classified as *primary, secondary,* or *tertiary* according to the number of hydrocarbon groups attached to the carbon atom bearing the hydroxyl groups. However, the nomenclature of the alcohols is somewhat more extensive than that encountered in other families of substances. **Common,** or **trivial, names** usually are employed for the simpler members having one to four carbon atoms. Such names are formed simply by naming the alkyl group bonded to the hydroxyl function, followed by the word *alcohol.* The following examples are illustrations of common names.

$$CH_3—OH \qquad CH_3—\overset{\displaystyle CH_3}{\underset{\displaystyle H}{\overset{|}{\underset{|}{C}}}}—OH \qquad CH_3—\overset{\displaystyle CH_3}{\underset{\displaystyle CH_3}{\overset{|}{\underset{|}{C}}}}—OH$$

Methyl alcohol Isopropyl alcohol *tert*-Butyl alcohol
(A primary alcohol) (A secondary alcohol) (A tertiary alcohol)

The IUPAC system of nomenclature is better adapted to naming the more complex members of the alcohol family, for which prefixes such as *secondary* and *tertiary* have little significance. Alcohols are named according to IUPAC rules by selecting and naming the longest carbon chain *including the hydroxyl group*. The terminal **e** of the parent hydrocarbon (alkane) is replaced by **ol.** As before, the chain is numbered to confer upon the functional group the smallest number (even if carbon-carbon double bonds or triple bonds are present). If more than one hydroxyl group appears on the chain, prefixes, such as **di, tri,** etc., are used. Alkyl side chains and other groups are named and their positions indicated. It would be profitable to consider a few examples.

C_2H_5—OH

Ethanol
(Ethyl alcohol)

$$\underset{(3)}{CH_3}-\underset{(2)}{\underset{|}{\underset{OH}{CH}}}-\underset{(1)}{CH_3}$$

2-Propanol
(Isopropyl alcohol)

$$\underset{(3)}{CH_3}-\underset{(2)}{\overset{\overset{(1)}{CH_3}}{\underset{\underset{CH_3}{|}}{\overset{|}{C}}}}-OH$$

2-Methyl-2-propanol
(*tert*-Butyl alcohol)

$$CH_3-CH_2-\overset{\overset{CH_3}{|}}{CH}-CH_2-OH$$

2-Methyl-1-butanol

$$CH_2{=}CH-CH_2-OH$$

2-Propen-1-ol
(Allyl alcohol)

$$\underset{OH}{\underset{|}{CH_3}}-\underset{OH}{\underset{|}{CH}}-CH_2$$

1,2-Propanediol
(Propylene glycol)

$$\underset{OH}{\underset{|}{CH_2}}-\underset{OH}{\underset{|}{CH}}-\underset{OH}{\underset{|}{CH_2}}$$

1,2,3-Propanetriol
(Glycerol)

The suffix **ol** is generic for compounds that contain hydroxyl groups. Although names such as cres**ol,** glycer**ol,** and cholester**ol** contain no clues to their structures, such names do indicate that each contains one or more hydroxyl groups.

A third system of nomenclature is called a **derived system.** According to this system, alcohols are considered to be derivatives of the simplest one—that is, hydroxy methane, or **carbinol.** Groups attached to the carbon atom bearing the hydroxyl are named as one word with the suffix carbinol. This system of nomenclature is seldom used except to name the phenyl substituted methanols.

Carbinol
(Methyl alcohol)

Diphenylcarbinol

Triphenylcarbinol

Draw structures for eight alcohols with a molecular formula $C_5H_{12}O$ and name them according to IUPAC rules. Classify each as primary, secondary, and tertiary.

Phenols are commonly named as derivatives of the parent substance, and simplest member of the family, **phenol.** Other substituents on the phenol ring are located by number or by *ortho, meta,* or *para* designations. For example,

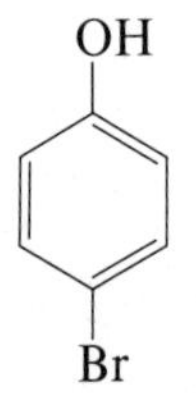

p-Bromophenol 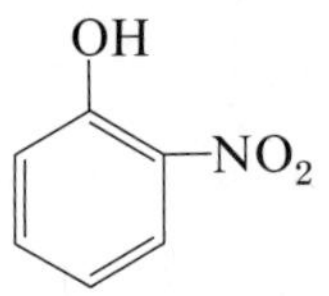   o-Nitrophenol 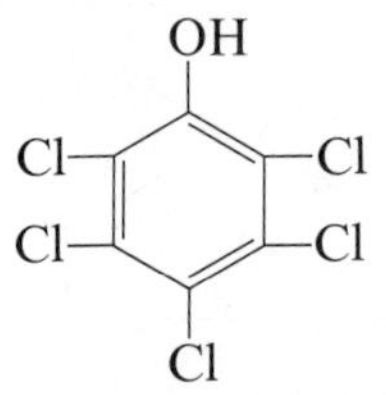   Pentachlorophenol

Sometimes phenols are named as hydroxy compounds or, as in the case of polyhydroxybenzenes, as *di-* or *triols*.

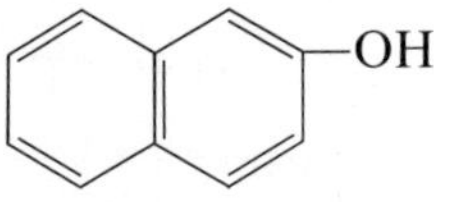

2-Hydroxynaphthalene
(β-Naphthol)

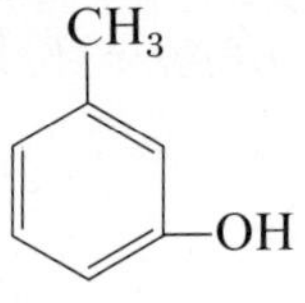

3-Hydroxytoluene
(m-Cresol)

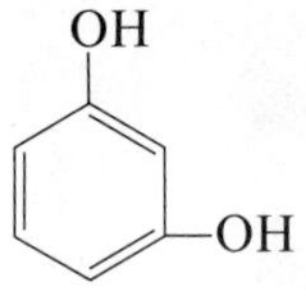

1,3-Benzenediol
(Resorcinol)

8.3 *Structure and Properties of the Alcohols and Phenols. Hydrogen Bonding*

Molecules in which hydrogen is bonded to a highly electronegative element such as fluorine, oxygen, or nitrogen may exhibit a dipole, the positive end of which is a relatively exposed hydrogen nucleus. Because of such exposure a strong attractive force develops between the hydrogen atom of one molecule and the electronegative element of a neighboring molecule. Water in the solid and liquid states exhibits such a dipole, with the hydrogen atoms of one molecule being attracted to the oxygen atoms of another. This type of intermolecular attraction between hydrogen and a donor atom is called **hydrogen bonding,** or **H-bonding,** and is represented between molecules by dotted lines. Liquids in which H-bonding occurs between molecules are called *associated* liquids and have properties unlike those of "normal" liquids.

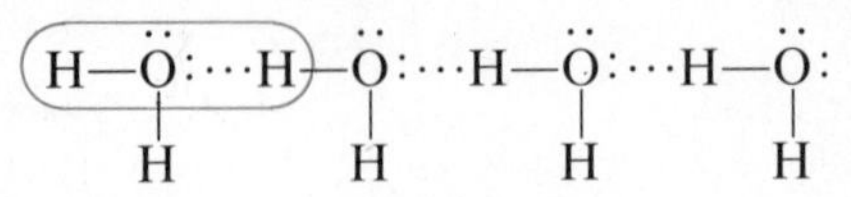

TABLE 8.1 *Boiling Points of Some Substituted Ethanes*

Compound	Molecular Wt.	B.P., °C
Ethyl alcohol	46	78.4
Ethyl chloride	64.5	12.2
Ethyl bromide	109	38.0
Ethyl iodide	156	72.4

The alcohol molecule, like water, also is highly polar and while the alcohol molecule has but one hydrogen attached to oxygen, it also is associated through hydrogen bond formation. Molecules capable of H-bonding are attracted to each other by the electrostatic interaction of their dipoles and the affinity of their hydrogens for a second electronegative atom.

$$\cdots\underset{R}{O}\text{—H}\cdots\underset{R}{O}\text{—H}\cdots\underset{R}{O}\text{—H}\cdots\underset{R}{O}\text{—H}\cdots\underset{R}{O}\text{—H}\cdots\underset{R}{O}\text{—H}\cdots$$

To pass from the liquid into the vapor form, a molecule of alcohol, like one of water, must receive sufficient energy to overcome the attraction of its neighbors. Hydrogen bonding thus explains why the boiling point of water is relatively high for such a simple molecule and also explains why the alcohols have considerably higher boiling points than their nonhydroxylic isomers. For example, the boiling point of ethyl alcohol, C_2H_5OH, is 78°C and that of its isomer, dimethyl ether, CH_3—O—CH_3 (incapable of H-bonding), is −24.9°C. Comparisons made in Table 8.1 illustrate how H-bonding, more than an increase in molecular weight, affects the boiling points of liquids.

The lower members of the alcohol family are like water in yet another respect. Inasmuch as the hydroxyl group comprises much of the molecular structure, the lower members of the alcohol family are miscible with water in all proportions. The alcohol molecule becomes more and more oil-like in character as the hydrocarbon segment becomes larger and its solubility in water diminishes markedly. A relationship between solubilities and the size and form of the alkyl groups is given in Table 8.2.

Phenol and other polyhydroxy benzenes are colorless, low melting solids only slightly soluble in water. Perhaps the most distinctive property of phenols, and one which sets them apart from the alcohols, is the acidic character they possess. Phenols (but not alcohols) dissolve readily in hydroxide bases to form salts called phenoxides.

Sodium phenoxide

TABLE 8.2 *Solubilities of C_1—C_6 Alcohols*

Name	Formula	Solubility (g/100 g H_2O)
Methanol	CH_3OH	completely miscible
Ethanol	C_2H_5OH	completely miscible
2-Propanol	$(CH_3)_2CHOH$	completely miscible
1-Propanol	$CH_3CH_2CH_2OH$	completely miscible
1-Butanol	$CH_3CH_2CH_2CH_2OH$	9 g
2-Methyl-1-propanol	$(CH_3)_2CHCH_2OH$	10 g
2-Butanol	$CH_3CH_2CH(OH)CH_3$	12.5 g
2-Methyl-2-propanol	$(CH_3)_3COH$	completely miscible
1-Pentanol	$CH_3(CH_2)_3CH_2OH$	2.7 g
2-Pentanol	$CH_3CH_2CH_2CH(OH)CH_3$	5.3 g
1-Hexanol	$CH_3(CH_2)_4CH_2OH$	0.6 g

The ease with which phenols form soluble salts may be attributable to the greater resonance stabilization of the phenoxide ion compared to that of phenol. Once the acidic hydrogen of phenol is removed as a proton, an unshared pair of electrons on the oxygen atom may become delocalized and enter into the resonance structure of the ring, as illustrated by the following structures.

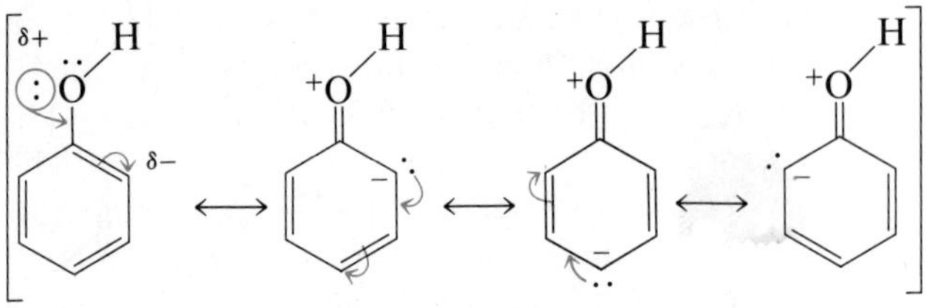

Resonance forms of phenol

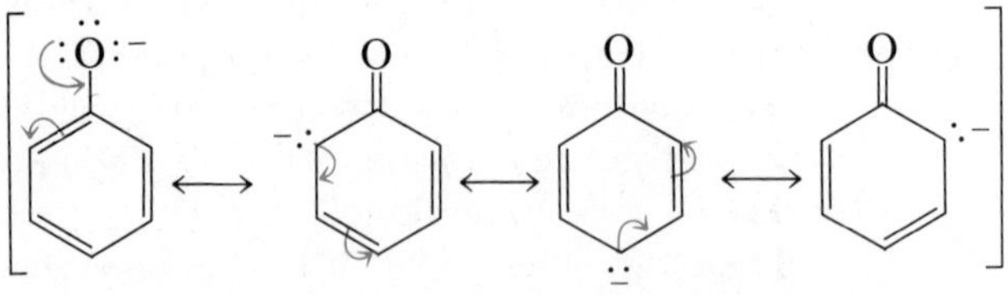

Resonance forms of phenoxide ion

Resonance confers greater stability on phenoxide ion than on phenol because there is no separation of positive and negative charges in the phenoxide resonance forms (Sec. 1.12). The separation of charges requires energy (to keep them apart) and is a destabilizing effect. There is no resonance stabilization of alkoxide ions, nor of alcohols; therefore, the alcohols are somewhat less acidic than water, principally because the alkyl groups of alkoxides prevent them from being solvated as well as hydroxide ion.

Phenols are not sufficiently acidic to react with sodium carbonate but will react with the strong bases sodium and potassium hydroxide to yield the phenoxides of these metals. This behavior allows other stronger organic acids which react with sodium carbonate to be separated as water-soluble salts from a mixture which might also include phenol. If strong electron-withdrawing groups are substituted in the *ortho* or *para* position of phenol, the substituted phenol is strongly acidic.

EXERCISE 8.2

Which of the following compounds would be most acidic? Least acidic? Why?
(a) Ethyl alcohol (b) *tert*-Butyl alcohol (c) Phenol (d) *o*-Nitrophenol
(e) *o*-Cresol (f) *p*-Chlorophenol

8.4 *Preparation of the Alcohols*

Before we consider general methods for the preparation of the alcohols, let us discuss briefly the very important first two alcohols of the aliphatic series.

Methyl alcohol, commonly called **wood alcohol,** was made prior to 1923 by the destructive distillation of hardwoods. Since then, synthetic methyl alcohol has been made much more economically by combining carbon monoxide and hydrogen under high pressure and in the presence of catalysts.

$$CO + 2 H_2 \xrightarrow[350°C,\ 200\ atm.]{ZnO\text{-}Cr_2O_3,} CH_3OH$$

Methanol

Methyl alcohol is a very poisonous substance which, if taken internally, causes visual impairment, complete blindness, or death. Death from the ingestion of as little as 30 ml of methyl alcohol has been reported. At one time methyl alcohol was extensively used as radiator antifreeze but has been replaced largely by ethylene glycol (Sec. 8.5). In order to protect people who recognize no distinction between the alcohols, methyl alcohol antifreeze usually was labelled by its manufacturers by its nonsuggestive IUPAC name, **methanol.**

Ethyl alcohol sometimes is called **grain alcohol** because starch from grain, when hydrolyzed to sugars and fermented by enzymes, produces ethyl alcohol and carbon dioxide. Starch from any source is a suitable starting material. The fermentation of sugar by yeast is a reaction practiced since antiquity and is the basis for the production of alcoholic spirits, and for the leavening action required in the baking process.

$$2\ (C_6H_{10}O_5)_n + n\ H_2O \xrightarrow[\text{in malt}]{\text{diastase}} n\ C_{12}H_{22}O_{11}$$

Starch   Maltose

$$C_{12}H_{22}O_{11} + H_2O \xrightarrow{\text{maltase}} 2\ C_6H_{12}O_6$$
$$\text{Glucose}$$

$$C_6H_{12}O_6 \xrightarrow{\text{zymase}} 2\ C_2H_5OH + 2\ CO_2$$
$$\text{Ethanol}$$

Inasmuch as the second step in the preceding sequence produces glucose, ethanol also may be prepared from this simple sugar directly. Grape juice, a rich source of glucose, will ferment to produce a wine with a maximum alcoholic content of approximately 12% by volume. The alcoholic content of liquors usually is designated as **proof spirit,** 100 proof indicating an alcoholic content of 50% by volume. The term "proof spirit" supposedly had its origin in an early and rather crude analytical procedure for determining the alcoholic content of whiskey. Whiskey of high alcoholic content, when poured onto a small mound of gunpowder and ignited, would burn with a flame sufficiently hot to ignite the powder. This was "proof" of spirit content. If the gunpowder failed to ignite, the presence of too much water was indicated and the powder became too wet to burn.

Ethyl alcohol used in the laboratory for solvent purposes seldom is pure alcohol. It usually is a mixture of 95% alcohol and 5% water. Ninety-five per cent represents the maximum purity obtainable when alcohol is distilled because this is the constant-boiling composition. A constant-boiling mixture of liquids, called an **azeotrope,** cannot be separated by fractional distillation. In order to obtain **absolute,** or 100% pure ethyl alcohol, the water must be removed by methods other than fractionation. One method is to distill a ternary mixture composed of alcohol, water, and benzene. These three liquids, in a composition of 18.5%, 7.4%, and 74.1%, respectively, also form an azeotrope with a constant-boiling temperature of 64.85°. Therefore, if sufficient benzene is added to 95% alcohol and the mixture distilled, the water is removed in the distillate along with benzene and some alcohol, but pure alcohol is left in the still pot.

In the laboratory the last traces of water may be removed by chemical combination. Calcium oxide, CaO, for example, reacts with water to produce calcium hydroxide, $Ca(OH)_2$, but does not react with the alcohol. Alcohol "dried" in this manner can be recovered by a simple distillation.

Alcoholic beverages always have been a prime source of tax revenue for most governments. Ethyl alcohol, when used as a solvent or as a reagent, is tax free. In order to prevent the use of tax-free alcohol for purposes other than scientific or industrial, the government requires that it be **denatured.** Denatured alcohol is alcohol rendered unfit for beverages by the addition of substances repugnant and difficult to remove. The reason for the government's vigilance over the manufacture and use of alcohol may be illustrated clearly by the following simple economics. The price of 95% (190-proof) alcohol in bulk lots is less than a dollar per gallon when purchased as a tax-free reagent. As 190-proof beverage spirit, the alcohol carries a $19.95 per gallon federal excise tax plus applicable state and local taxes.

A. Preparation of Alcohols from Alkenes. Large quantities of ethyl alcohol still are made by the fermentation of sugars found in blackstrap molasses, but its synthesis from ethylene has become a very important industrial process and provides most of the alcohol used in the chemical industry. Alkenes can be converted to alcohols by hydration which, in this case, is the reverse reaction we employed to produce an alkene from an alcohol (Sec. 3.8-C). In practice, the elements of water are added across the double bond indirectly. Sulfuric acid is added first and the alkyl sulfuric acid thus formed, when diluted with water and heated, undergoes hydrolysis to produce the alcohol according to the following equations which illustrate the preparation of ethanol from ethylene.

$$H-\overset{\overset{\displaystyle H}{|}}{C}=\overset{\overset{\displaystyle H}{|}}{C}-H + H^+\ \overset{(-)}{O}SO_3H \longrightarrow CH_3-\overset{\overset{\displaystyle H}{|}}{\underset{\underset{\displaystyle H}{|}}{C}}-OSO_3H$$

Ethyl sulfuric acid

$$CH_3-\overset{\overset{\displaystyle H}{|}}{\underset{\underset{\displaystyle H}{|}}{C}}-OSO_3H + H_2O \longrightarrow CH_3-CH_2-OH + H_2SO_4$$

Ethanol

Other members of the alcohol family can be prepared from alkenes by the same process.

$$R-\overset{\overset{\displaystyle H}{|}}{C}=CH_2 + \overset{+}{H}\ HSO_4^- \longrightarrow R-\overset{\overset{\displaystyle H}{|}}{\underset{\underset{\displaystyle OSO_3H}{|}}{C}}-CH_3 \overset{H_2O}{\longrightarrow} R-\overset{\overset{\displaystyle H}{|}}{\underset{\underset{\displaystyle OH}{|}}{C}}-CH_3$$

A secondary
alcohol

$$R-\overset{\overset{\displaystyle R}{|}}{C}=CH_2 \overset{H_2SO_4}{\longrightarrow} R-\overset{\overset{\displaystyle R}{|}}{\underset{\underset{\displaystyle OSO_3H}{|}}{C}}-CH_3 \overset{H_2O}{\longrightarrow} R-\overset{\overset{\displaystyle R}{|}}{\underset{\underset{\displaystyle OH}{|}}{C}}-CH_3$$

A tertiary
alcohol

The only primary alcohol that can be prepared by the hydration of an alkene in the above manner is ethanol from ethylene. Addition of ionic reagents to the double bond, you must remember, is in accordance with Markovnikov's rule (Sec. 3.8). In order to produce alcohols *via* an indirect hydration of alkenes in what appears to be an "anti" Markovnikov manner, a method called

hydroboration is practiced. The reagent used in this reaction is diborane, B_2H_6, which is formed from sodium borohydride and boron trifluoride according to the following equation.

$$3\,NaBH_4 + 4\,BF_3 \longrightarrow 2\,B_2H_6 + 3\,NaBF_4$$

Diborane is an unusual molecule and at first glance, as written above, may appear analogous to ethane. Actually diborane behaves as if it were a dimer of hypothetical borane, BH_3, and is often shown in reactions as $(BH_3)_2$. In its reaction with an alkene the electron-pair deficient boron of BH_3 becomes attached to the less substituted carbon of the double bond, apparently largely because of steric factors. At the same time, one of the hydrogen atoms of borane is donated to the other doubly-bonded carbon. Borane has three hydrogens to confer in this manner, and thus the end product becomes a trialkyl substituted boron. The latter is not isolated but is subsequently treated with an alkaline solution of hydrogen peroxide to yield the alcohol and boric acid. The net result is a *cis* addition of the elements of water across the carbon-carbon double bond to produce an alcohol in which the addition is "anti" Markovnikov. The reaction is illustrated using propylene.

$$CH_3-CH{=}CH_2 \longrightarrow CH_3CH_2CH_2-B\!\!<^{H}_{H}$$

n-Propyl boron hydride

$$CH_3CH_2CH_2-B\!\!<^{H}_{H} + 2\,CH_3-CH{=}CH_2 \longrightarrow (CH_3CH_2CH_2)_3B$$

Tri-*n*-propylborane

$$(CH_3CH_2CH_2)_3B + 3\,H_2O_2 \longrightarrow 3\,CH_3CH_2CH_2OH + B(OH)_3$$

n-Propyl alcohol

EXERCISE 8.3

Which alkenes must be your starting materials in the hydroboration reaction to produce (a) 3-methyl-2-butanol, (b) 3,3-dimethyl-1-butanol?

B. Hydrolysis of Alkyl Halides. Alcohols may be prepared by the hydrolysis of alkyl halides, but the reaction is of limited usefulness because alkene formation by *elimination* of halogen acid competes with the *substitution* reaction. Alkyl halides are hydrolyzed in a neutral, rather than in an alkaline, alcoholic solution in order to minimize the formation of alkenes, but some elimination may result even when a neutral hydrolysis is carried out. This is especially true when highly branched tertiary alkyl halides are hydrolyzed. Bulky groups, when attached to the same carbon atom that bears the halogen, not only inhibit a nucleophilic attack from the rear (S_N2), but appear to have a crowding effect upon each other. Strain, due to such crowding, may be relieved by the departure of the halide ion. The intermediate, tertiary carbonium ion now either can be attacked by the solvent at the face of the carbon atom, or at a β hydrogen. You will recognize the first choice as the S_N1, the second as the E1 reaction (Sec. 7.4). The elimination route appears to be favored, especially if an alkaline hydrolysis with a strong base is attempted. Where alkene formation is not possible, the hydrolysis of an alkyl halide results in the formation of an alcohol. A useful application of this method is the synthesis of benzyl alcohol from benzyl chloride.

| Toluene | Benzyl chloride | Benzyl alcohol |

C. Addition of Grignard Reagents to Aldehydes and Ketones.

The addition of Grignard reagents of the form RMgX or ArMgX to aldehydes or to ketones (Chap. 9) provides one of the best routes leading to the preparation of alcohols. An alkyl or aryl magnesium halide adds to the carbonyl group of an aldehyde or a ketone to produce a mixed salt of divalent magnesium. The electropositive magnesium seeks the carbonyl oxygen; the organic group joins the carbonyl carbon. The mode of this addition is illustrated by

The addition product is treated with dilute hydrochloric acid, or a saturated solution of ammonium chloride, and is smoothly converted into an alcohol and a magnesium dihalide.

The reaction is a versatile one by which alcohols of any class may be produced. Of the aldehydes, formaldehyde alone yields a primary alcohol when treated with a Grignard reagent. Other aldehydes give secondary alcohols, and ketones yield tertiary alcohols. General equations for the preparation of alcohols of each class from Grignard reagents are as follows:

$$RMgX + \underset{H}{\overset{H}{>}}C{=}O \xrightarrow{\text{followed by HX hydrolysis}} R{-}\underset{H}{\overset{H}{\underset{|}{\overset{|}{C}}}}{-}OH + MgX_2$$

Formaldehyde   A primary
alcohol

$$RMgX + \underset{R}{\overset{H}{>}}C{=}O \xrightarrow{\text{followed by HX hydrolysis}} R{-}\underset{R}{\overset{H}{\underset{|}{\overset{|}{C}}}}{-}OH + MgX_2$$

A secondary
alcohol

$$RMgX + \underset{R}{\overset{R}{>}}C{=}O \xrightarrow{\text{followed by HX hydrolysis}} R{-}\underset{R}{\overset{R}{\underset{|}{\overset{|}{C}}}}{-}OH + MgX_2$$

A tertiary
alcohol

An inspection of the structure of the desired alcohol will always indicate which Grignard reagent and which carbonyl compound may be combined to produce it. For example, the preparation of 3-methyl-2-butanol, a secondary alcohol, suggests an aldehyde for the carbonyl compound. Which aldehyde shall we use, and from what shall we prepare our Grignard reagent? By inspection, it may be seen that the portion of the structure indicated by the broken line can have its origin in acetaldehyde; that indicated by the solid line, in isobutyraldehyde.

$$CH_3{-}\underset{H}{\overset{CH_3}{\underset{|}{\overset{|}{C}}}}{-}\underset{OH}{\overset{H}{\underset{|}{\overset{|}{C}}}}{-}CH_3$$

(This part of the alcohol must have its
origin in the carbonyl compound.)

If acetaldehyde is our choice of a carbonyl group, it follows that the Grignard reagent used with it must be prepared from an isopropyl halide. On the other hand, if isobutyraldehyde is selected, the Grignard reagent must be prepared from methyl halide. The first set of reagents is more appealing than the second and may be used in the following manner.

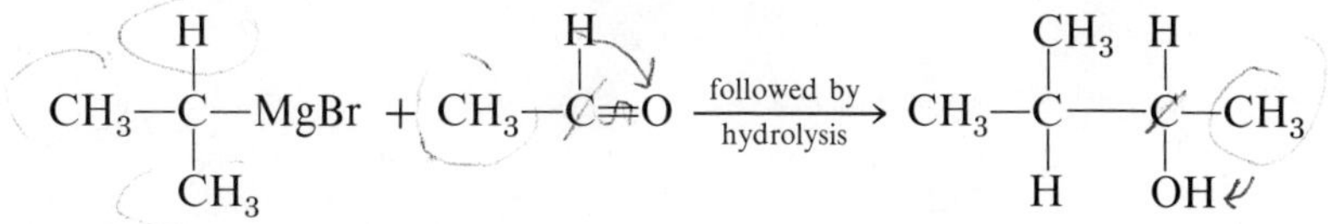

Isopropyl-magnesium Acetaldehyde   3-Methyl-2-butanol
bromide   (a secondary alcohol)

The isomeric, tertiary five-carbon alcohol must, of course, begin with a ketone.

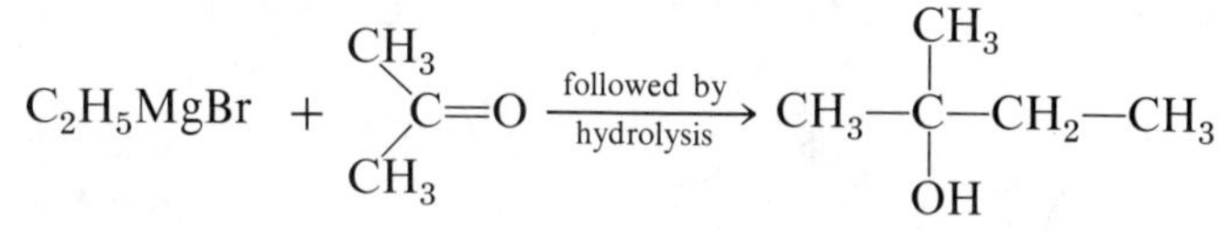

Ethylmagnesium Acetone   2-Methyl-2-butanol
bromide   (a tertiary alcohol)

Another synthetic route leading to a primary alcohol is the reaction of a Grignard reagent with ethylene oxide (Sec. 8.5). As in the carbonyl additions, the magnesium becomes attached to the oxygen atom. The reaction has the added advantage of lengthening the carbon chain by two carbons.

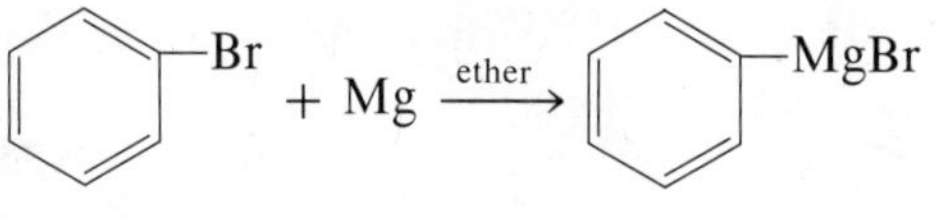

Phenylmagnesium
bromide

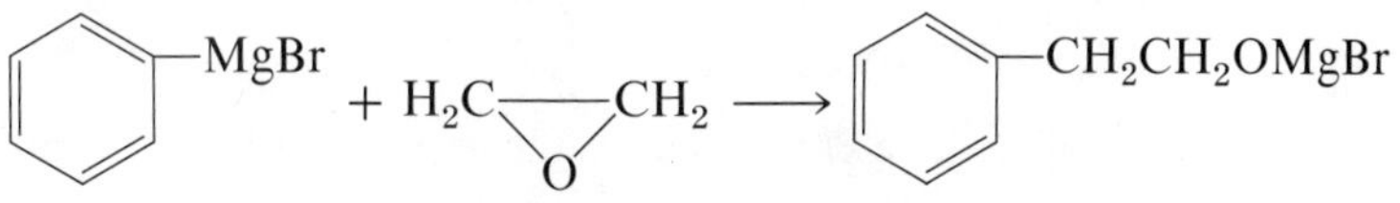

Ethylene
oxide

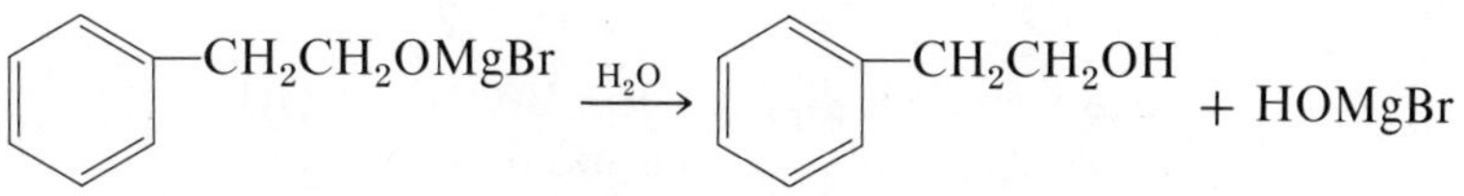

β-Phenylethyl alcohol
(2-Phenylethanol)

EXERCISE 8.4

Beginning with propylene and any other inorganic reagents that you might require, show how you might prepare (a) isopropyl alcohol, (b) *n*-propyl alcohol, (c) 2,3-dimethyl-2-butanol.

8.5   *Polyhydric Alcohols*

Polyhydric alcohols are those which contain more than one hydroxyl group. Compounds that have two hydroxyl groups are called **glycols. Ethylene glycol,** the principal component in permanent types of antifreeze, can be made from ethylene by several different methods. One of these proceeds by way of a chlorohydrin.

$$CH_2{=}CH_2 + HO^-Cl^+ \xrightarrow{(Cl_2+H_2O)} \underset{\substack{Cl \quad\; OH}}{CH_2{-}CH_2} \xrightarrow{H_2O,\ Na_2CO_3} \underset{\substack{OH \quad\; OH}}{CH_2{-}CH_2}$$

Ethylene · · · · · · · · · · · Ethylene · · · · · · · · · · · Ethylene
chlorohydrin · · · · · · · · · · · glycol

EXERCISE 8.5

The gauche conformation (Sec. 2.5) appears to be preferred for ethylene chlorohydrin. Can you suggest a reason why?

Another industrial method for the preparation of ethylene glycol is a simple hydration of ethylene oxide. Ethylene oxide has its origin in ethylene. Oxygen, in the presence of a silver catalyst, adds to the alkene.

$$2\,CH_2{=}CH_2 + O_2 \xrightarrow{Ag} 2\,\underset{O}{CH_2{-}CH_2}$$

$$\underset{O}{CH_2{-}CH_2} + H_2O \xrightarrow{HCl} \underset{\substack{OH \quad\; OH}}{CH_2{-}CH_2}$$

Ethylene · · · · · · · · · · · Ethylene oxide · · · · · · · · · · · Ethylene glycol

Ethylene glycol is miscible with water in all proportions. Its great solubility in water, along with its high boiling point (197.5°C), makes it an excellent antifreeze. Ethylene glycol is not readily lost by evaporation and offers far better winter protection to a water-cooled engine than does methanol (B.P., 64.6°C).

Propylene glycol, $CH_3CH(OH)CH_2OH$, is prepared in a manner similar to that for the preparation of ethylene glycol.

$$\underset{\text{Propylene}}{CH_3{-}\overset{H}{\underset{\;}{C}}{=}CH_2 + HO^-Cl^+} \longrightarrow \underset{\text{Propylene chlorohydrin}}{CH_3{-}\overset{H}{\underset{OH}{C}}{-}\overset{H}{\underset{Cl}{C}H}}$$

$$CH_3{-}\overset{H}{\underset{OH}{C}}{-}\overset{H}{\underset{Cl}{C}}{-}H + OH^- \longrightarrow CH_3{-}\overset{H}{\underset{OH}{C}}{-}\overset{H}{\underset{OH}{C}}{-}H + Cl^-$$

Propylene glycol

Propylene glycol has properties similar to those of ethylene glycol but is much less toxic. The low toxicity of propylene glycol permits its use as an emulsifying solvent and softening agent in cosmetics.

Glycerol ($HOCH_2CH(OH)CH_2OH$), or glycerine, as it is commonly called, is a trihydroxy alcohol, or triol. Like propylene glycol, glycerol also can be made from propylene.

$$CH_3-CH=CH_2 + Cl_2 \xrightarrow{600°C} Cl-CH_2-CH=CH_2 + HCl$$

Propylene   Allyl chloride

$$Cl-CH_2-CH=CH_2 + OH^- \longrightarrow HO-CH_2-CH=CH_2 + Cl^-$$

Allyl alcohol

$$HO-CH_2-CH=CH_2 + HO^-Cl^+ \longrightarrow HO-CH_2-\overset{\overset{\displaystyle H}{|}}{\underset{\underset{\displaystyle OH}{|}}{C}}-\overset{\overset{}{}}{\underset{\underset{\displaystyle Cl}{|}}{CH_2}}$$

Glycerol α-chlorohydrin

$$HO-CH_2-\overset{\overset{\displaystyle H}{|}}{\underset{\underset{\displaystyle OH}{|}}{C}}-CH_2Cl + OH^- \longrightarrow HOCH_2-CH(OH)-CH_2OH + Cl^-$$

Glycerol

You might wonder why, in the first step of the previous reaction sequence, chlorine replaces hydrogen of the methyl group and does not add across the double bond. Chlorine substitutes, rather than adds, because at the high temperature under which the reaction is carried out, substitution takes place by way of a free radical mechanism. Addition of halogen, you will recall, takes place by way of a polar mechanism (Sec. 3.7).

Large quantities of glycerol occur as a by-product in the manufacture of soaps from animal and vegetable fats and oils. It is a viscous, sweet-tasting liquid and, as might be expected with three hydroxyl groups present, is miscible with water in all proportions. It is an excellent humectant (moisture-retaining agent) and is used in the tobacco and cosmetic industries. Glycerol also is used for the preparation of "nitroglycerine," the high explosive component of "dynamite," in the production of plastics, synthetic fibers, and surface coatings.

$$\begin{array}{c}
\overset{\displaystyle H}{|} \\
H-C-OH \\
| \\
H-C-OH \\
| \\
H-C-OH \\
| \\
H
\end{array}
+ 3\,HNO_3 \xrightarrow{H_2SO_4}
\begin{array}{c}
\overset{\displaystyle H}{|} \\
H-C-O-NO_2 \\
| \\
H-C-O-NO_2 \\
| \\
H-C-O-NO_2 \\
| \\
H
\end{array}
+ 3\,H_2O$$

Glycerol   Glyceryl trinitrate
                            ("Nitroglycerine")

8.6 *Preparation of the Phenols*

Phenols cannot be prepared from aryl halides by displacement reactions unless the ring is substituted with powerful electron-withdrawing groups (Sec. 7.5). Simple aryl halides react under conditions suitable only for industrial processes by an elimination-addition mechanism thought to involve an unstable **benzyne** intermediate. In the **Dow process,** for example, phenol is produced from chlorobenzene by treatment with aqueous alkali at a very high pressure and temperature. Sodium phenoxide, the intial product, is converted to phenol by reaction with hydrochloric acid.

Benzyne   Sodium   Phenol
phenoxide

A more recent industrial process produces phenol from isopropylbenzene (cumene). Cumene is obtained from petroleum or may be synthesized from benzene and propene *via* a Friedel-Crafts reaction. Air, when forced through cumene in the presence of a trace of base, produces a hydroperoxide. The latter, on hydrolysis, rearranges and decomposes into phenol and the important by-product acetone.

Cumene

Cumene hydroperoxide   Acetone

Phenol may be prepared in the laboratory by the fusion of sodium benzenesulfonate (Sec. 4.7-D) with sodium hydroxide. The sodium phenoxide produced by this fusion is converted to free phenol by acid treatment.

Benzenesulfonic   Sodium
acid   benzenesulfonate

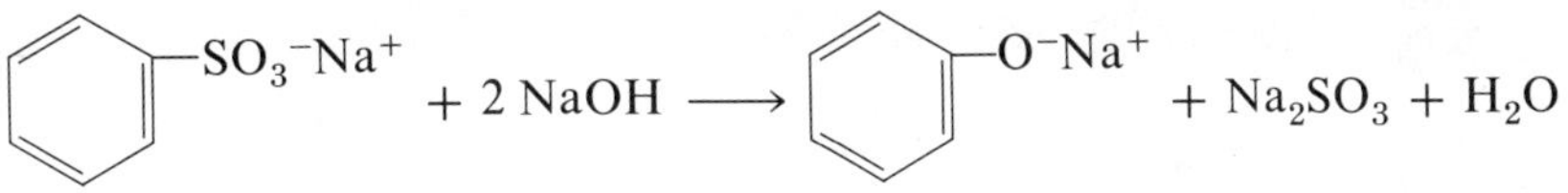

Sodium phenoxide

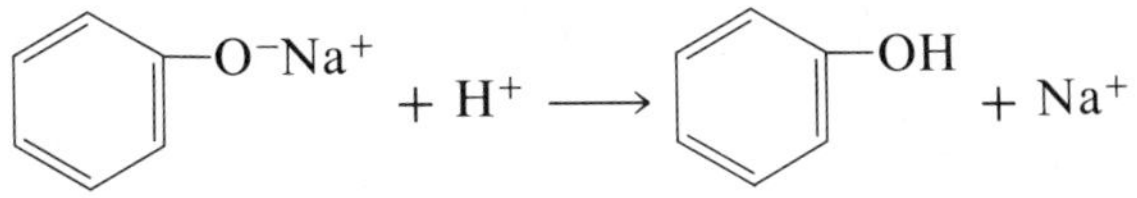

Phenol

8.7 *The Germicidal Properties of Phenols*

Phenol is one of the oldest of disinfectants. All phenolic compounds appear to have a germicidal power which is enhanced by the presence of alkyl groups on the ring. It appears that the optimum size of the side chain for maximum germicidal activity is six carbons, making **n-hexylresorcinol** a very fine antiseptic.

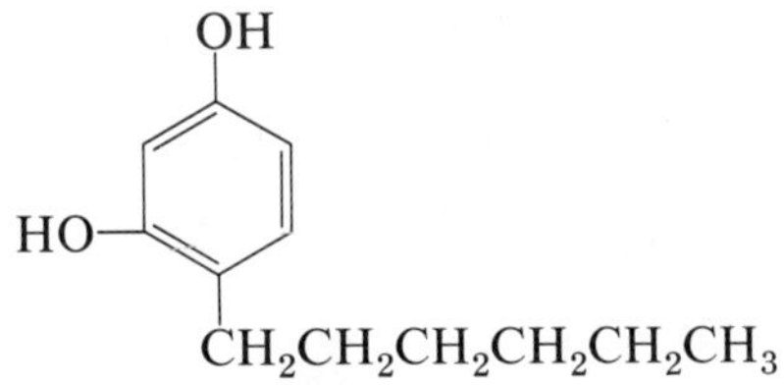

n-Hexylresorcinol

The killing power which an antiseptic has against microorganisms is measured against that of phenol. The germicidal efficiency of an antiseptic is measured in terms of an arbitrary unit called a **phenol coefficient.** For example, a germicide in a 1% solution that kills an organism in the same length of time as that required for a 5% phenol solution is assigned a phenol coefficient of 5.

Chlorine-substituted phenols are especially active against bacteria and fungi. **Pentachlorophenol,** for example, is an excellent fungicide and wood preservative and is widely used to protect against "dry rot" and termites. The chlorine substituted biphenyl phenolic compound 2,2'-dihydroxy-3,3',5,5', 6,6'-hexachlorodiphenylmethane, popularly known as "Hexachlorophene," has a phenol coefficient of about 125. In dilute form it has been used extensively in the manufacture of germicidal soaps, some toothpastes, and many deodorants, but products containing it have recently come under scrutiny as possibly causing systemic damage, especially when used on infants. The sale of products containing hexachlorophene is now restricted.

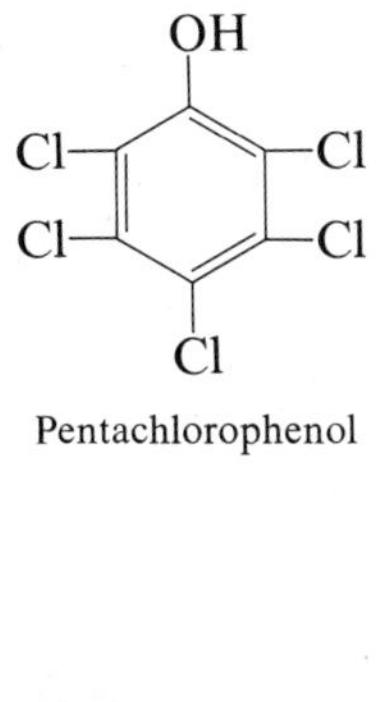

Pentachlorophenol

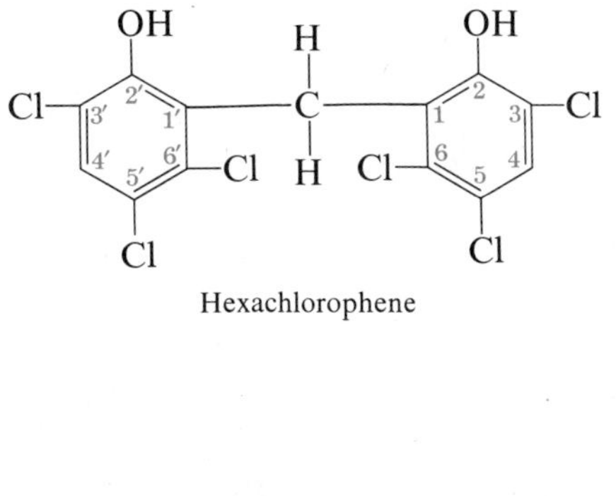

Hexachlorophene

Phenol and its homologs are toxic substances and have a caustic action on animal tissue. Care must be exercised in handling phenols; direct contact or inhalation of their vapors should be avoided.

EXERCISE 8.6

A germicide in a concentration of one part to 225 parts of water showed the same killing power against *Salmonella typhosa* in ten minutes as did a solution made of one part phenol in 100 parts of water during the same contact time. What is the phenol coefficient of the germicide in question?

(*Answer:* 2.25.)

8.8   *Reactions of the Alcohols and Phenols*

The reactions of the alcohols and phenols are of the following types:
(A) Reactions that result in O—H bond cleavage, RO╀H
(B) Reactions that result in C—O bond cleavage, R╀OH
(C) Reactions that result in oxidation of the carbinol carbon
(D) Reactions that involve the aromatic ring of phenols.

A-1. Cleavage of the O—H Bond. Salt Formation.   The aliphatic alcohols are weaker acids than water; therefore, the following equilibrium lies well to the left in aqueous solution.

$$R—OH + HO^-K^+ \rightleftharpoons R—O^-K^+ + H—OH$$

Although solutions of potassium hydroxide in methanol or ethanol contain considerable concentrations of alkoxide ion, the higher alcohols are less acidic and for all practical purposes do not yield useful concentrations of alkoxides on treatment with sodium or potassium hydroxide. Thus, the preferred preparation of the alkoxides (or alcoholates) is the reaction of the alcohol with active metals such as sodium or potassium, which takes place with the liberation of hydrogen gas.

$$2\,RO╀H + 2\,Na \longrightarrow 2\,RO^-Na^+ + H_2$$
Sodium
alkoxide

Primary alcohols exhibit a greater reactivity on treatment with sodium metal than do secondary alcohols, and the latter show a greater reactivity than tertiary alcohols. Of the alkali metals, potassium is more reactive toward any class of alcohols than is sodium. Thus, the reaction of potassium metal with *tert*-butyl alcohol gives potassium *tert*-butoxide, a very powerful base, widely used in both the laboratory and in industry.

The liberation of hydrogen gas from an unknown liquid sometimes is used as a test for alcohols. Of course, the liquid to be tested must be free of water.

A-2. Cleavage of the O—H Bond. Ether Formation.   The alkoxides or phenoxides of the alkali metals are strong bases (nucleophiles) that easily enter into S_N2 displacements of halogen from alkyl halides. This reaction, referred to as the **Williamson Ether Synthesis,** is best used to prepare mixed ethers—that is, ethers in which the two groups bridged by the oxygen atom are not the same.

$$(CH_3)_2CHO^-Na^+ + CH_3I \xrightarrow{S_N2} (CH_3)_2CH\!-\!O\!-\!CH_3 + NaI$$
Isopropyl methyl ether

If one of the groups in the ether is to be a branched structure, that part of the ether should have its origin in the alkoxide. If the alkyl halide is highly branched, elimination, rather than substitution, results.

$$CH_3O^-Na^+ + (CH_3)_2CHI \xrightarrow{E2} CH_3CH\!=\!CH_2 + CH_3OH + NaI$$
Propylene

A-3. Ester Formation.   Alcohols, when permitted to react with carboxylic acids (Chapter 10), produce esters. The reaction is catalyzed by strong mineral acids and, as indicated by the double arrow in the equation, is reversible.

Acid   Alcohol   Ester

The esterification reaction is discussed in greater detail in Section 10.5-G where the mechanism of the reaction shows why it is an O—H bond cleavage with respect to the alcohol.

Phenols, unlike the alcohols, cannot be esterified directly by reaction with a carboxylic acid but are best prepared by reaction with an acid chloride.

Acetyl chloride   Phenyl acetate

Phenol esters undergo an interesting rearrangement known as the **Fries rearrangement** when heated with aluminum chloride. The resulting products are the *ortho-* and *para-*phenolic ketones (Chapter 9).

o- and *p-*Hydroxyacetophenone

B. Reactions that Result in C—O Bond Cleavage. The reactions of alcohols in which the carbon-oxygen bond is cleaved fall into two familiar classes, substitution and elimination. The mechanisms for these reactions are similar to those for the S_N1, S_N2, E1, and E2 substitution and elimination reactions of alkyl halides (Sec. 7.4). Let us reconsider the dehydration of ethanol (Sec. 3.3) using a strong acid catalyst. The reaction is thought to take place in the following manner. In the first stage of the reaction, the acid (shown here as the hydrogen ion) protonates the electron-rich hydroxyl group of the alcohol molecule to form an alkyl substituted oxonium ion.

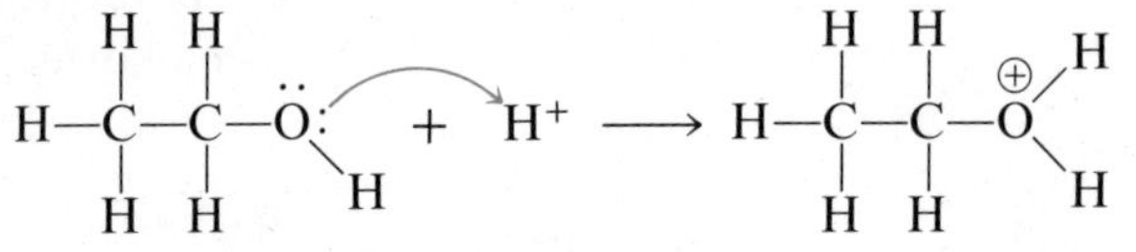

An alkyl oxonium ion

Next, some base present ($B:$ = bisulfate, HSO_4^-, or another molecule of alcohol) abstracts a proton from the β-carbon atom and a molecule of water departs to form the alkene by a typical E2 reaction mechanism.

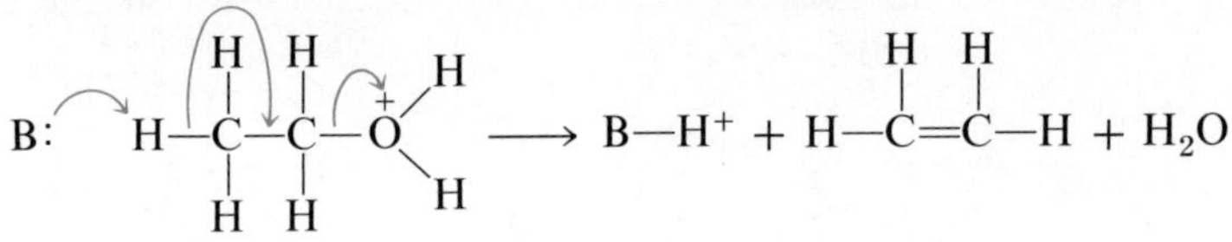

With tertiary alcohols the elimination proceeds *via* the E1 mechanism.

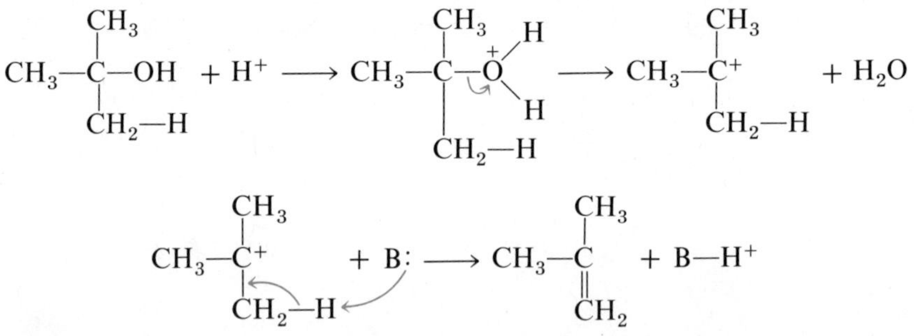

The important feature of each of these mechanisms is the role of the catalyst. If the alcohol were not protonated on the oxygen atom, the elimination would require the removal of hydroxide ion in the slow step of each reaction. However, hydroxide ion is a poor **leaving group** and is not easily removed in either substitution or elimination reactions. Protonation of the hydroxyl group to form the alkyloxonium ion changes the leaving group from hydroxide ion to water, a good leaving group, and facilitates both substitution and elimination. The use of acid catalysis to protonate hydroxyl (and alkoxyl) groups for the purpose of creating good leaving groups is fairly common practice in organic chemistry (see, for example, Sec. 4.7-C).

EXERCISE 8.7

Write mechanisms for the S_N2 reaction of *n*-butyl alcohol with concentrated hydrobromic acid to yield *n*-butyl bromide and for the S_N1 reaction of *tert*-butyl alcohol with cold concentrated hydrobromic acid to yield *tert*-butyl bromide.

Reactions of the foregoing types proceeding *via* E1 or S_N1 mechanisms *may* be complicated by the facility with which some carbonium ions rearrange to more stable carbonium ions. Rearrangements, when they occur, are often of the following types:

$$CH_3-\underset{\underset{H}{|}}{\overset{\overset{CH_3}{|}}{C}}-\underset{\underset{OH}{|}}{\overset{\overset{H}{|}}{C}}-CH_3 \underset{H^+Br^-}{\rightleftharpoons} CH_3-\underset{\underset{H}{|}}{\overset{\overset{CH_3}{|}}{C}}-\underset{\underset{OH_2^+}{|}}{\overset{\overset{H}{|}}{C}}-CH_3 \underset{-H_2O}{\rightleftharpoons} CH_3-\underset{\underset{H}{|}}{\overset{\overset{CH_3}{|}}{C}}-\overset{\overset{H}{|}}{\underset{+}{C}}-CH_3$$

3-Methyl-2-butanol

secondary carbonium ion

$\updownarrow$

$$CH_3-\underset{\underset{Br}{|}}{\overset{\overset{CH_3}{|}}{C}}-\underset{\underset{H}{|}}{\overset{\overset{H}{|}}{C}}-CH_3 \overset{Br^-}{\leftarrow} CH_3-\overset{\overset{CH_3}{|}}{\underset{+}{C}}-\underset{\underset{H}{|}}{\overset{\overset{H}{|}}{C}}-CH_3$$

2-Bromo-2-methylbutane

tertiary carbonium ion

$$CH_3-\underset{\underset{CH_3}{|}}{\overset{\overset{CH_3}{|}}{C}}-\underset{\underset{OH}{|}}{\overset{\overset{H}{|}}{C}}-CH_3 \underset{H_2SO_4}{\rightleftharpoons} CH_3-\underset{\underset{CH_3}{|}}{\overset{\overset{CH_3}{|}}{C}}-\underset{\underset{OH_2^+}{|}}{\overset{\overset{H}{|}}{C}}-CH_3 \underset{-H_2O}{\rightleftharpoons} CH_3-\underset{\underset{CH_3}{|}}{\overset{\overset{CH_3}{|}}{C}}-\overset{\overset{H}{|}}{\underset{+}{C}}-CH_3$$

3,3-Dimethyl-2-butanol

secondary carbonium ion

$\updownarrow$

$$CH_3-\underset{\underset{CH_3}{|}}{\overset{\overset{CH_3}{|}}{C}}=C-CH_3 \underset{-H^+}{\rightleftharpoons} CH_3-\overset{\overset{CH_3}{|}}{\underset{+}{C}}-\underset{\underset{CH_3}{|}}{\overset{\overset{H}{|}}{C}}-CH_3$$

2,3-Dimethyl-2-butene

tertiary carbonium ion

In each of the examples a group on a carbon next to the positively charged carbon migrated to the positive carbon, bringing with it a pair of electrons and creating a new, more stable carbonium ion. Rearrangements of this type are common in carbonium ion chemistry.

B-1. Replacement of the Hydroxyl Group by Acid Anions. The reactions of alcohols with phosphorus halides and thionyl chloride, $SOCl_2$, were reviewed under the preparation of the alkyl halides (Sec. 7.2-B). Most phenols do *not* undergo these reactions.

Treatment of an aliphatic alcohol with the **Lucas reagent** (a solution of zinc chloride in concentrated hydrochloric acid) produces an alkyl chloride.

$$R\text{—}OH + HCl \xrightarrow{ZnCl_2} R\text{—}Cl + H_2O$$

The alkyl chloride, when formed in this manner, is insoluble in the reagent and produces either a cloudy appearance or forms two layers. The Lucas reagent is used to distinguish between primary, secondary, and tertiary alcohols because a tertiary alcohol reacts immediately, the secondary after a few minutes, and the primary only when heated for some hours.

An alcohol treated with concentrated sulfuric acid at room temperature produces an alkyl hydrogen sulfate.

$$C_2H_5 \overbrace{(OH + H)}O\text{—}SO_2\text{—}OH \longrightarrow C_2H_5\text{—}O\text{—}SO_2\text{—}OH + H_2O$$
Ethyl hydrogen sulfate

The alkyl hydrogen sulfates still have one acidic hydrogen remaining, and are capable of forming salts. The sodium salts of the long chain alkyl hydrogen sulfates are excellent detergents (Sec. 11.8). Among the detergents with familiar trade names are Dreft, Cheer, Vel, and Trend.

$$CH_3\text{—}(CH_2)_{11}\text{—}O\text{—}SO_3H + NaOH \longrightarrow CH_3\text{—}(CH_2)_{11}\text{—}OSO_3^-Na^+ + H_2O$$
Lauryl hydrogen sulfate   Sodium lauryl sulfate

Alcohols react with nitric acid to yield alkyl nitrates. Perhaps the most widely used alkylnitrate is glyceryl trinitrate (commonly called "nitroglycerine"), the principal explosive ingredient in dynamite (Sec. 8.5). Glyceryl trinitrate has been employed medicinally as a vasodilator of short duration for the relief of angina pectoris in heart patients.

B-2. Cleavage of the C—O Bond. Dehydration. Ethanol may, in effect, be dehydrated to an olefin when its hydrogen sulfate ester is heated to 150° or higher.

$$CH_3CH_2\text{—}OSO_3H \xrightarrow{150°} H_2C\text{=}CH_2 + H_2SO_4$$

If the hydrogen sulfate ester is heated with an excess of alcohol at a temperature lower than 150°, the bisulfate ion is displaced by alcohol to produce a simple ether.

$$CH_3CH_2\text{—}\overset{\overset{\displaystyle H}{|}}{\underset{\cdot\cdot}{O}}\text{:}$$

$$CH_3CH_2\text{—}OSO_3H \xrightleftharpoons{130°} {}^-OSO_3H + CH_3CH_2\text{—}\overset{\overset{\displaystyle H}{|}}{\underset{\oplus}{O}}\text{—}CH_2CH_3$$

$$CH_3CH_2 \overset{\overset{\displaystyle H}{\displaystyle |}}{\underset{\oplus}{O}} CH_2CH_3 + {}^-OSO_3H \rightleftharpoons CH_3CH_2OCH_2CH_3 + H_2SO_4$$

Diethyl ether

C. Oxidation. When oxidized, primary and secondary alcohols give organic products containing the same number of carbon atoms. Primary alcohols can be oxidized to carboxylic acids (Sec. 10.4-A) and secondary alcohols to ketones (Sec. 9.4-A). Tertiary alcohols are not oxidizable without a rupture of carbon-carbon bonds. Reagents usually employed for the oxidation of alcohols are potassium dichromate in combination with concentrated sulfuric acid, or a hot, alkaline solution of potassium permanganate.

$$3\ CH_3CH_2OH + 2\ K_2Cr_2O_7 + 8\ H_2SO_4 \longrightarrow$$

Ethanol

$$3\ CH_3 - C\overset{\displaystyle O}{\underset{\displaystyle OH}{}} + 2\ K_2SO_4 + 2\ Cr_2(SO_4)_3 + 11\ H_2O$$

Acetic acid

$$3\ CH_3 - \overset{\overset{\displaystyle H}{\displaystyle |}}{\underset{\underset{\displaystyle OH}{\displaystyle |}}{C}} - CH_3 + 2\ KMnO_4 \longrightarrow 3\ CH_3 - \overset{\overset{\displaystyle O}{\displaystyle ||}}{C} - CH_3 + 2\ MnO_2 + 2\ KOH + 2\ H_2O$$

2-Propanol   Acetone

On Balancing Oxidation-Reduction Equations

It would seem in order at this point to illustrate how the equations for the preceding reactions were balanced.

Organic chemists when writing reaction equations frequently indicate to the right of the reaction arrow only the principal organic products obtained. The term "equation," however, implies "balance," and balanced oxidation-reduction equations may be written for organic reactions as well as for inorganic reactions—perhaps not as readily, but certainly with as much sport. Unfortunately, "oxidation number" as applied to carbon atoms in organic compounds does not have the same connotation as when applied to ions in inorganic salts and cannot always be determined by simple inspection. We usually think of an element undergoing oxidation as one losing electrons and one being reduced as gaining electrons. The carbon atom retains four electrons for sharing whether it is oxidized or reduced. The covalently bonded carbon atom *loses part of its normal* "hold" on the electron pair when oxidized—that is, when it is bonded to a more electronegative element. Conversely, it *gains* a greater "hold" on the electron pair when reduced—that is, when it is bonded to a more electropositive element. In other words, the carbon atom becomes *more* or *less* polar in an oxidation-reduction reaction. Therefore, by assigning polar values rather than oxidation numbers to carbon we can balance an oxidation-reduction equation involving carbon quite easily. We need to learn and practice but a few simple rules.

1. The oxidation number of hydrogen is $+1$ (except in the hydrides of active metals, e.g., $LiAlH_4$, where it is -1).
2. The oxidation number of oxygen is -2 (except in peroxides, e.g., H_2O_2, where it is -1).
3. The oxidation number of an element (or the polar number in the case of carbon) becomes more positive as it is oxidized, less positive when reduced.
4. Two carbon atoms bonded together effect no change in the polar number of either.
5. Oxidation must always be accompanied by an equivalent amount of reduction.

Before any equation can be balanced one must know beforehand the nature of the products obtained when certain reactants are brought together. One also must determine the degree of change in the oxidation or polar number of each element involved in an oxidation number change. For example,

The polar numbers -1 and $+3$ for the carbon atom in the alcohol and the acid, respectively, were determined according to Rules 1, 2, and 4 by adding algebraically all polar values conferred upon each carbon by the atoms bonded to it. You will note that the number two carbon atom in both alcohol and acid molecule undergoes no change and therefore remains the same throughout the reaction.

EXERCISE 8.8

Determine the polar number of the carbon atom in each of the following compounds:
(a) CH_4 (b) CH_3OH (c) $H_2C=O$, formaldehyde (d) HCOOH, formic acid, (e) CO_2

We are now ready to balance the equation

$$\underset{(reduction)}{\overset{(oxidation)}{CH_3CH_2OH + K_2Cr_2O_7 + H_2SO_4 \longrightarrow CH_3COOH + Cr_2(SO_4)_3}}$$
$$+ K_2SO_4 + H_2O$$

The oxidation step represents a loss of 4 in the polar number of carbon. The reduction step represents a gain of 3 in the oxidation number of each chromium atom for a total electron gain of 6 for both chromium atoms. In accordance with Rule 5, equivalence between oxidation and reduction is established by taking three times the amount of alcohol and its oxidation product, the acid, and twice the amount of oxidizing reagent and its reduction products, $Cr_2(SO_4)_3$ and K_2SO_4.

$$3 CH_3CH_2OH + 2 K_2Cr_2O_7 + H_2SO_4 \longrightarrow$$
$$3 CH_3COOH + 2 Cr_2(SO_4)_3 + 2 K_2SO_4 + H_2O$$

The oxidation-reduction step is now complete. All that remains to be done is to supply the necessary amount of sulfuric acid on the left side of the equation to provide for the sulfate salts on the other. A simple inspection will tell us that 8 moles of sulfuric acid are required. This amount of sulfuric acid automatically provides enough hydrogen (along with that obtained from the alcohol) to combine with the oxygen provided by the potassium dichromate in excess of that needed for the acetic acid. Excess hydrogen and oxygen then appear together as water on the right side of the equation. Let us now re-write the completed, balanced equation.

$$3 CH_3CH_2OH + 2 K_2Cr_2O_7 + 8 H_2SO_4 \longrightarrow$$
$$3 CH_3COOH + 2 Cr_2(SO_4)_3 + 2 K_2SO_4 + 11 H_2O$$

EXERCISE 8.9

Balance the equation for the oxidation of 2-propanol to acetone, $(CH_3)_2C=O$, using sodium dichromate in sulfuric acid solution as the oxidizing reagent.

D. Ring Substitution in Phenols. When phenol is nitrated with dilute HNO_3, both the *ortho* and the *para* isomeric nitrophenols are formed.

o-Nitrophenol   *p*-Nitrophenol

The *ortho* form can be separated from its *para* isomer by steam distillation. The *ortho* isomer is able to form a H-bond within itself, or *intra*molecularly, to produce a **chelated ring** (Gr., *chele,* claw). *Para* isomers must hydrogen bond *inter*molecularly to form an associated compound not volatile with steam.

Intramolecular H-Bonding (Chelation)   Intermolecular H-Bonding (Association)

Phenol, when nitrated directly with concentrated nitric acid, undergoes oxidation. For this reason, the high explosive 2,4,6-trinitrophenol, or picric acid, is obtained through a synthesis that begins with chlorobenzene. The first product, 2,4-dinitrochlorobenzene is then easily hydrolyzed to 2,4-dinitrophenol (Sec. 7.5) and the nitration continued to give picric acid in good yield.

Chlorobenzene

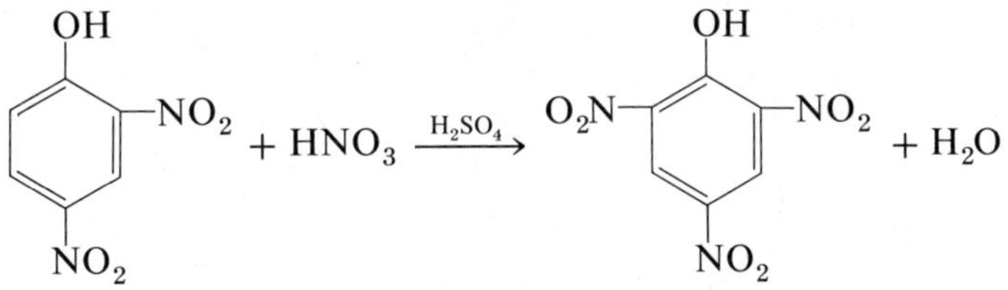

Picric acid
(2,4,6-Trinitrophenol)

Many organic substances form crystalline molecular complexes with picric acid. Such derivatives, called **picrates,** have sharp melting points and are very useful in the identification of "unknowns." Butesin picrate, the picrate of *n*-butyl-*p*-aminobenzoate, is a useful surface anesthetic for the treatment of burns. Picric acid is a bright yellow compound and also has been used as a direct dye on silk and wool (Sec. 19.3-A).

Phenol, when treated with sulfuric acid, yields both *ortho* and *para* phenol-sulfonic acids. The *ortho* isomer predominates at low temperatures, the *para* at high temperatures.

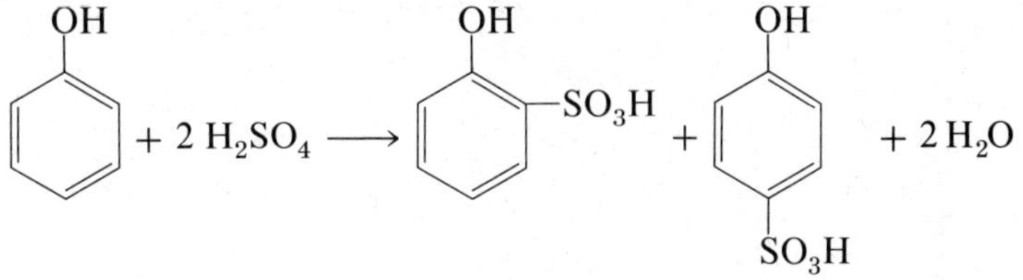

o-Phenol-   *p*-Phenol-
sulfonic acid   sulfonic acid

Phenol is very easily brominated. 2,4,6-Tribromophenol can be prepared simply by shaking an aqueous phenol solution with a saturated solution of bromine in water.

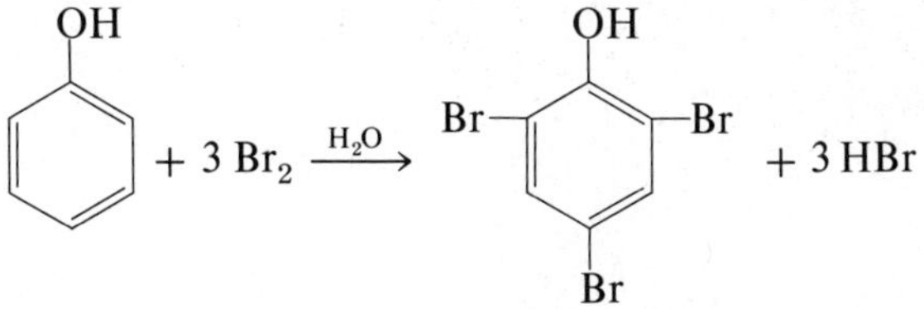

2,4,6-Tribromophenol

The catalytic reduction of phenol produces cyclohexanol.

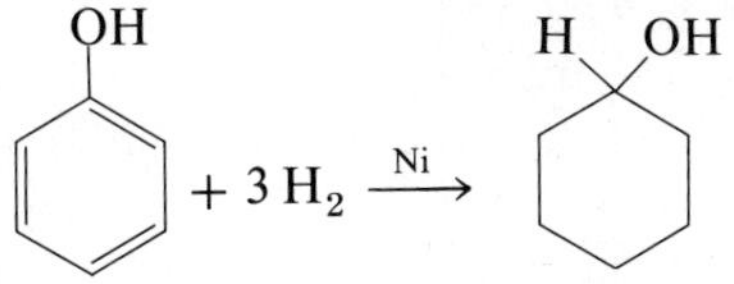

Cyclohexanol

8.9   *Dihydric Phenols and Quinones*

The three dihydric phenols can be prepared by standard procedures but, because of their industrial importance, are often made by special methods.

o-Dichlorobenzene	Catechol

o-Benzenedisulfonic acid	Resorcinol

Aniline	*p*-Benzoquinone	Hydroquinone

Both 1,2- and 1,4-dihydroxybenzene derivatives are characterized by easy oxidation to quinones and the easy reduction of the quinones back to the dihydric phenols.

Catechol	*o*-Benzoquinone	Hydroquinone	*p*-Benzoquinone

Dihydric and polyhydric phenols and their derivatives, as well as the quinones, are widely distributed in nature. The oxidation-reduction reactions of hydroquinone and quinone derivatives are important in certain biochemical processes. The quinones are not aromatic compounds, as may be readily seen by application of the Hückel rule (Sec. 4.2); rather they are α,β-unsaturated ketones (Sec. 9.12) and show the reactions expected of such substances. Among these reactions is the Diels-Alder reaction (Sec. 3.13), in which *p*-benzoquinone functions as an excellent dienophile.

8.10 *Alcohols and Phenols in Nature and in Industry*

Mono-, di-, and polyhydric alcohols and phenols are quite common in nature, and numerous examples of such naturally occurring alcohols and phenols are given in Chapters 14, 15, 16, and 18. Many of these examples have fairly complex structures, but even relatively simple alcohols and phenols play important roles in nature. For example, a number of the **pheromones** fall into these classes. The pheromones are substances secreted by animals and by insects for the purpose of intra-species communication. They may be used for such diverse purposes as marking a trail, warning of danger, attracting the opposite sex, or calling assembly. The use of natural sex attractants, or their synthetic analogs, to lure insects into traps may prove to be a relatively safe alternative to the use of chemical pesticides in the control of selected insect species, for it is known that as few as 30 molecules (perhaps even fewer!) will attract certain moths. Thus, the concentrations of chemicals required is incredibly small. Some typical pheromones are shown.

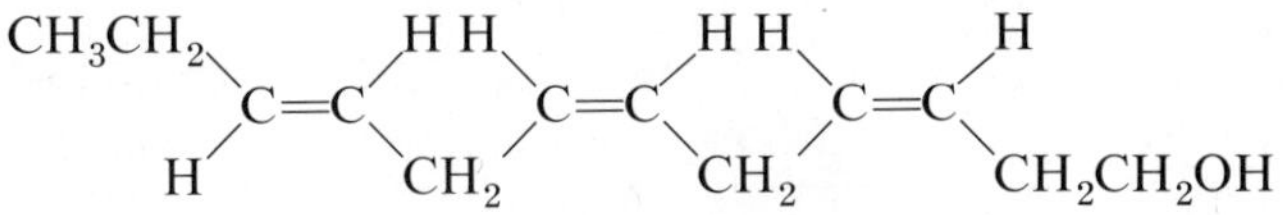

Trailmarker for termites

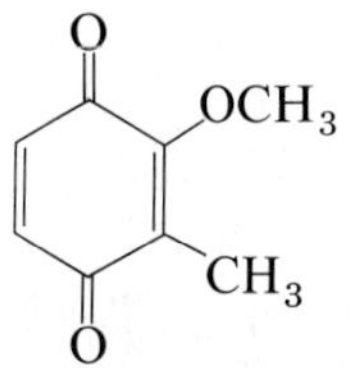

Chemical defense for millipedes

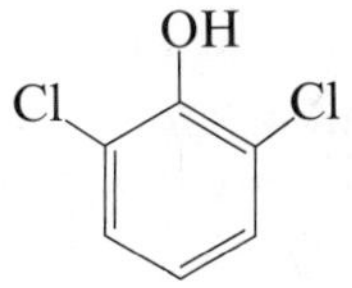

Sex attractant for lonestar tick

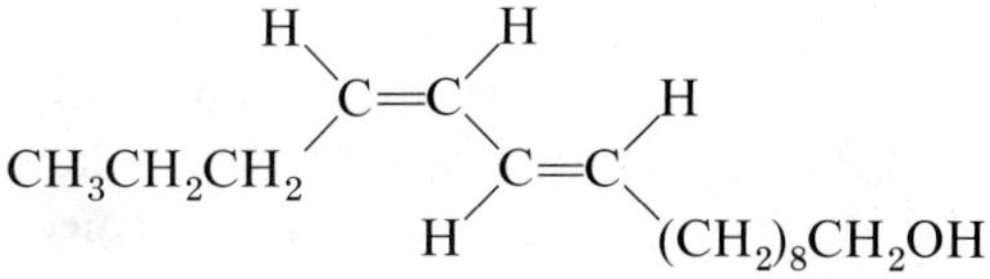

Sex attractant for silkworm moth

Dihydric phenols and the quinones serve in other important capacities in biochemical processes. Examples are the K vitamins (Sec. 17.11) and coenzyme Q, the latter being involved in electron-transport in the mitochondrial respiratory chain. The catechol moiety is represented in nature in the urushiols, the active ingredient in the allergenic oil of poison ivy.

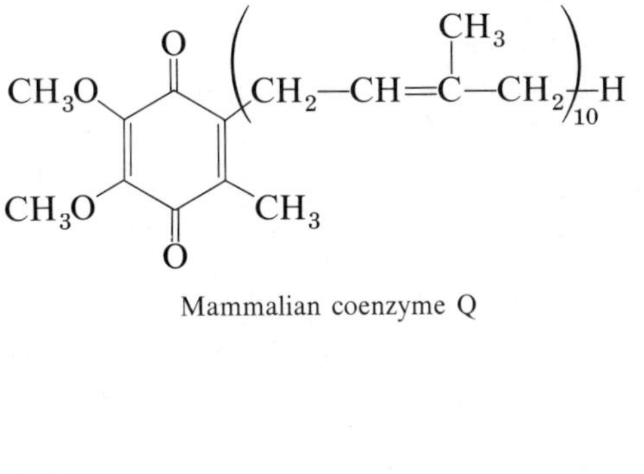

Mammalian coenzyme Q

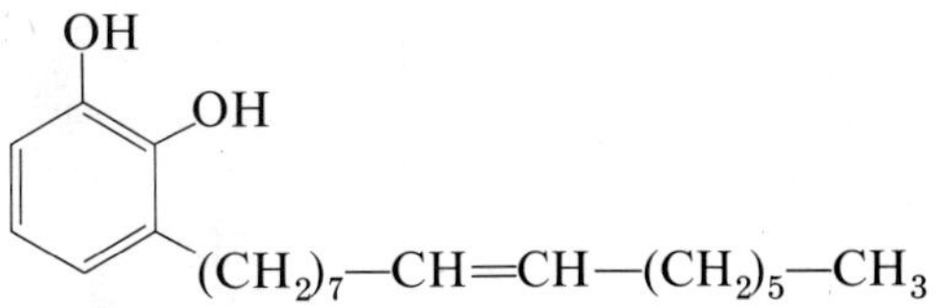

A urushiol

Phenol and its related compounds are of great industrial importance. They are starting materials for important pharmaceuticals such as **aspirin** (Sec. 12.6) and **epinephrine.** Certain derivatives of phenol are important photographic developers. The antiseptic power of certain phenolic compounds has already been mentioned. In the plastics industry phenol is used as starting material for **Bakelite** (Sec. 9.11) and **nylon.** The herbicides 2,4-dichlorophenoxyacetic acid and 2,4,5-trichlorophenoxyacetic acid, popularly known as **2,4-D** and **2,4,5-T,** respectively, have their origin in phenol. It is easy to see why a large percentage of the benzene produced each year is converted to phenol.

2,4-Dichlorophenoxyacetic acid
(2,4-D)

2,4,5-Trichlorophenoxyacetic acid
(2,4,5-T)

EXERCISE 8.10

What simple test would serve to distinguish a phenolic compound from cyclohexanol?

8.11 *Structure and Nomenclature of Ethers*

Ethers are compounds of the general formula R—O—R, or R—O—Ar, in which an oxygen bridge joins two hydrocarbon groups. Although the ethers are isomeric with the alcohols, their properties are vastly different.

Ethers may be named in one of two ways. According to the IUPAC system of nomenclature one of the alkyl groups bonded to the oxygen (the larger group if a mixed ether) is considered to be a substituted hydrocarbon. The smaller alkyl group with the oxygen is called an **alkoxy** substituent. In common nomenclature, both groups bridged by the oxygen are named and followed by the word ether. Sometimes the prefix "di" is employed if both groups are alike (symmetrical ethers), but this is not necessary. The following examples will illustrate these rules.

A simple ether   A mixed ether   Methyl ether   Ethyl ether   Methyl ethyl ether

Methyl phenyl ether
(Anisole)

Ethoxybenzene
(Ethyl phenyl ether)

1-Methoxypropane
(Methyl *n*-propyl ether)

8.12 *Properties of Ethers*

Ethers boil at much lower temperatures than do the alcohols from which they are derived because the oxygen atom now is attached only to carbon. Thus, H-bonding and association between many molecules no longer is possible. The boiling points of the ethers closely parallel those of the alkanes of the same molecular weight. You will note that the methylene group, $-(CH_2)-$, has almost the same formula weight (14) as an oxygen atom (16). Ethyl ether, with a molecular weight of 74, boils at 35°C; *n*-pentane, with a molecular weight of 72, boils at 36°C. The similarity may be illustrated also with *n*-propyl ether and *n*-heptane.

$$CH_3CH_2CH_2-O-CH_2CH_2CH_3 \qquad CH_3CH_2CH_2CH_2CH_2CH_2CH_3$$

n-Propyl ether   *n*-Heptane

M.W., 102; B.P., 91°C   M.W., 100; B.P., 98°C

The lower members of the aliphatic ethers are highly volatile and very flammable. Ethyl ether, the most important member of the family, is both an excellent organic solvent and a fine general anesthetic. Its high flammability, however, presents a hazard in the laboratory and in the operating room. The vapor of ethyl ether is heavier than air and has an annoying tendency to flow along the top of a laboratory bench and become ignited by a student's burner some distance away.

Ethyl ether, while not completely immiscible with water, is an excellent solvent to employ in extraction procedures involving aqueous solutions.[1]

8.13 *Preparation of Ethers*

A simple ether can be produced by the elimination of a molecule of water from two molecules of alcohol. The mechanism for this reaction is shown in Section 8.8-B2. In practice, a mixture of the alcohol and sulfuric acid is heated to a temperature of approximately 140°C. An additional volume of alcohol then is added as the ether distills.

$$C_2H_5\overbrace{-OH + H}-O-C_2H_5 \xrightarrow{H_2SO_4} C_2H_5-O-C_2H_5 + H_2O$$

Ethyl alcohol   Ethyl ether

Inasmuch as concentrated sulfuric acid may cause the elimination of a molecule of water intramolecularly, conditions must be carefully controlled to minimize this competing reaction. The several courses of reaction open to a mixture of alcohol and sulfuric acid have already been illustrated (Sec. 8.8-B). The sulfuric acid method is employed for the preparation of simple ethers. A method applicable to the preparation of mixed ethers involves the reaction of a

---

[1] At 25°C diethyl ether is soluble in water to the extent of 6%, and water is soluble in ether to the extent of approximately 1.5%. Ether dissolved in water, however, is easily removed by distillation.

metallic alkoxide and an alkyl halide (Williamson's Synthesis). The choice of reagents to employ in a Williamson ether synthesis, as was indicated in Section 8.8-A, must be made with some care.

Mixed ethers of the alkyl-aryl type also may be prepared by the Williamson method. The methyl and ethyl ethers of phenol are usually prepared by treating sodium phenoxide or phenol with the appropriate alkyl iodide or dialkyl sulfate in the presence of a base.

Methyl phenyl ether (Anisole)

| Methyl sulfate | | Methyl phenyl ether (Anisole) | Sodium methyl sulfate |

EXERCISE 8.11

A student wishes to prepare ethyl *tert*-butyl ether as a laboratory project. What reagents should he choose for this synthesis? Why?

8.14 *Reactions of the Ethers*

The ethers represent an extremely stable group of compounds and their reactions are few. The ethers are soluble in strong mineral acids because of the formation of **oxonium** salts. The ethereal oxygen provides unshared electron pairs for bond formation with an acid. Solubility in sulfuric acid is thus a convenient method for distinguishing ethers from hydrocarbons and alkyl halides.

$$C_2H_5-\overset{..}{\underset{|}{\overset{|}{O}}}: \,\, + \,\, H^+ \,\, A^- \longrightarrow C_2H_5-\overset{\oplus}{\underset{|}{\overset{..}{O}}}:H \,\, + \,\, A^-$$
$$\underset{C_2H_5}{} \qquad\qquad\qquad \underset{C_2H_5}{}$$

Oxonium ion

Ethers are resistant to attack by the usual chemical oxidizing agents; yet anhydrous ethyl ether, when exposed repeatedly to air over long periods of time, forms a highly explosive peroxide. Such ether peroxides are extremely dangerous, and anhydrous ether should be tested before it is distilled. In one test a sample of the ether is treated with an acidified solution of potassium iodide. If peroxides are present in the ether they will oxidize the iodide ion, I^-, to molecular iodine, I_2, yielding the characteristic brown color of iodine. Peroxides in ether may be removed by washing the ether with a ferrous sulfate solution. The peroxides oxidize ferrous ion to ferric and are thus eliminated.

Other reactions of the ethers involve a cleavage of the carbon-oxygen linkage. Concentrated hydriodic acid is an excellent reagent to employ for this cleavage. Each molecule of ether cleaved produces one equivalent of alkyl iodide and one equivalent of alcohol. If hydroiodic acid is used in excess, the alcohol initially formed also is converted to an alkyl iodide.

$$R\text{---}O\text{---}R' + HI \longrightarrow RI + R'OH$$

$$R\text{---}O\text{---}R' + HI \longrightarrow R\text{---}\overset{\oplus}{\underset{\underset{R'}{|}}{\ddot{O}\!:\!H}} \quad I^{\ominus} \xrightarrow[\text{(excess)}]{HI} RI + R'I + H_2O$$

If one group of an ether is alkyl, such as methyl or ethyl, and the other group is aryl, such as phenyl, then the iodide of one of the alkyl groups usually is one product of the cleavage, the other being a phenol.

Anisole

Cleavage of ethers also may be accomplished by the use of concentrated (48%) hydrobromic acid.

EXERCISE 8.12

A student storeroom assistant one morning found that the following labels had come undone from their bottles and were lying on the floor: Ethyl alcohol, Ethyl bromide, Ethyl ether.

What simple test could he have performed on a sample from each of the bottles in order to properly relabel them?

8.15 *Uses of Ethers*

Ethyl ether is perhaps most familiar to everyone as a general inhalation anesthetic. Halothane (Sec. 7.3) is less hazardous and has largely replaced ethyl ether for this purpose. In organic chemistry, ethers are excellent solvents for fats, waxes, oils, plastics, and lacquers. Ethyl ether is the solvent used in the preparation of the Grignard reagent (Secs. 2.7-A, 7.6).

Ethylene oxide, as was pointed out earlier, is an important intermediate in the manufacture of ethylene glycol. Ethylene oxide represents a highly strained ring and, unlike the dialkyl ethers, is a very reactive compound. When treated with alcohols, ethylene oxide converts them into monoalkyl ethers of ethylene glycol. These products are called **cellosolves** and combine the excellent solvent properties of both alcohols and ethers. Cellosolves are used as solvents for plastics and lacquers.

$$CH_2\!-\!\!CH_2 + C_2H_5OH \longrightarrow HOCH_2\!-\!CH_2\!-\!O\!-\!C_2H_5$$
$$\underset{O}{\diagdown\diagup}$$

Ethyl cellosolve

A cellosolve, when combined with a second molecule of ethylene oxide, produces monoalkyl ethers of diethylene glycol. These products are called **carbitols** and also are fine solvents.

$$HOCH_2\!-\!CH_2\!-\!O\!-\!C_2H_5 + CH_2\!-\!\!CH_2 \longrightarrow HO\!-\!(CH_2)_2\!-\!O\!-\!(CH_2)_2\!-\!O\!-\!C_2H_5$$
$$\underset{O}{\diagdown\diagup}$$

Ethyl carbitol
(Diethyleneglycol monoethyl ether)

Both the ether linkage and the hydroxyl group are found in many naturally occurring substances.

Summary

1. Alcohols may be classified as *primary, secondary,* and *tertiary;* ethers are classified as simple or mixed.

2. Alcohols associate through H-bond formation. H-bonding accounts for their abnormally high boiling points. Ethers are incapable of H-bonding and are not associated. They are low boiling, good solvents, and are rather inert chemically.

3. Phenols generally do not give the same reactions as do the aliphatic alcohols. Phenols have an acidic hydrogen. The hydroxyl group in phenols is a strong *ortho-para* director. Phenols are used in germicides, herbicides, fungicides, plastics, dyes, explosives, and many other useful everyday commodities.

4. **Nomenclature:**
   (A) Alcohols
      (a) (IUPAC) The suffix "ol" replaces terminal "e" of corresponding alkanes.
      (b) (Common) Group attached to —OH is named and followed by the word alcohol.

(c) (Derived) All groups attached to the hydroxylated carbon; are named as substituents of carbinol, e.g., CH_3OH.

(B) Phenols

(a) (IUPAC) Are named as hydroxy benzenes.

(b) (Common) Are named as phenol derivatives.

(C) Ethers

(a) (IUPAC) Are named as alkoxy alkanes.

(b) (Common) Are named simply as ethers. Both hydrocarbon groups are named.

5. **Preparation:**

(A) Alcohols can be prepared by one of the following methods:

(a) Synthetic methanol is prepared from CO and H_2.

(b) Ethanol is prepared by fermentation of sugars.

(c) Alcohols, in general, are prepared by the hydration of olefins (*via* H_2SO_4 or B_2H_6).

(d) Hydrolysis of alkyl halides.

(e) Addition of Grignard reagents to aldehydes and ketones.

(B) Phenol is prepared from

(a) Chlorobenzene (Dow process).

(b) Cumene (isopropylbenzene).

(c) Benzenesulfonic acid.

(C) Ethers are prepared by one of the following methods:

(a) Simple ethers from alcohols by the sulfuric acid method.

(b) Mixed ethers can be obtained *via* the Williamson synthesis.

6. **Reactions:** The reactions of alcohols, phenols, and ethers may be summarized as follows:

(A) Alcohols

(a) Cleavage of the O—H bond to produce alkoxides, ethers, and esters.

(b) Cleavage of the C—O bond to produce alkyl halides, sulfates, and nitrates.

(c) Dehydration to produce alkenes.

(d) Oxidation of primary alcohols to acids; secondary alcohols to ketones.

(B) Phenols

(a) Replacement of H by reaction with a strong base.

(b) Reduction of ring by catalytic hydrogenation.

(c) Ring substitution at *o*- and *p*-positions.

(C) Ethers

(a) Formation of oxonium salts in strong acids.

(b) Formation of peroxides.

(c) Cleavage with HI or HBr.

New Terms

absolute alcohol	Lucas reagent
associated liquid	oxonium salt
azeotrope	phenol coefficient
chelate	pheromone
Fries rearrangement	polar number
grain alcohol	"proof spirit"
H-bonding	Williamson synthesis
hydroboration	wood alcohol

Supplementary Exercises

EXERCISE 8.13 Assign an acceptable name to each of the following compounds:

(a) $(CH_3)_2CH-O-CH(CH_3)_2$

(b)

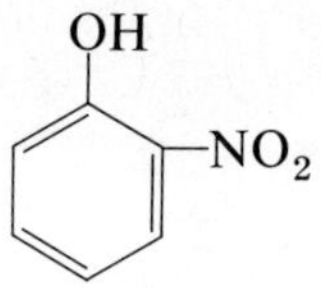

(c) $CH_3CH_2CH(OH)CH_3$

(d)

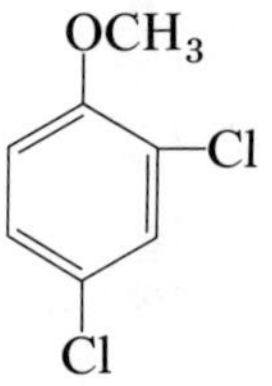

(e) $CH_3CH_2C(CH_3)_2OH$

(f)

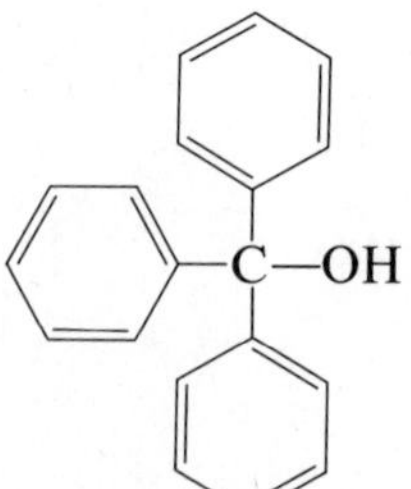

(g) $CH_3CH(OH)CH_2OH$

(h)

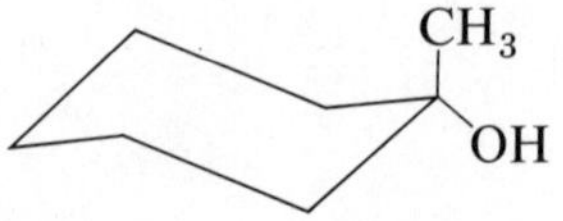

(i) 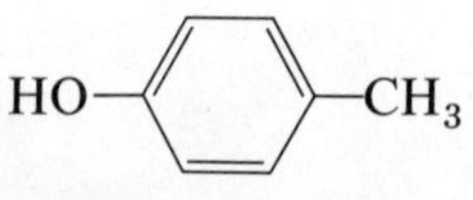 HO—⟨ ⟩—CH_3

(j)

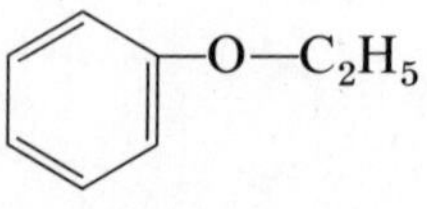

EXERCISE 8.14 Write structural formulas for the following.

(a) α-naphthol

(b) 2-phenyl-2-propanol

(c) benzyl alcohol

(d) 2,6-di-*tert*-butylphenol

(e) 1-phenylethanol

(f) 2-butene-1-ol

(g) dibenzyl ether

(h) *o*-nitrophenol

(i) sodium phenoxide

(j) potassium *tert*-butoxide

EXERCISE 8.15 Complete the following reactions.

(a) $CH_3—CH=CH_2 + HOCl \longrightarrow$

(b) $(CH_3)_2CHMgCl$ + ethylene oxide $\xrightarrow{\text{followed by hydrolysis}}$

(c) $(CH_3)_3COH + HCl \longrightarrow$

(d) $CH_3CH_2OH + H_2SO_4 \xrightarrow[140°C]{\Delta,}$

(e) 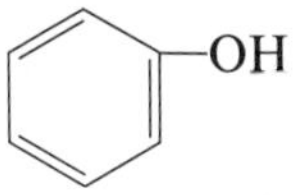 ⬡—OH $+ 3\,Br_2 \xrightarrow{H_2O}$

(f) 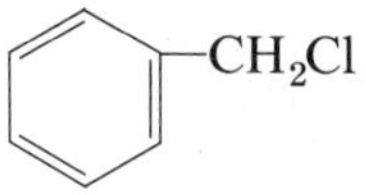 ⬡—CH_2Cl $+ Na^+OH^- \xrightarrow{\text{heat}}$

(g) *n*-propyl alcohol $+ PI_3 \xrightarrow{\text{heat}}$

(h) 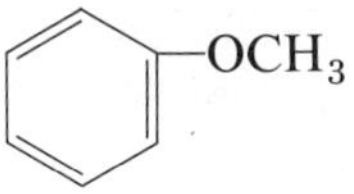 ⬡—OCH_3 $+ HI \xrightarrow{\text{heat}}$

(i) C_2H_5OH + sodium metal $\xrightarrow{\text{room temperature}}$

(j) *n*-propyl alcohol $+ K_2Cr_2O_7 + H_2SO_4 \xrightarrow{\text{heat}}$

(k) 1-butene $+ (BH_3)_2 \xrightarrow{H_2O_2,\ ^-OH}$

(l) 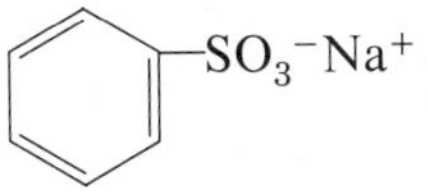 ⬡—$SO_3^-Na^+$ $+ NaOH \xrightarrow[\text{acidification}]{\text{fusion followed by}}$

(m) 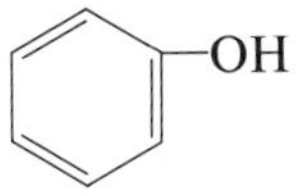 ⬡—OH $+ H_2 \xrightarrow[\text{heat}]{Ni,}$

EXERCISE 8.16   Using ethyl alcohol as your only organic starting material, write equations showing how you would prepare the following.

(a) ethylene   (f) ethylene glycol
(b) ethyl bromide   (g) ethylene oxide
(c) acetylene   (h) acetic acid
(d) 1-butyne   (i) iodoform
(e) ethylene chlorohydrin   (j) *n*-butyl alcohol

EXERCISE 8.17   Show two different reaction pathways whereby each of the following may be converted to the products shown.

(a) propene $\longrightarrow$ 1-propanol
(b) 1-butanol $\longrightarrow$ 1-bromobutane

EXERCISE 8.18   Arrange the compounds in the following series: (A) in an increasing order of solubility in water, (B) in an increasing order of reactivity toward Lucas reagent, and (C) in an increasing order of reactivity with sodium metal.

(a) 1-butanol   (d) 2-methyl-1-propanol
(b) 2-butanol   (e) ethanol
(c) 2-methyl-2-propanol   (f) 2,3-dimethyl-2-butanol

EXERCISE 8.19   Give a simple chemical test that would distinguish one from the other in each of the following pairs of compounds.

(a) *sec*-butyl alcohol and 1-hexyne
(b) isopropyl alcohol and isopropyl ether
(c) ethyl ether and *n*-pentane
(d) *tert*-butyl alcohol and *n*-butyl alcohol
(e) *sec*-butyl alcohol and *n*-butyl alcohol

EXERCISE 8.20   Draw the structures of both *cis* and *trans*-cyclopentane-1,2-diol. The *trans* isomer has a boiling point somewhat higher than that found for the *cis* isomer. Does this higher boiling point indicate a greater degree of association for the *trans* isomer? Why would the *cis* isomer be less inclined to form H-bonds with its neighbors?

EXERCISE 8.21   A compound containing only carbon, hydrogen, and oxygen was found to give no reaction with metallic sodium, PCl_5, or $SOCl_2$. When heated with an excess of HI only one product was isolated from the reaction mixture. On analysis the reaction product was found to contain 81.5% iodine. What could the original substance have been?

EXERCISE 8.22   The capacity of an automobile radiator is 20 quarts. If the density of ethylene glycol is 1.116 g/ml, how many quarts of the antifreeze must be used to protect the cooling system against freezing at $-40°C$ ($-40°F$)? (One quart = 960 ml. K_f for water = $1.86°C$/mole.)

EXERCISE 8.23 One of the classic rearrangements of carbonium ion chemistry is the *pinacol-pinacolone rearrangement.*

$$\underset{\text{Pinacol}}{\text{CH}_3\text{--}\underset{\underset{\text{CH}_3}{|}}{\overset{\overset{\text{OH}}{|}}{\text{C}}}\text{--}\underset{\underset{\text{CH}_3}{|}}{\overset{\overset{\text{OH}}{|}}{\text{C}}}\text{--CH}_3} \xrightarrow{\text{H}_2\text{SO}_4} \underset{\text{Pinacolone}}{\text{CH}_3\text{--}\underset{\underset{\text{CH}_3}{|}}{\overset{\overset{\text{CH}_3}{|}}{\text{C}}}\text{--}\overset{\overset{\text{O}}{||}}{\text{C}}\text{--CH}_3} + \text{H}_2\text{O}$$

Suggest a mechanism for this rearrangement. (*Hint:* Although you may need no information beyond that in this chapter, it may be helpful to review the four specific examples of resonance hybrids given in Sec. 1.12.)

EXERCISE 8.24 An optically active compound, $C_8H_{10}O$, gave an nmr spectrum consisting of a three-proton doublet at δ 1.33, a one-proton singlet at δ 2.75, a one-proton quartet at δ 4.67, and a five-proton singlet at δ 7.20. What are the name and structure of the compound?

EXERCISE 8.25 Show sequentially how each of the following conversions may be made.
   (a) propene $\longrightarrow$ 1-propanol
   (b) neopentyl alcohol $\longrightarrow$ 2-methyl-2-butene
   (c) 3,3-dimethyl-2-butanol $\longrightarrow$ 2-chloro-2,3-dimethylbutane
   (d) benzene $\longrightarrow$ β-phenylethyl alcohol
   (e) 3-methyl-1-butanol $\longrightarrow$ 2-methyl-2-butene

   (f) aniline $\longrightarrow$

EXERCISE 8.26 The product in each of the following acid-catalyzed reactions results from a typical reaction of a carbonium ion intermediate. Suggest a sequence of steps (mechanism) that could lead to each product.

(a)

(b)

(c)

EXERCISE 8.27   The infrared spectrum of a compound containing only carbon, hydrogen, and oxygen (molecular weight, 98) showed a broad, very strong band at 3400 cm^{-1}, a weak band at 2110 cm^{-1}, and the usual C—H bands at about 3000 cm^{-1}, but no other absorption above 1500 cm^{-1}. The nmr spectrum of the compound consisted of a one-proton singlet at δ 3.33, a one-proton singlet at δ 2.32, a two-proton quartet at δ 1.67, a three-proton singlet at δ 1.45, and a three-proton triplet at δ 0.95. What is the structure of the compound?

9

The Aldehydes and Ketones

Introduction

The aldehydes and ketones are compounds that contain the **carbonyl** group—a carbon-oxygen double bond, $-\overset{|}{C}{=}O$. The carbonyl is one of the most frequently encountered and, from the standpoint of synthetic organic chemistry, one of the most useful of the functional groups. Compounds which contain one or more carbonyl groups also are widely distributed in nature. For the most part, the aldehydes and ketones are of pleasant odor and are responsible for the active principles in a number of delightful-smelling, natural substances. For this reason, certain aldehydes and ketones are used as perfumes and as flavoring agents.

9.1 *The Structure of Aldehydes and Ketones*

The aldehydes and ketones often are referred to collectively as carbonyl compounds, but the two families differ in structure and in properties. The carbonyl carbon of an aldehyde is always bonded to *one hydrogen* atom, the remaining bond being shared with an alkyl or an aryl group. An exception is found in the case of formaldehyde (the simplest aldehyde), which has two hydrogens joined to the carbon atom of the carbonyl group. The carbonyl carbon of ketones, on the other hand, is bonded to *two organic* groups. Such groups may be identical or they may be different. Moreover, they may be either alkyl or aryl. The following examples serve to illustrate these structural variations.

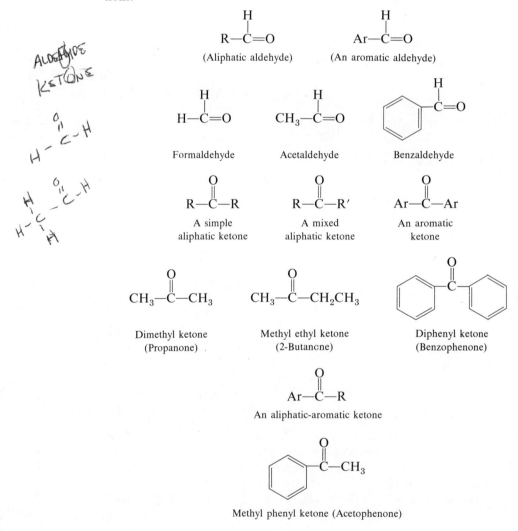

$$R—\overset{\overset{\displaystyle H}{|}}{C}=O$$
(Aliphatic aldehyde)

$$Ar—\overset{\overset{\displaystyle H}{|}}{C}=O$$
(An aromatic aldehyde)

$$H—\overset{\overset{\displaystyle H}{|}}{C}=O$$
Formaldehyde

$$CH_3—\overset{\overset{\displaystyle H}{|}}{C}=O$$
Acetaldehyde

Benzaldehyde

$$R—\overset{\overset{\displaystyle O}{||}}{C}—R$$
A simple aliphatic ketone

$$R—\overset{\overset{\displaystyle O}{||}}{C}—R'$$
A mixed aliphatic ketone

$$Ar—\overset{\overset{\displaystyle O}{||}}{C}—Ar$$
An aromatic ketone

$$CH_3—\overset{\overset{\displaystyle O}{||}}{C}—CH_3$$
Dimethyl ketone (Propanone)

$$CH_3—\overset{\overset{\displaystyle O}{||}}{C}—CH_2CH_3$$
Methyl ethyl ketone (2-Butanone)

Diphenyl ketone (Benzophenone)

$$Ar—\overset{\overset{\displaystyle O}{||}}{C}—R$$
An aliphatic-aromatic ketone

Methyl phenyl ketone (Acetophenone)

9.2 Nomenclature

Aldehydes are assigned common names derived from those of the carboxylic acids into which they are convertible by oxidation.

$$H-\overset{\overset{\displaystyle O}{\|}}{C}-H \xrightarrow{[O]} H-\overset{\overset{\displaystyle O}{\|}}{C}-OH$$

Formaldehyde   Formic acid

$$CH_3-\overset{\overset{\displaystyle O}{\|}}{C}-H \xrightarrow{[O]} CH_3-\overset{\overset{\displaystyle O}{\|}}{C}-OH$$

Acetaldehyde   Acetic acid CH_3COOH

Common names for the simple ketones are formed by naming both groups attached to the carbonyl carbon atom, then adding the word *ketone*. Trivial names, long in use, also are commonly employed. The IUPAC system of nomenclature follows established rules. The longest carbon chain (including the carbonyl carbon) is named after the parent hydrocarbon with *al* added as a suffix to designate an *al*dehyde. The carbonyl carbon atom of an aldehyde is always number one in the carbon chain and takes precedence over other functional groups that may be present. The carbonyl carbon atom of ketones, on the other hand, may appear at any point in the chain and must be located by number. The suffix *one* is used to designate a ket*one*. The examples illustrate these rules.

$$CH_3-\overset{\overset{\displaystyle H}{|}}{C}=O$$

Ethanal
(Acetaldehyde)

$$CH_3-\overset{\overset{\displaystyle H}{|}}{\underset{\underset{\displaystyle CH_3}{|}}{C}}-\overset{\overset{\displaystyle H}{|}}{C}=O$$

2-Methylpropanal
(Isobutyraldehyde)

$$CH_3-\overset{\overset{\displaystyle H}{|}}{\underset{(4)}{C}}=\overset{\overset{\displaystyle H}{|}}{\underset{(3)}{C}}-\overset{\overset{\displaystyle H}{|}}{\underset{(2)}{C}}=\underset{(1)}{O}$$

2-Butenal
(Crotonaldehyde)

$$CH_3-\overset{\overset{\displaystyle O}{\|}}{C}-CH_3$$

Propanone
(Acetone)

$$CH_3-\overset{\overset{\displaystyle O}{\|}}{\underset{(2)}{C}}-\underset{(3)}{CH_2}\underset{(4)}{CH_2}\underset{(5)}{CH_3}$$
(1)

2-Pentanone
(Methyl *n*-propyl ketone)

$$CH_3-CH_2-\overset{\overset{\displaystyle O}{\|}}{C}-CH_2-CH_3$$

3-Pentanone
(Diethyl ketone)

If the carbonyl group is attached to a benzene ring, ketones are given IUPAC names based on the names of the carboxylic acids from which they could be considered to be derived (see Sec. 9.4-D).

$$CH_3-\overset{\overset{\displaystyle O}{\|}}{C}-OH + H-\underset{}{\bigcirc} \xrightarrow[\text{required}]{\textit{usually several steps}} CH_3-\overset{\overset{\displaystyle O}{\|}}{C}-\underset{}{\bigcirc}$$

Acetic acid   Acetophenone

Benzoic acid + H— [benzene] → usually several steps required → Benzophenone

Usually the name is formed by replacing the *ic* or *oic acid* of the acid name by *ophenone* (exception: propionic acid → propiophenone).

In many naturally occurring substances both *al* and *one* suffixes frequently are employed in nonsystematic names to indicate the presence of aldehyde or ketone functions. The stem of the name frequently indicates the source of the substance. For example, *civetone,* a ketone found in a glandular secretion of the civet cat, has a cyclic structure. Civetone in a very dilute solution has a pleasant odor and is used in perfumery. *Citral,* an aldehyde found in oil of citrus fruits, is an unsaturated aldehyde used as a flavoring agent. The structures of both natural carbonyl compounds are shown following.

Civetone
(secretion of civet cat)
(9-Cycloheptadecene-1-one)

Citral (oil of lemon)
(3,7-Dimethyl-2,6-octadienal)

EXERCISE 9.1

The open chain aldehydes and ketones are isomeric. Draw structures for seven carbonyl compounds with the molecular formula $C_5H_{10}O$. Assign IUPAC names to each.

9.3 *Properties*

With the exception of formaldehyde, a gas, the lower molecular-weight aldehydes and ketones are liquids that have lower boiling points than alcohols of the same carbon content. The carbonyl compounds, lacking hydroxyl groups, do not hydrogen bond to themselves as do the alcohols, and therefore, are unassociated liquids. Table 9.1 compares the boiling points of some aldehydes and ketones with those of alcohols of the same carbon content.

The aldehydes and ketones, except the lower members which contain up to four carbon atoms, are practically insoluble in water. The lower members of the aldehyde family have sharp, irritating odors, but the higher molecular weight members, and nearly all members of the ketone family are fragrant. As was pointed out in the introductory section, certain of these are used in perfumery and as flavoring agents.

TABLE 9.1 *Boiling Points of Aldehydes and Ketones Compared to Those of Alcohols with the Same Carbon Chain*

Compound B.P., °C	Formula	Compound B.P., °C	Formula
Formaldehyde (−21)	$H-\overset{\displaystyle H}{\underset{}{C}}=O$	Methyl alcohol (64.6)	$H-\overset{\displaystyle H}{\underset{\displaystyle H}{C}}-OH$
Acetaldehyde (20.2)	$CH_3-\overset{\displaystyle H}{\underset{}{C}}=O$	Ethyl alcohol (78.3)	$CH_3-\overset{\displaystyle H}{\underset{\displaystyle H}{C}}-OH$
Propionaldehyde (48.8)	$CH_3CH_2\overset{\displaystyle H}{\underset{}{C}}=O$	n-Propylalcohol (97.8)	$CH_3CH_2\overset{\displaystyle H}{\underset{\displaystyle H}{C}}-OH$
Acetone (56.1)	$CH_3-\overset{\displaystyle O}{\overset{\|}{C}}-CH_3$	Isopropyl alcohol (82.5)	$CH_3-\overset{\displaystyle OH}{\underset{\displaystyle H}{C}}-CH_3$
n-Butyraldehyde (75.7)	$CH_3-CH_2CH_2-\overset{\displaystyle H}{\underset{}{C}}=O$	n-Butyl alcohol (117.7)	$CH_3CH_2CH_2\overset{\displaystyle H}{\underset{\displaystyle H}{C}}-OH$
Methyl ethyl ketone (79.6)	$CH_3-CH_2-\overset{\displaystyle O}{\overset{\|}{C}}-CH_3$	sec-Butyl alcohol (99.5)	$CH_3-CH_2-\overset{\displaystyle OH}{\underset{\displaystyle H}{C}}-CH_3$

9.4 Preparation of Aldehydes and Ketones

A-1. Oxidation of Alcohols (Dehydrogenation). The word *aldehyde* is a composite name originally given to this family of compounds to describe them as products of *al*cohol *dehy*drogenation. The oxidation of primary and secondary alcohols through dehydrogenation is a method for the preparation of aldehydes and ketones, respectively. Dehydrogenation is accomplished by passing the alcohol vapors over a heated copper or silver catalyst.

$$R-\underset{\boxed{H\ \ H}}{\overset{H}{C}}-O \xrightarrow[250°C]{Cu,} H_2 + R-\overset{H}{C}=O \qquad R-\underset{\boxed{H\ \ H}}{\overset{R}{C}}-O \xrightarrow[250°C]{Cu,} H_2 + R-\overset{R}{C}=O$$

Chemical oxidation of a primary alcohol is one of the most direct methods for the preparation of the simple aldehydes. Formaldehyde and acetaldehyde, perhaps the most important members of the aliphatic series of aldehydes, may be produced by oxidation of methyl and ethyl alcohols, respectively. The oxidation is generally done in air with the help of a catalyst.

$$2\,H\!-\!\overset{\displaystyle H}{\underset{\displaystyle H}{\overset{|}{\underset{|}{C}}}}\!-\!OH + O_2\ (\text{air}) \xrightarrow[250°C]{Cu,} 2\ H\!-\!\overset{\displaystyle H}{\overset{|}{C}}\!=\!O + 2\,H_2O$$

Formaldehyde

In aqueous solution the aldehyde hydrate, which is in equilibrium with the aldehyde, is especially susceptible to further oxidation. An aldehyde formed by the oxidation of an alcohol, therefore, will be converted directly into an acid unless some provision is made for its removal or the oxidation is done in a nonaqueous medium.

$$R\!-\!\overset{\displaystyle H}{\underset{\displaystyle H}{\overset{|}{\underset{|}{C}}}}\!-\!OH \xrightarrow{[O]} \left[R\!-\!\overset{\displaystyle OH}{\underset{\displaystyle H}{\overset{|}{\underset{|}{C}}}}\!-\!OH \right] \underset{+H_2O}{\overset{-H_2O}{\rightleftharpoons}} R\!-\!\overset{\displaystyle H}{\overset{|}{C}}\!=\!O$$

| Primary alcohol | Aldehyde hydrate | Aldehyde |

$$[O]\ \Big\downarrow\ \text{fast}$$

$$R\!-\!\overset{\displaystyle O}{\overset{\|}{C}}\!-\!OH$$

Acid

$$3\,CH_3CH_2OH + Cr_2O_7{}^{2-} + 8\,H^+ \xrightarrow{50°C} 3\,CH_3\overset{\displaystyle H}{\overset{|}{C}}\!=\!O + 2\,Cr^{3+} + 7\,H_2O$$

Ethyl alcohol (B.P. 78.3°C)   Acetaldehyde (B.P. 20.8°C)

$$3\,CH_3\!-\!\overset{\displaystyle H}{\overset{|}{C}}\!=\!O + Cr_2O_7{}^{2-} + 8\,H^+ \longrightarrow 3\,CH_3\!-\!\overset{\displaystyle O}{\overset{\|}{C}}\!-\!OH + 2\,Cr^{3+} + 4\,H_2O$$

Acetic acid (B.P. 118.1°C)

The boiling points of the aldehydes not only are much lower than those of the alcohols from which they may be obtained (Table 9.1), but also are considerably lower than the boiling points of the corresponding acids they would yield on further oxidation. This difference in physical properties is fortunate, for it makes possible the removal of the lower molecular weight aldehydes as they are formed by simple distillation. For less volatile aldehydes the complex between chromium trioxide and the heterocyclic amine, pyridine (Sec. 17.4) may be used as the oxidant in methylene chloride solution.

$$CH_3(CH_2)_4 \overset{H}{\underset{H}{-C-}} OH \xrightarrow[\text{CH}_2\text{Cl}_2]{\text{CrO}_3(\text{C}_5\text{H}_5\text{N})_2} CH_3(CH_2)_4 -C \overset{O}{\underset{H}{\diagdown}}$$

1-Hexanol   Hexanal

Ketones also may be prepared by oxidation methods using secondary alcohols as starting material. Ketones, unlike the aldehydes, are not easily oxidized (carbon-carbon bonds must be broken), and can be made in high yield by this method.

$$CH_3 \overset{H}{\underset{OH}{-C-}} CH_3 \xrightarrow{[O]} CH_3 \overset{O}{\underset{\|}{-C-}} CH_3$$

2-Propanol   2-Propanone
(Isopropyl alcohol)   (Acetone)

A-2. Oxidation of Alkenes (Ozonization). The oxidation of alkenes as a possible route to carbonyl compounds is discussed in some detail in Section 3.10-C.

B. Hydrolysis of *gem*-Dihalides. Dihalides in which the two halogen atoms are attached to the same carbon atom (*geminal* or *gem*-dihalides) may be hydrolyzed to carbonyl compounds.

The hydrolysis of *gem*-dihalides, as a preparative method, is employed most advantageously for the preparation of aromatic aldehydes. Beginning with toluene (readily available from petroleum), the side chain on the benzene ring can easily be chlorinated to benzal chloride. The latter is then hydrolyzed to benzaldehyde.

Toluene   Benzal
chloride

Benzaldehyde

C. Hydration of Alkynes. Acetaldehyde may be made by hydration of acetylene. This reaction is discussed in Section 3.19-B.

$$H-C\equiv C-H + HOH \xrightarrow[\text{HgSO}_4]{\text{H}_2\text{SO}_4,} \left[H-\overset{\overset{\displaystyle H}{|}}{C}=\overset{\overset{\displaystyle H}{|}}{C}-OH \right] \longrightarrow CH_3-\overset{\overset{\displaystyle H}{|}}{C}=O$$

Acetylene   Unstable enol   Acetaldehyde
intermediate

Large quantities of acetaldehyde prepared commercially by this method are used in the production of acetic acid. An alkyl-substituted acetylene, when hydrated, yields a methyl ketone. Note that the mode of addition is in accordance with Markovnikov's Rule (Sec. 3.8).

$$R-C\equiv C-H + H_2O \xrightarrow[\text{H}_2\text{SO}_4]{\text{Hg}^{2+},\ \text{SO}_4{}^{2-},} \left[R-\overset{\overset{\displaystyle }{|}}{\underset{\underset{\displaystyle OH}{|}}{C}}=CH_2 \right] \longrightarrow R-\overset{\overset{\displaystyle O}{\|}}{C}-CH_3$$

D. Friedel-Crafts Acylation. The preparation of benzaldehyde from toluene already has been illustrated (Sec. 9.4-B). Certain aromatic aldehydes may be prepared by the direct introduction of the aldehyde group into the benzene nucleus through a modified Friedel-Crafts reaction known as **formylation.**

Formyl chloride, $H-\overset{\overset{\displaystyle O}{\|}}{C}-Cl$, is a unstable compound and decomposes to carbon monoxide and hydrogen chloride. A combination of these two gases in the presence of anhydrous aluminum chloride accomplishes the same result as if the acid chloride were used. The reaction is known as the Gatterman-Koch reaction.

$$HCl + CO \longrightarrow \left[Cl-C{\overset{\displaystyle H}{\diagdown_{\displaystyle O}}} \right]$$

Aromatic ketones also may be prepared by the Friedel-Crafts acylation reaction using either acid chlorides or acid anhydrides. In either case, an acyl group, $R-\overset{\overset{\displaystyle O}{\|}}{C}-$, is attached directly to the aromatic ring. The method is illustrated in the preparation of acetophenone using acetyl chloride, the acid chloride of acetic acid.

$$CH_3-C\underset{Cl}{\overset{O}{\diagup}} + AlCl_3 \longrightarrow CH_3-C\overset{O}{\underset{\oplus}{\diagdown}}\cdots ClAlCl_3^{(-)}$$

Acetyl chloride

$$\text{Acetophenone} \quad + HCl + AlCl_3$$

E. Other Methods Based on Carboxylic Acids or Their Derivatives. In a number of useful procedures besides the Friedel-Crafts acylation for the preparation of aldehydes and ketones, carboxylic acids or their derivatives may be used as starting materials. Some of these procedures are described in Sec. 10.5.

F. Industrial Procedures. Industrially, **acetone** is prepared by the dehydrogenation of isopropyl alcohol and, to a lesser degree, as a by-product in the Weizmann fermentation of carbohydrates. The Weizmann fermentation, a process discovered by Chaim Weizmann[1] during World War I, is a bacterial fermentation. Sugar from either corn mash or blackstrap molasses is inoculated with the bacterium, *Clostridium acetobutylicum.* Under anaerobic ("without air") conditions the fermentation produces *n*-butyl alcohol, acetone, and ethyl alcohol in approximate yields of 60%, 30%, and 10%, respectively.

Acetone is used principally as a solvent and as an intermediate in organic synthesis.

Furfuraldehyde, $C_5H_4O_2$, usually called **furfural,** is a very important industrial aldehyde obtained from agricultural wastes such as corn cobs and oat hulls (Sec. 17.2). It is a colorless liquid when freshly distilled, but on exposure to air becomes oxidized to a deep brown or black liquid. Treatment of furfural (a cyclic ether as well as a carbonyl compound) with a mineral acid results in a ring cleavage. In neutral or basic solutions furfural gives all the reactions of benzaldehyde. Furfural is used in the petroleum industry for the refining of lubricants, as starting material in the manufacture of some nylon, and in synthetic resins. The structure of furfural is shown.

$$\begin{array}{c} H-C\text{------}C-H \\ \| \qquad\quad \| \\ H-C \qquad C-CHO \\ \diagdown\;\;\diagup \\ O \end{array}$$

Furfuraldehyde
(Furfural)

---

[1] Chaim Weizmann, 1874–1952. University of Manchester. First president of newly founded Israel.

9.5   *The Carbonyl Group*

It would be profitable to discuss briefly the nature of the carbonyl group before we consider the reactions of the aldehydes and ketones. The unsaturated carbonyl carbon atom, like the doubly bonded carbon atoms of the alkenes, is joined by three σ bonds to three other atoms. All lie in the same plane and form angles of 120° with each other. The planarity of the carbonyl group is illustrated using formaldehyde as an example.

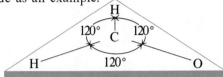

The fourth bond, a π bond, is made up of a pair of shared electrons, one each from the carbonyl carbon and the oxygen atom. These π electrons, analogous to the π electrons of the carbon-carbon double bond, confer upon the carbonyl group a special reactivity. The oxygen atom, you will remember, is far more electronegative (electron-attracting) than is the carbon atom. This attraction for electrons on the part of the oxygen atom produces an unequal distribution of electrons between the nuclei of the carbon and the oxygen atoms. The carbon-oxygen bond is thus a highly polarized one—the oxygen electron-rich and the carbon electron-deficient.

$$\left[\overset{\delta+}{\underset{}{>}}C \overset{\delta-}{=\!\!\!\longrightarrow} \ddot{O} : \longleftrightarrow \overset{+}{\underset{}{>}}C - \ddot{O} :^{-} \right]$$

In the addition reactions illustrated in the following sections, bases and nucleophiles (Sec. 1.11) attack the electron-deficient carbonyl carbon atom, and acids and electrophiles (Sec. 1.11) attack the basic oxygen atom. Thus, with most reagents, carbonyl addition reactions show the same overall course: addition of the negative (nucleophilic) part of the reagent to the carbon atom and addition of the positive (electrophilic) part of the reagent to the oxygen atom.

$$\overset{\delta+}{\underset{}{>}}C \overset{\delta-}{=\!\!\!\longrightarrow} O$$
$$\underset{\ddot{B}^{-} \quad H^{+}}{}$$

EXERCISE 9.2

Of the two structures shown, which carbonyl carbon atom will be more readily attacked by the hydroxide ion? Why?

$$CH_3 - \overset{\overset{\displaystyle O}{\|}}{C} - CH_3 \qquad CH_3 - \overset{\overset{\displaystyle O}{\|}}{C} - CCl_3$$

A carbonyl group attached to a benzene nucleus deactivates the ring through its electron-attracting tendency and, in turn, is deactivated by the ring. This effect not only makes the ring more resistant to further attack by electrophilic groups but also directs entering groups to the *meta* position when substitution does take place. Furthermore, carbonyl groups attached to aromatic rings are less reactive in addition reactions than aliphatic and alicyclic carbonyl compounds.

EXERCISE 9.3

Write structures for the reasonable resonance forms for acetophenone, showing the interaction of the carbonyl group and the ring. On the basis of your structures explain why acetone is more reactive than acetophenone in most addition reactions.

Another important characteristic of carbonyl compounds is the acidity of hydrogen atoms on carbon atoms *alpha* to the carbonyl group (called α-hydrogens). We first encountered carbon acids in the alkynes (Sec. 3.16); however, acetone is about 100,000 times as strong an acid as acetylene (based on their respective ionization constants). Because of the reactivity of the α-hydrogens, in solution aldehydes and ketones may exist as equilibrium mixtures of two isomeric forms, a **keto** form and an **enol** form.

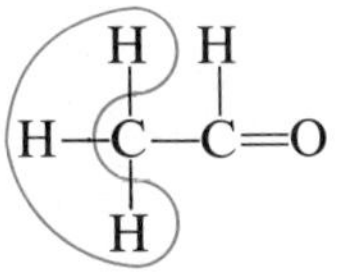

Acetaldehyde
(three α-hydrogens)

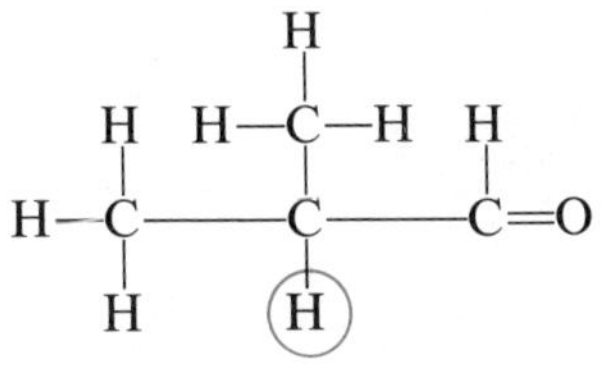

Isobutyraldehyde
(one α-hydrogen)

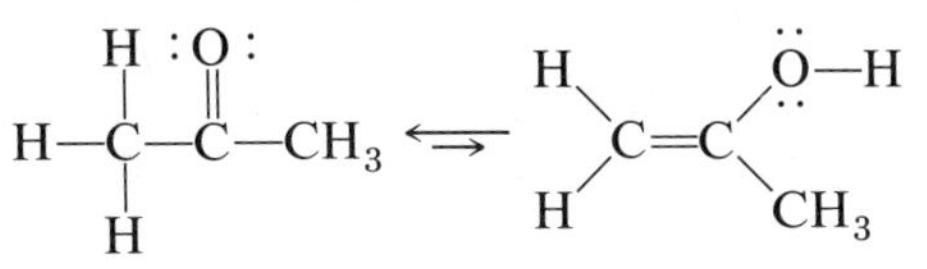

Keto form
of acetone

Enol form
of acetone

This type of isomerism is called **tautomerism** and the isomers are known as **tautomers.** At equilibrium for most simple carbonyl compounds very little enol is present, but the equilibrium is established very rapidly; therefore, enols are important intermediates in many of the reactions of carbonyl compounds. Enols may be stabilized by conversion of the hydroxyl group to ethers or esters (Sec. 10.5-G). An example is the important, high-energy metabolic intermediate, phosphoenolpyruvate (PEP) shown in the acid form.

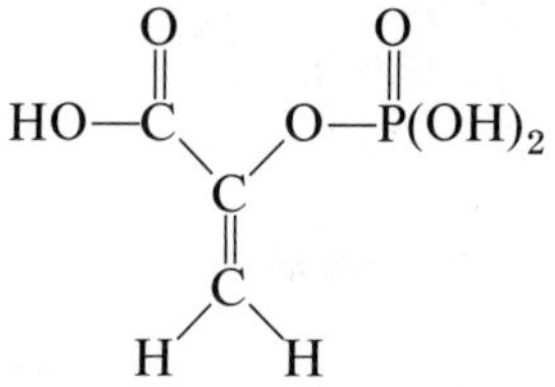

$$\text{HO-}\overset{\overset{\displaystyle O}{\|}}{C}\text{ }\overset{\overset{\displaystyle O}{\|}}{\underset{\underset{\displaystyle C}{|}}{C}}\text{-O-}\overset{\overset{\displaystyle O}{\|}}{P}(OH)_2$$

Phosphoric acid ester
of the enol

Enolization is catalyzed by both acids and bases as is shown briefly in the following equations:

Base-catalyzed enolization:

(1) $HO:^- + H-CH_2-C\overset{O:}{\underset{CH_3}{}} \rightleftharpoons \left[^-:CH_2-C\overset{O:}{\underset{CH_3}{}} \longleftrightarrow CH_2=C\overset{O:^-}{\underset{CH_3}{}} \right] + H_2O$

Enolate anion

(2) $\left[^-:CH_2-C\overset{O:}{\underset{CH_3}{}} \longleftrightarrow CH_2=C\overset{O:^-}{\underset{CH_3}{}} \right] + H-OH \rightleftharpoons CH_2=C\overset{O-H}{\underset{CH_3}{}} + :OH^-$

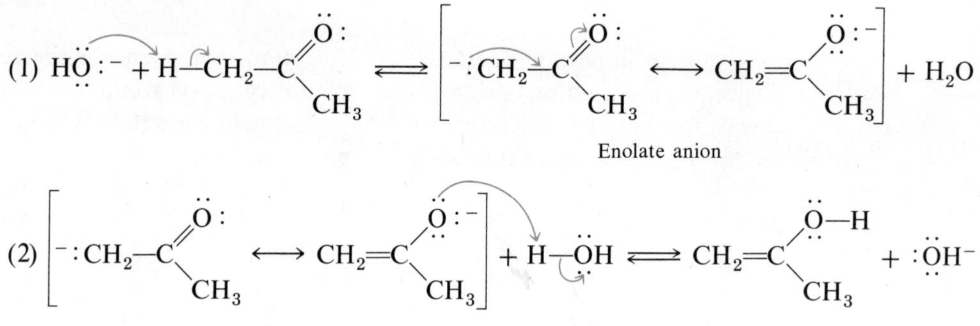

Acid-catalyzed enolization:

(1) $H-CH_2-C\overset{O:}{\underset{CH_3}{}} + H^+ \rightleftharpoons \left[H-CH_2-C\overset{\overset{+}{O}-H}{\underset{CH_3}{}} \longleftrightarrow H-CH_2-\overset{+}{C}\overset{O-H}{\underset{CH_3}{}} \right]$

Protonated acetone (oxonium ion)

(2) $\left[H-CH_2-C\overset{\overset{+}{O}-H}{\underset{CH_3}{}} \longleftrightarrow H-CH_2-\overset{+}{C}\overset{O-H}{\underset{CH_3}{}} \right] \rightleftharpoons CH_2=C\overset{O-H}{\underset{CH_3}{}} + H^+$

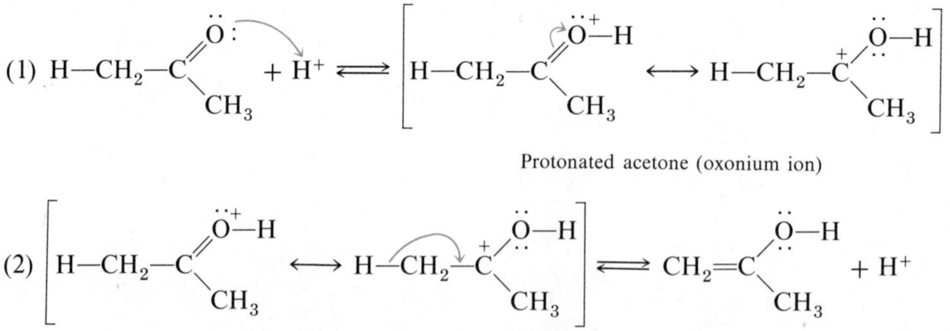

EXERCISE 9.4

Explain why acetone is a stronger acid than acetylene. (*Hint:* Look at the mechanism for the base-catalyzed enolization.)

9.6 Addition Reactions

A. Addition of Hydrogen Cyanide. The elements of hydrogen cyanide, HCN, add to aldehydes and ketones to yield **cyanohydrins** when the reaction is carried out with a basic catalyst.

$$HCN + \overset{(-)}{OH} \longrightarrow H_2O + \overset{(-)}{CN}$$

$$\underset{}{\overset{}{>}}C\!\!=\!\!O + \overset{(-)}{CN} \longrightarrow \underset{}{\overset{}{>}}C\overset{O^-}{\underset{CN}{<}}$$

$$\underset{}{\overset{}{>}}C\overset{O^-}{\underset{CN}{<}} + H_2O \longrightarrow \underset{}{\overset{}{>}}C\overset{OH}{\underset{CN}{<}} + \overset{(-)}{OH}$$

Hydrogen cyanide, a highly toxic gas, is rarely handled directly, but is produced at the site of a reaction by treating sodium or potassium cyanide with a mineral acid. However, the amount of acid used must never be enough to react with all the cyanide ion; otherwise the basic conditions favorable to the reaction will be lacking.

$$\underset{CH_3}{\overset{CH_3}{>}}C\!\!=\!\!O + NaCN + H_2SO_4 \xrightarrow{10\text{-}20°C} \underset{CH_3}{\overset{CH_3}{>}}C\overset{OH}{\underset{CN}{<}} + NaHSO_4$$

Acetone   Acetone
                                          cyanohydrin

Aldehydes, aliphatic methyl ketones, and cyclic ketones form cyanohydrins. Inasmuch as these addition compounds can be hydrolyzed to acids (Sec. 10.4-B) the cyanohydrins are valuable intermediates in organic synthesis. The conversion of acetaldehyde to lactic acid is an example.

$$CH_3\!-\!\overset{H}{\underset{}{C}}\!\!=\!\!O + HCN \longrightarrow CH_3\!-\!\overset{H}{\underset{OH}{C}}\!-\!CN$$

Acetaldehyde   Acetaldehyde
                                    cyanohydrin

$$CH_3\!-\!\overset{H}{\underset{OH}{C}}\!-\!CN + HCl + 2\,H_2O \longrightarrow CH_3\!-\!\overset{H}{\underset{OH}{C}}\!-\!\overset{O}{\overset{\|}{C}}\!-\!OH + NH_4Cl$$

Lactic acid
(a hydroxy acid)

B. Addition of Sodium Bisulfite. A saturated solution of sodium bisulfite, when mixed with aldehydes, cyclic ketones, and with some methyl ketones, forms addition products. These are crystalline solids which may be separated from the mixture by filtration. Treatment of the addition product with a mineral acid regenerates the original carbonyl compound.

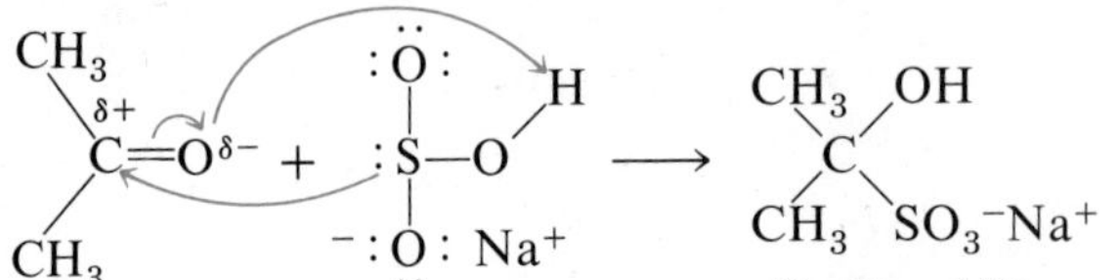

Bisulfite addition
compound of acetone

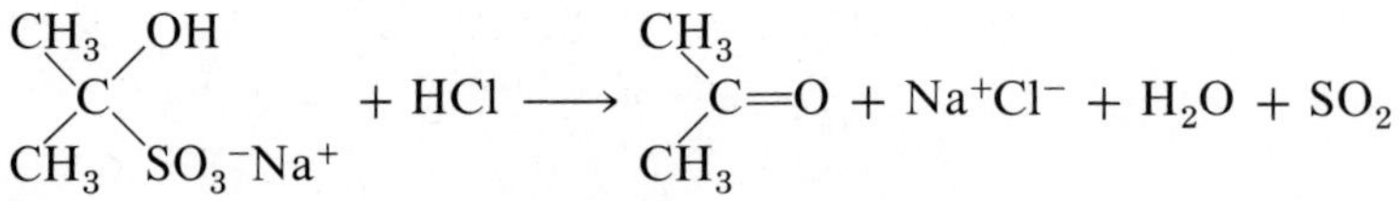

The reaction provides a useful method for the separation of aldehydes and certain ketones from mixtures. Aromatic ketones and aliphatic ketones having both alkyl groups larger than methyl show little or no tendency to form bisulfite addition compounds.

EXERCISE 9.5

Suggest a reason why cyclopentanone forms a sodium bisulfite addition compound but diethyl ketone does not.

C. Addition of Water and Alcohols. The aldehydes and ketones show little tendency to form *stable* hydrates. In a few compounds, in which the carbonyl group is attached to other strong electron-attracting groups, a hydrate sufficiently stable to be isolated is formed. One of the best examples of such a compound is trichloroacetaldehyde, or **chloral,** which forms a stable hydrate used medicinally as a soporific.

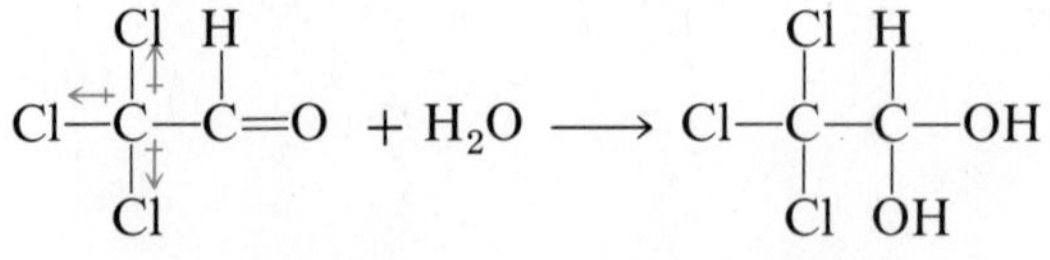

Chloral   Chloral hydrate

On the other hand, alcohols will add to aldehydes in the presence of acid catalysts. An unstable addition product called a **hemiacetal,** formed in the first stage of the reaction, reacts with a second molecule of alcohol to yield a stable compound called an **acetal.**

$$CH_3-\overset{\overset{\displaystyle H}{|}}{C}=O + C_2H_5OH \underset{H^+A^-}{\rightleftharpoons} CH_3-\overset{\overset{\displaystyle H}{|}}{\underset{\underset{\displaystyle OC_2H_5}{}}{C}}{\diagdown}{OH}$$

A hemiacetal
(unstable)

$$CH_3-\overset{\overset{\displaystyle H}{\diagup}}{\underset{\underset{\displaystyle OC_2H_5}{\diagdown}}{C}}{-}OH + C_2H_5OH \xrightarrow{H^+A^-} CH_3-\overset{\overset{\displaystyle H}{\diagup}}{\underset{\underset{\displaystyle OC_2H_5}{\diagdown}}{C}}{-}OC_2H_5 + H_2O$$

Acetaldehyde diethyl acetal
(stable)

The reaction is reversible and the acetal, when hydrolyzed in an acid solution, readily regenerates the original alcohol and aldehyde.

The equilibrium for the formation of higher molecular weight acetals and most **ketals** (from the reaction of ketones with alcohols) is unfavorable; therefore, the water formed is removed (by azeotropic distillation (Sec. 8.4)) to push the equilibrium to the right. Cyclic acetals and ketals with 5- and 6-membered rings form very readily and are widely employed in carbohydrate chemistry. Acetone is the ketone most frequently used with carbohydrates, giving what is called an isopropylidene derivative.

$$CH_3-\overset{\overset{\displaystyle O}{\|}}{C}-CH_3 + HO-CH_2-CH_2-OH \underset{H^+}{\rightleftharpoons}$$

Acetone   Ethylene glycol   A cyclic ketal

D. **Addition of Grignard Reagents.** The addition of Grignard reagents (Sec. 7.6) of the form RMgX or ArMgX to aldehydes or to ketones as a route to the alcohols has already been presented in detail in Section 8.4-C.

EXERCISE 9.6

Using propene and any inorganic reagents that you might need, show how you could obtain the necessary Grignard reagent and carbonyl compound needed to prepare 2,3-dimethyl-2-butanol.

9.7 *Addition Reactions with Loss of Water*

Reactions in which two reactants combine with the loss of a molecule of water (or alcohol) are called **condensations.**

A. Condensation with Ammonia Derivatives.   Certain derivatives of ammonia that contain the primary amino group, —NH_2, add to aldehydes and ketones to form unstable intermediates. The initial addition product loses the elements of water (condensation) to form a carbon-nitrogen double bond.

$$\begin{array}{c}R\\[-2pt]{\diagdown}\\[-4pt]\\[-4pt]R{\diagup}\end{array}C{=}O + H_2NR' \rightleftharpoons \begin{array}{c}R\\[-2pt]{\diagdown}\\[-4pt]\\[-4pt]R{\diagup}\end{array}C\begin{array}{c}OH\\ H\\ N\\ {\diagdown}\\ R'\end{array} \longrightarrow \begin{array}{c}R\\[-2pt]{\diagdown}\\[-4pt]\\[-4pt]R{\diagup}\end{array}C{=}N\begin{array}{c}\\ {\diagdown}\\ R'\end{array} + H_2O$$

Many of these condensation products are crystalline solids with sharp melting points. For this reason they frequently are employed for the preparation of aldehyde and ketone derivatives needed in identification work. Ammonia derivatives commonly used and their condensation products follow.

Carbonyl Compound	Ammonia Derivative	Condensation Product
$\underset{\text{A ketone*}}{R{-}\overset{\displaystyle R}{\overset{\mid}{C}}{=}O}$	$+$ $\underset{\text{Hydroxylamine}}{\overset{\displaystyle H}{\underset{\displaystyle H}{>}}N{-}OH}$	$\xrightarrow{H^+}$ $\underset{\text{An oxime}}{R{-}\overset{\displaystyle R}{\overset{\mid}{C}}{=}N{-}OH} + H_2O$
$\underset{\text{A ketone}}{R{-}\overset{\displaystyle R}{\overset{\mid}{C}}{=}O}$	$+$ $\underset{\text{Phenylhydrazine}}{\overset{\displaystyle H}{\underset{\displaystyle H}{>}}N{-}\overset{\displaystyle H}{\overset{\mid}{N}}{-}C_6H_5}$	$\xrightarrow{H^+}$ $\underset{\text{A phenylhydrazone}}{R{-}\overset{\displaystyle R}{\overset{\mid}{C}}{=}N{-}\overset{\displaystyle H}{\overset{\mid}{N}}{-}C_6H_5} + H_2O$
$\underset{\text{A ketone}}{R{-}\overset{\displaystyle R}{\overset{\mid}{C}}{=}O}$	$+$ $\underset{\text{Semicarbazide}}{\overset{\displaystyle H}{\underset{\displaystyle H}{>}}N{-}\overset{\displaystyle H}{\overset{\mid}{N}}{-}\overset{\displaystyle O}{\overset{\|}{C}}{-}NH_2}$	$\xrightarrow{H^+}$ $\underset{\text{A semicarbazone}}{R{-}\overset{\displaystyle R}{\overset{\mid}{C}}{=}N{-}\overset{\displaystyle H}{\overset{\mid}{N}}{-}\overset{\displaystyle O}{\overset{\|}{C}}{-}NH_2} + H_2O$

* If an R were H, the compound would be an aldehyde.

The carbonyl derivative formed in each case simply is designated as the **aldoxime,** the **ketoxime,** the **phenylhydrazone,** or the **semicarbazone** of the carbonyl compound from which it was prepared.

$$CH_3{-}\overset{\displaystyle H}{\overset{\mid}{C}}{=}O + H_2N{-}OH \longrightarrow CH_3{-}\overset{\displaystyle H}{\overset{\mid}{C}}{=}N\diagup^{OH} + H_2O$$
$$\text{Acetaldoxime (M.P. 47°C)}$$

$$\begin{array}{c}CH_3CH_2\\[-2pt]{\diagdown}\\[-4pt]\\[-4pt]CH_3CH_2{\diagup}\end{array}C{=}O + H_2NOH \longrightarrow \begin{array}{c}CH_3CH_2\\[-2pt]{\diagdown}\\[-4pt]\\[-4pt]CH_3CH_2{\diagup}\end{array}C{=}N\diagup^{OH} + H_2O$$
$$\text{Diethylketoxime (M.P. 69°C)}$$

EXERCISE 9.7

Draw two possible structures for the oxime obtainable from 2-butanone. To what type of stereoisomers are these related?

The semicarbazones usually are higher melting solids than are the oximes or phenylhydrazones. Should the oxime or the phenylhydrazone of an unknown carbonyl compound be a low melting solid or an oil, as is sometimes the case,

TABLE 9.2 *Derivatives Useful for the Identification of Aldehydes and Ketones*

Compound	Formula	B.P., °C	Oxime M.P., °C	Phenylhy-drazone M.P., °C	Semi-carbazide M.P., °C	2,4-Dini-trophenyl-hydrazone M.P., °C
Acetone	$CH_3-\overset{\overset{O}{\|\|}}{C}-CH_3$	56	59	42	187	126
Acetaldehyde	$CH_3-\overset{\overset{H}{\|}}{C}=O$	20	47	63	162	147
α-Methyl-*n*-butyraldehyde	$CH_3CH_2CH(CH_3)\overset{\overset{H}{\|}}{C}=O$	93	oil	oil	103	120
Benzaldehyde	$\overset{\overset{H}{\|}}{C}=O$ (phenyl)	179	35	158	222	237
Methyl ethyl ketone	$C_2H_5-\overset{\overset{CH_3}{\|}}{C}=O$	80	oil	oil	146	117
Diethyl ketone	$(C_2H_5)_2C=O$	102	69	oil	139	156
Di-*n*-propyl ketone	$(C_3H_7)_2C=O$	145	oil	oil	133	75
Isopropyl methyl ketone	$(CH_3)_2CH\overset{\overset{CH_3}{\|}}{C}=O$	94	oil	oil	113	117
Cyclopentanone	(cyclopentanone structure)	131	56	50	205	142

the semicarbazone usually will serve to identify it. Frequently, the higher the molecular weight of a carbonyl derivative, the higher its melting point. Thus, 2,4-*dinitrophenylhydrazine* often is used in place of phenylhydrazine for the preparation of carbonyl derivatives.

$$H—N—NH_2$$

2,4-Dinitrophenylhydrazine

Table 9.2 illustrates the wide divergence in melting points of oximes, phenylhydrazones, and semicarbazones of some common carbonyl compounds.

B. The Aldol Condensation.   In the resonance-stabilized enolate ion (Sec. 9.5), the negative charge is largely on the more electronegative oxygen atom, as is indicated by the name, *enolate*. However, as was stated in Sec. 3.12, resonance hybrids may show the reactions that would be expected of more than one contributing form. Thus, enolate ions can and do behave as carbanions (carbon nucleophiles) in many reactions. When an enolate ion adds (as if it were a carbanion) to another molecule of aldehyde or ketone, the reaction is called an **aldol[2] condensation.** An aldol condensation involving acetaldehyde reacting with itself in the presence of a basic catalyst is shown as an example. Because the aldol condensation is reversible, the reaction often gives better results if it's followed by one with a more favorable equilibrium (such as dehydration) or by an essentially irreversible reaction (such as the Cannizzaro reaction), as shown in other examples in this section. The reversibility of the aldol condensation is important in biological processes for it is involved in both the biosynthesis (anabolism) and degradation (catabolism) of carbohydrates, with enzymes called *aldolases* serving as the catalysts.

_____

[2] Aldol, a composite word for *ald*ehyde + alcoh*ol*.

$$CH_3-\underset{\underset{\displaystyle :\overset{..}{\underset{..}{O}}:^-}{|}}{\overset{\overset{\displaystyle H}{|}}{C}}-CH_2-\overset{\overset{\displaystyle H}{|}}{C}=O + H_2O \; \rightleftharpoons \; CH_3-\underset{\underset{\displaystyle OH}{|}}{\overset{\overset{\displaystyle H}{|}}{C}}-CH_2-\overset{\overset{\displaystyle H}{|}}{C}=O + OH^-$$

Aldol

(3-Hydroxybutanal or

β-Hydroxybutyraldehyde)

Aldols readily lose water when heated to give **α,β-unsaturated carbonyl compounds.**

$$CH_3-\underset{\underset{\displaystyle OH}{|}}{\overset{\overset{\displaystyle H}{|}}{C}}-\underset{\underset{\displaystyle H}{|}}{\overset{\overset{\displaystyle H}{|}}{C}}-\overset{\overset{\displaystyle H}{|}}{C}=O \; \xrightarrow{\;heat\;} \; H_2O + CH_3-\overset{\overset{\displaystyle H}{|}}{C}=\overset{\overset{\displaystyle H}{|}}{C}-\overset{\overset{\displaystyle H}{|}}{C}=O$$

2-Butenal

(Crotonaldehyde)

Ketones containing α-hydrogen also are capable of an aldol-type condensation.

EXERCISE 9.8

Write the structures for the four condensation products possible when a mixture of acetaldehyde and *n*-butyraldehyde is warmed with dilute sodium hydroxide.

An aldehyde that lacks an α-hydrogen can enter into an aldol condensation with another aldehyde only by serving as the carbanion acceptor. An excellent illustration of this kind of situation is found in the preparation of pentaerythritol, $(HOCH_2)_4C$, a polyhydric alcohol used in the preparation of explosives. The first step in the preparation of pentaerythritol is the condensation of formaldehyde with acetaldehyde to produce trimethylolacetaldehyde.

$$O=\underset{\underset{\displaystyle H}{}}{\overset{\overset{\displaystyle H}{}}{C}} \quad \overset{\overset{\displaystyle O}{\|}}{\underset{}{C}} \quad + \; H-\underset{\underset{\displaystyle H}{|}}{\overset{\overset{\displaystyle H}{|}}{C}}-\overset{\overset{\displaystyle H}{|}}{C}=O \; \xrightarrow{Ca(OH)_2} \; (HOCH_2)_3C-\overset{\overset{\displaystyle H}{|}}{C}=O$$

Formaldehyde   Acetaldehyde   Trimethylolacetaldehyde

The second step is a "crossed" Cannizzaro reaction (Sec. 9.9-B) between two aldehydes neither of which has α-hydrogen.

$$(HOCH_2)_3C-\overset{\overset{\displaystyle H}{|}}{C}{=}O + CH_2O + H_2O \xrightarrow{\text{NaOH}} (HOCH_2)_4C + H-\overset{\overset{\displaystyle O}{\|}}{C}-OH$$

Pentaerythritol   Formic acid
        (as sodium salt)

Benzaldehyde reacts with acetaldehyde to produce cinnamaldehyde, an α,β-unsaturated aromatic aldehyde used as a flavoring agent.

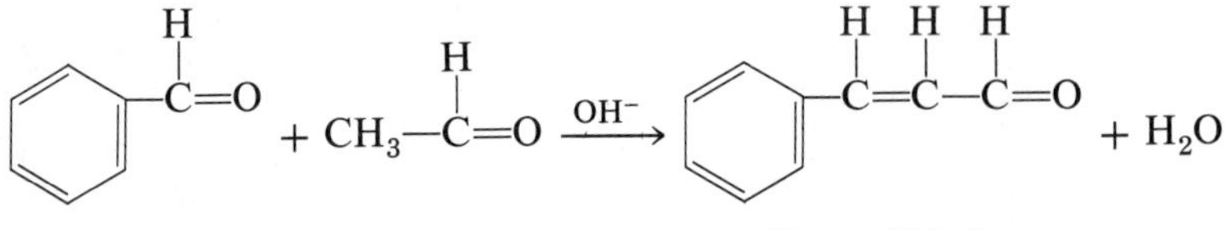

Benzaldehyde   Cinnamaldehyde

EXERCISE 9.9

The following compounds can be made by the aldol condensation, perhaps followed by another reaction. Write the equations for the preparation of these substances.

(1)

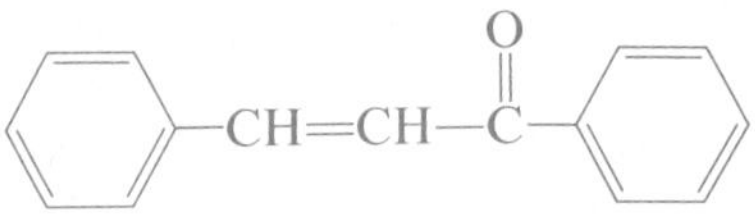

(2) $CH_3-\overset{\overset{\displaystyle CH_3}{|}}{\underset{\underset{\displaystyle OH}{|}}{C}}-CH_2-\overset{\overset{\displaystyle O}{\|}}{C}-CH_3$

(3)

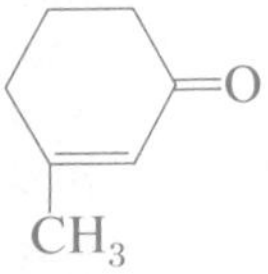

C. The Wittig Reaction. Phosphorus **ylids,**[3] which are prepared as shown below, react readily with aldehydes and ketones to give alkenes with precisely determined substitution on the double bond; that is, there will be no isomers except *cis,trans*-isomers. At present, this very general reaction, known as the **Wittig reaction,** is one of the most useful methods for the synthesis of alkenes.

(1) $(C_6H_5)_3P\colon\!\!\overset{\frown}{+}CH_3\!\!-\!\!\overset{..}{\underset{..}{Br}}\colon \longrightarrow CH_3-\overset{+}{P}(C_6H_5)_3\ Br^-$

  Triphenyl-   Methyltriphenyl-
  phosphine   phosphonium bromide

---

[3] Ylids are internal salts formed by removing a proton from a carbon atom adjacent to a positively charged heteroatom (usually phosphorus or sulfur).

(2) $B:^- + \overset{+}{C}H_2 - \overset{+}{P}(C_6H_5)_3 \ Br^- \longrightarrow$ H

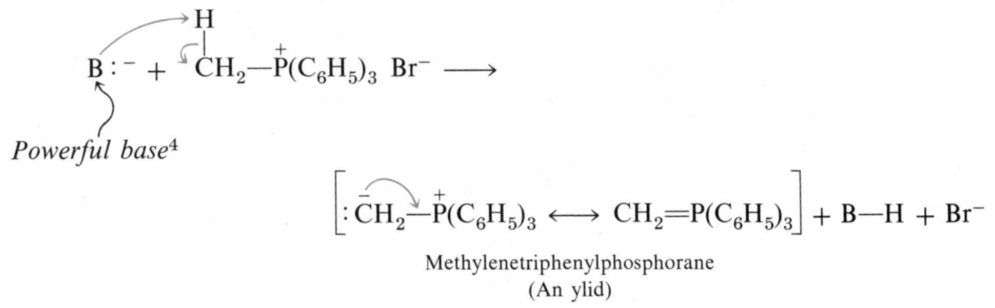

Powerful base[4]

$$\left[:\bar{C}H_2 - \overset{+}{P}(C_6H_5)_3 \longleftrightarrow CH_2{=}P(C_6H_5)_3 \right] + B{-}H + Br^-$$

Methylenetriphenylphosphorane
(An ylid)

(3) $:\bar{C}H_2 - \overset{+}{P}(C_6H_5)_3 + \overset{\delta^-}{O}{=}\overset{\delta^+}{C}\overset{CH_3}{\underset{CH_2CH_3}{\diagup}} \longrightarrow \left[\begin{array}{c} CH_2{-}C\overset{CH_3}{\underset{CH_2CH_3}{\diagdown}} \\ (C_6H_5)_3P{-}O \end{array} \right]$

an unstable oxaphosphetane

$$CH_2{=}C\overset{CH_3}{\underset{CH_2CH_3}{\diagup}} + (C_6H_5)_3P{=}O$$

2-Methyl-1-butene   Triphenylphosphine oxide

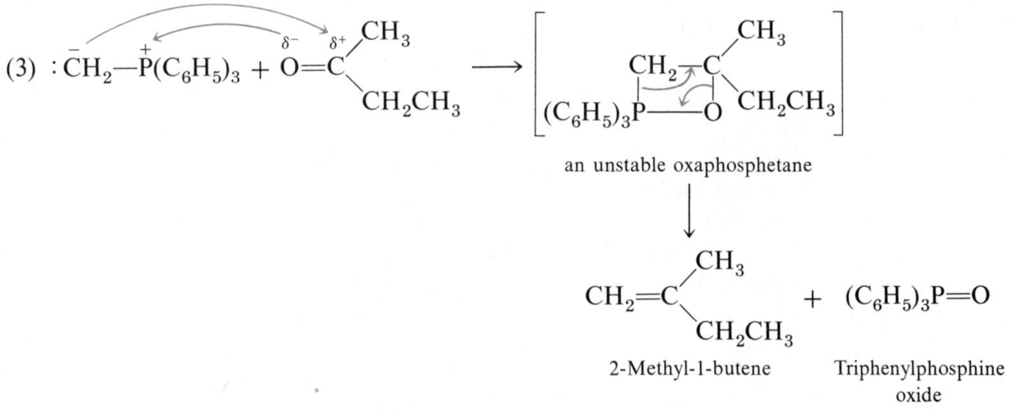

Most aldehydes and ketones can be used in this reaction sequence, and alkyl halides that react well by the S_N2 mechanism can be substituted for methyl bromide.

9.8 Replacement of α-Hydrogen by Halogen (Haloform Reaction)

Aldehydes and ketones are much more readily halogenated than the alkanes (Sec. 2.8-B). The reaction, which is ionic rather than free radical in nature, invariably results in replacement of an α-hydrogen to give an α-halo

---

[4] The preferred strong base for this reaction is a carbanion prepared from and in the very useful solvent, dimethyl sulfoxide (DMSO), by treatment with sodium hydride.

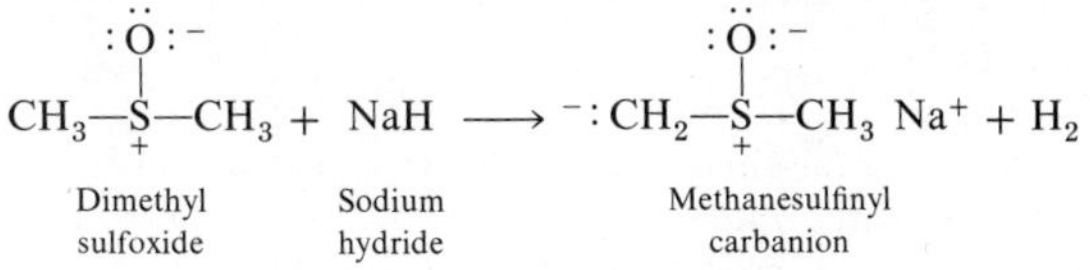

$$CH_3{-}\overset{:\ddot{O}:^-}{\underset{+}{S}}{-}CH_3 + NaH \longrightarrow {}^-{:}CH_2{-}\overset{:\ddot{O}:^-}{\underset{+}{S}}{-}CH_3 \ Na^+ + H_2$$

Dimethyl   Sodium   Methanesulfinyl
sulfoxide   hydride   carbanion

Other strong bases can be used, such as *n*-butyllithium.

ketone and is catalyzed by both acids and bases. Acid-catalyzed halogenation proceeds through the enol form of the aldehyde or ketone and tends to stop after the introduction of one halogen atom.

$$
\text{Acetophenone} \quad + \text{Cl}_2 \xrightarrow[\text{(HCl)}]{} \quad \alpha\text{-Chloroacetophenone}
$$

Acetophenone

α-Chloroacetophenone
(Phenacyl chloride)

Phenacyl chloride is a potent **lachrymator** (tear gas), which is used widely in law enforcement (for example, as a constituent of MACE).

Base-catalyzed halogenation proceeds through the enolate ion and tends to give polyhalogenated products. Thus, when acetaldehyde or a methyl ketone is warmed with an alkaline solution of chlorine, bromine, or iodine, the product is chloroform, bromoform, or iodoform, respectively. This reaction, called the **haloform** reaction, appears to take place in two stages. In the first stage, the three hydrogens on the α-carbon atom are successively replaced by halogen, each additional hydrogen more easily than the one before because of the strong **inductive effect**[5] of the halogen atom(s) that makes the hydrogen atom more acidic.

$$
\text{CH}_3-\text{C} \underset{\text{H}}{\overset{\text{O}}{\Big\langle}} + \text{Cl}_2 + \text{NaOH} \longrightarrow \text{Cl}-\underset{\text{H}}{\overset{\text{H}}{\text{C}}}-\text{C} \underset{\text{H}}{\overset{\text{O}}{\Big\langle}} + \text{H}_2\text{O} + \text{NaCl}
$$

Acetaldehyde

$$
\text{Cl}-\underset{\text{H}}{\overset{\text{H}}{\text{C}}}-\text{C} \underset{\text{H}}{\overset{\text{O}}{\Big\langle}} + 2\,\text{Cl}_2 + 2\,\text{NaOH} \longrightarrow \text{Cl}_3\text{C}-\text{C} \underset{\text{H}}{\overset{\text{O}}{\Big\langle}} + 2\,\text{H}_2\text{O} + 2\,\text{NaCl}
$$

---

[5] When atoms of different electronegativities are bonded, there will be a shift of electrons toward the more electronegative atom and away from the less electronegative atom, resulting in the formation of a dipole (Sec. 5.7). The electrostatic effect of this dipole is called the inductive effect. The effect is referred to as a minus I ($-$I) effect if electrons are withdrawn from groups attached to the dipole and a plus I ($+$I) effect if electrons are donated to such groups. The effect of the chlorine atoms on the acidity of the α-hydrogens of an aldehyde or ketone can be seen from the structure of the enolate ion:

$$
\left[\text{Cl} \overset{\curvearrowleft}{\longleftarrow} \overset{..}{\underset{\text{H}}{\text{C}}}-\text{C} \underset{\text{R}}{\overset{\overset{..}{\text{O}}}{\Big\langle}} \longleftrightarrow \text{Cl} \overset{\curvearrowleft}{\longleftarrow} \text{C}=\text{C} \underset{\text{R}}{\overset{\overset{..}{\text{O}}:^-}{\Big\langle}} \right]
$$

Since the chlorine atom tends to draw electrons away from the α-carbon atom, the negative charge on the carbon atom is somewhat diminished and the anion is stabilized. Thus, the $-$I effect of the chlorine can be said to enhance the acidity of the α-hydrogen by stabilizing the enolate ion that results when the α-hydrogen is removed by a base.

In the second stage of the reaction the molecule is cleaved by the base as a result of an addition-elimination mechanism.

$$Cl_3C-\overset{\delta+}{\underset{H}{C}}\overset{\overset{\delta-}{\ddot{O}}:}{\diagdown} \;+\; {}^{-}:\ddot{O}H \longrightarrow Cl_3C-\overset{:\overset{..}{O}:^{-}}{\underset{H}{\overset{|}{C}}}-\ddot{O}H \longrightarrow Cl_3C:^{-} + H-\overset{\overset{..}{\ddot{O}}:}{C}\diagdown_{\overset{..}{O}H}$$

$$Cl_3C:^{-} + H-\overset{\overset{..}{\ddot{O}}:}{C}\diagdown_{\overset{..}{O}H} \longrightarrow Cl_3C-H + H-\overset{\overset{..}{\ddot{O}}:}{C}\diagdown_{\overset{..}{O}:^{-}}$$

Chloroform   Formate ion

A structural requirement for any compound that gives a positive haloform reaction is that it contain an **acetyl group,** $CH_3-\overset{\overset{O}{\|}}{C}-$ or one oxidizable to an acetyl group. The first requirement is met by acetaldehyde and all methyl ketones. The second requirement, of course, is met by all methyl carbinols of the structure, $CH_3-\overset{\overset{H}{|}}{\underset{|}{C}}-OH$. Of the primary alcohols, ethanol alone gives the haloform reaction because it is oxidizable to acetaldehyde. The haloform reaction is useful not only as a preparative method for the haloforms (Sec. 7.3), but also as a diagnostic test for the presence of the groupings indicated. In practice, a solution of iodine is added to an unknown compound in an aqueous alkaline solution. A positive reaction will yield iodoform, CHI_3, a bright yellow solid (chloroform and bromoform are liquids) which may be identified by its sharp pungent odor and its melting point.

9.9 *Oxidation Reactions*

A. Oxidation by Tollens' Reagent, Fehling's, and Benedict's Solutions.

Aldehydes are so easily oxidized that even the mildest of oxidizing reagents will serve to bring about their conversion to acids. Ketones, on the other hand, are fairly resistant to oxidation. The oxidation of ketones, when forced by the use of strong oxidizing reagents and heat, results in a rupture of carbon–carbon bonds to produce acids.

$$CH_3-\overset{\overset{O}{\|}}{C}-CH_2CH_3 \xrightarrow[\text{heat}]{KMnO_4,\ H^+,} CH_3-\overset{\overset{O}{\|}}{C}\diagdown_{OH} \text{ and/or } CH_3-CH_2-\overset{\overset{O}{\|}}{C}\diagdown_{OH}$$

An exception to this general rule may be found in the oxidation of ketones by selenium dioxide, SeO_2. A methylene group adjacent to the carbonyl is oxidized by the use of this reagent to another carbonyl group. Methyl ethyl ketone, for example, can be oxidized to diacetyl.

$$CH_3-CH_2-\overset{\overset{\displaystyle O}{\|}}{C}-CH_3 + SeO_2 \longrightarrow CH_3-\overset{\overset{\displaystyle O}{\|}}{C}-\overset{\overset{\displaystyle O}{\|}}{C}-CH_3 + H_2O + Se$$

2-Butanone   Butanedione

(Methyl ethyl ketone)   (Diacetyl)

The ease with which an oxidation takes place provides a simple method for distinguishing between aldehydes and ketones. Mild oxidizing agents may be used for this purpose. **Tollens' reagent,** an ammoniacal solution of silver oxide, $Ag(NH_3)_2OH$, **Fehling's solution,** an alkaline solution of cupric ion complexed with sodium potassium tartrate (Rochelle salt), and **Benedict's solution,** an alkaline solution of cupric ion complexed with sodium citrate, are three reagents commonly used to detect the presence of an aldehyde group. When Tollens' reagent is used to oxidize an aldehyde, the silver ion is reduced to the metallic form and, if the reaction is carried out in a clean test tube, deposits as a mirror.

$$R-\overset{\overset{\displaystyle H}{|}}{C}=O + 2\,Ag(NH_3)_2OH \longrightarrow R-\overset{\overset{\displaystyle O}{\|}}{C}-O^-NH_4^+ + 2\,Ag + H_2O + 3\,NH_3$$

When Fehling's and Benedict's solutions are used to oxidize an aldehyde, the complexed cupric ion (deep blue) is reduced to cuprous oxide (red).

$$R-\overset{\overset{\displaystyle H}{|}}{C}=O + 2\,Cu(OH)_2 + NaOH \longrightarrow R-\overset{\overset{\displaystyle O}{\|}}{C}-O^-Na^+ + Cu_2O + 3\,H_2O$$

Aromatic aldehydes react with Tollens' reagent but do not react with either Fehling's or Benedict's solutions. A means of distinguishing between aliphatic and aromatic aldehydes is thus provided by this difference in reactivity with the two types of reagents.

B. Autooxidation and Reduction (Cannizzaro Reaction). Aldehydes which lack an α-hydrogen, when heated with a concentrated sodium or potassium hydroxide solution, undergo an intermolecular oxidation-reduction. One molecule of such an aldehyde is reduced to an alcohol at the expense of another molecule. The second molecule is oxidized to an acid. The mechanism for this interesting disproportionation, known as the **Cannizzaro reaction,** is illustrated using formaldehyde.

$$\underset{H}{\overset{H}{>}}C=O + N\overset{+}{a}\overset{-}{O}H \rightleftharpoons \underset{H}{\overset{H}{>}}C\underset{OH}{\overset{O^-}{<}} + Na^+$$

Formaldehyde

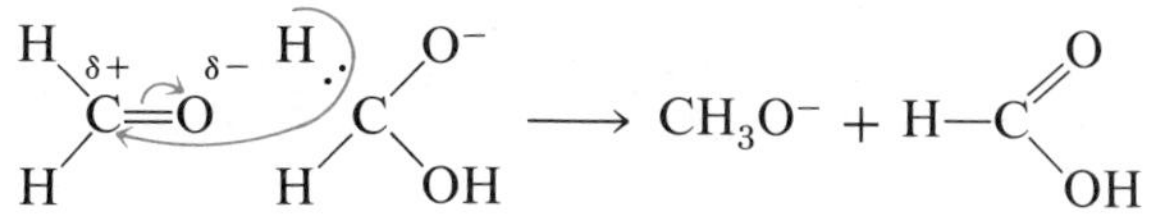

Hydride ion
transfer
(slow)

Acid-base
reaction
(fast)

Methyl
alcohol

Sodium
formate

Two different aldehydes, each lacking α-hydrogen, engage in a **crossed Cannizzaro reaction** when heated in an alkaline solution. This type of Cannizzaro reaction was illustrated in the preparation of pentaerythritol (Sec. 9.7-B).

9.10 *Reduction*

The carbonyl group of aldehydes and ketones may be reduced to primary and secondary alcohols respectively. This transformation can be accomplished either catalytically with hydrogen (hydrogenation) or by means of a chemical reducing agent such as lithium aluminum hydride, $LiAlH_4$.

$$R-\overset{\overset{H}{|}}{C}=O + 2\ H_2 \xrightarrow[\text{pressure}]{\text{Pt or Ni}} R-CH_2-OH$$

$$R-\overset{\overset{R}{|}}{C}=O \xrightarrow{\text{LiAlH}_4} R-\overset{\overset{R}{|}}{\underset{\underset{H}{|}}{C}}-OH$$

Another reducing agent specific for the reduction of carbonyl groups is sodium borohydride, $NaBH_4$. This reagent is less reactive than $LiAlH_4$ and may be used in water or alcoholic solutions. Lithium aluminum hydride, on the other hand, reacts violently with such hydroxylic solvents.

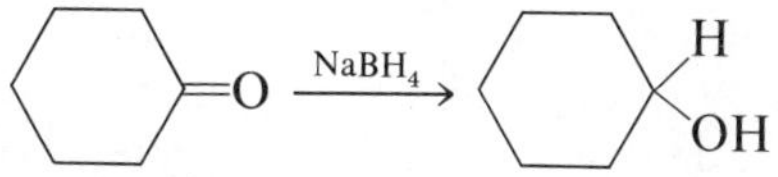

Cyclohexanone

Cyclohexanol

EXERCISE 9.10

Cyclohexanone (b.p. 156°) may be prepared by the oxidation of cyclohexanol (b.p. 161°). If you had prepared cyclohexanone by this method, how could you isolate it uncontaminated by traces of unreacted cyclohexanol?

The carbonyl group of a ketone can be reduced to a methylene group when refluxed with concentrated hydrochloric acid in the presence of amalgamated zinc. The reaction, known as the **Clemmensen reduction,** provides a method for converting carbonyls to hydrocarbons, and often is used as a sequel to the Friedel-Crafts acylation of the aromatic ring. The net result, in this case, is the monosubstitution of the ring by an unbranched alkyl group—an objective not always possible in the Friedel-Crafts alkylation.

Propiophenone
(Ethyl phenyl ketone)

n-Propylbenzene

9.11 *Polymerization of Aldehydes*

Acetaldehyde, as you have noted, is rather unstable toward oxidation. This inherent instability, along with a boiling point of only 20°C, makes acetaldehyde a difficult compound to store and use. Fortunately, when treated with acid at a low temperature, acetaldehyde undergoes self-addition to give the cyclic trimer,[6] **paraldehyde** (b.p. 125°C).

Acetaldehyde

Paraldehyde

In this form the aldehyde is not only stable to oxidation but no longer is easily lost by evaporation. Paraldehyde, when warmed, depolymerizes to regenerate acetaldehyde. Acetaldehyde, when warmed in concentrated alkaline solution, appears to undergo repeated aldol condensation accompanied by dehydration and polymerization to yield viscous, resinous products of undetermined structure. Acetaldehyde (as paraldehyde) has been used medicinally, as a soporific, but its most important use is as a starting material in many organic syntheses.

Formaldehyde is not marketed in its gaseous form, but either as **formalin,** a 37–40% aqueous solution, or as the polymer, **paraformaldehyde,** $HO(CH_2O)_n H$,

---

[6] Monomer: Gr., *mono,* one; *meros,* part. The simplest structural unit in a polymer. A trimer is composed of three molecules of monomer.

with *n* having an average value of 30. Paraformaldehyde is an amorphous white solid prepared by slowly evaporating formalin under reduced pressure. The polymerization involves the self-addition of many molecules of formaldehyde.

$$\underset{H}{\overset{H}{>}}C=O + H_2O \rightleftharpoons [HOCH_2OH]$$

$$[HOCH_2OH] + n\ HCHO \longrightarrow HO-\left(\overset{H}{\underset{H}{\overset{|}{\underset{|}{C}}}}-O\right)_{n+1}H$$

Paraformaldehyde

Depolymerization of paraformaldehyde, as in the case of paraldehyde, is brought about by heating. This ready change of state from solid to gaseous allows formaldehyde to be easily stored and used.

A high molecular weight linear polymer named "Delrin" has been prepared from formaldehyde. It is a remarkable plastic and shows promise of becoming a very useful one for structural materials where strength and resiliency are important.

Formaldehyde is perhaps the most important member of the aldehyde family. Its industrial importance lies principally in its ability to **copolymerize**[7] with phenol, with urea $[(H_2N)_2C=O]$, and with melamine ($C_3H_6N_6$, a cyclic triamino compound) to produce hard, electrical nonconducting, infusible resins of the "Bakelite" and "Melmac" type. Structural units of these useful plastics are shown.

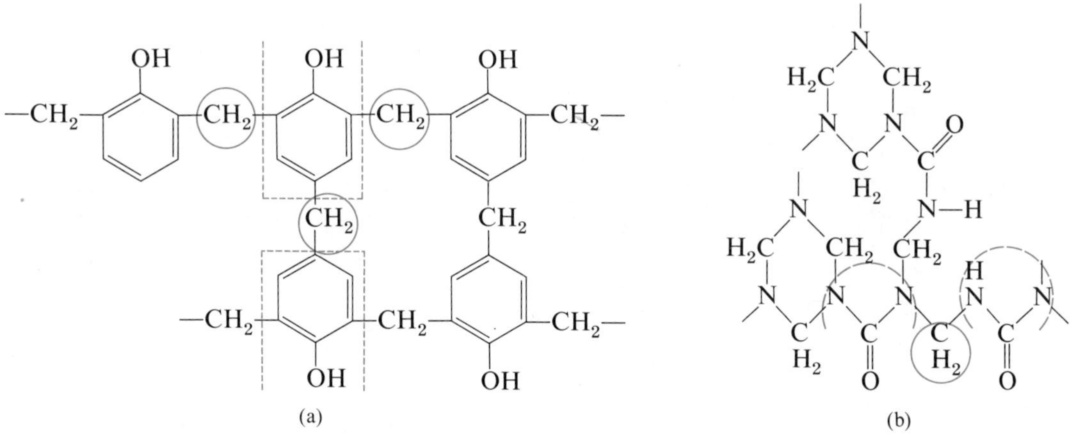

Structural units in (a) Bakelite and (b) urea-formaldehyde resins

---

[7] Copolymerization: a reaction in which two or more unlike monomers polymerize with each other.

Formaldehyde reacts with proteins to harden them and makes them less susceptible to putrefaction. For this reason it is used in the preservation of biological specimens and in embalming agents. Formaldehyde is toxic to insects and many microorganisms, and in the form of paraformaldehyde "candles" is conveniently used as a fumigant.

9.12 α,β-Unsaturated Carbonyl Compounds

When a carbonyl group is conjugated with a double bond, as in the α,β-unsaturated aldehydes and ketones, the conjugated system tends to show both 1,2- (normal) and 1,4- (conjugate) addition reactions (Sec. 3.12). Thus, hydrogen bromide adds in the 1,4-manner to methyl vinyl ketone.

Methyl vinyl ketone   *two of the three principal forms*

unstable enol   4-Bromo-2-butanone

There are two important observations that should be made concerning additions to conjugated systems. First, the numbers 1,2- and 1,4- do not refer to the carbon-chain numbering schemes used in nomenclature. They refer to the conjugated system; and, as shown for methyl vinyl ketone, the numbering *usually* begins with the hetero atom (oxygen), if such an atom is present. Second, although the addition of hydrogen bromide takes place in a 1,4-manner, because the intermediate enol is unstable and rearranges to the ketone, the final product *appears* to be that formed by *anti*-Markovnikov 3,4-addition. Thus, the terms 1,2- and 1,4- refer to the mechanism of addition, not to the structures of the final products. In 1,2-addition, two adjacent atoms are the only atoms required by the mechanism. In 1,4-addition, the mechanism requires addition across a 1,4-system.

Nucleophilic reagents, such as organometallic reagents or carbanions, may add 1,2- or 1,4- or both, depending on the nature of the conjugated system and the reagent. Grignard reagents tend to give variable results; and, for this reason, it is customary to use alkyllithiums (Sec. 2.7-B) for 1,2-addition and lithium-dialkyl cuprates (Sec. 2.7-B) for 1,4-addition, since these reagents tend to add in a specific fashion.

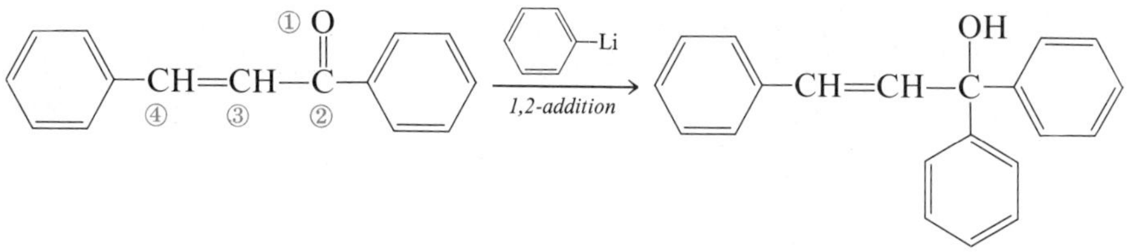

Benzalacetophenone
(Chalcone)

1,1,3-Triphenyl-2-propen-1-ol
(Diphenylstyrylcarbinol)

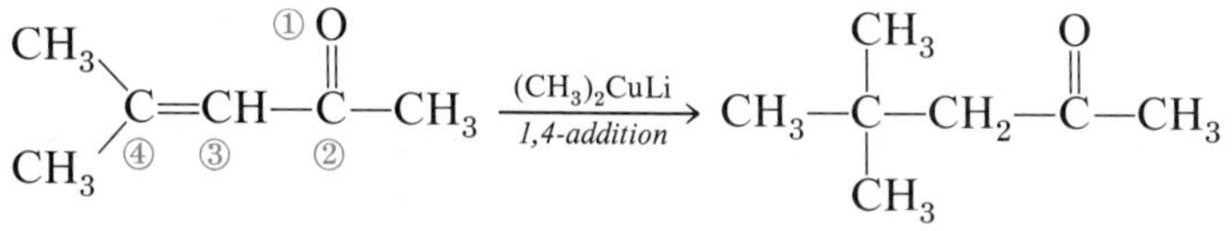

Mesityl oxide
(4-Methyl-3-buten-2-one)

4,4-Dimethyl-2-butanone
(Methyl neopentyl ketone)

EXERCISE 9.11

Draw the important resonance structures for methyl vinyl ketone.

Summary

1. **Structure**

   The functional group of both the aldehydes and ketones is the carbonyl
   group, $\diagdown C{=}O$. The general formula for the aldehydes is $R{-}\overset{H}{\underset{}{C}}{=}O$; for the
   ketones, $R{-}\overset{R}{\underset{}{C}}{=}O$.

2. **Nomenclature**

   IUPAC nomenclature uses the suffix **al** to name an aldehyde; **one** to designate a ketone. Systematic nomenclature is governed by IUPAC rules previously outlined.

3. **Properties**

   (A) The simple aldehydes and ketones are nonassociated, low boiling
   liquids.

   (B) Only aldehydes and ketones of low molecular weight are soluble in
   water.

(C) The simple aldehydes have sharp, irritating odors.

(D) The aromatic aldehydes and nearly all the ketones are fragrant.

(E) The electron-withdrawing effect of the carbonyl oxygen is transmitted to the α-carbon atom and bestows acidic properties upon the hydrogen atoms bonded to it.

4. **Preparation**

(A) General methods for the preparation of the aldehydes are
   (a) The catalytic dehydrogenation or oxidation of primary alcohols.
   (b) The hydrolysis of geminal dihalides (must be 1,1-dihalides).

(B) Special methods of preparation of aldehydes are
   (a) The Gatterman-Koch reaction for aromatic aldehydes.
   (b) The hydration of acetylene for preparing acetaldehyde.

(C) General methods for the preparation of the ketones are
   (a) The oxidation of secondary alcohols.
   (b) The hydration of alkynes other than acetylene.

(D) Special methods for the preparation of ketones are
   (a) The Friedel-Crafts acylation reaction for aromatic ketones.
   (b) The Weizmann fermentation for acetone.

5. **Reactions of Carbonyl Compounds**

(A) Oxidation.
Aldehydes are easily oxidized to acids even by mild oxidizing reagents; ketones generally cannot be oxidized without a rupture of the carbon chain.

(B) Addition of nucleophilic reagents—i.e., bases, electron-pair donors.
   (a) The following reagents give "straight" addition products.
      (i) HCN adds to aldehydes and ketones to form cyanohydrins.
      (ii) $NaHSO_3$ adds to aldehydes, methyl ketones, and cyclic ketones to give sodium bisulfite salts.
      (iii) Hydrogen adds to the carbonyl group of aldehydes and ketones to produce primary and secondary alcohols respectively.
      (iv) Grignard reagents, when treated with aldehydes, lead to the preparation of *secondary* alcohols; when treated with ketones, Grignard reagents lead to the preparation of *tertiary* alcohols. The only aldehyde capable of forming a primary alcohol with a Grignard reagent is formaldehyde.
      (v) Water adds to the carbonyl group of aldehydes and ketones to give unstable hydrates. Alcohols add to give stable acetals and ketals.
   (b) Certain nitrogen-containing nucleophilic reagents add to aldehydes and ketones to produce unstable addition products. Subsequent loss of water (condensation) produces a carbon-nitrogen

double bond. *Hydroxylamine* reacts with aldehydes and ketones to yield *oximes;* hydrazine and its derivatives give the corresponding *hydrazones*. *Semicarbazide* reacts to produce *semicarbazones*. All three types of derivatives are useful in the identification of "unknowns."

(C) Reactions of α-Hydrogen.

    (a) An aldol condensation results when an aldehyde or a ketone with α-hydrogen is treated with a dilute base. The product (an aldol) readily loses water to yield an α,β-unsaturated carbonyl.

    (b) Acetaldehyde and methyl ketones undergo replacement of all three α-hydrogen atoms when warmed with an alkaline solution of the halogens to yield *haloforms* and salts of acids.

(D) Disproportionation.

Aldehydes lacking α-hydrogen atoms undergo the Cannizzaro reaction when treated with concentrated sodium or potassium hydroxide. Ketones do not give the Cannizzaro reaction.

(E) The Wittig reaction.

Aldehydes and ketones react with phosphorus ylids to give alkenes.

(F) Polymerization.

Formaldehyde is capable of self-addition to yield polymers of varying molecular weights. Acetaldehyde trimerizes to paraldehyde. Formaldehyde copolymerizes with phenol (Bakelite), and with urea, and melamine (Melmac). These are useful plastics with desirable thermal and electrical properties.

(G) Conjugate addition.

α,β-Unsaturated aldehydes and ketones may undergo 1,2- or 1,4-addition or both.

New Terms

acetal	inductive effect
acyl group	ketal
aldol condensation	keto form
carbonyl group	oxime
cyanohydrin	phenylhydrazone
enol form	semicarbazone
enolate anion	Tollens' reagent
Fehling's solution	α,β-unsaturated carbonyl compound
hemiacetal	ylid

Supplementary Exercises

EXERCISE 9.12   Assign an acceptable name to each of the following compounds.

(a) 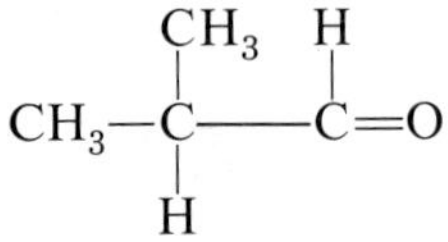 $CH_3-\overset{\overset{\displaystyle CH_3}{|}}{\underset{\underset{\displaystyle H}{|}}{C}}-\overset{\overset{\displaystyle H}{|}}{C}=O$

(b) [cyclopentanone structure]

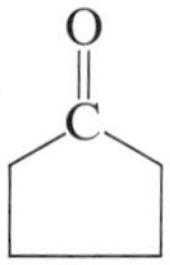

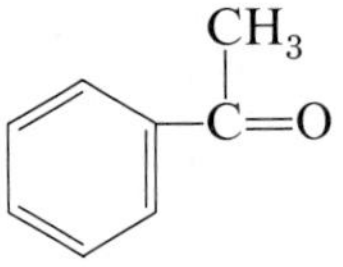

(c) [phenyl with $\overset{\overset{\displaystyle CH_3}{|}}{C}=O$ group]

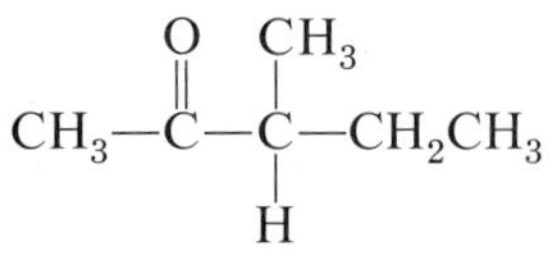

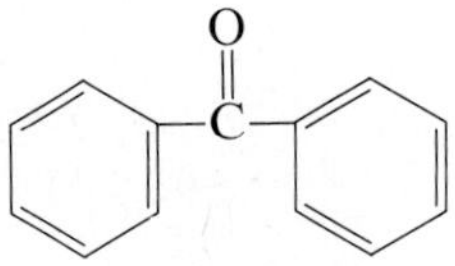

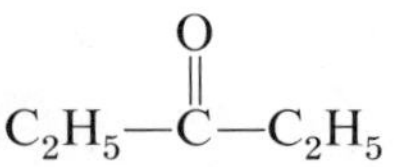

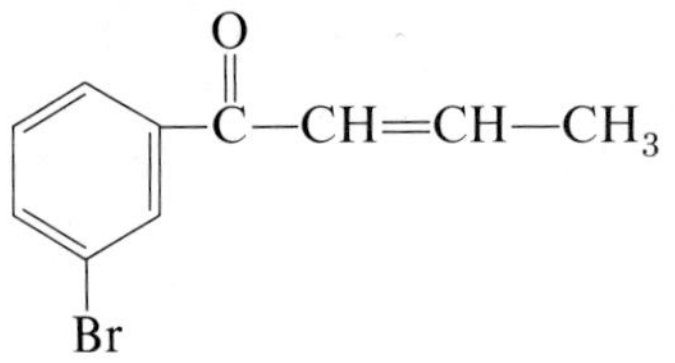

(d) $CH_3-\overset{\overset{\displaystyle O}{\|}}{C}-\overset{\overset{\displaystyle CH_3}{|}}{\underset{\underset{\displaystyle H}{|}}{C}}-CH_2CH_3$

(e) [benzophenone structure, two phenyl rings with $\overset{O}{\|}{C}$]

(f) $C_2H_5-\overset{\overset{\displaystyle O}{\|}}{C}-C_2H_5$

(g) [phenyl with Br substituent and $\overset{\overset{\displaystyle O}{\|}}{C}-CH=CH-CH_3$]

(h) [cyclohexenone ring with CH_3 substituent]

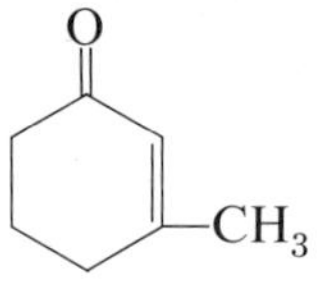

EXERCISE 9.13   Write structures for the following.

(a) propionaldehyde
(b) diisopropyl ketone
(c) 2-methylpropanal
(d) 2-octanone

(e) α-bromoacetophenone
(f) cinnamaldehyde
(g) furfural
(h) 4-phenyl-3-buten-2-one

EXERCISE 9.14   Complete the following reactions by supplying the necessary but missing component. This could be either a reactant, a catalyst, or a product.

(a) $(?) + I_2$ in aqueous $KI + NaOH \longrightarrow CH_3COO^- Na^+ + CHI_3$

(b) Acetone $+ (?) \longrightarrow (CH_3)_2C(OH)CN$

(c) Acetaldehyde $+ C_2H_5OH \xrightarrow{H^+}$

(d) $(?) + C_2H_5MgBr \xrightarrow[\text{hydrolysis}]{\text{followed by}} (C_2H_5)_2C(OH)CH_3$

(e) Acetophenone $+ H_2NOH \xrightarrow{H^+}$

(f) *n*-Butyraldehyde + NaHSO$_3$ $\longrightarrow$

(g) Acetone $\xrightarrow{\text{10\% NaOH}}$

(h) Benzaldehyde + KOH $\longrightarrow$

(i) Methanol $\xrightarrow{\text{CuO, heat}}$

(j) (?) + Ag(NH$_3$)$_2$OH $\longrightarrow$ $\underline{\text{Ag}}$ + CH$_3$COO$^-$NH$_4$$^+$

EXERCISE 9.15 Write structures for the products formed when each of the following is treated with ethylmagnesium bromide, followed by acid hydrolysis.

(a) H—C=O (with H above C)

(b) CH$_3$—C=O (with H above C)

(c) CH$_3$—C=O (with CH$_3$ above C)

(d) C$_2$H$_5$—C=O (with CH$_3$ above C)

(e) benzene ring with —C=O (H above C)

(f) benzene ring with —C—CH$_3$ (O double bond above C)

(g) benzene ring with —CH=CH—C—CH$_3$ (O double bond above C)   (two products)

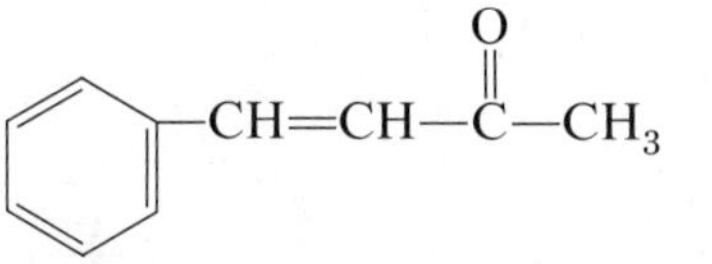

EXERCISE 9.16 Which of the following compounds will give a positive iodoform reaction?

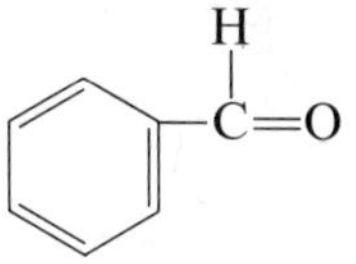

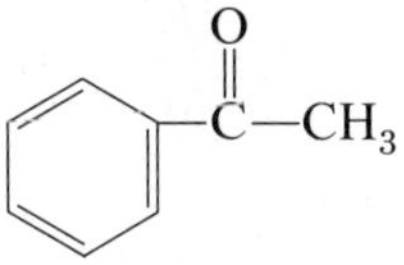

(a) (CH$_3$)$_2$CHOH

(b) C$_2$H$_5$—C=O (with CH$_3$ above C)

(c) CH$_3$—C—OH (with CH$_3$ above and CH$_3$ below C)

(d) benzene ring with —C=O (H above C)

(e) CH$_3$CH$_2$OH

(f) benzene ring with —C—CH$_3$ (O double bond above C)

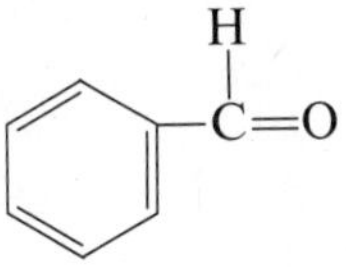

(g) CH$_3$CH$_2$—C—CH$_3$ (with OH above and H below C)

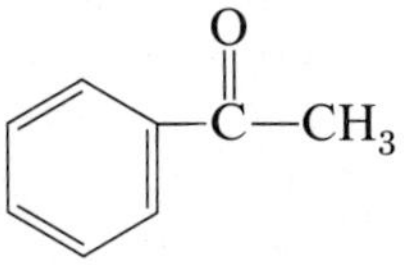

(h) CH$_3$CH$_2$—C=O (with H above C)

EXERCISE 9.17 Using acetylene as your only organic starting material show how you would synthesize each of the following compounds. You may use whatever inorganic reagents you consider necessary.

(a) acetaldehyde   (e) 2-butanol
(b) ethyl bromide   (f) methyl ethyl ketone
(c) iodoform   (g) 2-butenol
(d) acetic acid   (h) propionic acid, CH_3CH_2COOH

EXERCISE 9.18 Identify the aldehyde or ketone that will react with ethylmagnesium bromide to produce

(a) 2-butanol   (d) 2-methyl-2-butanol
(b) 3-methyl-3-pentanol   (e) 3-methyl-3-octanol
(c) *n*-propyl alcohol

EXERCISE 9.19 A student attempted to prepare 1-(4-hydroxyphenyl)-ethanol by the following series of reactions:

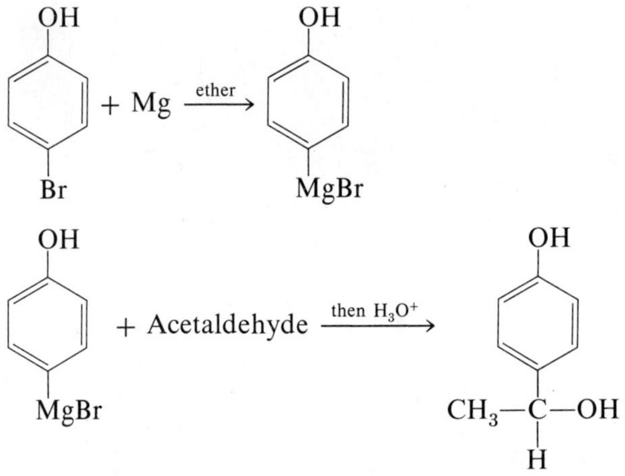

The expected product was not obtained. Why did this reaction fail? Outline a reaction sequence that will yield the desired product free of any *ortho* isomer. (*Hint: See Section 8.8-D.*)

EXERCISE 9.20 Indicate simple test tube reactions which would serve to distinguish between members of the following pairs of carbonyl compounds without the need for taking boiling points or melting points, or preparing solid derivatives.

(a) acetaldehyde and acetone
(b) 2-octanone and benzaldehyde
(c) 2,2-dimethylpropanal and 2-butanone
(d) 3-pentanone and cyclopentanone
(e) acetophenone and benzaldehyde

EXERCISE 9.21 A student was given an "unknown" carbonyl compound from among those listed in Table 9.2. He obtained a boiling point of 90–92°. Consulting Table 9.2

he noted that his unknown could possibly be one of two different compounds. He proceeded to prepare a 2,4-dinitrophenylhydrazone of his compound but realized too late that this derivative would still leave the identity of his "unknown" in question. Without preparing a second derivative, what simple test tube reaction would help him to establish the identity of his unknown? What derivative should he then prepare?

EXERCISE 9.22   Compound (A), $C_5H_{12}O$, when refluxed with potassium dichromate and sulfuric acid was converted to (B), $C_5H_{10}O$. Compound (B) formed a derivative with 2,4-dinitrophenylhydrazine but gave neither a positive haloform reaction nor would it form an addition product with sodium bisulfite. Give the structure and name of the original compound.

EXERCISE 9.23   An alkene, C_6H_{12}, after ozonization and subsequent hydrolysis, yielded two products. One of these gave a positive iodoform reaction but a negative Tollens' test. The other product gave a positive Tollens' test, but a negative iodoform reaction. What is the structure and name of the alkene?

EXERCISE 9.24   An organic compound, $C_5H_{10}O$, showed strong absorption at 1700 cm^{-1} in the ir region. With only this much information it is possible to draw seven different structures that fit the above molecular formula, but only one will be in agreement with the nmr spectrum of the compound, which consisted of only a quartet at δ 2.45 and a triplet at δ 0.97. The ratio of the areas was quartet : triplet = 2 : 3. What is the name of the compound in question?

EXERCISE 9.25   A colorless liquid, $C_6H_{12}O_3$, absorbed strongly in the ir at 1720 cm^{-1} and gave the following nmr spectrum: 1-proton triplet at δ 4.71, a 6-proton singlet at δ 3.29, a 2-proton doublet at δ 2.64, and a 3-proton singlet at δ 2.11. Suggest a structure for the compound.

EXERCISE 9.26   Periodic acid, HIO$_4$, is a reagent used to cleave the carbon-carbon bond in 1,2-glycols, 1,2-dicarbonyl compounds, and α-hydroxy carbonyl compounds. In the course of the cleavage, hydroxyl groups are converted to carbonyl groups and carbonyl groups to carboxylic acid groups, as shown.

When compound A, $C_6H_{14}O_2$, was treated with HIO$_4$, it gave a single compound B, C_3H_6O. The latter showed a strong ir absorption band at 1720 cm^{-1} and reacted immediately with a solution of iodine in aqueous sodium hydroxide solution to produce iodoform. When compound A was treated with sulfuric acid, a new compound C, $C_6H_{12}O$, resulted, which also gave a positive iodoform reaction. The nmr spectrum of C showed only two singlets at δ 1.10 and δ 2.50 in a proton ratio of 3:1. What are the structures and names of compounds A, B, and C?

EXERCISE 9.27   Outline sequences of reactions that would be suitable for bringing about the following transformations. Use any other reagents required.

(a) $CH_3CH_2CH_2CH=CH_2 \longrightarrow$

$$\underset{H}{\overset{CH_3CH_2}{\diagdown}} C = C \underset{CH_3}{\overset{CH_2CH_2CH_3}{\diagup}}$$

(b) $CH_3 - \overset{\overset{\displaystyle O}{\|}}{C} - CH_3 \longrightarrow CH_3 - \overset{\overset{\displaystyle CH_3}{|}}{\underset{\underset{\displaystyle CH_3}{|}}{C}} - C \overset{\displaystyle O}{\underset{\displaystyle OH}{\diagup}}$

(c) $CH_3CH_2CH_2CH=CH_2 \longrightarrow CH_3CH_2CH_2CH_2C \overset{\displaystyle O}{\underset{\displaystyle H}{\diagup}} \longrightarrow$

$$CH_3CH_2CH_2\underset{\diagdown}{\overset{\diagup}{CH}} \overset{CH_2OH}{\underset{CH_2OH}{}} - CH_2OH$$

(d) $HC\equiv CH \longrightarrow CH_3CH_2CH_2CH_2\overset{\overset{\displaystyle O}{\|}}{C}CH_3$

(e) ⬡ $\longrightarrow$ ⬡$-CH_2CH_2CH_2CH_2CH_2CH_3$

10

The Carboxylic Acids and Their Derivatives

Introduction

Substances which contain a **carboxyl group,** $-C{\overset{\displaystyle O}{\underset{\displaystyle OH}{}}}$ (also written simply as —COOH, or as $-CO_2H$), make up a large family of compounds known as **carboxylic acids.** The carboxyl group represents the highest oxidation state of a carbon atom when bonded to another carbon and is a group found as a part of the structure of many natural products. It is not surprising to learn that acids occur naturally when one considers the fact that we live in an oxidizing atmosphere. Many carboxylic acids have been known for a long time.

Acetic acid, CH_3COOH, for example (the sour principle of vinegar from which it takes its name), is a substance known since antiquity and frequently is referred to in the Bible. A number of acids play vital roles in body functions and are important intermediates in the metabolic processes. A large number of acids, or their derivatives, are useful everyday commodities.

10.1 Formulas and Nomenclature

Carboxylic acids of the aliphatic series frequently are referred to as "fatty" acids because those containing an even number of carbon atoms (four or greater) exist in a combined form with glycerol as fats and oils. Acids which contain but one carboxyl group (monocarboxylic) have the general formula R—COOH (R = alkyl) or Ar—COOH (Ar = aryl).

$$H-C\overset{\displaystyle O}{\underset{\displaystyle OH}{}} \qquad CH_3-C\overset{\displaystyle O}{\underset{\displaystyle OH}{}} \qquad CH_3CH_2CH_2-C\overset{\displaystyle O}{\underset{\displaystyle OH}{}}$$

Formic acid   Acetic acid   n-Butyric acid

The carboxyl group in aromatic acids is attached directly to the benzene ring. However, the phenyl group, C_6H_5—, may appear as a substituent on any carbon atom of an aliphatic acid. Such acids should be classified as aryl-substituted aliphatic acids, because their reactions essentially are those given by acids of the aliphatic series.

Benzoic acid   p-Bromobenzoic acid   Phenylacetic acid

Many of the carboxylic acids had been known before rules for systematic nomenclature were devised. As a result, they usually are called by their common names. Many of these have their origin in Greek and Latin, and their names often indicate the original source of the acid. There are no easy rules for remembering common names, and they simply must be learned. IUPAC nomenclature follows general rules. The final **e** of the hydrocarbon stem is replaced by **oic**, followed by the word **acid**. The names, formulas, and derivations of some of the more common carboxylic acids are given in Table 10.1.

The carbon atom of the carboxyl group is always carbon 1 in systematic nomenclature, other numbers locating the position of substituents. Greek letters, α-, β-, γ-, δ-, etc., are used to locate substituents in common names. The α-carbon in an acid is always the carbon atom adjacent, or joined, to the carboxyl group.

$$\overset{\delta}{\underset{(5)}{C}}-\overset{\gamma}{\underset{(4)}{C}}-\overset{\beta}{\underset{(3)}{C}}-\overset{\alpha}{\underset{(2)}{C}}-\overset{}{\underset{(1)}{C}}\overset{\displaystyle O}{\underset{\displaystyle OH}{}}$$

TABLE 10.1 *Some Common Carboxylic Acids*

Name		Formula	Derivation
(Common)	**(IUPAC)**		
Formic	Methanoic	H—COOH	L. *formica*, ant
Acetic	Ethanoic	CH_3COOH	L. *acetum*, vinegar
Propionic	Propanoic	CH_3CH_2COOH	Gr. *protos* first; *pion* fat
n-Butyric	Butanoic	$CH_3CH_2CH_2COOH$	L. *butyrum*, butter
n-Valeric	Pentanoic	$CH_3(CH_2)_3COOH$	L. *valere*, powerful
Caproic	Hexanoic	$CH_3(CH_2)_4COOH$	L. *caper*, goat
Caprylic	Octanoic	$CH_3(CH_2)_6COOH$	L. *caper*, goat
Capric	Decanoic	$CH_3(CH_2)_8COOH$	L. *caper*, goat
Lauric	Dodecanoic	$CH_3(CH_2)_{10}COOH$	Laurel
Palmitic	Hexadecanoic	$CH_3(CH_2)_{14}COOH$	Palm oil
Stearic	Octadecanoic	$CH_3(CH_2)_{16}COOH$	Gr., *stear*, tallow
Benzoic	Benzenecar-boxylic acid	⬡—COOH	(gum benzoin)

In a few cases, certain acids are more conveniently named as derivatives of acetic acid or as alkanecarboxylic acids. The following examples will illustrate the various methods for naming substituted fatty acids.

$$\underset{(3)}{\overset{\beta}{}}\underset{(2)}{\overset{\alpha}{}}\underset{(1)}{}$$

$$CH_3-\underset{|}{CH}-COOH$$
$$\quad\;\; Br$$

α-Bromopropionic acid
(2-Bromopropanoic acid)

$$\underset{(4)}{\overset{\gamma}{}}\underset{(3)}{\overset{\beta}{}}\underset{(2)}{\overset{\alpha}{}}\underset{(1)}{}$$

$$CH_3-\underset{|}{CH}-CH_2-COOH$$
$$\quad\;\; OH$$

β-Hydroxybutyric acid
(3-Hydroxybutanoic acid)

$$\overset{CH_3}{\underset{|}{}}$$
$$CH_3-\underset{|}{C}-COOH$$
$$\quad\;\; CH_3$$

Trimethylacetic acid
(2,2-Dimethylpropanoic acid)
(Pivalic acid)

$$CH_2{=}CH-CH_2COOH$$

Vinylacetic acid
(3-Butenoic acid)

Cyclohexanecarboxylic acid

Aromatic acids usually are designated by common names, or named as derivatives of the parent acid—benzoic, C_6H_5COOH. Substituents on the benzene ring are designated either by number, or by the prefixes *ortho-*(*o-*), *meta-*(*m-*), or *para-*(*p-*). The ring carbon atom that bears the carboxyl group is always number 1 when an aromatic acid is named as a substituted benzoic acid.

Salicylic acid
(*o*-Hydroxybenzoic acid)

2,6-Dimethylbenzoic acid

Mandelic acid
(α-Hydroxyphenylacetic acid)

10.2 *Acidity and Structure*

The carboxylic acids are much weaker acids than the mineral acids (HCl, H_2SO_4, HNO_3), but they are more acidic than the phenols. The carboxylic acids do not show a strong tendency to dissociate into protons and their corresponding acid anions. Indeed, the equilibrium between the ionized and unionized forms of a carboxylic acid lies far to the left. The reaction of acetic acid with water may be used as an illustration.

Acetic acid

Acetate ion

Hydronium ion

Acetic acid in a 0.1 M aqueous solution has been found to dissociate only to the extent of 1.34% at room temperature. Statistically, this means that less than two molecules of acetic acid out of every hundred undergo ionization at any one time. The hydronium ion concentration may be related to that of other components in the acid solution by means of an equilibrium constant (in this case an ionization constant, K_a). A small numerical value for K_a indicates a weak acid in which the bulk of the acid molecules remain in the undissociated form.

$$K_a = \frac{\text{concentration of hydronium ion} \times \text{concentration of acetate ion}}{\text{concentration of undissociated acetic acid}^1}$$

$$= \frac{[H_3O^+][CH_3COO^-]}{[CH_3COOH]}$$

$$= \frac{(0.1 \times 0.0134)^2}{0.1 \times (1 - 0.0134)} = 1.8 \times 10^{-5}$$

---

[1] The concentration of water has been omitted from the above expression for K_a because the number of water molecules involved in the formation of hydronium ions is negligible with respect to the total number of water molecules present. It may be assumed that the concentration of water remains nearly constant.

An acid stronger than acetic acid would be dissociated to a greater extent. Thus, for such an acid the numerator in the equation for K_a would be larger, and the denominator would be smaller, leading to a value of K_a larger than that for acetic acid. Therefore, the stronger the acid, the larger its ionization constant, K_a.

Because most organic acids are relatively weak, their ionization constants are very small numbers expressed in negative powers of 10. Thus, it is convenient and common practice to convert these rather cumbersome numbers into numbers easier to handle. This is done by taking the *negative* logarithm of K_a and calling the result the pK_a value for the acid.

$$pK_a = -\log K_a$$

Thus, for acetic acid

$$pK_a = -\log (1.8 \times 10^{-5}) = 4.74$$

Note that the *smaller* the value of pK_a, the *stronger* the acid will be. The pK_a of a water-soluble acid can be readily measured with a simple pH meter. The weak acid is 50 per cent neutralized with a strong base (sodium or potassium hydroxide solution), and the pH of the solution is measured. The following approximation is then sufficiently accurate for most purposes.

$$pK_a \simeq pH_{half-neutralization}$$

Biochemists, in particular, prefer to use pK_a values because of their relationship to the amounts of the acid and its salt present in solution at a given pH: at one pH unit above the acid's pK_a value the ratio of salt to acid is $10:1$; at two pH units above the pK_a the ratio is $100:1$.[2]

EXERCISE 10.1

The pK_a values for some important acidic substances (not all carboxylic acids) with which you may be familiar, at least by name, are: ascorbic acid (vitamin C), 4.30; Barbital (a "sleeping tablet"), 7.98; citric acid (first of three ionization constants), 3.13; lactic acid, 3.86. Rank these acidic substances in order of increasing acidity. Which, if any, of these acidic substances would be converted 95% or more into their salts in the following fluids: beer (pH, 4.5), intracellular muscle fluids (pH, 6.1), cow's milk (pH, 6.6), intestinal juice (pH, 8.0). (*Hint:* $\log (95/5) = 1.3$.)

---

[2] For the weak acid, HA, and its salt, A^-, the Henderson-Hasselbalch equation may be used to calculate such ratios:

$$pH = pK_a + \log \frac{[A^-]}{[HA]}$$

Although the tendency of carboxylic acids to dissociate is but slight, two factors play a part in the process. The first factor, **resonance stabilization,** helps to explain why the carboxylic acids are acids at all. The gain in stability afforded through resonance of the acid anion serves as the driving force which helps to promote the ionization process.

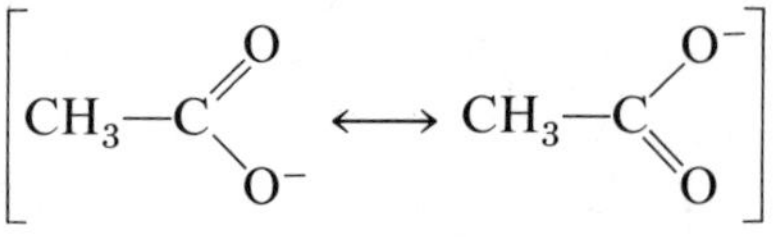

$$\left[CH_3-C\underset{O^-}{\overset{O}{<}} \longleftrightarrow CH_3-C\underset{O}{\overset{O^-}{<}} \right]$$

(a)   (b)

The resonance forms of the acetate ion actually appear to be that of a hybrid intermediate to (a) and (b) and is more correctly represented by the structure below.

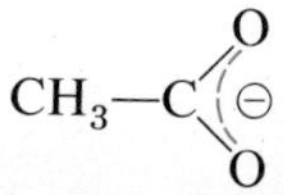

$$CH_3-C\underset{O}{\overset{O}{<}}\ominus$$

The second factor is due to an **inductive effect.** Electrophilic groups on the hydrocarbon portion of an acid have the effect of promoting ionization. Such groups withdraw electrons from the carboxyl group, promote the departure of the positive proton, and help stabilize the anion. For example, chloroacetic acid, $Cl-CH_2COOH$, is far more acidic than acetic acid due to the inductive effect of the chlorine atom.

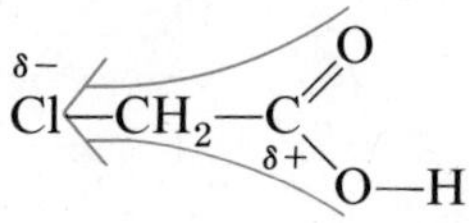

$$\underset{\delta-}{Cl}\leftarrow CH_2-\underset{\delta+}{C}\underset{O-H}{\overset{O}{<}}$$

A second chlorine atom increases the acidity still further, and three chlorine atoms on the α-carbon produce an acid about 13,000 times as strong as acetic acid (based on the relative K_a values). The inductive effects diminish when electron-withdrawing groups are farther removed from the carboxyl group and are strongest when such substituents are positioned on the α-carbon. Inductive effects also vary with the electronegativities of the elements bonded to the α-carbon atom. Reference to Table 10.2 will show that fluoroacetic acid is nearly four times as strong an acid as iodoacetic acid. The inductive effects of electron-releasing groups on the α-carbon atom destabilize the anion by intensifying the negative charge, thus making proton departure more difficult and the acid weaker.

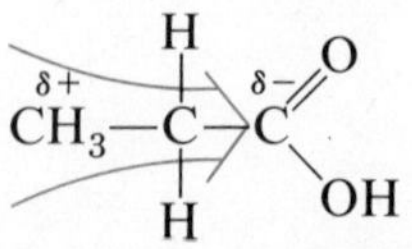

$$CH_3-\underset{\delta+}{\overset{H}{\underset{H}{C}}}\rightarrow\underset{\delta-}{C}\overset{O}{\underset{OH}{<}}$$

Ring-deactivating groups (Sec. 4.9), when substituted in the *ortho* or *para* positions of benzoic acid, make the substituted acid stronger than benzoic. On the other hand, ring-activating groups, only when substituted in the *para* position, have an acid-weakening effect. Table 10.2 compares the strengths of a number of carboxylic acids and illustrates the effects of substituents and their positions upon acid strength.

TABLE 10.2 *Relative Strengths of Some Organic Acids* (25°C)

Name of Acid	Structure	pK_a
Water*	H—OH	15.74
Formic	H—COOH	3.75
Acetic	CH_3COOH	4.76
Monochloroacetic	$ClCH_2COOH$	2.86
Dichloroacetic	$Cl_2CHCOOH$	1.30
Trichloroacetic	Cl_3CCOOH	0.64
Bromoacetic	$BrCH_2COOH$	2.90
Iodoacetic	ICH_2COOH	3.18
Fluoroacetic	FCH_2COOH	2.59
Trimethylacetic	$(CH_3)_3CCOOH$	5.05
Propionic	CH_3CH_2COOH	4.87
α-Chloropropionic	$CH_3\underset{\underset{Cl}{\mid}}{C}HCOOH$	2.80
β-Chloropropionic	$ClCH_2CH_2COOH$	4.00
Benzoic	C_6H_5COOH	4.21
Phenylacetic	⬡—CH_2COOH	4.31
p-Chlorobenzoic	Cl—⬡—COOH	3.99
p-Nitrobenzoic	O_2N—⬡—COOH	3.44
p-Methoxybenzoic	CH_3O—⬡—COOH	4.49

* While water does not belong in the above category, it is included for reference value. The value of pK_a is given for the equation: $K = [H^+][OH^-]/[H_2O]$, where $[H_2O]$ is 55.5 moles per liter.

EXERCISE 10.2

Without consulting Table 10.2 arrange the acids in the following set in order of diminishing acidity: (a) acetic acid, (b) propionic acid, (c) trimethylacetic acid, (d) α-chloropropionic acid, (e) β-chloropropionic acid.

10.3   Properties

The lower molecular weight members of the aliphatic series of carboxylic acids are liquids with sharp or disagreeable odors. Those with four to ten carbon atoms are particularly obnoxious. The odor of rancid butter, limburger cheese, and stale sweat vividly exemplify this unpleasant property. The higher members are waxlike solids and almost odorless. The boiling points of fatty acids increase regularly by an approximate 20°C increment per methylene unit. The abnormally high boiling points of the fatty acids are accounted for by hydrogen bond formation between acid molecules. Hydrogen bonding also accounts for the fact that molecular weight measurements reveal the fatty acids to be largely "dimeric," or double, molecules. Both linear and cyclic dimers are present, with the latter predominating.

$$R-C\begin{smallmatrix} O\cdots H-O \\ \\ O-H\cdots O \end{smallmatrix}C-R$$

Cyclic dimer

Carboxylic acids, with the exception of the first five members of the aliphatic series, are not very soluble in water. The aromatic acids usually are crystalline solids, also sparingly soluble in cold water.

10.4   Preparation of Acids

A number of straight-chain aliphatic acids and some aromatic acids are available as natural products. Others can be prepared by one of the following methods.

A. Oxidation Methods.   Direct oxidation of primary alcohols provides one of the most direct routes to the corresponding aliphatic acids. Potassium or sodium dichromate in combination with concentrated sulfuric acid is frequently used to bring about this change.

$$3\,R-CH_2OH + 2\,Cr_2O_7^{2-} + 16\,H^+ \longrightarrow 3\,R-C\begin{smallmatrix}O\\\\OH\end{smallmatrix} + 4\,Cr^{3+} + 11\,H_2O$$

The benzene carboxylic acids are obtained by the oxidation of alkyl derivatives of benzene. The benzene ring itself appears to be rather resistant to oxidation but facilitates the oxidation of the carbon atom attached to it. Should more than one alkyl group be attached to the ring, all become oxidized to carboxyl groups. Thus, toluene and other monoalkylated benzenes yield benzoic acid and the three isomeric xylenes, $C_6H_4(CH_3)_2$, yield the corresponding phthalic acids.

$$\text{Toluene} + 3\,[O] \xrightarrow{Na_2Cr_2O_7,\ H_2SO_4} \text{Benzoic acid (COOH)} + H_2O$$

Toluene   Benzoic acid

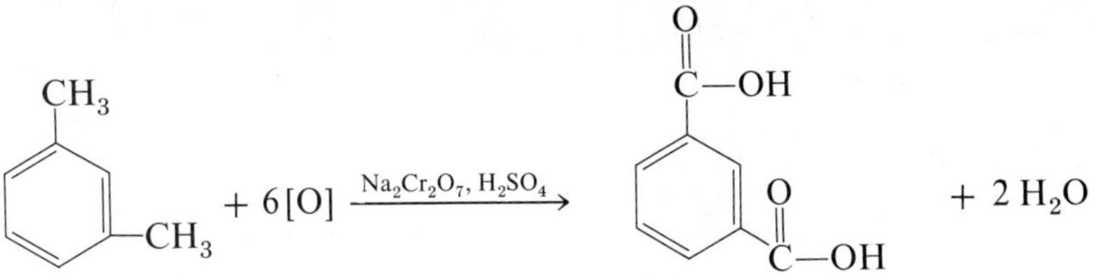

m-Xylene

1,3-Benzenedicarboxylic acid
(Isophthalic acid)

Although the length of the side chain on the benzene nucleus may vary, it is degraded to the last or ring-attached carbon atom. All other carbon atoms in the side chain are oxidized to carbon dioxide.

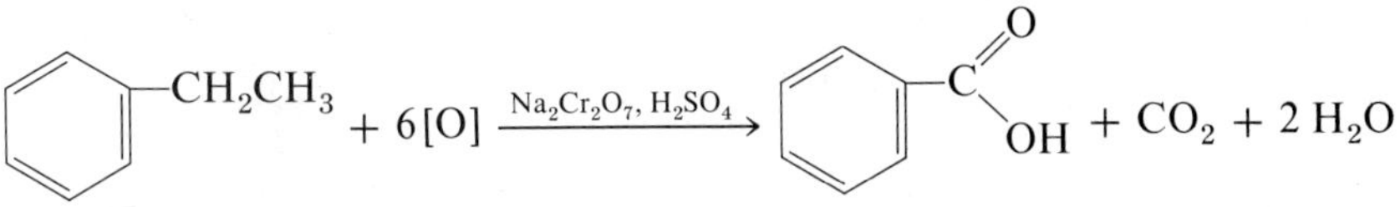

Ethylbenzene

Benzoic acid

EXERCISE 10.3

Balance the equation for the oxidation of ethylbenzene. (*Hint: See Section 8.8-C.*)

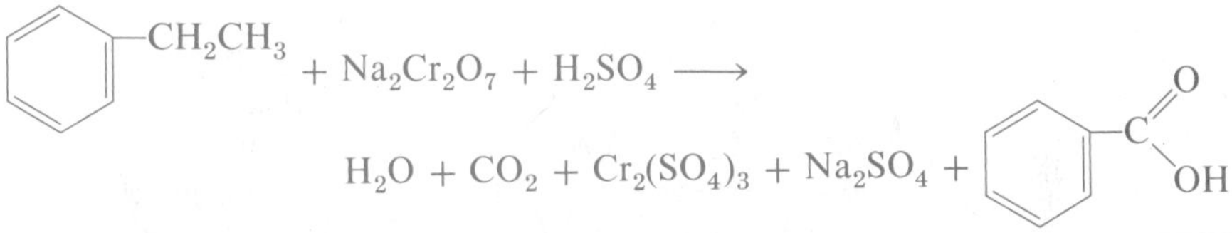

Naphthalene, obtainable from coal tar, is a source of phthalic acid. Phthalic anhydride, formed when phthalic acid is heated in excess of 200°C, is an important industrial material used in the manufacture of glyptal resins for surface coatings (Sec. 12.4).

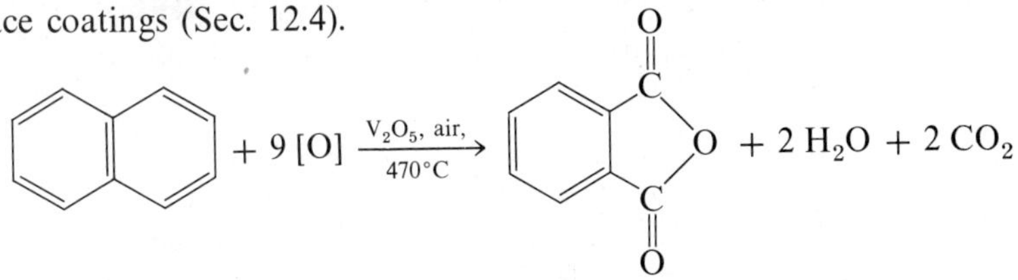

Naphthalene

Phthalic anhydride

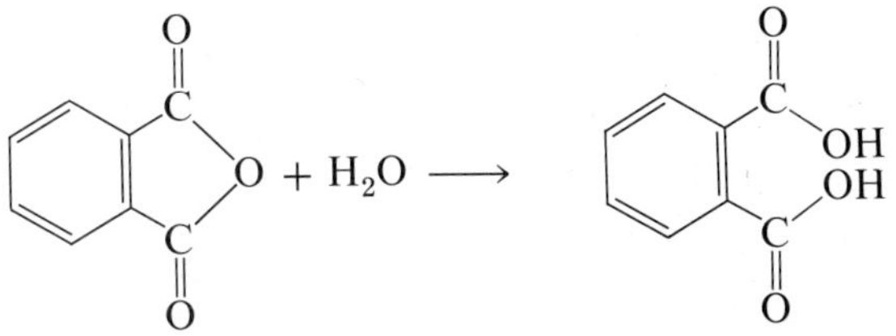

Phthalic acid
(1,2-Benzenedicarboxylic acid)

B. Hydrolysis Methods. The nitriles (prepared from alkyl halides and potassium cyanide) may be hydrolyzed to carboxylic acids by refluxing with aqueous acid or alkali.

$$C_2H_5Br + KCN \longrightarrow C_2H_5{-}CN$$
Ethyl cyanide
(Propionitrile)

Acid hydrolysis

$$C_2H_5{-}CN + 2\,H_2O + HCl \longrightarrow C_2H_5{-}C\overset{\displaystyle O}{\underset{\displaystyle OH}{<}} + NH_4{}^+Cl^-$$
Propionic acid

Alkaline hydrolysis

$$C_2H_5{-}CN + 2\,H_2O + NaOH \longrightarrow C_2H_5{-}C\overset{\displaystyle O}{\underset{\displaystyle O^-Na^+}{<}} + NH_3 + H_2O$$
Sodium propionate

The nitriles are named according to the acids they yield when hydrolyzed. Thus, methyl cyanide, CH_3CN, is named *acetonitrile;* phenyl cyanide, C_6H_5CN, *benzonitrile,* etc. The formation of nitriles from alkyl halides, when followed by hydrolysis, provides a method for increasing the length of the carbon chain. The same result can be achieved by the addition of hydrogen cyanide to an appropriate organic compound. For example, aldehydes add hydrogen cyanide to yield cyanohydrins and thus provide a route to the α-hydroxy acids.

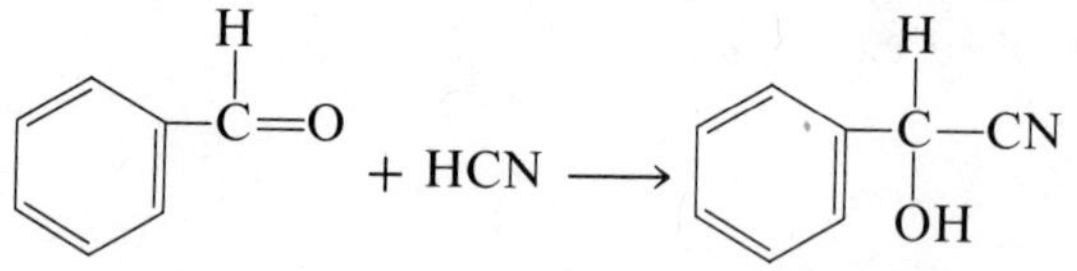

Benzaldehyde   Benzaldehyde cyanohydrin (Mandelonitrile)

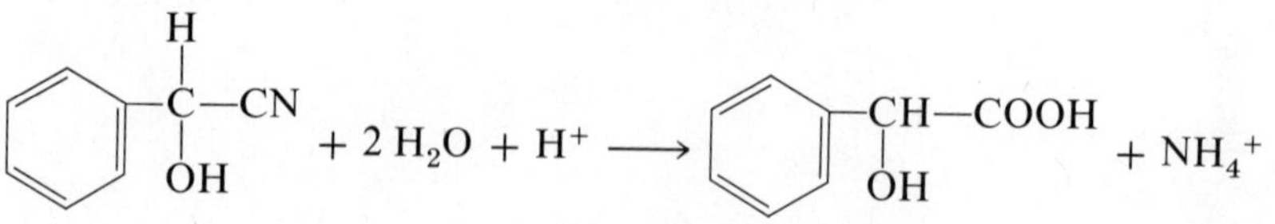

Mandelic acid

Benzotrichloride, produced by the chlorination of toluene in sunlight, can be converted by hydrolysis to benzoic acid as shown in the following reaction.

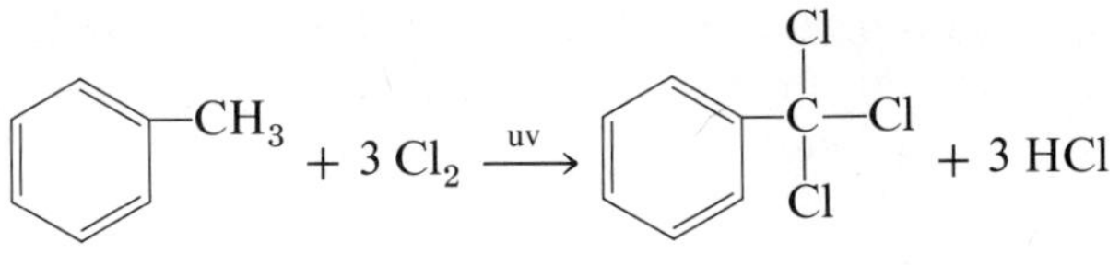

Toluene   Benzotrichloride

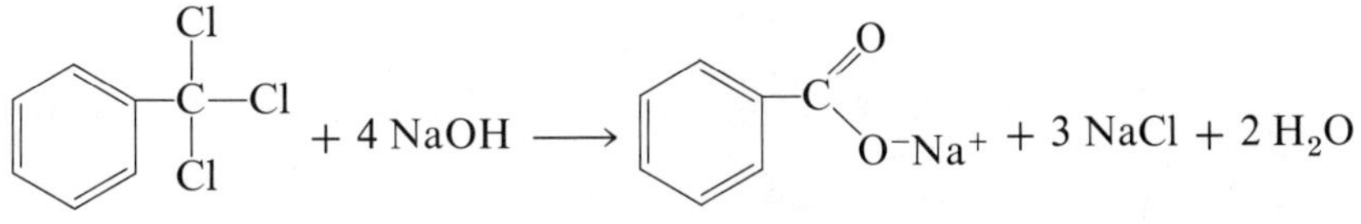

Sodium benzoate

The free organic acid can be precipitated from the salt solution by acidification with hydrochloric acid.

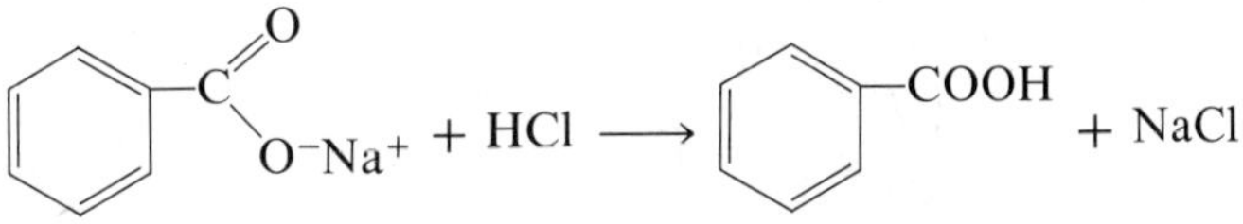

Benzoic acid

C. The Carbonation of Grignard Reagents.

Grignard reagents are especially useful for the preparation of acids, either aliphatic or aromatic. One of the best general methods for the preparation of carboxylic acids is to treat the appropriate Grignard reagent with anhydrous carbon dioxide.

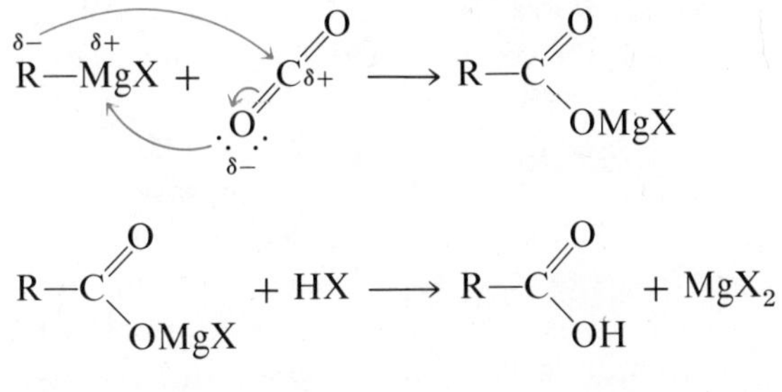

You will note that the carbonation of a Grignard reagent, like the hydrolysis of a nitrile, is a reaction that also increases by one the length of the carbon chain.

10.5 *Reactions of the Carboxylic Acids and Their Derivatives*

Many of the reactions engaged in by carboxylic acids give products which correctly can be said to have been derived from acids. The term "acid derivatives," however, usually is reserved to describe those in which some other structure has replaced the hydroxyl group and from which the original acid can be regained by hydrolysis. In the following sections are outlined a number of the more general reactions of the acids. Following these are the specific reactions of the derivative. It soon will become apparent that in a number of instances a reaction involving one acid derivative is a route leading to the preparation of another.

A. Preparation of Salts. The carboxylic acids, although feebly acidic when compared to mineral acids, are relatively strong when compared to water (Table 10.2). They may be neutralized quantitatively by carbonates, bicarbonates, and hydroxide bases to produce salts and water.

$$2\ R-C\overset{O}{\underset{OH}{\diagup}} + Na_2CO_3 \longrightarrow 2\ R-C\overset{O}{\underset{O^-Na^+}{\diagup}} + CO_2 + H_2O$$

$$R-C\overset{O}{\underset{OH}{\diagup}} + NaHCO_3 \longrightarrow R-C\overset{O}{\underset{O^-Na^+}{\diagup}} + CO_2 + H_2O$$

$$R-C\overset{O}{\underset{OH}{\diagup}} + NaOH \longrightarrow R-C\overset{O}{\underset{O^-Na^+}{\diagup}} + H_2O$$

Salts of organic acids are named as one would name salts of inorganic acids—that is, the cation is named first followed by the name of the acid anion. The latter is derived by dropping the *ic* of the acid and adding *ate*. For example,

$$CH_3-C\overset{O}{\underset{O^-Na^+}{\diagup}}$$

$$CH_3-(CH_2)_{16}-C\overset{O}{\underset{O^-Li^+}{\diagup}}$$

Sodium acetate   Lithium stearate

The equivalent weight of an acid can be determined by titrating it with a standardized base. The **neutralization equivalent** of an acid, abbreviated as **N.E.**, gives a measure of its molecular weight and often is useful in helping to establish the acid's identity. The neutralization equivalent may be defined as the weight of acid in grams that is required to neutralize one gram equivalent of standardized base. If an unknown acid is monoprotic—that is, has but one proton to lose—its neutralization equivalent and its molecular weight is one and the same. A dicarboxylic acid, of course, has a neutralization equivalent one-half its molecular weight, a tricarboxylic acid, one-third, etc. You will want to ascertain for yourself the correctness of the N.E. values for the examples below.

$$CH_3C\overset{O}{\underset{OH}{\diagup}}$$

Acetic acid   N.E. = 60

Phthalic acid   N.E. = 83

EXERCISE 10.4

The proper identity of an acid was in doubt. It could be either *d,l*-mandelic acid or *d,l*-tropic acid.

$$
\begin{array}{cc}
\text{C}_6\text{H}_5\text{—}\overset{\displaystyle \text{H}}{\underset{\displaystyle \text{CH}_2\text{OH}}{\text{C}}}\text{—COOH}
&
\text{C}_6\text{H}_5\text{—}\overset{\displaystyle \text{H}}{\underset{\displaystyle \text{OH}}{\text{C}}}\text{—COOH}
\end{array}
$$

d,l-Tropic acid   *d,l*-Mandelic acid
(M.P. 117°)   (M.P. 118°)

A 0.228-g sample of the acid required 15 ml of 0.1 N sodium hydroxide when titrated to the phenolphthalein end point. Which acid was it?

B. Reactions of Salts. The ammonium salts of carboxylic acids, when heated strongly, lose the elements of water to form acid amides.

$$
\text{CH}_3\text{—C}\overset{\displaystyle \text{O}}{\underset{\displaystyle \overset{-}{\text{O}}\overset{+}{\text{N}}\text{H}_4}{\Big\langle}} \xrightarrow{\text{heat}} \text{CH}_3\text{—C}\overset{\displaystyle \text{O}}{\underset{\displaystyle \text{NH}_2}{\Big\langle}} + \text{H}_2\text{O}
$$

Ammonium acetate   Acetamide

Alkali metal salts of the long chain fatty acids are called **soaps** (Chapter 11). Lithium stearate, Sec. 10.5-A, and other heavy metal salts, when blended with oils, form lubricating greases. Calcium propionate, $(\text{CH}_3\text{—CH}_2\text{—C}\overset{\text{O}}{\text{—}}\text{O—})_2\text{Ca}$, is a commonly used additive in bread to retard spoilage (molding). Certain of the zinc salts of the higher fatty acids also are excellent fungicides and are used in the treatment of skin diseases such as athlete's foot. One of these, zinc undecylenate (from 10-undecenoic acid), is particularly effective. Its structure is shown.

$$
\left(\text{CH}_2{=}\overset{\displaystyle \text{H}}{\text{C}}\text{—(CH}_2)_8\text{—}\overset{\displaystyle \text{O}}{\text{C}}\text{—O} \right)_{\!2} \!\! \text{Zn}
$$

Zinc undecylenate

The cupric salts of naphthenic acids (cyclic, nonaromatic acids derived from petroleum) are used as wood preservatives.

C. Preparation of Acid Halides. The acid halides also are referred to as the **acyl halides.** The name of an acyl group, $\text{R—C}\overset{\text{O}}{\Big\langle}$, is derived from an aliphatic acid by dropping the *ic* ending of the acid and adding *yl*. The aromatic acid

halides (aroyl halides) are formed and named in the same manner. The following examples will illustrate the usage of the term as a group name.

Acet*ic* acid	Acet*yl* chloride	Acetylacetone
(Ethano*ic* acid)	(Ethano*yl* chloride)	(2,4-Pentanedione)

Benzoic acid	Benzoyl chloride	*o*-Benzoylbenzoic acid
		(Benzophenone-
		o-carboxylic acid)

Acid halides other than the chlorides can be made, but the chlorides are
more easily and economically prepared. The acyl chlorides are very reactive
compounds and usually are employed where acyl halides are required. They are
easily prepared by the reaction of the appropriate acid with either phosphorus
trichloride or phosphorus pentachloride.

Acetic acid	Phosphorus	Acetyl chloride
	*tri*chloride	

Thionyl chloride, often used in the preparation of acid chlorides, is a superior
reagent to the phosphorus halides for two reasons: (1) the by-products of the
reaction are gases and are easily removed, (2) the reagent is a low boiling liquid.
The second advantage permits any excess reagent to be easily removed by
distillation.

Benzoic acid	Thionyl	Benzoyl
	chloride	chloride

D. Reactions of the Acid Halides.   The acid halides are low boiling liquids of
irritating odor. They are extremely reactive compounds because the inductive
effect of the halogen atom further diminishes the electron density on the
carbonyl carbon atom. Attack on the carbonyl carbon by nucleophiles (Sec.
1.11) is enhanced.

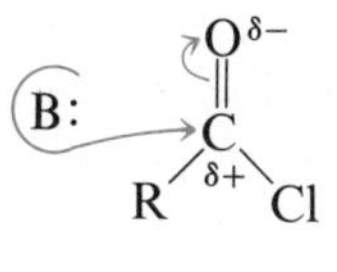

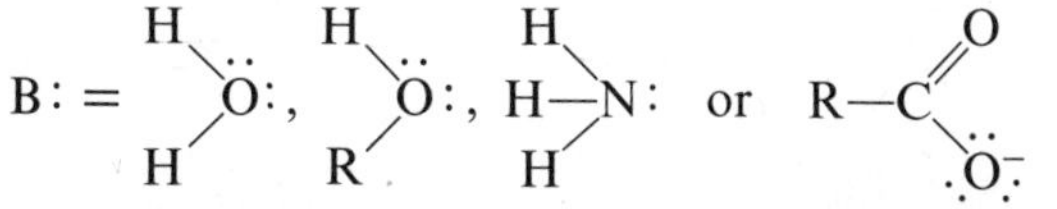

The reaction of an acyl halide with water is called **hydrolysis** and results in the displacement of halogen by hydroxyl to reform the original organic acid.

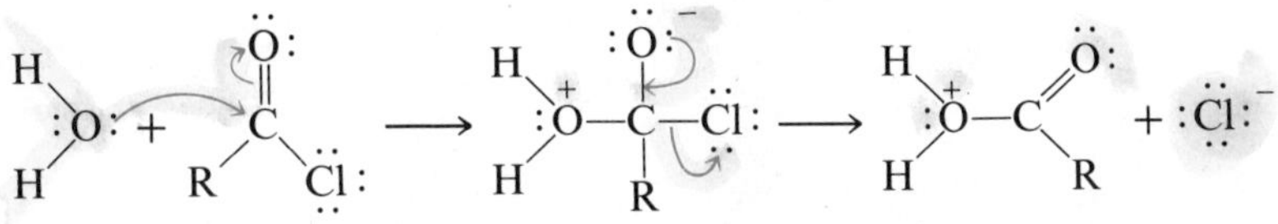

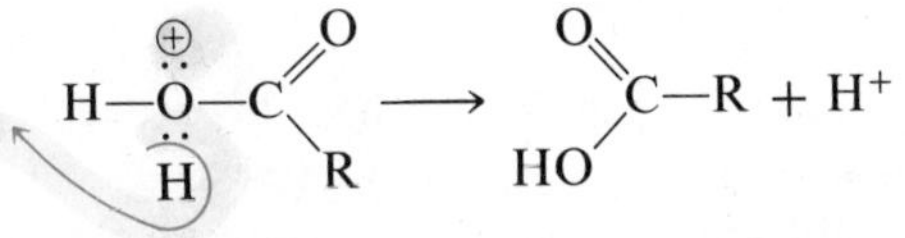

The net equation may be shown as

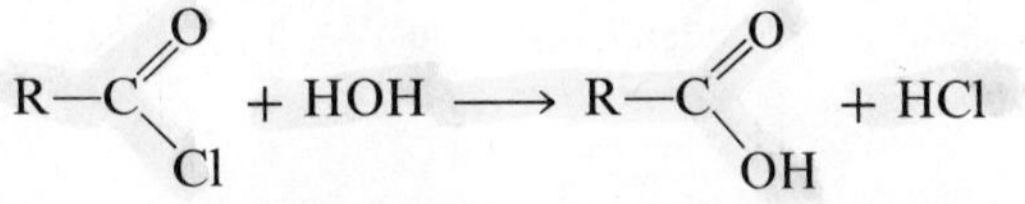

The reaction of an acyl halide with an alcohol is called **alcoholysis** and results in the displacement of halogen by an alkoxy group, —OR, to produce an **ester** (Sec. 10.5-G).

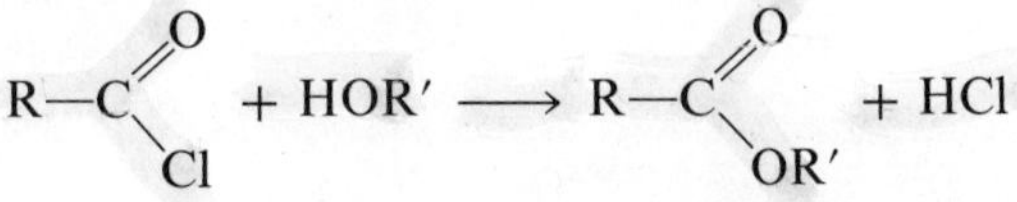

An ester

The reaction of an acyl halide with ammonia is called **ammonolysis,** and results in the displacement of halogen by an amino group, —NH$_2$, to produce an **amide** (Sec. 10.5-I).

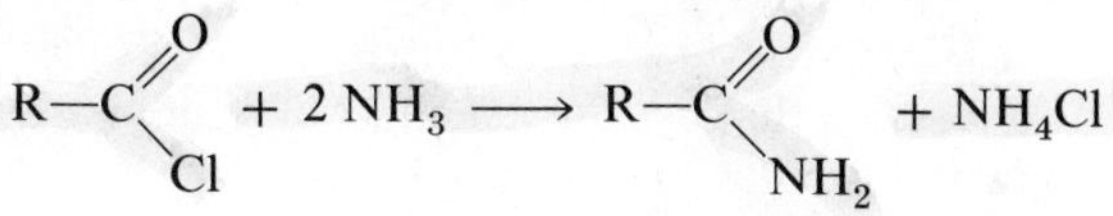

An amide

Acyl halides react with salts of acids to yield acid **anhydrides.**

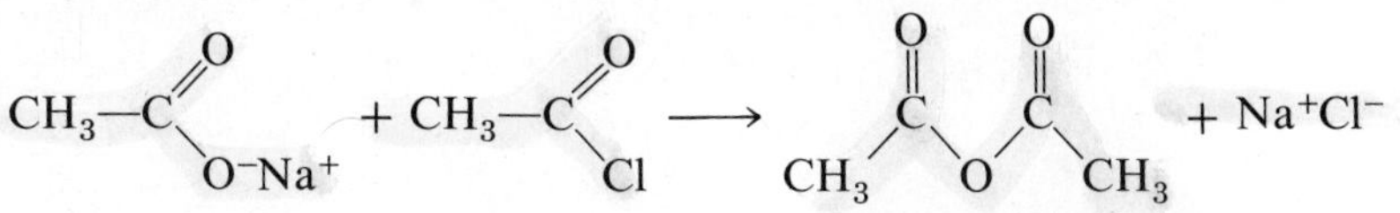

Sodium acetate   Acetyl chloride   Acetic anhydride

Acyl halides form enols more readily than acids and thus undergo halogenation in the α-position more readily than do acids and provide an easy route to α-halogenated acids. In practice, α-bromoacids usually are prepared. The appropriate acid is treated with red phosphorus and bromine to produce the acyl bromide as the initial product.

$$6\ CH_3C\underset{OH}{\overset{O}{\diagup}} + 2\ P + 3\ Br_2 \longrightarrow 6\ H\overset{H}{\underset{H}{-C-}}C\underset{Br}{\overset{O}{\diagup}} + 2\ P(OH)_3$$

Acetyl bromide

$$CH_3-C\underset{Br}{\overset{O}{\diagup}} + Br_2 \longrightarrow BrCH_2-C\underset{Br}{\overset{O}{\diagup}} + HBr$$

α-Bromoacetyl
bromide

Subsequent reaction of the α-bromoacetyl bromide with unreacted acid results in a transfer of the more reactive acetyl bromine, not the α-bromine. The reaction thus continues until all the original acid is converted into the α-bromo derivative. This method for preparing the α-halogen acids is called the **Hell-Volhard-Zelinsky reaction.**

$$Br\overset{H}{\underset{H}{-C-}}C\underset{Br}{\overset{O}{\diagup}} + CH_3-C\underset{OH}{\overset{O}{\diagup}} \rightleftharpoons Br\overset{H}{\underset{H}{-C-}}C\underset{OH}{\overset{O}{\diagup}} + CH_3-C\underset{Br}{\overset{O}{\diagup}}$$

α-Bromoacetic acid

EXERCISE 10.5

What simple test tube reaction would serve to distinguish between bromoacetic acid and acetyl bromide?

E. Preparation of Acid Anhydrides.

The preparation of an acid anhydride is a reaction that has the net effect of removing a molecule of water from between two molecules of the acid.

$$CH_3-C\underset{(OH\ \ H)O}{\overset{O}{\diagup}} \diagdown C-CH_3 \longrightarrow \begin{matrix} CH_3-C\overset{O}{\diagup} \\ \diagdown O \\ CH_3-C\diagdown_O \end{matrix} + H_2O$$

Acetic anhydride

Direct dehydration, however, is seldom practiced. Dehydration usually is accomplished indirectly by reaction between the sodium salt of the acid with its acid chloride (Sec. 10.5-D). Two acyl groups are thus bridged by an oxygen atom by a method much like that employed in the Williamson ether synthesis (Sec. 8.8A-2). Acetic anhydride is by far the most important acid anhydride. Industrially it is prepared from the very reactive "anhydride", ketene. Acetic acid adds to ketene to produce acetic anhydride.

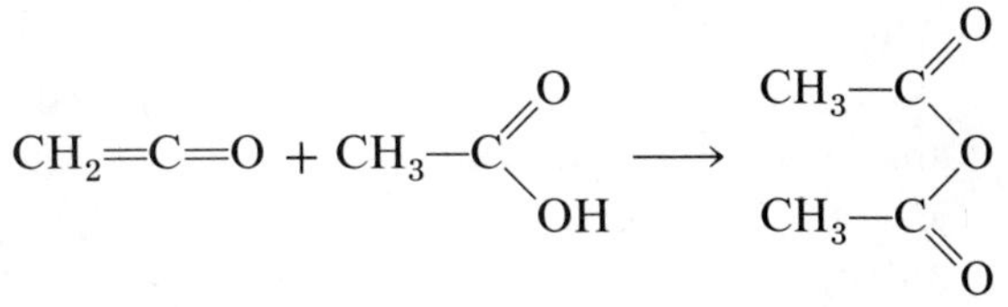

$$CH_2{=}C{=}O + CH_3{-}C\overset{O}{\underset{OH}{\big\langle}} \longrightarrow \begin{matrix} CH_3{-}C\overset{O}{\diagdown} \\ O \\ CH_3{-}C\diagup \\ O \end{matrix}$$

Ketene   Acetic acid   Acetic anhydride

F. Reactions of the Acid Anhydrides. The reactions of the acid anhydrides with water, alcohol, and ammonia parallel those already shown for the acyl halides. As shown by the following reactions, at least one molecule of the organic acid is a product of each reaction.

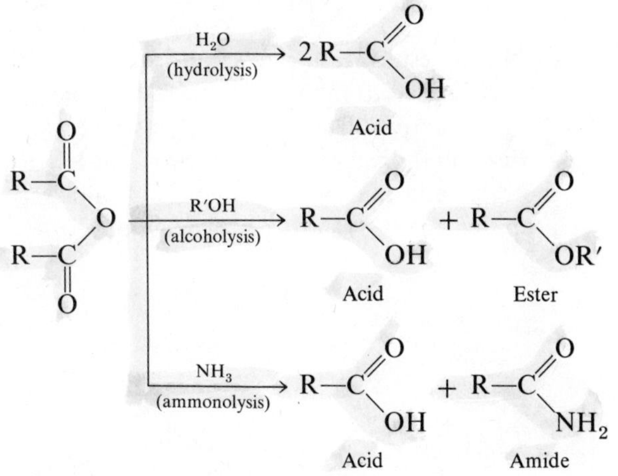

Acid anhydrides may be used in place of acyl halides as acylating reagents in the Friedel-Crafts reaction (Sec. 9.4-D).

G. Preparation of Esters. The reaction of an acid with an alcohol in the presence of a mineral acid catalyst produces an ester. A direct esterification of alcohols and acids in this manner is known as the **Fischer esterification.**

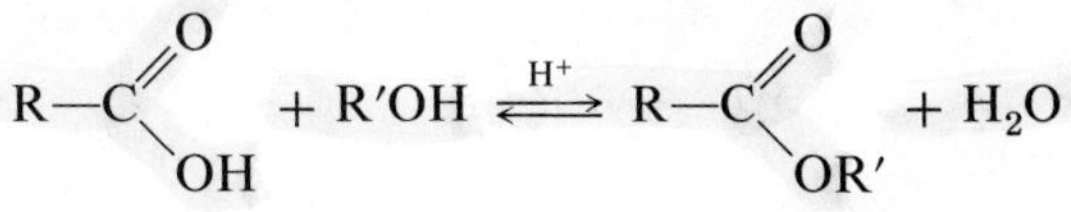

$$R{-}C\overset{O}{\underset{OH}{\big\langle}} + R'OH \underset{}{\overset{H^+}{\rightleftarrows}} R{-}C\overset{O}{\underset{OR'}{\big\langle}} + H_2O$$

Direct esterification is a reversible reaction, and at equilibrium appreciable amounts of unreacted acid and alcohol may be present. Thus, if we start with equimolar quantities of acetic acid and ethyl alcohol and allow them to react until equilibrium is established, two-thirds of the acid and the alcohol will have been converted to ester and one-third will remain as unreacted acid and alcohol.

$$K_e = \frac{\text{ethyl acetate} \times \text{water}}{\text{acetic acid} \times \text{ethyl alcohol}} = \frac{(2/3)^2}{(1/3)^2} = 4$$

The principle of LeChatelier may be applied to shift the equilibrium to the right (promote esterification) either by increasing the concentration of the alcohol or the acid, or by removing the ester or the water as it is formed. The reverse reaction (hydrolysis), on the other hand, can be made complete only when carried out in an alkaline solution (Sec. 10.5-H).

Esters are named in a manner similar to that used in naming the salts of the carboxylic acids, in that the group attached to the oxygen is named first.

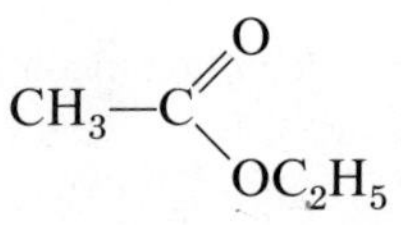

Ethyl acetate 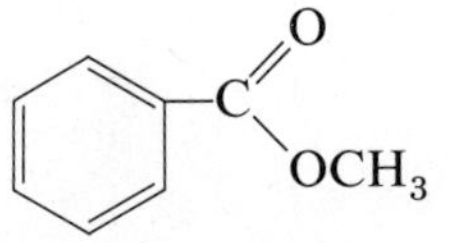   Methyl benzoate

In the esterification reaction the oxygen atom in the alkoxy group of the ester is derived from the alcohol, and the oxygen atom in the molecule of water formed is derived from the acid. Thus, when alcohols containing the ^{18}O isotope are used in the esterification reaction, the **labeled** oxygen is found exclusively in the ester and not in the water. The following reaction mechanism has been proposed for the esterification reaction.

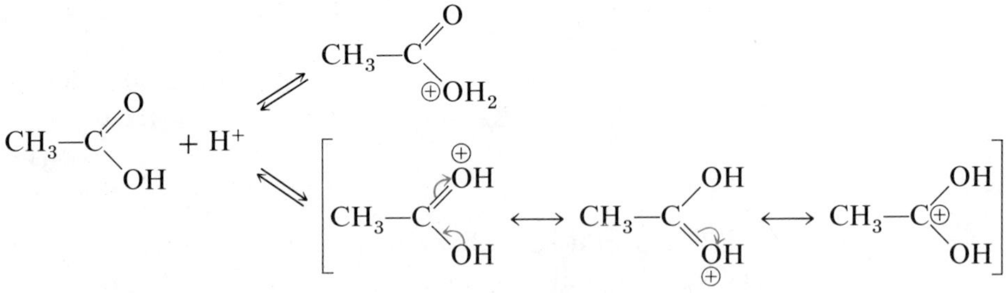

The acid is protonated on both oxygens, but only the oxonium ion formed by protonation on the carbonyl oxygen is involved in the reaction with alcohol.

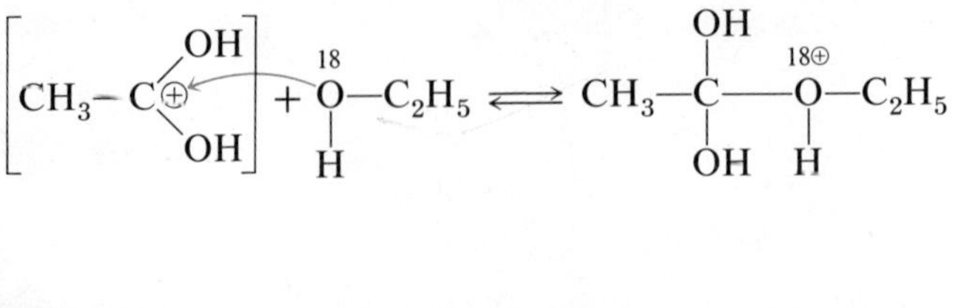

$$CH_3-\overset{\displaystyle OH}{\underset{\displaystyle OH \;\; H}{\overset{|}{\underset{|}{C}}}}-\overset{18\oplus}{O}-C_2H_5 \;\rightleftharpoons\; CH_3-\overset{\displaystyle OH}{\underset{\displaystyle \overset{O\oplus}{\underset{H \quad H}{}}}{\overset{|}{\underset{}{C}}}}-\overset{18}{O}-C_2H_5$$

$$CH_3-\overset{\displaystyle OH}{\underset{\displaystyle \overset{O\oplus}{\underset{H \quad H}{}}}{\overset{|}{\underset{}{C}}}}-\overset{18}{O}-C_2H_5 \;\rightleftharpoons\; H_2O + \left[CH_3-\overset{\displaystyle OH}{\underset{\displaystyle \oplus}{\overset{|}{\underset{}{C}}}}-\overset{18}{O}-C_2H_5 \;\longleftrightarrow\; CH_3-\overset{\displaystyle \oplus OH}{\overset{\|}{C}}-\overset{18}{O}-C_2H_5 \right]$$

$$\left[CH_3-\overset{\displaystyle \oplus OH}{\overset{\|}{C}}-\overset{18}{O}-C_2H_5 \right] \;\rightleftharpoons\; H^+ + CH_3-\overset{O}{\overset{\diagup\!\!\diagdown}{C}}\overset{}{\underset{18\;\;\;\;}{}} \;\; \underset{OC_2H_5}{}$$

Esters also may be prepared by heating the salt of an acid with an alkyl halide. This reaction is not reversible.

$$R-C\overset{\displaystyle O}{\underset{\displaystyle O^-Ag^+}{\diagup\!\!\diagdown}} + R'I \longrightarrow AgI\downarrow + R-C\overset{\displaystyle O}{\underset{\displaystyle OR'}{\diagup\!\!\diagdown}}$$

The esters are liquids of rather pleasant odor. A number are responsible (at least in part) for the odors of certain fruits and other plant parts, and are used in artificial flavorings and in perfumes. The esters are excellent solvents for lacquers and plastics.

H. Reactions of the Esters.

The acid hydrolysis of esters has been shown to be a reaction in equilibrium with the esterification reaction. In order to hydrolyze an ester irreversibly—that is, to have it quantitatively reform the acid (as a salt) and the alcohol, hydrolysis must be carried out in an alkaline solution. Alkaline hydrolysis of an ester is called **saponification** because soaps are prepared by the alkaline hydrolysis of fats and oils, esters of glycerol (Sec. 11.3). The acid component of an ester, when the latter is saponified, always forms a salt of the organic acid. Subsequent treatment of the salt with mineral acid regenerates the organic acid.

$$CH_3-C\overset{\displaystyle O}{\underset{\displaystyle OC_2H_5}{\diagup\!\!\diagdown}} + Na^+OH^- \longrightarrow CH_3-C\overset{\displaystyle O}{\underset{\displaystyle O^-Na^+}{\diagup\!\!\diagdown}} + C_2H_5OH$$

$$\text{Ethyl acetate} \hspace{6cm} \text{Sodium acetate}$$

$$CH_3-C\overset{\displaystyle O}{\underset{\displaystyle O^-Na^+}{\diagup\!\!\diagdown}} + HCl \longrightarrow CH_3COOH + Na^+Cl^-$$

$$\hspace{5cm} \text{Acetic acid}$$

Ammonolysis of an ester forms an alcohol and an **amide.**

$$R-C{\overset{\displaystyle O}{\underset{\displaystyle OR'}{\big\langle}}} + NH_3 \longrightarrow R-C{\overset{\displaystyle O}{\underset{\displaystyle NH_2}{\big\langle}}} + R'OH$$

An amide

Esters may be reduced to alcohols. The reduction can be accomplished chemically by using either sodium metal and an alcohol or by the use of lithium aluminum hydride, $LiAlH_4$.

$$R-C{\overset{\displaystyle O}{\underset{\displaystyle OR'}{\big\langle}}} + 4\,Na + 2\,R'OH \longrightarrow R-CH_2O^-Na^+ + 3\,R'O^-Na^+$$

$$R-CH_2O^-Na^+ + H_2O \longrightarrow R-CH_2-OH + Na^+OH^-$$

$$4\,R-C{\overset{\displaystyle O}{\underset{\displaystyle OR'}{\big\langle}}} + 2\,LiAlH_4 \xrightarrow[\text{ether}]{\text{anhydrous}} LiAl(OCH_2R)_4 + LiAl(OR')_4$$

$$LiAl(OCH_2R)_4 + 4\,HCl \longrightarrow LiCl + AlCl_3 + 4\,RCH_2OH$$

$$LiAl(OR')_4 + 4\,HCl \longrightarrow LiCl + AlCl_3 + 4\,R'OH$$

Esters react with Grignard reagents to form tertiary alcohols in which two of the three alkyl groups come from *the Grignard reagent.* In the tertiary alcohols made by the reaction of a ketone with a Grignard reagent, you will recall, two of the three alkyl groups come from *the ketone* and the third from the Grignard reagent (Sec. 8.4-C).

$$R-C{\overset{\displaystyle \ddot{O}}{\underset{\displaystyle OR'}{\big\langle}}} + \overset{\delta- \;\; \delta+}{R'MgX} \longrightarrow R-\overset{\displaystyle :\!\ddot{O}:^- \;^+MgX}{\underset{\displaystyle R''}{\overset{\displaystyle |}{\underset{\displaystyle |}{C}}}}-O-R'$$

$$R-\overset{\displaystyle :\!\ddot{O}:^- \;^+MgX}{\underset{\displaystyle R''}{\overset{\displaystyle |}{\underset{\displaystyle |}{C}}}}-OR' \longrightarrow \overset{\displaystyle R}{\underset{\displaystyle R''}{C}}{=}O + R'-O^- \;^+MgX$$

$$\overset{\displaystyle R}{\underset{\displaystyle R''}{C}}{=}O + R''MgX \longrightarrow R-\overset{\displaystyle R''}{\underset{\displaystyle R''}{\overset{\displaystyle |}{\underset{\displaystyle |}{C}}}}-OMgX$$

$$R-\overset{\displaystyle R''}{\underset{\displaystyle R''}{\overset{\displaystyle |}{\underset{\displaystyle |}{C}}}}-OMgX + HX \longrightarrow R-\overset{\displaystyle R''}{\underset{\displaystyle R''}{\overset{\displaystyle |}{\underset{\displaystyle |}{C}}}}-OH + MgX_2$$

EXERCISE 10.6

What simple test tube reaction would serve to distinguish between acetic anhydride and ethyl acetate?

The α-hydrogen atoms of an ester such as ethyl acetate (pK_a 24.5)[3] are somewhat less acidic than those of a ketone such as acetone (pK_a 20); however, they will react with a strong base such as sodium ethoxide to produce the ester enolate anion. The ester anion is a strong nucleophile which then may attack the carbonyl carbon atom of a second ester molecule. Elimination of ethoxide ion results in the formation of the β-keto ester, ethyl acetoacetate in the case of ethyl acetate. The reaction, illustrated in the following sequence, is called the **Claisen condensation**[4] and bears a resemblance to the aldol condensation (Sec. 9.7-B) except that it is an addition-elimination rather than a simple addition reaction. You may have noted by this time that many of the reactions of acid derivatives proceed by addition-elimination mechanisms, that is, the addition of one nucleophile followed by the elimination of another. Examples given thus far include: hydrolysis of an acid chloride, esterification, and the reaction of Grignard reagents with esters.

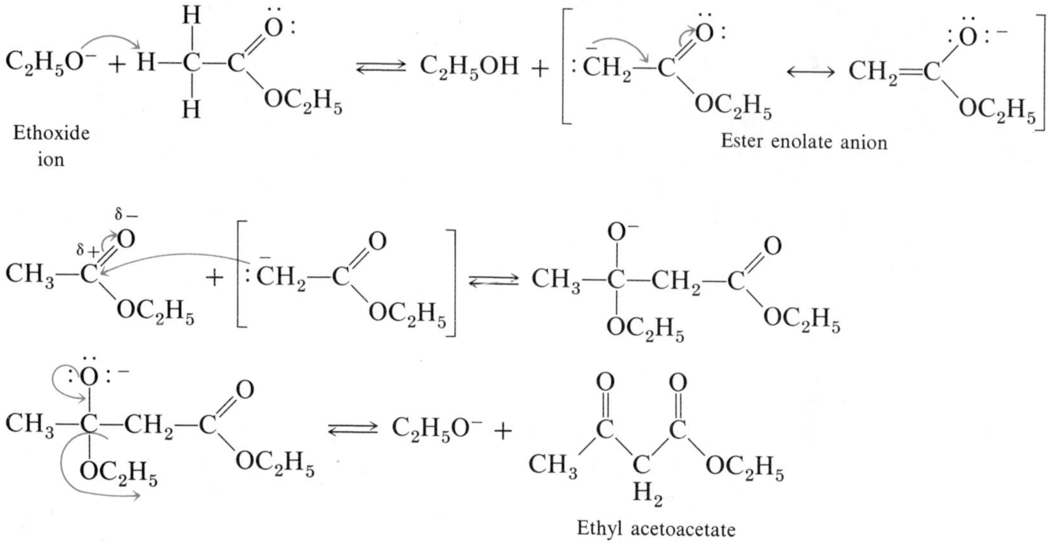

The Claisen condensation is also an important biological reaction involved in the degradation and biosynthesis of the fatty acids (Sec. 11.10), although the conditions for the biochemical reactions are quite different from those employed in the laboratory.

---

[3] The pK_a's of *very* weak acids ($pK_a > 14$) are measured by indirect methods.

[4] Ludwig Claisen (1851–1930) was professor of chemistry at the University of Kiel.

The β-keto esters are useful in a number of organic syntheses that lead to ketones and carboxylic acids that are difficult or impossible to obtain by other methods. These reactions are possible, in part, because the β-keto esters such as ethyl acetoacetate (pK_a 11) are stronger acids than either the ketones or the esters and form well-defined enolate anions that can be used in a variety of nucleophilic displacements (such as the S_N2 substitution of alkyl halides). To describe these in detail is beyond the scope of this text; however, an example of a reaction of this type is given in Sec. 12.7. The greater acidity of the β-keto esters is also reflected in the greater stability of the enol form relative to the keto form (Sec. 9.5). The stability of the enol tautomer may be attributed in part to an intramolecular H-bond.

Keto form	Enol form
In hexane: 51%	*In hexane: 49%*
In water: 90%	*In water: 10%*
Reacts with $C_6H_5NHNH_2$	*Reacts with bromine and gives a red color with $FeCl_3$ (a test reaction for phenols and enols)*

I. Preparation of Amides. Amides are conveniently prepared in the laboratory by the ammonolysis of acyl halides (Sec. 10.5-D) or acid anhydrides.

Amides are named after their parent acids by replacing *ic* or *oic* with *amide*. Examples of both common and systematic names are shown.

Benzamide	Propionamide (Propanamide)

With the exception of formamide, $H-\overset{O}{\underset{||}{C}}-NH_2$, the amides are solids, with sharp melting points. This property makes them very useful derivatives for the identification of acids.

J. Reactions of the Amides. The amides can be hydrolyzed in acid or in alkaline solution. Hydrolysis carried out in an acid solution produces the free organic acid and an ammonium salt. Hydrolysis carried out in a basic solution produces the free base (NH_3), and the salt of the organic acid. Both types of procedures are illustrated by the following reaction equations.

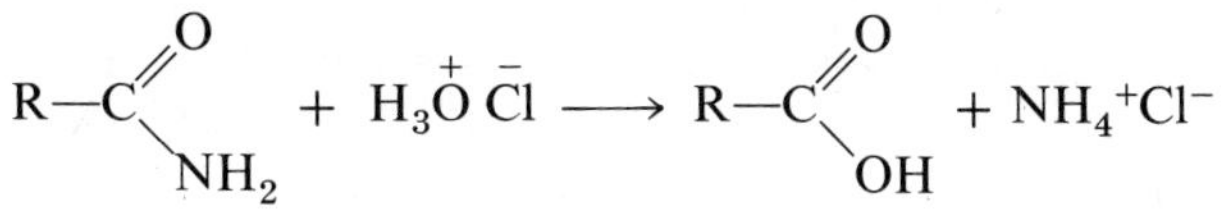

$$R-C\overset{O}{\underset{NH_2}{\big\langle}} + H_3\overset{+}{O}\,\overset{-}{Cl} \longrightarrow R-C\overset{O}{\underset{OH}{\big\langle}} + NH_4{}^+Cl^-$$

Acid hydrolysis

$$R-C\overset{O}{\underset{NH_2}{\big\langle}} + Na\overset{+}{}\overset{-}{OH} \longrightarrow R-C\overset{O}{\underset{O^-Na^+}{\big\langle}} + NH_3$$

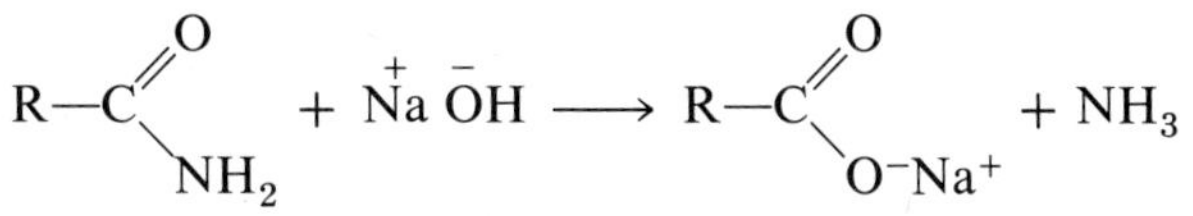

Alkaline hydrolysis

The amides form nitriles when heated in the presence of a strong dehydrating agent such as phosphorus pentoxide.

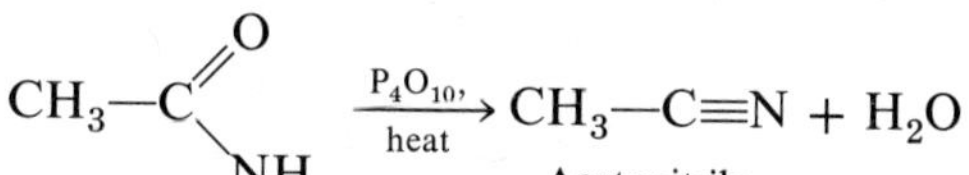

$$CH_3-C\overset{O}{\underset{NH_2}{\big\langle}} \xrightarrow[\text{heat}]{P_4O_{10},} CH_3-C\equiv N + H_2O$$

Acetonitrile

EXERCISE 10.7

What simple test will serve to distinguish benzamide (M.P., 128°) from benzoic acid (M.P., 122°)?

K. Reduction of Carboxylic Acids. Carboxylic acids may be reduced with lithiumaluminum hydride or with diborane (Sec. 8.4-A) but *not* with sodium borohydride (Sec. 9.10).

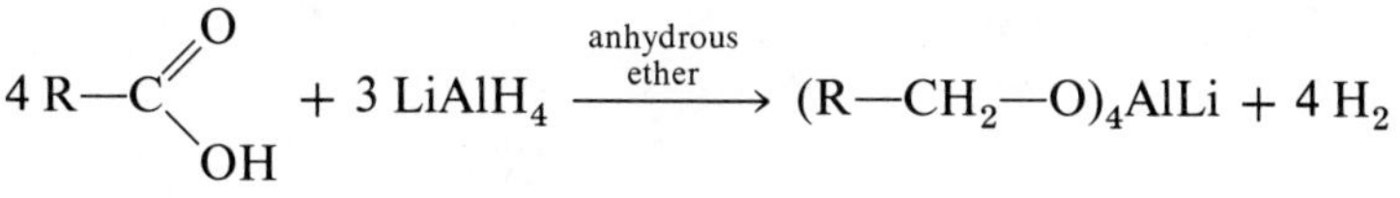

$$4\,R-C\overset{O}{\underset{OH}{\big\langle}} + 3\,LiAlH_4 \xrightarrow{\text{anhydrous ether}} (R-CH_2-O)_4AlLi + 4\,H_2$$

$$(R-CH_2-O)_4AlLi \xrightarrow[H^+]{H_2O} 4\,R-CH_2-OH$$

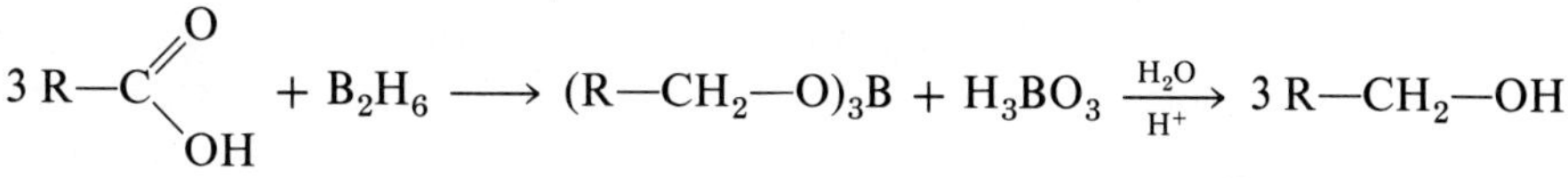

$$3\,R-C\overset{O}{\underset{OH}{\big\langle}} + B_2H_6 \longrightarrow (R-CH_2-O)_3B + H_3BO_3 \xrightarrow[H^+]{H_2O} 3\,R-CH_2-OH$$

L. Addition of Organolithium Reagents to Acids. The addition of Grignard reagents to esters (Sec. 10.5-H) cannot be stopped at the intermediate ketone stage because the ketone is more reactive than the ester. Acids react with Grignard reagents to liberate the hydrocarbon (corresponding to the alkyl group of the reagent, Sec. 2.7-A) and to form the relatively insoluble magnesium halide salt of the acid. Usually no further reaction occurs. However, alkyllithiums react with carboxylic acids to give soluble salts to which a second mole of alkyllithium adds to form a stable dilithium salt. Hydrolysis of the latter gives a ketone.

$$2\ CH_3Li + R-C\!\!\begin{array}{c}O\\\\OH\end{array} \longrightarrow CH_4 + R-C\!\!\begin{array}{c}O\\\\O^-Li^+\end{array} \xrightarrow{CH_3Li} R-\overset{\displaystyle O^-Li^+}{\underset{\displaystyle O^-Li^+}{C}}-CH_3$$

$$R-\overset{\displaystyle O^-Li^+}{\underset{\displaystyle O^-Li^+}{C}}-CH_3 \xrightarrow{H_2O} R-\overset{\displaystyle OH}{\underset{\displaystyle OH}{C}}-CH_3 \xrightarrow{-H_2O} R-\overset{\displaystyle O}{\underset{\displaystyle \ }{C}}-CH_3$$

EXERCISE 10.8

Lithiumdialkyl cuprates prefer not to add 1,2 to carbonyl groups (Sec. 9.12); therefore, they would appear to be useful organometallic reagents for converting some type of acid derivative to ketones. On the basis of what you have learned about these reagents (Secs. 2.7-B and 9.12), suggest an acid derivative that might be used in this way.

10.6 Properties of the Aromatic Acids

The reactions of the aromatic carboxylic acids, in general, are very similar to those of the aliphatic acids. The benzene ring, when attached directly to the carboxyl group, increases the acidic properties of the latter. Electron-withdrawing substituents *ortho* or *para* to the carboxyl group further increase the acidity of benzoic acid (Table 10.2). Ring substituents *ortho* to the carboxyl group can inhibit or completely prevent the direct esterification of an aromatic acid by an alcohol. Conversely, an ester of benzoic acid with substituents in the *ortho* positions is not easily hydrolyzed. This type of interference, attributable to the spatial requirements of groups, is called **steric hindrance.** Such steric effects, when caused by substituents on positions ortho to the reactive group, sometimes are specifically referred to as *ortho* effects. The carboxyl group, you will recall, is a *meta-director* and prevents other substituents from entering *ortho* positions. *Ortho, para*-substituted benzoic acids, therefore, must be prepared indirectly. For example, *p*-nitrobenzoic acid is not prepared from benzoic acid but is obtained by the oxidation of *p*-nitrotoluene.

The synthesis of an organic compound requires not only the use of the proper reactions but also the use of these reactions in the proper order or sequence. Sometimes such routes are lengthy and circuitous but may be the only ones leading to the desired product. At other times the orientation of substitution may be such so that devising a synthesis is relatively simple.

10.7 *Formic Acid*

The structure and properties of formic acid are somewhat different from those of other members of the acid series and deserve special mention. In the first place, the structure of formic acid is that of both an aldehyde and an acid. As such it is easily oxidized by mild oxidizing agents. Tollens' reagent (Sec. 9.9), for example, oxidizes it to carbon dioxide and water.

$$H-C\underset{O^-}{\overset{O}{\diagdown}} + 2\,Ag^+ + 3\,OH^- \longrightarrow 2\,H_2O + CO_3^{2-} + 2\,Ag$$

Formate ion

The acid chloride of formic acid, unlike those of its homologs, is not stable at ordinary temperatures. All attempts to prepare formyl chloride yield only carbon monoxide and hydrogen chloride. However, in combination these two gases sometimes enter into a Friedel-Crafts reaction as formyl chloride (Sec. 9.4-D).

Formic acid is prepared industrially by heating carbon monoxide and sodium hydroxide.

$$CO + NaOH \xrightarrow[\text{6-7 atmospheres}]{150°C,} H-C\underset{O^-Na^+}{\overset{O}{\diagdown}}$$

Sodium formate

The free acid is liberated from its sodium salt by reaction with sulfuric acid.

$$H-C\underset{O^-Na^+}{\overset{O}{\diagdown}} + H_2SO_4 \longrightarrow H-C\underset{OH}{\overset{O}{\diagdown}} + NaHSO_4$$

Formic acid

Formic acid also is obtained as a by-product in the production of pentaerythritol (Sec. 9.7-B).

EXERCISE 10.9

At 25°C formic acid ionizes to the extent of 4.3% in a 0.10 M solution. (a) What is the hydrogen ion concentration of this solution, $[H^+]$? (b) How much greater is the $[H^+]$ of a 0.1 M formic acid solution than that of a 0.1 M acetic acid solution?

Answer: (a) 0.0043 M, (b) 3.2 times greater.

10.8   *Acetic Acid*

Acetic acid, although classified as one of the fatty acids, is not a component acid of fats and oils. It is the sour principle of vinegar—a 5% solution of acetic acid. Synthetic acetic acid brought to this dilution is "white" vinegar whereas "brown" vinegar is the natural product of apple juice fermentation. Vinegars also may be made from other fruit juices. The fermentation process produces ethyl alcohol as the first product. Certain enzymes, if present, then catalyze the further oxidation of ethyl alcohol to acetic acid.

$$\underset{\text{Fruit sugar}}{C_6H_{12}O_6} \xrightarrow[\text{of yeast}]{\text{enzymes}} 2\,C_2H_5OH + 2\,CO_2$$

$$C_2H_5OH + O_2\,(\text{air}) \xrightarrow[\text{acetobacter}]{\text{enzymes of}} CH_3-C\underset{OH}{\overset{O}{\diagdown}} + H_2O$$

Pure acetic acid, m.p. 16.7°C, is referred to as "glacial" acetic acid because it appears as an icelike solid at lower than room temperature. Acetic acid is by far the most important of the monocarboxylic acids.

EXERCISE 10.10

Vinegar has a density of 1.0055 g/ml. If this represents a 5% (by weight) acetic acid solution, what is the concentration of vinegar in terms of molarity?

Answer: 0.8378 M.

Summary

1. **Structure**

   The carboxyl group, —COOH, is the functional group of the carboxylic acids. The aliphatic acids are referred to as *fatty acids* because many appear in a combined form with glycerol as fats.

2. **Nomenclature of Acids**

   The carboxylic acids usually are assigned common names. Systematic nomenclature follows general rules already learned. The suffix *oic* is added to the alkane stem followed by the word *acid*. Substituents in the hydrocarbon segment of an acid may be located by Greek letters, by numbers, or (in the case of aromatic acids) by *ortho, para,* and *meta* prefixes.

   Removal of the hydroxyl group of an acid forms the *acyl*

$$\left(R-C\!\!\!\diagup^{\displaystyle O}_{\displaystyle \diagdown} \right) \text{ or the } aroyl \left(Ar-C\!\!\!\diagup^{\displaystyle O}_{\displaystyle \diagdown} \right) \text{ group.}$$

3. **Properties of the Acids**

   The carboxylic acids are weak acids. They are capable of associating (H-bonding) and are high boiling liquids or crystalline solids sparingly soluble in water. H-bonding also makes possible the formation of double molecules or dimers.

4. **Preparation of the Acids**

   The carboxylic acids may be prepared by any of the following methods.
   - (a) Oxidation of primary alcohols, methyl ketones (haloform reaction, Sec. 9.8) or side chains on the aromatic nucleus.
   - (b) Hydrolysis of nitriles.
   - (c) Carbonation of Grignard reagents.

5. **Reactions**

   Aliphatic and aromatic acids give reactions that are almost identical. Any differences are of degree rather than of kind. These reactions include:
   - (a) Replacement of the ionizable hydrogen by another cation (salt formation).
   - (b) Replacement of the hydroxyl by halide to yield **acyl** or

   aroyl halides, $R-C\!\!\!\diagup^{\displaystyle O}_{\displaystyle \diagdown X}$ or $Ar-C\!\!\!\diagup^{\displaystyle O}_{\displaystyle \diagdown X}$

   - (c) Replacement of the hydroxyl by an alkoxyl, —OR, to yield

   esters, $R-C\!\!\!\diagup^{\displaystyle O}_{\displaystyle \diagdown OR}$

(d) Replacement of the hydroxyl by an amino group to yield

amides, R—C $\begin{array}{c} \nearrow O \\ \searrow NH_2 \end{array}$

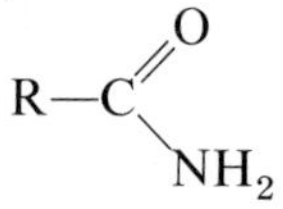

(e) Replacement of the hydroxyl by carboxylate,

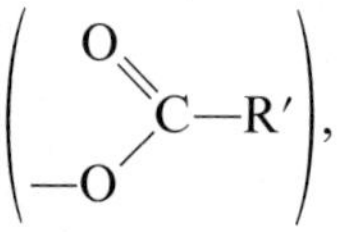

$\left(\begin{array}{c} O \\ \diagdown \\ -O \end{array} C-R' \right)$, to yield **anhydrides,** $\begin{array}{ccc} O & & O \\ \parallel & & \parallel \\ C & & C \\ R & O & R' \end{array}$

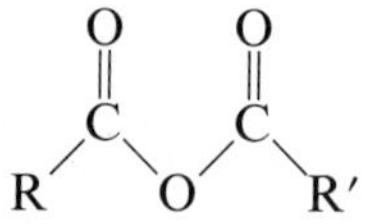

(f) Reduction to alcohols. Reduction of an acid may be accomplished directly by the use of $LiAlH_4$. Reduction of an acid may be accomplished *indirectly* via its ester—that is, the ester of an acid is reduced to produce two alcohols.

(g) The carboxyl group attached directly to the benzene ring orients incoming substituents to the *meta* position.

(h) The acids (with the exception of the first member) are stable to oxidation.

New Terms

acid dimers

acyl group

aroyl group

alcoholysis

ammonolysis

esterification

glacial acetic acid

hydrolysis

neutralization equivalent

pK_a

saponification

steric hindrance

tautomerism

Supplementary Exercises

EXERCISE 10.11 Draw structural formulas for each of the following compounds.

(a) valeric acid

(b) α,β-dimethylcaproic acid

(c) 3-butenoic acid

(d) ethyl bromoacetate

(e) propanoyl chloride

(f) cyclobutanecarboxylic acid

(g) phthalic anhydride

(h) butyramide

(i) trimethylacetic acid

(j) methyl salicylate

(k) calcium propionate

(l) *n*-valeronitrile

EXERCISE 10.12 Identify each lettered product in the following reaction sequences.

(a) $CH_3-COOH \xrightarrow{SOCl_2}$ (A) $\xrightarrow{NH_3}$ (B) $\xrightarrow{P_2O_5}$ (C)

(b) $CH_3CN \xrightarrow[\text{reflux}]{Na^+OH^-,}$ (A) + (B);

$\quad\quad$ A + HCl $\longrightarrow$ CH_3COOH + NaCl

(c) $n\text{-}C_4H_9-Br \xrightarrow{NaCN}$ (A) $\xrightarrow[\text{HCl}]{H_2O}$ (B) $\xrightarrow{LiAlH_4}$ (C)

(d) $C_2H_5Br \xrightarrow[\text{ether}]{\text{Mg, anhydrous}}$ (A) $\xrightarrow[\text{(2) hydrolysis}]{\text{(1) } CO_2}$ (B) $\xrightarrow{PCl_3}$ (C) $\xrightarrow{C_2H_5OH}$

$\quad\quad\quad\quad\quad\quad\quad\quad\quad\quad\quad\quad\quad\quad\quad$ (D) $\xrightarrow[\text{heat}]{Na, C_2H_5OH,}$ (E) + (F);

$\quad\quad$ (F) + I_2 + NaOH $\longrightarrow$ HCI_3

EXERCISE 10.13 Calculate the pK_a values for the following acids whose acid ionization constants are given in parentheses following their names: (a) methoxyacetic acid (2.9×10^{-4}), (b) 3-butenoic acid (4.5×10^{-5}), (c) 3-butynoic acid (4.8×10^{-4}).

EXERCISE 10.14 Calculate the acid ionization constants for the following acids whose pK_a values are given in parentheses following their names: (a) trifluoroacetic acid (0.23), (b) acetyl chloride ($\sim$16 for hydrogens of methyl group), (c) ethyl alcohol (17), (d) phenol (9.99), (e) propionic acid (4.87).

EXERCISE 10.15 Without referring to a table of acidity constants arrange the acids in the following set in an order of diminishing acidity.
(a) benzoic acid, (b) *p*-nitrobenzoic acid, (c) phenylacetic acid, (d) *p*-methoxybenzoic acid.

EXERCISE 10.16 Complete the following reaction equations and name the organic product(s) obtained.

(a) 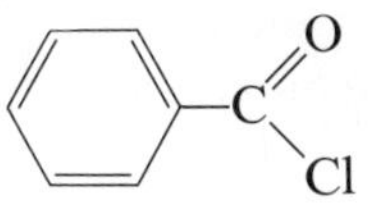 + $NH_3 \longrightarrow$

(b) C_2H_5I + KCN $\longrightarrow$

(c) 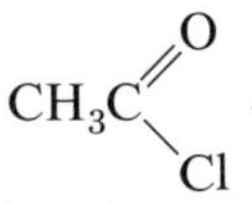 $CH_3C\overset{O}{\underset{Cl}{\big\langle}}$ + $(CH_3)_2CHOH \longrightarrow$

(d) 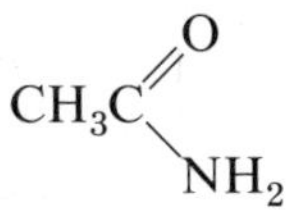 $CH_3C\overset{O}{\underset{NH_2}{\big\langle}} \xrightarrow{P_4O_{10}}$

(e) CH$_3$—CH$_2$—$\overset{\overset{\displaystyle CH_3}{|}}{C}$=O + Br$_2$ + NaOH $\longrightarrow$

(f) CH$_3$—C$\overset{\displaystyle O}{\underset{\displaystyle NH_2}{\diagup}}$ + NaOH $\xrightarrow{\text{reflux}}$

(g) CH$_3$—C$\overset{\displaystyle O}{\underset{\displaystyle OH}{\diagup}}$ + SOCl$_2$ $\longrightarrow$

(h) Product of (b) + H$_2$O + NaOH $\xrightarrow{\text{reflux}}$

(i) Salicylic acid + Br$_2$ $\xrightarrow{\text{Fe}}$

(j) Ethyl acetate + methylmagnesium iodide $\longrightarrow$

(k) *p*-Nitrotoluene + K$_2$Cr$_2$O$_7$ + H$_2$SO$_4$ $\longrightarrow$

EXERCISE 10.17 Using limestone (CaCO$_3$) and coke (C) as your only carbon-containing starting material and any other inorganic reagents you may need, show how you might synthesize the following compounds. (*Note:* The product of one reaction may be used as starting material for the preparation of another.)

(a) acetylene
(b) acetaldehyde
(c) ethyl bromide
(d) 2-butanol
(e) iodoform
(f) propionic acid
(g) acetone

(h) isopropyl alcohol
(i) 2-bromopropane
(j) *n*-propyl bromide
(k) acetic acid
(l) acetic anhydride
(m) isopropyl acetate
(n) isobutyric acid

EXERCISE 10.18 Outline a procedure for the removal of acetic acid (b.p., 118°) from a mixture which also contains *n*-butyl alcohol (b.p., 117°) and *n*-butyl acetate (b.p., 126°).

EXERCISE 10.19 Using a minimum number of steps, outline a reaction sequence for the conversion of a carboxylic acid to the next higher homolog. Use a readily available acid as your starting acid.

EXERCISE 10.20 Beginning with toluene as your only organic starting material outline a reaction or reaction sequence that would lead to the preparation of the following compounds.

(a) *p*-bromotoluene
(b) 4-methylbenzoic acid
(c) *p*-bromobenzamide
(d) *p*-bromophenylacetic acid

(e) terephthalic acid
(*p*-C$_6$H$_4$(CO$_2$H)$_2$)
(f) benzyl phenyl ketone

EXERCISE 10.21 An unknown acid was believed to be either *o*-bromobenzoic acid or 2,4-dibromobenzoic acid. A 0.2412-g sample of the acid neutralized 12.4 ml of 0.098 N sodium hydroxide solution. Which acid was it?

EXERCISE 10.22 Compound (A), $C_6H_{14}O$, showed the following reactions:

$$C_6H_{14}O \xrightarrow{\text{NaOH} + I_2} CHI_3 + (B) [C_5H_9O_2Na]$$

$$(A) \xrightarrow{H_2SO_4} (C) [C_6H_{12}] \xrightarrow{Br_2 \text{ in } CCl_4} (D) [C_6H_{12}Br_2]$$

$$(C) + O_3 \xrightarrow[\substack{\text{reductive} \\ \text{hydrolysis}}]{\text{followed by}} (E) [C_3H_6O] \text{ (the only product)}$$

$$(E) \xrightarrow{\text{NaOH} + I_2} CHI_3 + (F) [C_2H_3O_2Na]$$

Draw structures and assign acceptable names to compounds (A) through (F).

EXERCISE 10.23 A neutral compound which contained only carbon, hydrogen, and oxygen was hydrolyzed to yield two new products A and B. Compound A was found to be an acid with a neutralization equivalent of 60 ± 1. Compound B gave a positive iodoform test reaction, reacted readily with the Lucas reagent, and on treatment with concentrated H_2SO_4 gave propylene. Give the name and structure of the original compound.

EXERCISE 10.24 A liquid, $C_7H_{14}O_2$, on hydrolysis gave compounds A and B. Compound A was found to be an acid with a neutralization equivalent of 88 ± 1. Compound B gave a positive iodoform reaction and when treated with PBr_3 formed an alkyl bromide. A Grignard reagent, when prepared from the alkyl bromide and caused to react with carbon dioxide, produced an acid identical to compound A. Give the structure and name of the original compound.

EXERCISE 10.25 The following compounds have the molecular formula $C_9H_{10}O_2$, and the structural feature $-C\overset{\displaystyle O}{\underset{\displaystyle O-}{\big\Vert}}$, in common. Draw a structure for each isomer that is consistent with the following sets of data:

Compound A: A sweet-smelling liquid that was insoluble in water and in cold dilute base. Strong oxidation of the compound produced benzoic acid.

Compound B: A white, odorless solid that was insoluble in water but soluble in dilute sodium hydroxide. Vigorous oxidation produced phthalic acid.

Compound C: A sweet-smelling liquid that was insoluble in water and in cold, dilute sodium hydroxide. Treatment with Tollens' reagent produced a silver mirror. Strong oxidation produced phthalic acid.

Compound D: A colorless solid that was insoluble in water but soluble in dilute sodium hydroxide. Treatment with hot potassium permanganate produced benzoic acid.

EXERCISE 10.26 Which of the compounds in Exercise 10.25 would be expected to have an nmr spectrum consisting of a 3-hydrogen singlet at δ 2.00, a 2-hydrogen singlet at δ 5.05, and a 5-hydrogen singlet at δ 7.29?

EXERCISE 10.27 An organic compound, $C_6H_{12}O_2$, gave an infrared spectrum with strong absorption at 1740 and 1715 cm^{-1} and an nmr spectrum consisting of a 9-proton singlet at δ 1.08, a 2-proton singlet at δ 2.20, and a 1-proton singlet at δ 11.24. Propose a structure for the compound.

EXERCISE 10.28 An organic compound, $C_6H_{11}BrO_2$, showed strong absorption in the ir at 1733 cm^{-1} and gave an nmr spectrum consisting of a 3-proton triplet at δ 1.19, a 6-proton singlet at δ 1.92, and a 2-proton quartet at δ 4.20. Propose a structure for the compound.

EXERCISE 10.29 Outline sequences of reactions that would be suitable for bringing about the following transformations.

(a) $=CH_2 \longrightarrow$ $-CH_2COOH$

(b) $HOOC-(CH_2)_5-COOH \longrightarrow$ $=O$

$\overset{|}{C}OOC_2H_5$

(c) $CH_3CH=C\overset{\displaystyle CH_3}{\underset{\displaystyle CH_3}{\diagdown}} \longrightarrow CH_3CH_2\overset{\overset{\displaystyle CH_3}{|}}{\underset{\underset{\displaystyle CH_3}{|}}{C}}-COOH \longrightarrow$

$CH_3CH_2\overset{\overset{\displaystyle CH_3}{|}}{\underset{\underset{\displaystyle CH_3}{|}}{C}}\overset{\overset{\displaystyle O}{||}}{-C}-CH_3$

(d) $CH_3CH_2CH_2CH_2COOH \longrightarrow CH_3CH_2CH_2COOH$

11

Fats, Oils, Waxes; Soap and Detergents; Prostaglandins

Introduction

Natural products which are soluble in ether, chloroform, carbon tetrachloride, and other water-immiscible organic solvents, but insoluble in water, are known as **lipids.** The lipids are important constituents of all plant and animal tissue. About 40–50% of most membranes is composed of lipids of various types. In this chapter our attention will be focused on the saponifiable lipids derived from the fatty acids: the fats, oils,[1] and waxes. The important biochemical regulators, the prostaglandins, are also derived from the fatty acids and are included in our discussion. In addition to the fats and oils (complex

---

[1] Oils, as the term is used in the present chapter, refers to glycerides (see Sec. 11.2) which are liquid at room temperature and *not* to mineral oils or petroleum products.

lipids), there are other fatlike substances, the terpenes and steroids, which are known as simple lipids and are discussed in Chapter 17. Not only do the edible fats and oils make up approximately 40% of the American diet, but they also provide the raw material for the preparation of numerous important commodities.

11.1   *Waxes*

Waxes are esters of long-chain, unbranched fatty acids and long-chain alcohols. Both the acid and the alcohol that combine to form a wax may be sixteen to thirty carbons in length. The general formula of a wax is essentially that of a simple ester, $R-\overset{\overset{\textstyle O}{\|}}{C}-O-R'$.

Both plants and animals produce natural waxes. Waxes usually are mixtures of esters and contain, in addition, small amounts of free acids, alcohols, and even hydrocarbons. The waxes melt over a wide range of temperature ($35°-100°C$), have a satiny "waxy" feel, and are very insoluble in water. On the other hand, they are quite soluble in many organic solvents and may be compounded into a number of useful everyday commodities. Such wax solutions generally are used as protective coatings. They are often protective coatings in nature, too.

Beeswax is largely ceryl myristate, $C_{13}H_{27}\overset{\overset{\textstyle O}{\|}}{C}-O-C_{26}H_{53}$, along with some esters of cerotic acid, $C_{25}H_{51}COOH$, and a few per cent of hydrocarbons. It melts between $62-65°C$ and is used in the preparation of shoe polishes, candles, and paper coatings.

Carnauba wax, a plant wax found as a coating on the leaf of the Brazilian palm, is largely myricyl cerotate, $C_{25}H_{51}\overset{\overset{\textstyle O}{\|}}{C}-O-C_{31}H_{63}$. Carnauba wax melts between $80-87°C$ and is used in polishes and as a coating on mimeograph stencils.

Spermaceti wax consists mainly of cetyl palmitate, $C_{15}H_{31}\overset{\overset{\textstyle O}{\|}}{C}-O-C_{16}H_{33}$ along with some free cetyl alcohol, $C_{16}H_{33}OH$. It is obtained from the head cavity of the sperm whale. Spermaceti wax has a melting range of $42-50°C$ and is used primarily as an emollient in ointments and in cosmetics. It is also used in the manufacture of candles.

Fats and Oils

11.2   *Structure and Composition of Fats and Oils*

Fats and oils differ from waxes in that they are glycerides, or esters of glycerol, a trihydroxy alcohol (Sec. 8.5).

$$
\begin{array}{c}
H \\
| \\
H-C-O-\overset{\displaystyle O}{\overset{\|}{C}}-R \\
| \\
H-C-O-\overset{\displaystyle O}{\overset{\|}{C}}-R \\
| \\
H-C-O-\overset{\displaystyle O}{\overset{\|}{C}}-R \\
| \\
H
\end{array}
$$

A simple glyceride

A simple glyceride is one in which all R groups in the previous general formula are identical. If R in the general formula represents an aliphatic group, $C_n H_{2n+1}$, then the number of carbons in the group usually is an odd number from 3 to 17. Such chains are saturated. If R is an unsaturated alkyl group, of the form $C_n H_{2n-1}$, $C_n H_{2n-3}$, or $C_n H_{2n-5}$, then n usually is 17.

$$
\begin{array}{c}
H \\
| \\
H-C-O-\overset{\displaystyle O}{\overset{\|}{C}}-(CH_2)_{14}-CH_3 \\
| \\
H-C-O-\overset{\displaystyle O}{\overset{\|}{C}}-(CH_2)_{14}-CH_3 \\
| \\
H-C-O-\overset{\displaystyle O}{\overset{\|}{C}}-(CH_2)_{14}-CH_3 \\
| \\
H
\end{array}
$$

Glyceryl tripalmitate (a simple glyceride)

The natural fats and oils usually are not simple glycerides. The three acid residues produced when fats and oils are hydrolyzed usually vary not only in length but also in the degree of unsaturation. The principal structural difference between oils and fats lies in the degree of saturation of the acid residues, and this accounts for the differences in both the physical and chemical properties of these two classes of glycerides.

The fats are glyceryl esters in which long-chain *saturated* acid components predominate. They are solids or semi-solids and are principally animal products. **Lauric,** $CH_3(CH_2)_{10}COOH$; **palmitic,** $CH_3(CH_2)_{14}COOH$; and **stearic,** $CH_3(CH_2)_{16}COOH$ are the most important acids obtained by the hydrolysis of fats. Such long-chain carboxylic acids usually are called *fatty acids* because they are obtained from fats. They are, for the most part, insoluble in water but soluble in organic solvents. It is entirely possible that all three of the fatty acids named above could be present in a mixed glyceride such as that shown by the formula on page 340.

$$H-\underset{\underset{\displaystyle H}{|}}{\overset{\overset{\displaystyle H}{|}}{C}}-O-\overset{\overset{\displaystyle O}{\|}}{C}-(CH_2)_{10}-CH_3$$

$$H-C-O-\overset{\overset{\displaystyle O}{\|}}{C}-(CH_2)_{14}-CH_3$$

$$H-\underset{\underset{\displaystyle H}{|}}{C}-O-\overset{\overset{\displaystyle O}{\|}}{C}-(CH_2)_{16}-CH_3$$

Glyceryl lauropalmitostearate
(a mixed glyceride)

Oils, on the other hand, are glyceryl esters in which long-chain *unsaturated* acid components predominate. They are liquids and are largely of vegetable origin.

$$H-\underset{\displaystyle |}{\overset{\overset{\displaystyle H}{|}}{C}}-O-\overset{\overset{\displaystyle O}{\|}}{C}-(CH_2)_7-CH{=}CH-(CH_2)_7CH_3$$

$$H-C-O-\overset{\overset{\displaystyle O}{\|}}{C}-(CH_2)_7-CH{=}CH-(CH_2)_7CH_3$$

$$H-\underset{\underset{\displaystyle H}{|}}{C}-O-\overset{\overset{\displaystyle O}{\|}}{C}-(CH_2)_7-CH{=}CH-(CH_2)_7CH_3$$

Glyceryl trioleate (Triolein)[2]
(an oil)

The presence of unsaturation in the acid component of a fat tends to lower its melting point. The most important unsaturated acids obtainable by the hydrolysis of oils are the C_{18} acids—**oleic, linoleic,** and **linolenic,** of which the structures are given below.

$$\underset{(18)}{CH_3}-(CH_2)_7-\underset{(10)}{CH}{=}\underset{(9)}{CH}-(CH_2)_7-\underset{(1)}{\overset{\overset{\displaystyle O}{\diagup}}{C}}\diagdown_{OH}$$

Oleic acid
((Z)-9-Octadecenoic acid)

$$\underset{(18)}{CH_3}-(CH_2)_4-\underset{(13)}{CH}{=}\underset{(12)}{CH}-CH_2-\underset{(10)}{CH}{=}\underset{(9)}{CH}-(CH_2)_7-\underset{(1)}{\overset{\overset{\displaystyle O}{\diagup}}{C}}\diagdown_{OH}$$

Linoleic acid
((Z,Z)-9,12-Octadecadienoic acid)

---

[2] If all fatty acid residues of a glyceride are the same, the name is shortened by dropping "glyceryl" and changing "tri . . . ate" to "tri . . . in."

$$CH_3 \underset{(18)}{-}CH_2 \underset{(17)}{-}CH \underset{(16)}{=}CH \underset{(15)}{-}CH_2 \underset{(14)}{-}CH \underset{(13)}{=}CH \underset{(12)}{-}CH_2 \underset{(11)}{-}CH \underset{(10)}{=}CH \underset{(9)}{-}(CH_2)_7 \underset{(1)}{-}C \overset{O}{\diagdown}_{OH}$$

Linolenic acid
((Z,Z,Z)-9,12,15-Octadecatrienoic acid)

You will note that the first double bond in each of the above structures is found in the middle of the carbon chain and that other double bonds are farther removed from the carboxyl group. Methylene ($-CH_2-$) units separate one double bond from another, and the unsaturated hydrocarbon portion of these three acids does not represent a conjugated system (Sec. 3.12). Some natural oils do contain acids with systems of alternating single and double bonds. One of these is **tung oil,** which on hydrolysis yields **eleostearic** as the principal acid.

$$CH_3 \underset{(14)}{-}CH_2CH_2CH_2CH \underset{(13)}{=}CH \underset{(12)}{-}CH \underset{(11)}{=}CH \underset{(10)}{-}CH \underset{(9)}{=}CH-(CH_2)_7 -C \overset{O}{\diagdown}_{OH}$$

Eleostearic acid
((Z,E,E)-9,11,13-Octadecatrienoic acid)

Unsaturation in a fatty acid also makes possible *cis,trans*-isomerism depending on the configuration of the hydrogen atoms attached to the doubly-bonded carbon atoms. Oleic acid has the *cis* configuration whereas its isomer, elaidic acid, has the *trans* configuration. When more than one double bond is present in the fatty acid molecule, of course more than two *cis,trans*-isomers are possible. Generally speaking, the *cis* isomers are the forms found naturally occurring in the unsaturated acid component of food fats and oils. As may be seen from Table 11.1 the most abundant of all saturated acids found in fats is palmitic; the most abundant of the unsaturated acids found in edible oils is oleic. See page 342 for Table 11.1.

EXERCISE 11.1

How many *cis,trans*-isomers are possible for linoleic acid?

11.3 *Saponification and Iodine Values*

The composition of fats and oils is variable and depends not only upon the plant or animal species involved in their production, but also upon climatic and dietetic factors as well. The approximate molecular weights of the acid components in fats and oils may be obtained from **saponification values.** Saponification is a term specifically applied to the hydrolysis of an ester when the action is carried out in alkaline solution. The saponification value of a fat or an oil is an arbitrary unit, defined as the **number of milligrams of potassium hydroxide**

TABLE 11.1 *Fatty Acid Components of Some Common Fats and Oils*

	Component Acids (percent)*							Saponification Values	Iodine Values
	Myristic C_{14}	Palmitic C_{16}	Stearic C_{18}	Oleic	Linoleic	Linolenic	Eleostearic		
Fats									
Butter	7–10	24–26	10–13	28–31	1.0–2.5	0.2–0.5		210–230	30–40
Lard	1–2	28–30	12–18	40–50	7–13	0–1		195–203	46–70
Tallow	3–6	24–32	20–25	37–43	2–3			190–200	30–48
Edible Oils									
Olive oil		9–10	2–3	73–84	10–12	trace		187–196	79–90
Corn oil	1–2	8–12	2–5	19–49	34–62	trace		187–196	109–133
Soybean oil		6–10	2–5	20–30	50–60	5–11		189–195	127–138
Cottonseed oil	0–2	20–25	1–2	23–35	40–50	trace		190–198	105–114
Peanut oil		8–9	2–3	50–65	20–30			188–195	84–102
Safflower oil		6–7	2–3	12–14	75–80	0.5–0.15		188–194	140–156
Nonedible Oils									
Linseed oil		4–7	2–4	25–40	35–40	25–60		187–195	170–185
Tung oil		3–4	0–1	4–15			75–90	190–197	163–171

* Totals less than 100% indicate the presence of lower or higher acids in small amounts.

required to saponify one gram of the fat or oil. Because there are three ester linkages in a glyceride to hydrolyze, three equivalents of potassium hydroxide are required to saponify one molecular weight of any fat or oil. The following equation and sample calculation show how a saponification value is determined.

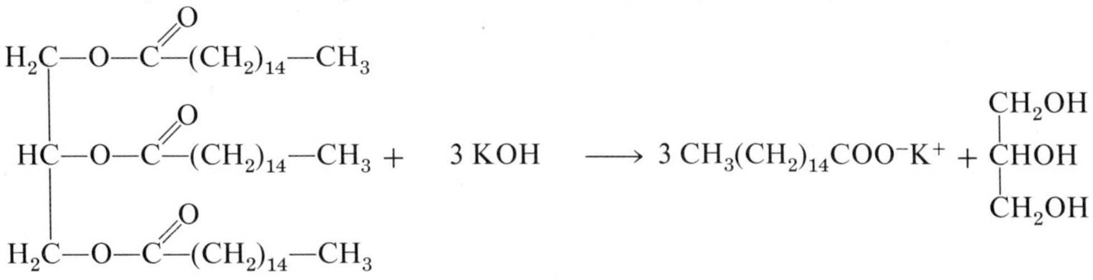

| Glyceryl tripalmitate (tripalmitin) | (3 × gram formula weight) | Potassium palmitate (a soap) | Glycerol |

M.W. = 806

$$\text{Saponification value of tripalmitin} = \frac{168{,}000 \text{ mg KOH/mole fat}}{806 \text{ g fat/mole fat}}$$

$$= 208 \text{ mg KOH/g fat}$$

A small saponification value for a fat or an oil indicates a high molecular weight.

The extent of unsaturation in a fat or oil is expressed in terms of its **iodine value.** The iodine value is defined as the **number of grams of iodine which will add to 100 grams of fat or oil.** The iodine value of tripalmitin (above) with no unsaturation would, of course, be zero. The determination of an iodine value may be illustrated by using triolein as an example.

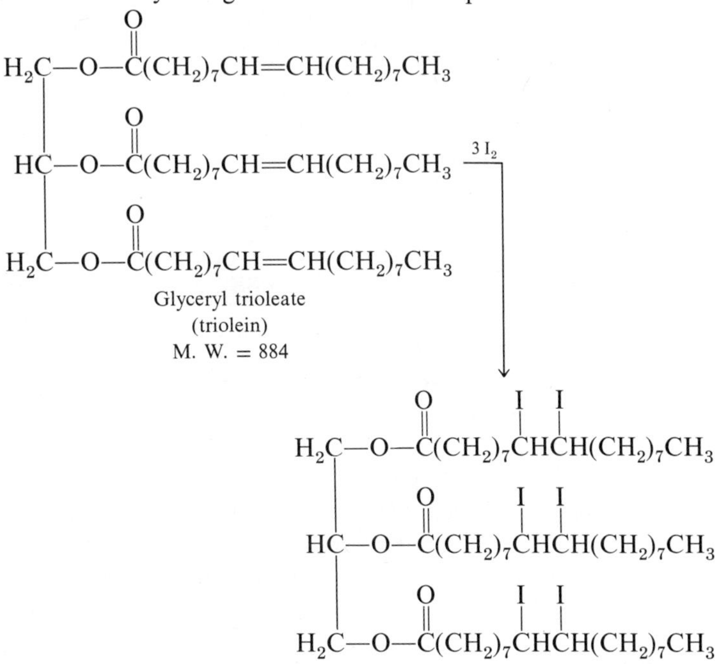

Glyceryl trioleate
(triolein)
M. W. = 884

The above equation reveals that 761.4 g (3 moles) of iodine will add to 884 g of the oil. The number of grams of iodine that will add to 100 g of oil will be $100/884 \times 761.4$ g of iodine, or 86 g of iodine. (Atomic weight of iodine = 126.9)

$$\text{Iodine value} = \frac{6 \times 126.9 \times 100}{884} = 86$$

A high iodine value indicates a high degree of unsaturation. While the above equation shows the addition of molecular iodine, in practice iodine monobromide (IBr) is the reagent actually used.

Table 11.1 indicates the composition of some of the more important fats and oils along with their saponification and iodine values.

EXERCISE 11.2

If 90% or more of the fatty acid components of castor oil is ricinoleic acid (see Exercise 8-f, Chapter 5), what are the saponification and iodine values for castor oil?

Answer: Sap. value = 176–187; Iodine value = 81–90.

Reactions of the Fats and Oils

11.4   *Saponification. (Soap Preparation)*

The glycerides, like other esters, may be hydrolyzed by heating with a solution of sodium or potassium hydroxide. The hydrolysis products are glycerol and the alkali metal salts of long-chain fatty acids (Sec. 10.5-H) The latter are called **soaps,** and alkaline hydrolysis is called **saponification,** whether the term is applied to fats, oils, or simple esters.

The best soaps are those in which the hydrocarbon segment is saturated and is from twelve to eighteen carbon atoms in length. Sodium soaps are hard soaps and commonly are used as cake soaps. Potassium soaps are soft soaps, produce a finer lather, and usually are employed in shaving creams. Natural or "hardened" fats (Sec. 11.5) are saponified by heating them in open kettles with a solution of sodium hydroxide. The kettles are quite large and can accommodate over fifty tons of fat per batch. When the reaction is complete, the thick curds of soap are precipitated by the addition of sodium chloride (common ion effect). The water layer is drawn off and the glycerol it contains is recovered by concentration and distillation. The crude soap, which contains salt, alkali, and some residual glycerol, is heated with water to dissolve these impurities. The soap is again precipitated by the addition of salt. The washing and precipitation procedure is repeated several times, Finally, the soap is heated with sufficient water to give a smooth mixture, which on standing, separates into a homogeneous upper layer of kettle soap. Kettle soap may be used without further treatment or it may be blended with other ingredients to produce soaps for special purposes. Pumice is incorporated for scouring soaps, perfumes for toilet soaps, and antiseptics (phenol, hexachlorophene) for medicated or deodorizing soaps. When air is blown into molten soap, the specific gravity of the solidified product is lowered to 0.8–0.9. This treatment produces a floating soap.

Sodium and potassium soaps are soluble in water and are the common household soaps. The long-chain fatty acid salts of heavier metals are not water soluble and may be blended with mineral oils to form lubricating greases.

11.5 *Hydrogenation. ("Hardening" of Oils)*

Unsaturation in a fat or oil may be diminished by catalytic hydrogenation.

$$H_2C-O-\overset{O}{\overset{\|}{C}}-(CH_2)_7-CH=CH-(CH_2)_7CH_3$$

$$H-\overset{\displaystyle|}{C}-O-\overset{O}{\overset{\|}{C}}-(CH_2)_7-CH=CH-(CH_2)_7CH_3 + 3\,H_2 \xrightarrow[\text{catalyst}]{\text{Ni}}$$

$$H_2\overset{\displaystyle|}{C}-O-\overset{O}{\overset{\|}{C}}-(CH_2)_7-CH=CH-(CH_2)_7CH_3$$

Triolein

$$H_2C-O-\overset{O}{\overset{\|}{C}}-(CH_2)_{16}CH_3$$

$$H-\overset{\displaystyle|}{C}-O-\overset{O}{\overset{\|}{C}}-(CH_2)_{16}CH_3$$

$$H_2\overset{\displaystyle|}{C}-O-\overset{O}{\overset{\|}{C}}-(CH_2)_{16}CH_3$$

Tristearin

The process is controllable and is used to convert low melting fats or oils to higher melting fats of any desired consistency. Because the melting point of a fat increases with saturation, oils may be converted into semi-solid fats by hydrogenation. This hydrogenation process is known as **hardening.** Margarines are made by hardening oils to the consistency of butter. When churned with skim milk, fortified with vitamin A, and artificially colored, these butter substitutes have not only the flavor and color of butter, but its nutritional advantages as well. Butter is approximately 80% fat, and margarine, by federal regulation, must also contain not less than 80% fat.

Many of the cooking fats now available have their origin in vegetable oils. When hydrogenation is carried to completion, glycerol and long-chain alcohols are produced. The latter are used in the manufacture of synthetic detergents (Sec. 11.8).

$$H_2C—O—\overset{\displaystyle O}{\overset{\|}{C}}—(CH_2)_{16}CH_3$$

$$H—\overset{|}{\underset{|}{C}}—O—\overset{\displaystyle O}{\overset{\|}{C}}—(CH_2)_{16}CH_3 + 6\ H_2 \xrightarrow{\text{catalyst}} \overset{\displaystyle CH_2OH}{\underset{\displaystyle CH_2OH}{\overset{|}{\underset{|}{CHOH}}}} + 3\ CH_3(CH_2)_{16}CH_2OH$$

$$H_2C—O—\overset{\displaystyle O}{\overset{\|}{C}}—(CH_2)_{16}CH_3$$

| Glyceryl tristearate | Glycerol | 1-Octadecanol |

11.6 *Oxidation. (Rancidity)*

Edible unsaturated fats and oils, when exposed to air and light for long periods of time, not only are subject to slow hydrolysis but the acid components produced also are subject to oxidative cleavage at the site of unsaturation.

$$CH_3—(CH_2)_7—CH\overset{\xi}{=\!\!\!\!\!=}CH—(CH_2)_7COOH \xrightarrow{O_2(air)}$$

Oleic acid

$$CH_3(CH_2)_7COOH + HO—\overset{\displaystyle O}{\overset{\|}{C}}—(CH_2)_7—\overset{\displaystyle O}{\overset{\|}{C}}—OH$$

Pelargonic acid   Azelaic acid

The lower molecular weight and more volatile acids that are produced by this exposure impart an offensive odor to fats. This condition is known as **rancidity.** If the volatile acids are produced by hydrolysis, the resultant rancidity is known as hydrolytic rancidity. Butter, especially, when left uncovered and out of the refrigerator, easily becomes rancid through hydrolytic rancidity. A substantial portion of the fatty acid components of butterfat is made up of butyric, caproic, caprylic, and capric acids. These are liberated when butter is hydrolyzed and are responsible for the unpleasant odor of rancid butter. Oxidation leading to rancidity (oxidative rancidity) in fats and oils is catalyzed by the presence of certain metallic salts. Proper packaging of foods, therefore, is of the utmost importance. The addition of **antioxidants** will stabilize edible fats for long periods of storage. These inhibitors, or interceptors as they sometimes are called, are themselves easily oxidizable substances. Some are natural products such as δ-tocopherol (Sec. 17.11-c) and the lecithins. The lecithins, as evident in their formula, are themselves glycerides.

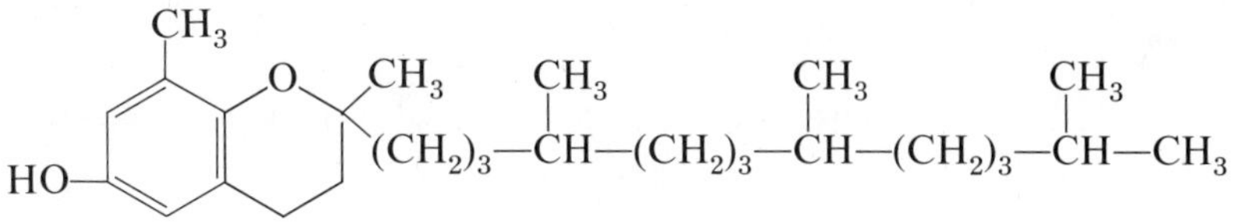

δ-Tocopherol

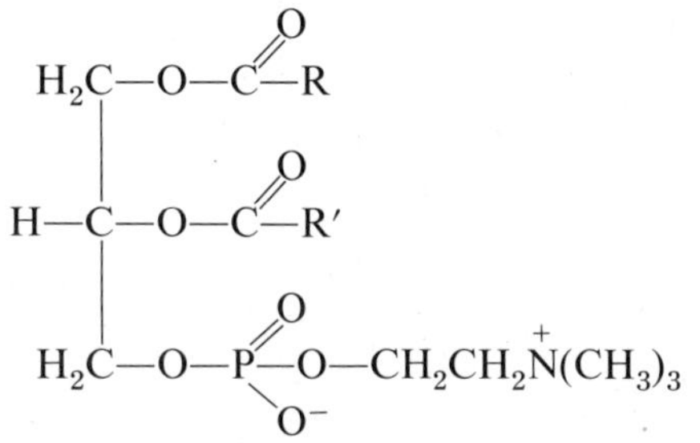

A lecithin

Other antioxidants employed to stabilize fats contain a modified phenolic structure, as does δ-tocopherol (above). Two of these, 3-*tert*-butyl-4-hydroxyanisole (BHA), and 2,6-di-*t*-butyl-4-methylphenol (BHT), are used to stabilize cooking oils at the high temperatures required for the preparation of potato and corn chips and other foods. BHA is also used to stabilize lubricating oils, rubber, and gasoline—all substances containing carbon-carbon unsaturated bonds.

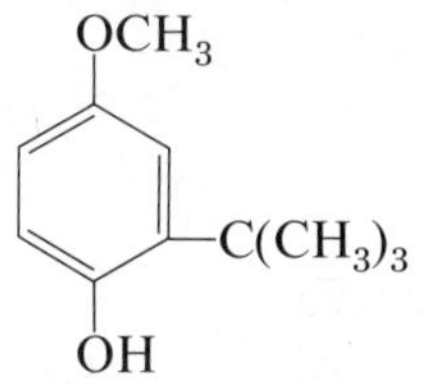

3-*tert*-Butyl-4-hydroxyanisole

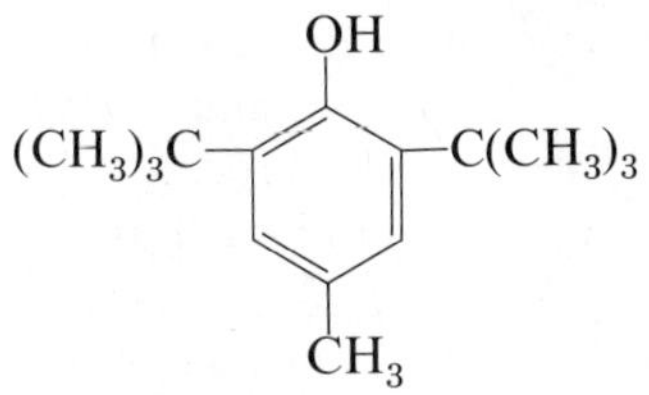

2,6-di-*t*-butyl-4-methylphenol

Antioxidants effectively suppress rancidity when used in only minute amounts (0.01–0.001%).

11.7 *Oxidation. (Drying)*

The highly unsaturated linseed and tung oils (Table 11.1) are especially reactive. The hydrogen atoms of the methylene group, flanked by doubly-bonded carbon atoms in both linoleic and linolenic acids, are reactive enough to form peroxides, which then undergo polymerization.

$$CH_3-(CH_2)_4-\overset{\overset{\displaystyle H}{|}}{C}=\overset{\overset{\displaystyle H}{|}}{C}-\overset{\overset{\displaystyle H}{|}}{\underset{\underset{\displaystyle H}{|}}{C}}-\overset{\overset{\displaystyle H}{|}}{C}=\overset{\overset{\displaystyle H}{|}}{C}-(CH_2)_7-\overset{\overset{\displaystyle O}{||}}{C}-OH \quad \text{Linoleic acid}$$

Active methylene group

$\downarrow O_2(\text{air})$

$$-CH=CH-\overset{\overset{\displaystyle H}{|}}{\underset{\underset{\displaystyle O-O-H}{|}}{C}}-CH=CH-$$

$\downarrow$ polymerization

$$-----CH-CH-\overset{|}{\underset{\underset{\displaystyle O}{|}}{C}}-CH-CH-----$$

$$-----CH-CH-\overset{|}{\underset{|}{C}}\quad CH-CH-----$$

A cross-linked unit in a polymerized drying oil

Hydroperoxide formation, when accompanied by polymerization, converts highly unsaturated oils into a vast network of interlinked units. Such units form a hard, dry, tough film when exposed to air. Unsaturated oils are for this reason used in the vehicle—i.e., the liquid medium, of paints and varnishes as **drying oils.** A paint usually contains catalysts (lead and cobalt salts) to hasten the drying process. Other ingredients are finely ground pigments and volatile thinners such as turpentine.

Linoleum and **oilcloth** are made by applying a viscous mixture of drying oils, pigments, and fillers as a thick coating onto heavy fabric or cloth. When "dried" these products make very durable surface coverings. Because the oxidation of unsaturated vegetable oils is an exothermic reaction, oily rags used in painting are subject to spontaneous combustion. Such oily waste should be destroyed after use by burning or should be denied oxygen by keeping it in a closed container. This type of autoignition cannot occur if oily rags are spread out so that the heat generated by oxidation cannot build up to a kindling temperature.

11.8 *Detergents. (Soaps and Syndets)*

Detergents (L., *detergere,* to wipe off) are cleansing agents and include both soaps and syndets (*syn*thetic *det*ergents). The molecular structures of synthetic detergents and soaps are similar. Both types of cleansing agents are characterized by the presence of a long, nonpolar, hydrophobic (water-hating) hydrocarbon tail and a polar, hydrophilic (water-loving) head.

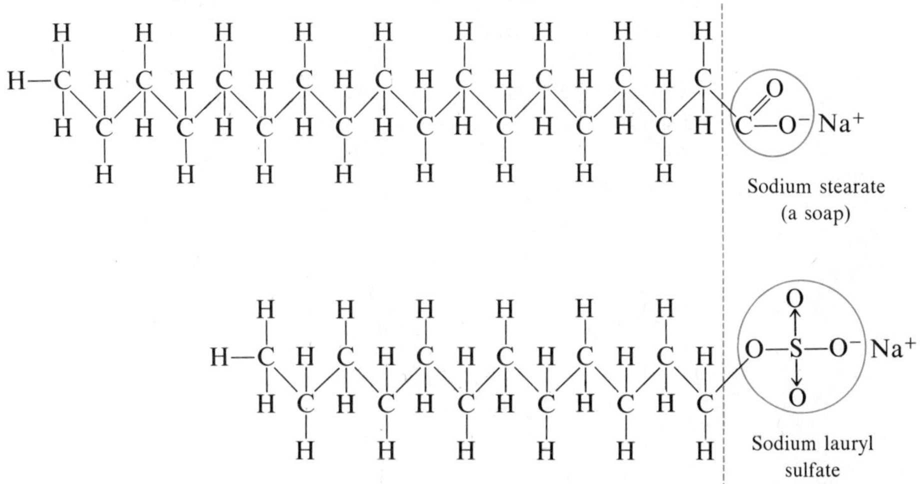

Sodium stearate
(a soap)

Sodium lauryl
sulfate
(a syndet)

Oil-soluble tail   Water-soluble head

Soaps, when used in hard water, precipitate as insoluble calcium, magnesium, and iron salts of the long-chain fatty acids, but synthetic detergents are not precipitated in the presence of these ions. Solubility in hard water is one of the principal advantages which a synthetic detergent has over a soap. On the other hand, the large scale use of syndets in this country within the past decade has presented a serious disposal problem. Many of these detergents have an aromatic ring in the hydrocarbon segment of the molecule. This type is generally classified as alkyl benzene sulfonate (ABS) detergents.

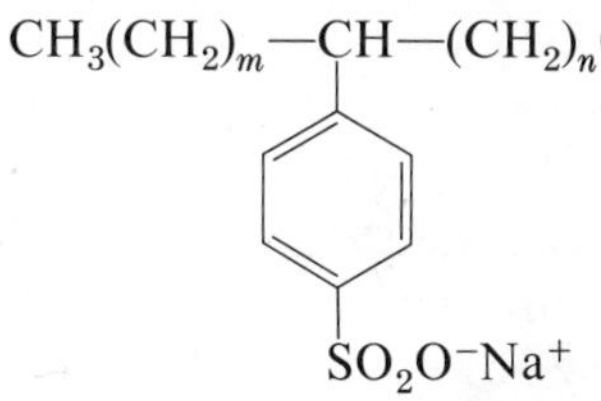

$$CH_3(CH_2)_m—CH—(CH_2)_nCH_3$$

$$m + n = 9\text{-}12$$

$$SO_2O^-Na^+$$

An alkyl benzene sulfonate

The straight-chain fatty acid salts with an even number of carbon atoms are biodegradable—that is, easily degraded by microorganisms. Unfortunately some of the early ABS types of detergents had highly branched alkyl side chains which resisted degradation (breakdown) and reappeared in some fresh water sources quite active and sudsy. Present day ABS derivatives have the biodegradable structure shown with branching confined to carbon atom adjacent to the ring. The long-chain hydrocarbon segments of both soap and synthetic detergent molecules shown above are embodied in the acid anion portion of sodium salts. This does not mean that a detergent, like a soap, must be a sodium or a potassium salt. On the contrary, the oil-soluble tail or "business end" of some detergents is cationic. Solubilizing power, not sudsing action, is the principal criterion for a good detergent. For this reason certain detergents have been developed, especially for use in automatic clothes and dish washers, that are neither anionic nor cationic but are simply long-chain polar molecules not unlike the carbitols (Sec. 8.15). Examples of all three types of detergents are shown.

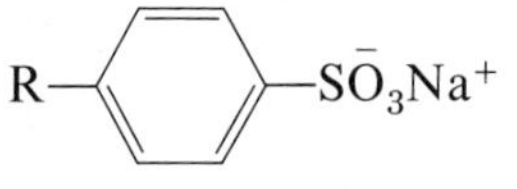

$$R-\!\!\left\langle\bigcirc\right\rangle\!\!-\overset{-}{S}O_3Na^+ \qquad R-\overset{+}{N}(CH_3)_3Cl^-$$

Anionic	Cationic
$R = C_{12}-C_{18}$	$R = C_{16}$

$$R-\!\!\left\langle\bigcirc\right\rangle\!\!-O-(CH_2CH_2O)_xCH_2CH_2OH$$

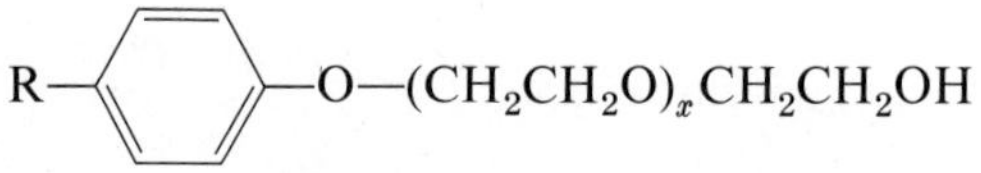

Nonionic
$R = C_8-C_{10}; \; x = 8-12$

11.9 *Detergents and the Ecology*

Within recent years the detergent industry has been severely criticized by environmentalists for contributing to the pollution of our lakes and streams. Many detergents, while biodegradable, have a high phosphate content to enhance the cleaning power of the detergents. When discharged into the sewer systems of our cities the phosphates are unchanged, and, once in the river or lake into which sewage is discharged, they nourish algae and other plant life to excess, causing undesireable growth. As a result, eutrophication proceeds, in which algal growth is unrestrained, the water becomes depleted of oxygen, and the lake or stream begins to "die"—that is, it becomes barren of aquatic life that depends on oxygen for survival.

Sodium nitrilotriacetate [$N(CH_2COO^-Na^+)_3$], **NTA,** at one time showed great promise as a satisfactory replacement for phosphates in detergents, and the chemical industry geared itself to produce it in quantity. Unfortunately, NTA was found to possess a carcinogenic activity in laboratory animals and therefore has been banned as a detergent additive. Ironically, recent experiments have shown that it is both the inorganic and organic matter from human waste, from garbage disposal units, from feed lots, and from agricultural fertilizers,—not phosphates from detergents—that is mainly responsible for using up the oxygen in our streams and lakes.

The mechanics of detergent action seem to be the same for both soaps and syndets. In the removal of soil from clothing and grease from dishes, the hydrocarbon segment of the detergent molecule dissolves in the oily or greasy

FIGURE 11.1 *The Solvating Action of Soap on a Droplet of Water*

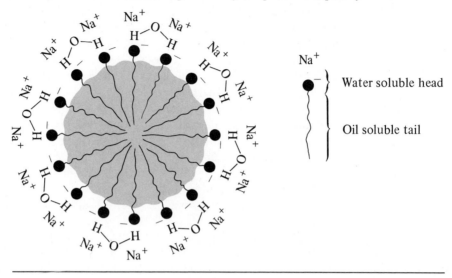

layer (like dissolves like). The oily layer then is dislodged as small globules by the mechanical action of rubbing, tumbling, or stirring. Once loosened, each globule becomes emulsified, or suspended in water, because the polar end of the detergent molecule is attracted to the polar water molecule. Such tiny droplets of "oil-in-water" emulsion do not coagulate, or condense, into huge globules because the electrical nature of the outer film surrounding each droplet is the same. A particle that has adjacent to its surface a double layer of charges of opposite sign is called a **micelle.** One micelle thus tends to repel the other. The role that a detergent or a soap molecule plays in the removal of oil is illustrated (in part) by Figure 11.1.

EXERCISE 11.3

Could a satisfactory soap be made from butter? Why?

11.10 *Digestion and Metabolism of Fats*

Enzymes called *lipases*[3] are active in the hydrolysis of fats. Gastric lipase, found in the stomach, catalyzes the hydrolysis of fats but to a small extent. Fats

---

[3] Enzymes usually are named by appending **ase** to their functions. Thus an **oxidase** catalyzes chemical combination with oxygen; **dehydrogenase,** the removal of hydrogen. The fats and oils belong to a general classification of natural substances known as *lipids*. Thus a **lipase** is an enzyme whose function is specific for a lipid, in this case a fat.

are hydrolyzed mainly in the small intestine where the environment is slightly alkaline. In this part of the digestive tract the fat first is emulsified by the bile and then hydrolyzed by the action of pancreatic lipase (steapsin), an enzyme made in the pancreas. The hydrolysis products, glycerol and fatty acids, are absorbed through the wall of the intestinal tract where recombination to form glycerides (fats) occurs. Fats are transported from the intestinal wall to various parts of the body to be stored as depot fat, used in the formation of protoplasm, or oxidized to supply energy.

The body can make both saturated and monounsaturated fatty acids by modifying dietary fats or synthesizing them from carbohydrates or proteins. However, certain polyunsaturated fatty acids, referred to as "essential" fatty acids, cannot be made and must be supplied in the diet.

Fats produce approximately 9.5 kcal of heat per gram when oxidized in the body to carbon dioxide and water. This number of calories is more than twice the energy obtained from the oxidation of an equivalent weight of either protein or carbohydrate.

Fats, once deposited at a storage site, do not remain there for long. The catabolism (degradation) and anabolism (synthesis) of fats represents a dynamic state. Most recent evidence indicates that the essential unit in the synthesis or in the degradation of a fat is the acetyl group, $CH_3-\overset{\overset{\displaystyle O}{\|}}{C}-$. The discovery of coenzyme A (A = acetylation) by Lipmann[4] (1947) showed that this substance plays a principal role in metabolism *via* 2-carbon units related to acetic acid. Coenzyme A is a complex molecule composed of four separately identifiable groups.

| Adenylic acid residue | Pyrophosphate | Pantothenic acid residue (Vitamin B$_5$) | β-Mercapto-ethylamine |

Coenzyme A (CoA)

---

[4] Fritz Lipmann (1899–   ). Member and Professor of the Rockefeller Institute, New York. Shared the Nobel prize (1953) in medicine and physiology with H. A. Krebs.

The terminal sulfhydryl (—SH) group is a very important part of the coenzyme A structure inasmuch as it is involved in transferring (donating or accepting) the acetyl group.

In addition to coenzyme A, two other species are required for the **biosynthesis** of fatty acids—acyl carrier protein (ACP) and an enzyme, β-ketoacyl-ACP-synthase. The structure of ACP resembles that of coenzyme A but differs in the replacement of the adenylic acid residue and one phosphate group by a polypeptide (Sec. 15.7).

| Polypeptide chain | Phosphate | Pantothenic acid residue | β-Mercapto-ethylamine |

Acyl carrier protein (ACP)

We will not attempt to portray the structure of the enzyme, β-ketoacyl-ACP-synthase except to point that it too has a terminal sulfhydryl group; therefore, we will represent its structure by *synthase*—SH. Both the biosynthetic and degradative pathways require acyl derivatives in which the hydrogen attached to the sulfur atom has been replaced by an acyl group in coenzyme A, ACP, and β-ketoacyl-ACP-synthase. These species are interconnected by the following equilibria.

FIGURE 11.2   *Degradation and Biosynthesis of Fatty Acids*

Tricarboxylic acid cycle (TCA)

$$CO_2 + H_2O$$

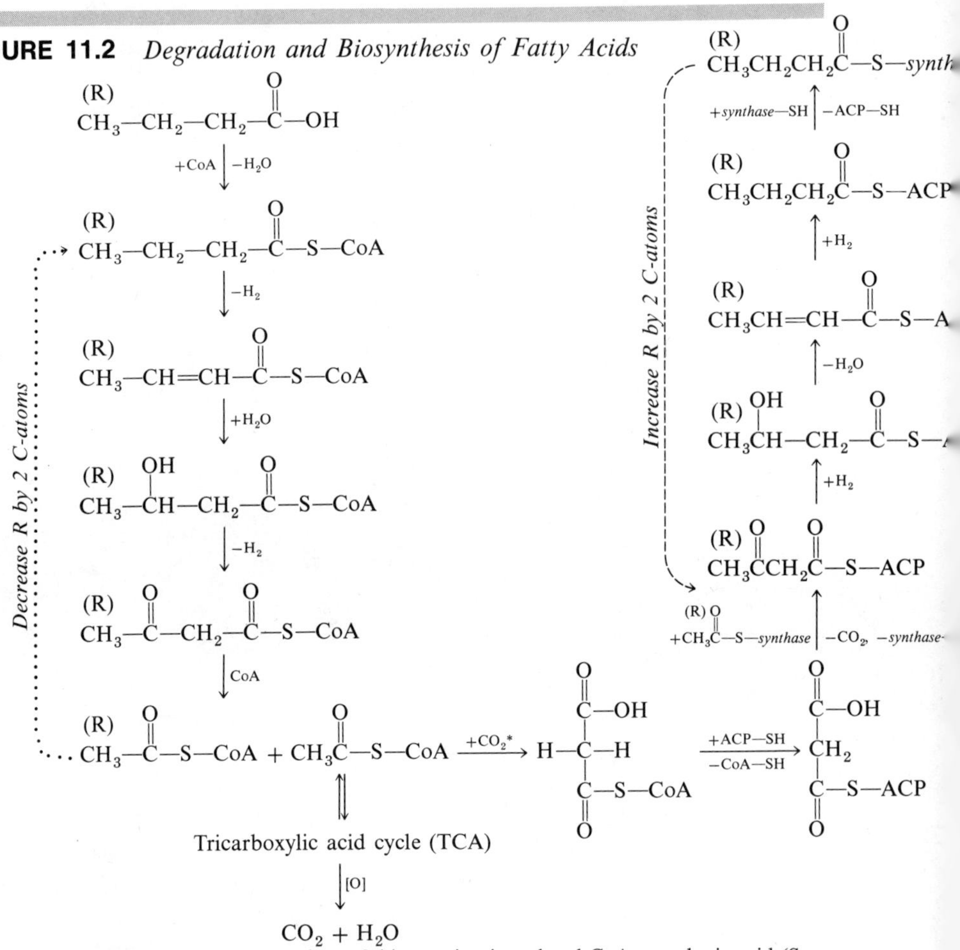

* As bicarbonate ion, HCO_3^-. The product of this reaction is malonyl CoA, a malonic acid (Sec. 12.2 and 12.3) derivative.

Each of the acetyl or acetoacetyl derivatives are simply thioesters (sulfur analogs of esters) with a complex substituent in place of the usual alkyl group. The first equilibrium is the biochemical analog of the Claisen condensation, and the second two equilibria are **transesterification** reactions, that is, reactions in which the "alkyl" group of the thioester is exchanged for a different "alkyl" group.

In Figure 11.2 the presently accepted sequences of reactions involved in the degradation and biosynthesis of fatty acids are shown for the breakdown and the synthesis of butyric acid. For the *degradation* of a longer-chain fatty acid, we can substitute R for the methyl group in butyric acid and follow the cycle shown by the dotted line. In each pass through the cycle two carbon atoms are lost from the right-hand end of the molecule (and R becomes shorter by two carbon atoms). For the *synthesis* of a longer-chain acid we follow the cycle shown by the dashed line, substituting the acyl-S-*synthase* at the top for the one at the bottom (and R becomes longer by two carbon atoms). Note the simplicity of the individual steps, reductions and oxidations (dehydrogenations), hydrations and dehydrations. Of course, the actual reactions are much more complex, involving a number of enzymes. If you go on to study biochemistry, you will find that the reactions are beautifully designed to minimize the activational energies for each step, to assure that the overall course of reaction is exothermic, to utilize available sources of energy, and to minimize interaction between synthetic and degradative processes.

11.11 *The Prostaglandins*

The prostaglandins are a relatively large group of 20-carbon atom acids containing a 5-membered ring that have been isolated from a variety of animal tissues. At least 14 prostaglandins occur in human seminal plasma and numerous others have been found in many types of cell fluids or synthesized in the laboratory during the past 15 years. Although present in only trace amounts, the prostaglandins have a number of important biological roles of a hormonal or regulatory nature such as the regulation of metabolism, stimulation of the contraction of smooth muscles, depression of blood pressure, inhibition of gastric secretion, and dilation of the bronchial vessels. It has even been suggested that aspirin (Sec. 12.6) functions, in part, by inhibiting prostaglandin production in the body. The prostaglandins are formed by the biosynthetic cyclization and oxidation of 20-carbon atom unsaturated fatty acids, such as arachidonic acid, which is derived in turn from the essential fatty acid, linoleic acid. An abbreviated biosynthetic pathway is shown in the following equations for the biological preparation of prostaglandin E_2, which along with prostaglandin $F_{2\alpha}$ has seen limited clinical use to induce labor or terminate pregnancy.

Arachidonic acid
(5,8,11,14-Eicosatetraenoic acid)

Prostaglandin E_2 (PGE$_2$)

As shown, oxygen functions as a diradical in this biosynthesis, and a resonance stabilized allylic radical is formed in the first step by an enzyme-controlled oxidation. The inhibitory effect of aspirin on prostaglandin production may be due to its antioxidant properties (a property aspirin should share with the other phenolic antioxidants (Sec. 11.6)). In some of the other prostaglandins the carbonyl group is reduced to form the *cis*-diol or the cyclopentanol hydroxyl group is eliminated by dehydration to form the unsaturated cyclopentenone derivative. The shorthand designations PGE_2 and $PGF_{2\alpha}$ refer to the hydroxy ketone and the cyclopentanediol series, respectively. The subscript number refers to the number of carbon-carbon double bonds present in the side chains, and a subscript α or β indicates the *cis* or *trans* configuration of the diol, respectively.

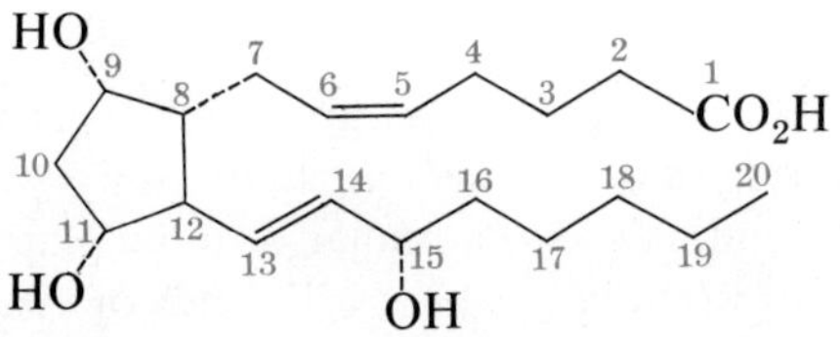

$PGF_{2\alpha}$

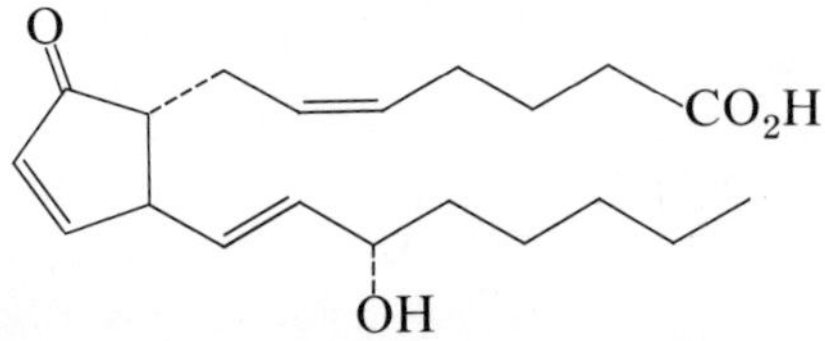

PGA_2

Summary

Structure and Classification

1. Fats, oils and waxes belong to a general classification of natural esters known as lipids.

2. Waxes are simple esters in which both the acid and alcohol components are long-chain structures.

3. Fats and oils are long-chain fatty acid esters of glycerol called glycerides. Glycerides may be simple (all fatty acids alike) or mixed (fatty acids different).

4. Fatty acids may be predominantly saturated (as in fats) or unsaturated (as in oils).

5. Saponification numbers provide an index to the molecular weight of a fat or oil.

6. Iodine numbers reveal the degree of unsaturation of a fat or oil.

7. The reactions of fats and oils may be summarized as follows.
   (a) Alkaline hydrolysis (saponification) of a glyceride yields glycerol and soaps.
   (b) Hardening (hydrogenation) converts oils to fats. Hydrogenation, if complete, converts a fat to glycerol and long-chain alcohols.
   (c) Hydrolysis and oxidation of *edible* oils and fats results in rancidity. Rancidity in fat and oil food products can be retarded by the addition of antioxidants.
   (d) Oxidation of unsaturated *inedible* oils with air forms peroxides and polymers. Such oils are used in paints and lacquers as drying oils.

8. Fats and oils have more than twice the caloric value of carbohydrates and proteins.

9. Detergents are cleansing agents and include both soaps and synthetic detergents. Synthetic detergents are called syndets.

10. Soluble soaps are sodium or potassium salts of long-chain fatty acids; heavy metal salts are water insoluble.

11. A syndet may be anionic—i.e., a sodium salt of a long-chain alcohol sulfate or an alkyl substituted aryl sulfonate; cationic—i.e., a long-chain substituted ammonium salt, or a long-chain nonionic structure.

12. The mechanics of cleansing with soaps or with syndets are the same.

13. Soaps precipitate in hard water; syndets do not.

14. The catabolism (degradation) and anabolism (synthesis) of fats in the animal system appear to involve a sequence of 2-carbon units related to acetic acid.

15. Prostaglandins are cyclic derivatives of 20-carbon atom fatty acids which function as biological regulators.

New Terms

acyl carrier protein (ACP)	iodine value
anionic detergent	lipase
antioxidant	lipid
biosynthesis	margarine

cationic detergent

coenzyme A (CoA)

drying oil

emulsion

glyceride

hardening of oils

micelle

nonionic detergent

prostaglandin

rancidity

saponification value

syndet

Supplementary Exercises

EXERCISE 11.4 Name the following compounds:

(a) $C_{25}H_{51}-\overset{\overset{O}{\|}}{C}-O-C_{12}H_{25}$ (b) $C_{15}H_{31}-\overset{\overset{O}{\|}}{C}-O^-Na^+$

(c)
$$H_2C-O-\overset{\overset{O}{\|}}{C}-C_{13}H_{27}$$
$$HC-O-\overset{\overset{O}{\|}}{C}-C_{15}H_{31}$$
$$H_2C-O-\overset{\overset{O}{\|}}{C}-C_{17}H_{33}$$

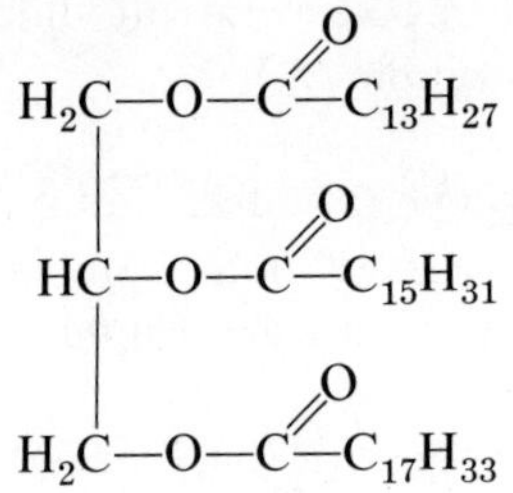

(d)
$$H_2C-O-\overset{\overset{O}{\|}}{C}-C_{17}H_{33}$$
$$HC-O-\overset{\overset{O}{\|}}{C}-C_{17}H_{33}$$
$$H_2C-O-\overset{\overset{O}{\|}}{C}-C_{17}H_{33}$$

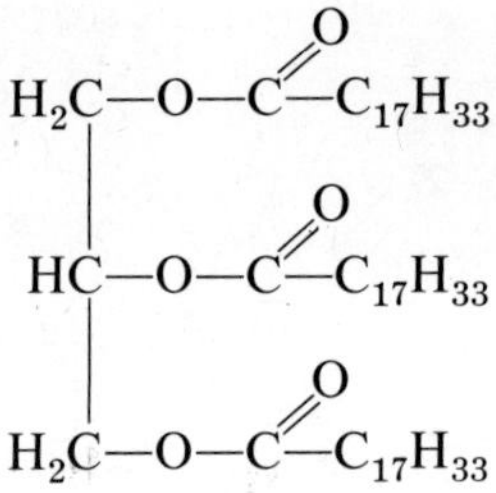

(e)

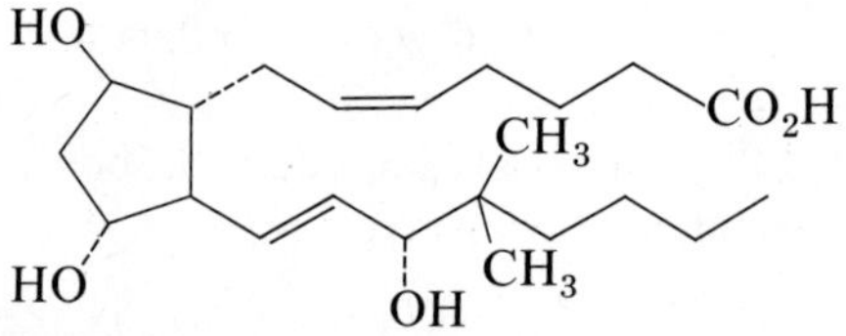

EXERCISE 11.5 Draw structures for the following:

(a) oleic acid

(b) glyceryl butyropalmitostearate

(c) a soap

(d) a prostaglandin

(e) a sulfate detergent

(f) a sulfonate detergent

(g) a cationic detergent

(h) acetyl CoA

EXERCISE 11.6 Draw structures of glycerides which might possibly be those of individual molecules found in

(a) butter

(b) tallow

(c) linseed oil

(d) tung oil

(e) safflower oil

(*Note: Consult Table 11.1.*)

EXERCISE 11.7 Calculate the saponification number and the iodine value for Compound (c), Exercise 11.6. Check your values to see if they come within the ranges given in Table 11.1.

EXERCISE 11.8 By means of equations illustrate chemical reactions which, when carried out with a fat or an oil, result in:
(a) formation of soap
(b) "hardening" of vegetable oil
(c) drying of paint
(d) rancidity in butter

EXERCISE 11.9 A grease spot made by hot butter on a linen napkin may be removed by laundering. If the spot had been made on a woolen flannel skirt or trousers it should be removed by "dry" cleaning. Explain the mechanics involved in each case.

EXERCISE 11.10 Why does not soap lather easily in most deep-well water?

EXERCISE 11.11 Explain the difference between each of the following.
(a) hydrolysis and saponification
(b) a drying and a nondrying oil
(c) a fat and an oil
(d) a soap and a syndet
(e) hydrolytic and oxidative rancidity
(f) a fat and a wax

EXERCISE 11.12 Why do most of the long-chain fatty acids contain an even number of carbon atoms?

EXERCISE 11.13 The saponification number of a fat is 200 and its iodine value is 30. Which of the saturated and unsaturated fatty acids probably predominate in this glyceride? What general conclusions may be made concerning this fat?

EXERCISE 11.14 The β-hydroxybutyric acid formed in the first reductive step of the biosynthesis of fatty acids has the D-configuration, while the β-hydroxybutyric acid formed in the hydration of crotonic acid in the degradation of fatty acids has the L-configuration. Draw the Fischer projections of these enantiomers.

EXERCISE 11.15 When the prostaglandin PGE$_2$ is treated with sodium borohydride, two products result. The ir and nmr spectra of the products show that the carboxyl group and carbon-carbon double bonds have not been reduced. One of the products is PGF$_{2\alpha}$. What is the probable structure of the other product?

EXERCISE 11.16 The configuration of most of the double bonds in the unsaturated fatty acids is *cis*, while the configuration of the α,β-unsaturated acids formed in the degradation of fatty acids is *trans*. The enzyme that catalyzes the hydration of α,β-unsaturated acids will function with both *cis-* and *trans-*α,β-unsaturated acids (but not with β,γ-unsaturated acids); however, with the *cis-*acid it gives the D-β-hydroxy acid rather than the L-β-hydroxy acid *required* by the enzyme which catalyzes the oxidation of the β-hydroxy acid to the β-keto acid in the next step. Therefore, when unsaturated acids are degraded, two additional enzymes are required beyond those needed for the degradation of saturated fatty acid. One of these *isomerases* converts β,γ-unsaturated acids into *trans-*α,β-unsaturated acids and the other converts D-β-hydroxy acids into L-β-hydroxy acids (as the CoA derivatives, of course). Using linoleic acid as an example, show at what points in the degradation these *isomerases* are required. Do this schematically by the successive elimination of 2-carbon units from the simplified structure given, as shown for the first three cycles. Draw the simplified structure of the acid residue after each cycle and describe in words or structures the points where the *isomerases* must function.

Linoleic acid

12

Bifunctional Acids

Introduction

Compounds discussed in previous chapters usually had but one functional group. Occasionally, as in the case of the diols and the dihalides, more than one functional group was encountered, but in these instances the groups were of the same kind. In the present chapter the compounds considered are primarily acids but have, in addition to the carboxyl, another functional group. The chemical behavior of a bifunctional acid, for the most part, is the behavior characteristic of each group separately. However, the dual nature of bifunctional acids confers special properties upon these compounds. Such changes in properties are particularly significant when the functional groups are in close proximity to each other.

Dicarboxylic Acids

12.1 *Nomenclature and Properties*

The aliphatic dicarboxylic acids are naturally occurring, colorless, crystalline solids often referred to by common names usually of Latin or Greek derivation. Such names frequently indicate a natural source of the acid. IUPAC nomenclature follows established rules. One simply adds the suffix *dioic* to the parent hydrocarbon of the same number of carbon atoms. For example, oxalic acid (HOOC—COOH), the simplest member of the dicarboxylic acid series, is named **ethanedioic acid.** Malonic acid (HOOC—CH$_2$—COOH), the next member of the family, is named **propanedioic acid,** etc. Table 12.1 lists the common names and structures, along with the acidic properties, of the dicarboxylic acids containing 2–10 carbon atoms.

The first two members of the dicarboxylic acid family are much stronger acids than acetic acid. The presence of a second carboxyl group adjacent to, or near, another appears to withdraw electrons from the first. This inductive effect becomes weaker if the carboxyl functions are farther separated by intervening methylene groups. An electron withdrawal on the part of one carboxyl group tends to promote ionization of the other carboxyl group by stabilizing the negatively charged mono-anion. Ionization of the second carboxyl group is more difficult because of electrostatic repulsion in the dianion between the two negatively charged carboxylate groups that tends to destabilize the dianion. The value of the second ionization constant, for this reason, is generally much smaller than that of the first.

12.2 *Preparation and Reactions of the Dicarboxylic Acids*

Dicarboxylic acids usually can be prepared by adapting methods used for the preparation of the monocarboxylic acids. Functional groups easily convertible to carboxyl groups usually provide a route to the dicarboxylic acids. For example, the oxidation of unsaturated acids and the hydrolysis of nitriles are methods frequently used. The preparation of malonic and azelaic acids by such methods is illustrated in the following reactions.

Oxidation of an unsaturated acid

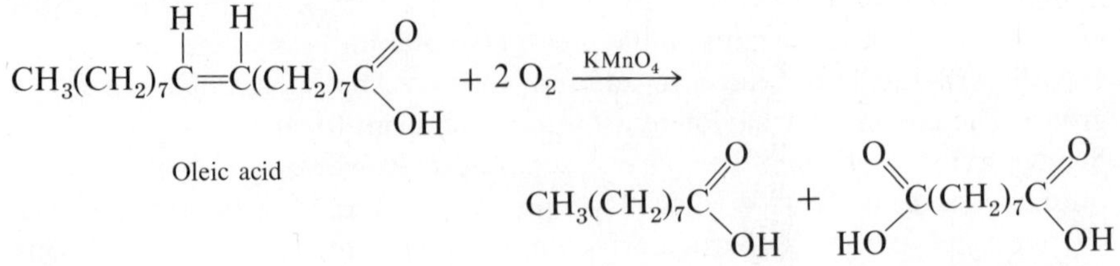

Oleic acid

Nonanoic acid   Azelaic acid

TABLE 12.1 *Dicarboxylic Acids: Names, Structures, Melting Points, and Acid Properties*

Name	Formula	M.P. (°C)	pK_1 (25°)	pK_2 (25°)
Acetic* (L., *acetum*, vinegar)	CH_3COOH	16.6	4.76	
Oxalic (Gr., *oxys*, sharp)	HOOC—COOH	187	1.27	4.27
Malonic (L., *malum*, apple)	HOOC—CH_2—COOH	135 (dec.)	2.86	5.70
Succinic (L., *succinum*, amber)	HOOC—$(CH_2)_2$—COOH	185	4.21	5.64
Glutaric (Composite of *glutamic* + *tartaric*)	HOOC—$(CH_2)_3$—COOH	97.5	4.34	5.42
Adipic (L., *adipis*, fat)	HOOC—$(CH_2)_4$—COOH	151	4.43	5.41
Pimelic (Gr., *pimele*, fat)	HOOC—$(CH_2)_5$—COOH	105	4.50	5.42
Suberic (L., *suber*, cork)	HOOC—$(CH_2)_6$—COOH	142	4.52	5.41
Azelaic (Fr., *azote*, nitrogen: Gr., *elaion*, olive oil)	HOOC—$(CH_2)_7$—COOH	106	4.53	5.40
Sebacic (L., *sebum*, tallow)	HOOC—$(CH_2)_8$—COOH	134	4.70	5.42
Phthalic (Composite of *naphthalene* + *ic*)		191	2.95	5.41
Maleic		130	1.94	6.22
Fumaric (L., *fumus*, smoke)		287	3.02	4.38

* Included for reference value.

Hydrolysis of a nitrile

$$ClCH_2COOH + NaHCO_3 \longrightarrow ClCH_2COO^-Na^+ + H_2O + CO_2$$

Chloroacetic acid   Sodium chloroacetate

$$ClCH_2COO^-Na^+ + K^+CN^- \longrightarrow N{\equiv}C{-}CH_2COO^-Na^+ + KCl$$

Sodium cyanoacetate

$$N{\equiv}C{-}CH_2COO^-Na^+ + 2\,HCl + 2\,H_2O \longrightarrow$$

$$HOOC{-}CH_2{-}COOH + NH_4Cl + NaCl$$

Malonic acid

The derivatives of the dicarboxylic acids are, in general, the same as those produced from the monocarboxylic acids and include salts, esters, amides, and acid halides. The behavior of the dicarboxylic acids, when heated, is unique, however, and deserves special mention.

Oxalic acid, when strongly heated, undergoes decomposition according to the following equation.

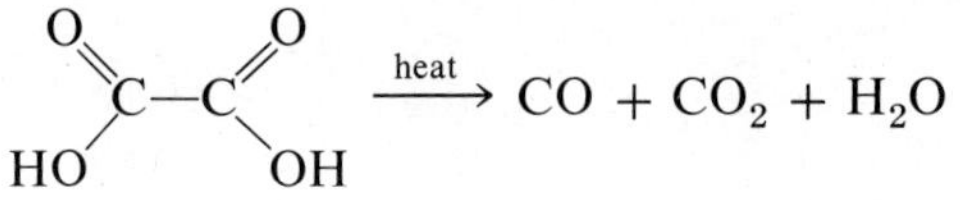

Oxalic acid

Malonic acid is so easily decomposed by heat that it is decarboxylated at its melting point (137°C).

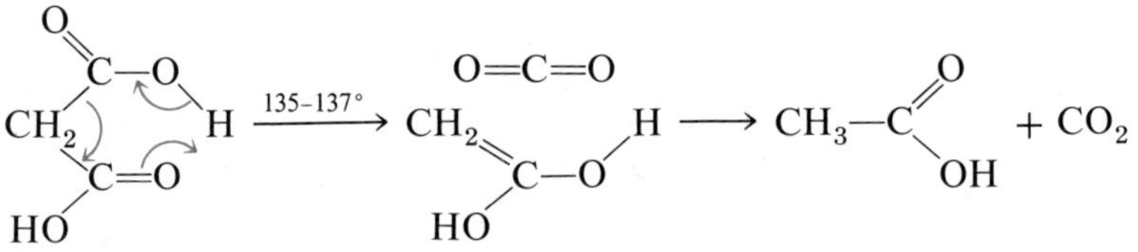

Malonic acid   Enol of acetic acid   Acetic acid

Succinic and glutaric acids, when heated, lose water to produce their corresponding cyclic anhydrides.

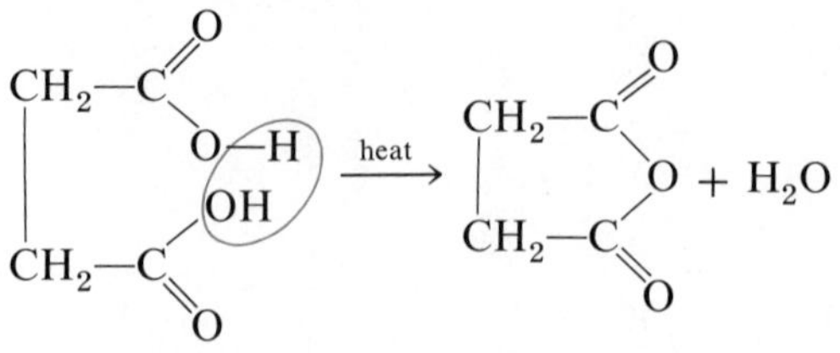

Succinic acid   Succinic anhydride

Glutaric acid   Glutaric anhydride

EXERCISE 12.1

Why is it not possible to convert a halogen acid into a Grignard reagent and then, by carbonation of the Grignard reagent with anhydrous CO_2, form a second carboxyl group?

12.3 *The Malonic Ester Synthesis*

Substituted malonic acids, like malonic acid, also are easily decarboxylated. The ease with which carbon dioxide can be split out from only one carboxyl group of malonic acid makes it a very useful acid in a number of organic syntheses. A synthesis which involves a substituted malonic acid begins with its ethyl ester, diethyl malonate, or "malonic ester" as it usually is called. The methylene carbon of ethyl malonate is flanked on both sides by electron-withdrawing carbonyl groups. This dual influence causes the α-hydrogens to be much more acidic (pK_a 13) than those in a simple ester such as ethyl acetate (pK_a 24.5). Thus, these hydrogens are easily removed by a basic reagent, such as sodium ethoxide. The anion of sodiomalonic ester, when formed, is nucleophilic and reacts readily with alkyl halides.

Diethyl malonate
(Malonic ester)

Diethyl sodiomalonate

Many synthetic organic compounds may be prepared *via* a substituted malonic ester. Indeed, the "malonic ester synthesis" often is the only route open to the preparation of a number of desirable compounds. The usefulness of the malonic ester synthesis is illustrated by the following reaction sequence, which illustrates the preparation of one of the barbiturates (malonylureas) commonly used in medicine as hypnotics (sleeping tablets).

The Preparation of 5,5-Diethylbarbituric Acid

Step 1

Step 2

Diethyl ethylmalonate

Step 3

Diethyl diethylmalonate

Step 4

Urea

5,5-Diethylbarbituric acid
(Barbital)

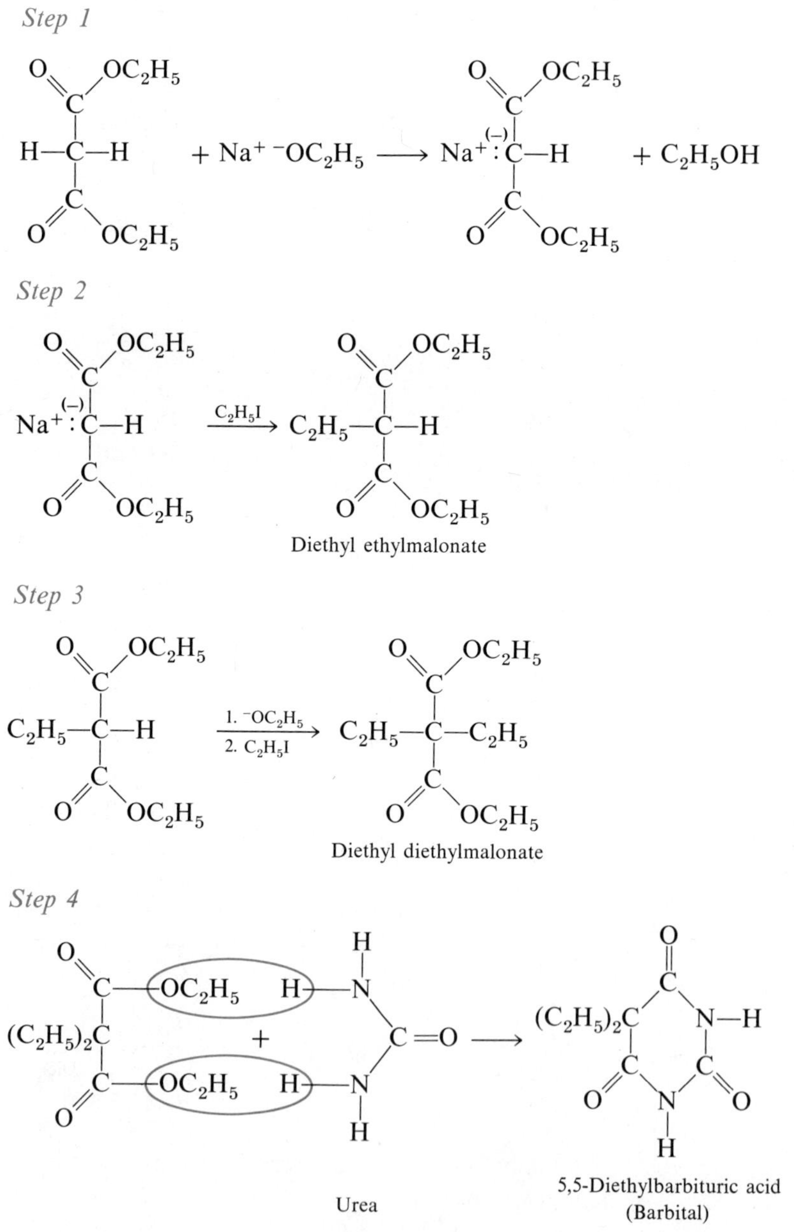

In the laboratory the malonic ester synthesis is widely used for the synthesis of substituted acetic acids, $R-CH_2-CO_2H$ and $\begin{array}{c} R \\ R' \end{array}\!\!\!>\!CH-CO_2H$, where R and R' are derived from the alkyl halides used to alkylate malonic ester and the acetic acid moiety is derived from malonic acid (by decarboxylation). To illustrate this useful procedure we will interrupt the preparation of 5,5-diethyl-barbituric acid after Step 2 and prepare butyric acid.

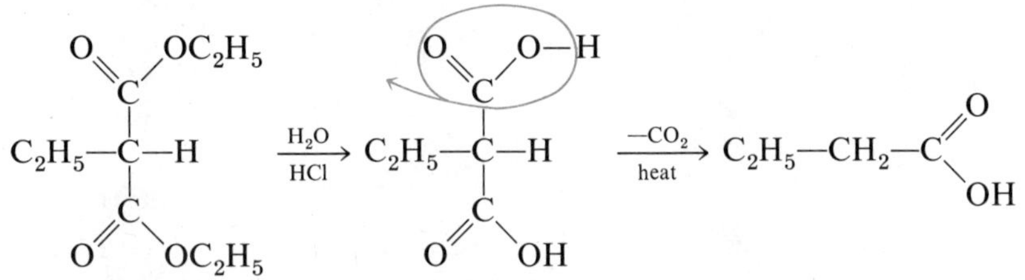

Diethyl ethyl-malonate → Diethylmalonic acid → Butyric acid

EXERCISE 12.2

Show by means of equations how the malonic ester synthesis could be used to prepare (a) propionic acid, (b) 2-methylpropanoic acid (dimethylacetic acid), (c) succinic acid.

12.4 *Other Important Dicarboxylic Acids*

Adipic acid, used in large quantities for the production of nylon (Sec. 17.2), is an important industrial chemical. One commercial method for the production of adipic acid uses cyclohexanol as starting material. Cyclohexanol is oxidized to adipic acid according to the following reaction.

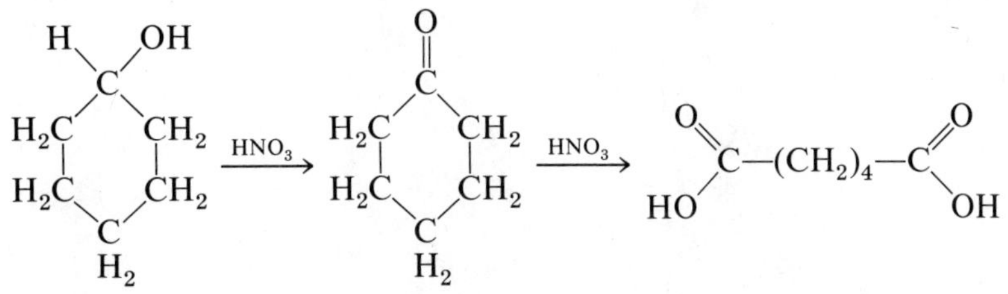

Cyclohexanol → Cyclohexanone → Adipic acid

Corn cobs and oat hulls provide another abundant source of raw material for the preparation of adipic acid. Furfural, an aromatic aldehyde obtainable from such agricultural wastes, can be converted into adipic acid through a series of reactions illustrated in Sec. 13.5.

Phthalic acid, or 1,2-benzenedicarboxylic acid, is one of the more useful aromatic dicarboxylic acids. Phthalic acid is obtained by the vigorous oxidation of naphthalene (Secs. 10.4-A and 4.7). The *para* isomer, terephthalic acid, is prepared by the oxidation of the methyl groups of *p*-xylene.

p-Xylene   Terephthalic acid
(1,4-Benzene-
dicarboxylic acid)

Phthalic and terephthalic acids, when esterified with polyhydric alcohols, produce high molecular weight polyesters. The reaction of phthalic acid with ethylene glycol or with glycerol produces the **glyptal resins** (*gly*cerol + *phthalic* acid) that are now widely used as synthetic auto finishes. The synthetic fiber known as "Dacron" or "Terylene" is a polyester of ethylene glycol and terephthalic acid. Dacron is not produced by the direct esterification of terephthalic acid, but rather by an ester interchange **(transesterification)** between methyl terephthalate and ethylene glycol.

Dacron

The same polyester, when produced in the form of a film, is marketed under the trade name "Mylar."

Hydroxy and Halogen Acids

12.5 *Structure and Nomenclature*

Certain of the hydroxy acids (see Table 12.2) are very common naturally occurring substances and usually are referred to by common names. The halogen acids, on the other hand, are mostly synthetic and are named systematically. Hydroxy and halogen substituents on the carbon chain can be located either by numbers or by Greek letters.

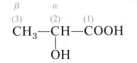

β α
(3) (2) (1)
$CH_3—CH—COOH$ 2-Hydroxypropanoic acid
       | α-Hydroxypropionic acid
       OH   (Lactic acid)

β α
(3) (2) (1)
$CH_3—CH—COOH$ 2-Bromopropanoic acid
       | α-Bromopropionic acid
       Br

TABLE 12.2 *Naturally Occurring Hydroxy Acids*

Common Name	Structure	pK_a (25°)
Glycolic (*Glycerol*)	$HOCH_2COOH$	3.82
Lactic (L. *lactis*, milk)	$CH_3—CH—COOH$ $\qquad\quad$| $\qquad\quad OH$	3.86
Malic (L. *malus*, apple)	$\qquad$H $\qquad$| $HO—C—COOH$ $\qquad$$CH_2COOH$	pK_1 3.40 pK_2 5.05
Tartaric	$\qquad$H $\qquad$| $HO—C—COOH$ $HO—C—COOH$ $\qquad$| $\qquad$H	pK_1 3.22 pK_2 4.81
Citric (L. *citrum*, lemon, lime)	$\qquad$$CH_2—COOH$ $\qquad$\| $HO—C—COOH$ $\qquad$$CH_2—COOH$	pK_1 3.13 pK_2 4.76 pK_3 6.40
Salicylic (L. *salix*, willow)	OH benzene ring—COOH	pK_1 3.00 pK_2 12.38
Mandelic (Gr. *mandel*, almond)	H O | || benzene ring—C—C $\qquad\quad$|$\quad$\\OH $\qquad\quad$OH	3.41

12.6 *Preparation and Reactions of Substituted Acids*

The chemistry of substituted acids is interrelated. A method that leads to the preparation of one substituted acid usually involves a reaction of another acid already bearing a different replaceable substituent.

Acids that have a halogen atom on the α-carbon atom are important starting materials for a number of substituted acids. Such α-halogen acids are easily prepared *via* the Hell-Volhard-Zelinsky reaction (Sec. 10.5-D). The reactions exhibited by halogen acids are similar in nature to those reviewed for the alkyl halides. For example, dehydrohalogenation with alcoholic potassium hydroxide converts a halogen acid to an unsaturated acid.

$$\underset{\text{α-Bromopropionic acid}}{H-\overset{\overset{\displaystyle H}{|}}{\underset{\underset{\displaystyle H}{|}}{C}}-\overset{\overset{\displaystyle H}{|}}{\underset{\underset{\displaystyle Br}{|}}{C}}-\overset{\displaystyle O}{\underset{\displaystyle OH}{C}}} \quad\xrightarrow[\text{KOH}]{\text{alcoholic}}\quad \underset{\text{Acrylic acid}}{CH_2=\overset{\overset{\displaystyle H}{|}}{C}-COOH} + H_2O + K^+Br^-$$

The addition of hydrogen bromide to acrylic acid will not reform α-bromopropionic acid but will yield instead β-bromopropionic acid by 1,4-addition (Sec. 9.12).

$$\underset{\text{Acrylic acid}}{H-\overset{\overset{\displaystyle H}{|}}{C}=\overset{\overset{\displaystyle H}{|}}{C}-COOH} \quad\xrightarrow[1,4]{\text{HBr}}\quad \underset{\text{β-Bromopropionic acid}}{H-\overset{\overset{\displaystyle H}{|}}{\underset{\underset{\displaystyle Br}{|}}{C}}-CH_2-COOH}$$

Hydrolysis of a halogen acid with dilute aqueous alkali produces the corresponding hydroxy acid.

$$CH_3-\overset{\overset{\displaystyle H}{|}}{\underset{\underset{\displaystyle Br}{|}}{C}}-COOH + Na^+OH^- \quad\longrightarrow\quad \underset{\text{Lactic acid}}{CH_3-\overset{\overset{\displaystyle H}{|}}{\underset{\underset{\displaystyle OH}{|}}{C}}-COOH} + Na^+Br^-$$

Hydroxy acids also may be prepared by the hydrolysis of cyanohydrins (Sec. 9.6-A).

The hydroxy acids, when heated strongly, show a tendency to lose water. Strong heating converts α-hydroxy acids into cyclic diesters known as lactides. In the intermolecular action of two molecules of α-hydroxy acid to form a lactide, each molecule supplies both an alcohol and an acid function for the esterification of the other.

$$\underset{\text{Lactic acid (two molecules)}}{\text{H}_3\text{C}-\text{C}-\text{O}-\text{H} \ \ \text{H}-\text{O}-\text{C}-\text{H}} \xrightarrow{\text{heat}} \underset{\text{A lactide}}{\text{H}_3\text{C}-\text{C}-\text{O}-\text{C}-\text{H}} + 2\,\text{H}_2\text{O}$$

Lactic acid (two molecules)   A lactide

The β-hydroxy acids, when heated strongly, form α,β-unsaturated acids,

$$\underset{\beta\text{-Hydroxybutyric acid}}{\text{CH}_3-\text{CH}-\overset{\text{H}}{\underset{\text{OH} \ \ \text{H}}{\text{C}}}-\text{COOH}} \xrightarrow{\text{heat}} \text{H}_2\text{O} + \underset{\substack{\text{2-Butenoic acid}\\\text{(Crotonic acid)}}}{\text{CH}_3-\overset{\text{H}}{\text{C}}=\overset{\text{H}}{\text{C}}-\text{COOH}}$$

β-Hydroxybutyric acid

2-Butenoic acid
(Crotonic acid)

The γ- and the δ-hydroxy acids can react intramolecularly to produce five- and six-atom cyclic inner esters known as **γ-** and **δ-lactones.**

$$\underset{\gamma\text{-Hydroxybutyric acid}}{\begin{array}{c}\text{CH}_2\text{---}\text{CH}_2\\ | \qquad \qquad \text{C}=\text{O}\\ \text{H}_2\text{COH} \ \ \text{HO}\end{array}} \xrightarrow{\text{heat}} \underset{\gamma\text{-Butyrolactone}}{\begin{array}{c}\text{CH}_2-\text{CH}_2\\ | \qquad \qquad \text{C}=\text{O}\\ \text{CH}_2-\text{O}\end{array}} + \text{H}_2\text{O}$$

γ-Hydroxybutyric acid   γ-Butyrolactone

If, instead of the hydroxyl group, the amino group, ($-\text{NH}_2$), is a substituent in the γ or δ position, heating produces analogous compounds in which the nitrogen atom becomes part of the ring. Such compounds are called **γ-** and **δ-lactams.**

$$\begin{array}{c}\text{H}_2\text{C}\text{---}\text{CH}_2\\ \text{H}_2\text{C} \qquad \text{C}\\ \diagdown \ \ \diagup \ \ \diagdown\\ \text{N} \qquad \text{O}\\ |\\ \text{H}\end{array}$$

γ-Butyrolactam

One of the most important aromatic hydroxy acids, perhaps, is salicylic, or *o*-hydroxybenzoic, acid. The importance of salicylic acid and its derivatives lies in the fact that they are widely used as antipyretics and analgesics.

Salicylic acid

Salicylic acid is manufactured commercially in large quantities by heating sodium phenoxide with carbon dioxide under high pressure. The free acid is obtained from its sodium salt by treatment with mineral acid.

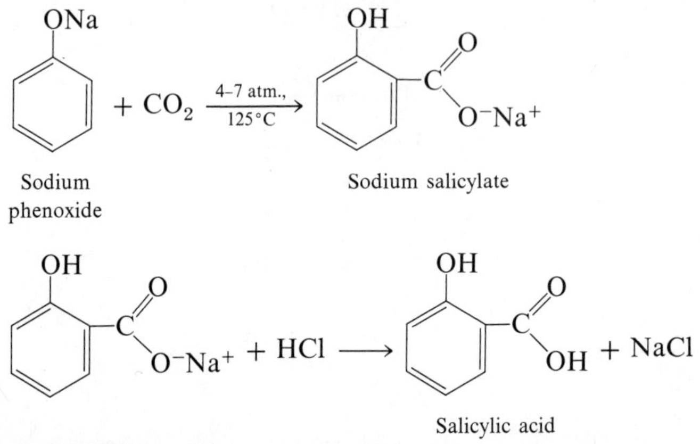

Sodium
phenoxide

Sodium salicylate

Salicylic acid

Salicylic acid is a stronger acid than benzoic due to the presence of the adjacent phenolic group. The ionization of salicylic acid is promoted because the salicylate ion, once formed, is stabilized by H-bonding between the phenolic hydrogen and one or the other of the oxygen atoms of the carboxylate group.

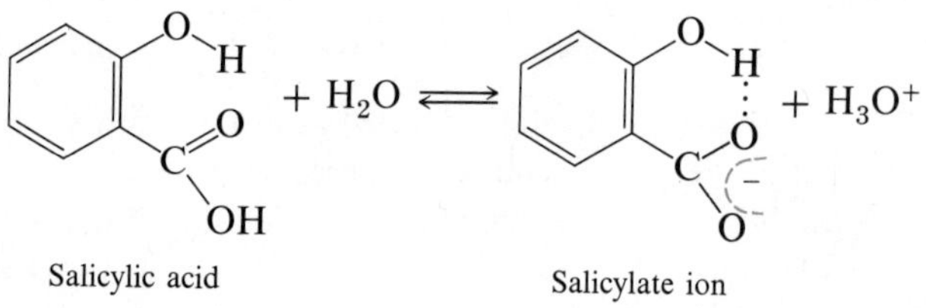

Salicylic acid

Salicylate ion

Salicylic acid is the starting material for the preparation of several very useful esters. In some of these the carboxyl group is involved, in others only the phenolic group. Esters formed from only the carboxyl group of salicylic acid are called *salicylates*. Methyl salicylate, a flavoring agent widely used in confections and in toothpastes, and referred to as *oil of wintergreen,* is an example of such an ester. It is prepared by the direct esterification of salicylic acid by methyl alcohol.

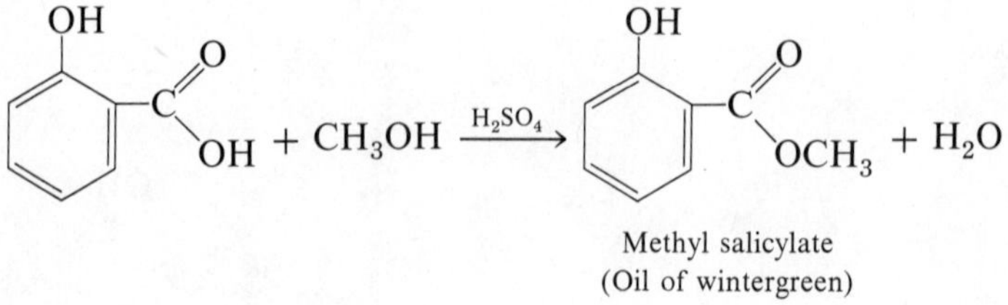

Methyl salicylate
(Oil of wintergreen)

Aspirin is by far the best example of an ester involving only the phenolic group of salicylic acid. Aspirin, or acetylsalicylic acid, is prepared by acetylation of the hydroxy group by acetic anhydride.

Acetic anhydride   Acetyl salicylic acid
(Aspirin)

Phenyl salicylate, or **Salol,** is used as an enteric coating for pills and tablets. An orally administered medicinal, if uncoated, is subject to hydrolysis in the acid environment of the stomach and either its effectiveness may be lost or a gastric disturbance may result. The coating of salol ensures safe passage into the intestine where, upon reaching an alkaline environment, the coating is hydrolyzed and the medicinal released.

Phenyl salicylate   Phenol   Salicylic acid
(Salol)

12.7 Keto Acids

Certain of the keto acids play vital roles as metabolic intermediates in biological oxidation and reduction reactions. Pyruvic acid, $CH_3\overset{\text{O}}{\underset{\|}{C}}-\overset{\text{O}}{\underset{\|}{C}}-OH$, and acetoacetic acid are especially important. Pyruvic acid is a principal intermediate in the aerobic metabolism of carbohydrates and the precursor to oxalacetic acid, without which the tricarboxylic acid cycle will not operate (see Fig. 14.1). Inspection of the scheme outlined for the metabolism of fatty acids (Sec. 11.9) shows the CoA ester of acetoacetic acid also as an intermediate in the metabolic process. The abnormal metabolism of fatty acids by diabetics, if uncontrolled, releases an excess of acetoacetic acid into the blood stream. The tendency of

β-keto acids to decarboxylate results in the release of acetone. The accumulation of acetoacetic acid and acetone in the blood may reach concentrations sufficient to cause death.

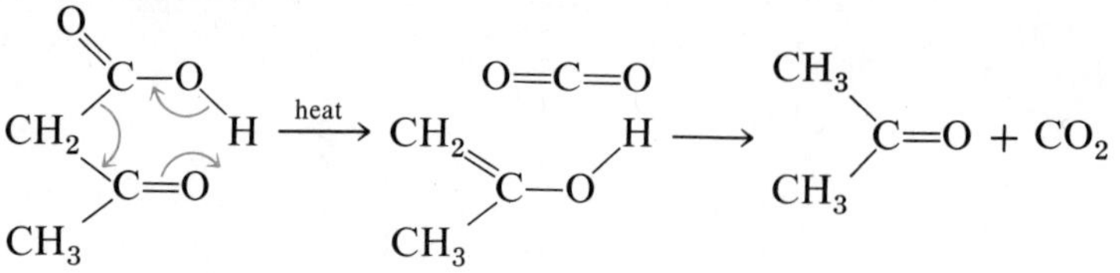

| Acetoacetic acid | Enol form of acetone | Acetone |

EXERCISE 12.3

Compare the mechanisms given for the decarboxylation of malonic acid and of acetoacetic acid and suggest a reason for the experimental observation that β-keto acids such as acetoacetic acid decarboxylate much more readily when heated than do malonic acid and its derivatives.

The readiness with which a β-keto acid decarboxylates, when heated, makes such acids useful intermediates in a number of organic syntheses. The α-hydrogens of ethyl acetoacetate are even more acidic (pK_a 11) than those of malonic ester (pK_a 13); therefore, they may be replaced in the same manner used for malonic ester (Sec. 12.3). The alkylated β-keto ester may be hydrolyzed with dilute base or with mineral acid to form the β-keto acid. Decarboxylation of the latter by heating yields a substituted acetone derivative,

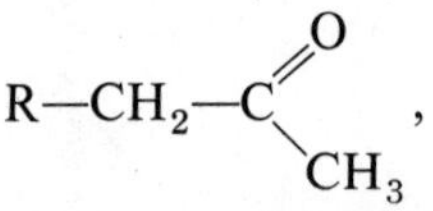

$$R-CH_2-C\overset{\displaystyle O}{\underset{\displaystyle CH_3}{}}$$, in which the alkyl group is derived from the alkyl halide

used in the alkylation step, and the acetone moiety is derived from the acetoacetic ester. This sequence of reactions, called the acetoacetic ester synthesis, is illustrated in the preparation of 2-hexanone.

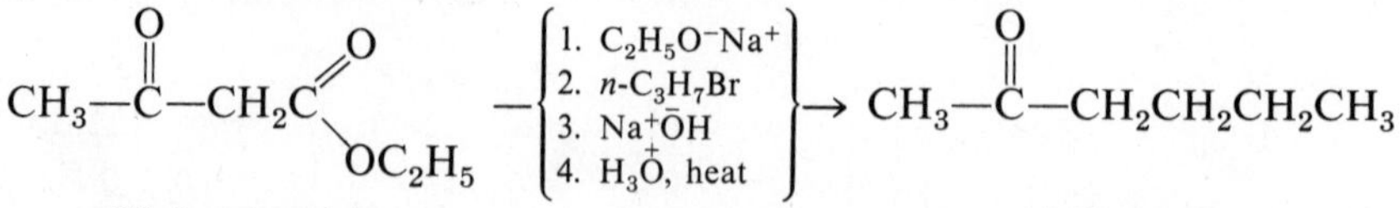

Ethyl acetoacetate   2-Hexanone

EXERCISE 12.4

In step (2) of the previous reaction sequence a tertiary alkyl halide may not be used. Why?

12.8 *Unsaturated Acids. cis,trans-Isomerism*

Malic acid, when strongly heated, loses the elements of water from adjacent carbon atoms to yield maleic and fumaric acids—two isomeric unsaturated dicarboxylic acids.

| Malic acid | Maleic acid (*cis*-Butenedioic acid) M.P. 130°C | Fumaric acid (*trans*-Butenedioic acid) M.P. 287°C |

Maleic acid when heated loses water to form an anhydride. Fumaric acid, on the other hand, is incapable of anhydride formation. However, when heated to a sufficiently high temperature (300°C), fumaric acid rearranges into maleic acid, which then yields the anhydride.

Maleic acid   Maleic anhydride

Both acids, when hydrogenated, yield succinic acid.

Maleic acd   Succinic acid   Fumaric acid

Maleic anhydride is synthesized commercially from benzene by an oxidative process similar to that used for phthalic anhydride (Secs. 4.7-E and 10-4-A). Maleic anhydride is one of the most powerful dieneophiles known (Sec. 3.13).

Summary

1. The bifunctional acids contain, in addition to a carboxyl group, one or more other functions. These may be
   (a) a halogen
   (b) a hydroxyl
   (c) a carbon-carbon double bond
   (d) a second carboxyl group
   (e) a keto group

2. The chemical behavior of bifunctional acids, generally, is either that of the monocarboxylic acids or that usually associated with the second functional group.

3. The presence of a second functional group in an acid molecule may lead to cyclization, unsaturation, loss of carbon dioxide, or polymerization.

4. Dicarboxylic acids may be prepared by the same general methods used for the preparation of monocarboxylic acids. Such methods include
   (a) the oxidation of unsaturated acids
   (b) hydrolysis of nitriles

5. Dicarboxylic acids usually are referred to by common names.

6. Adipic acid is used in the manufacture of "Nylon." Phthalic acid is used in the manufacture of *glyptal* resins, and terephthalic acid is used in the manufacture of "Dacron."

7. The chemistry of the halogen and hydroxy acids is interrelated.

8. Halogen acids usually are prepared by
   (a) the direct halogenation by the use of halogen and red phosphorus—e.g., the Hell-Volhard-Zelinsky reaction
   (b) the action of an inorganic halide upon a hydroxy acid
   (c) the oxidation of a halohydrin
   (d) the addition of hydrogen halide to an unsaturated acid

9. Hydroxy acids may be prepared by
   (a) the alkaline hydrolysis of the corresponding halogen acid
   (b) the hydrolysis of a cyanohydrin

10. Malic acid, when dehydrated, produces a pair of isomeric, unsaturated dicarboxylic acids called **fumaric** and **maleic.** These acids are *cis,trans*-isomers.

New Terms

acetoacetic ester synthesis

analgesic

antipyretic

barbiturates

enteric coating

glyptal resin

hypnotic

lactam

lactone

malonic ester synthesis

salol

transesterification

Supplementary Exercises

EXERCISE 12.5 Write structural formulas for each of the following.

(a) adipyl chloride

(b) methyl succinate

(c) terephthalic acid

(d) ethylmalonic acid

(e) ethyl cyanoacetate

(f) δ-valerolactone

(g) α-methylglutaric acid

(h) aspirin

EXERCISE 12.6 Name each of the following structures.

(a) 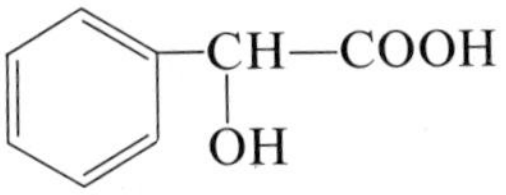 [benzene ring]—CH—COOH, with OH on the CH

(c) 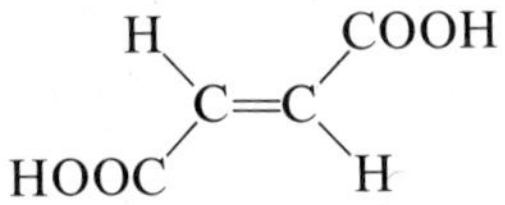 H, $COOH$ / $C=C$ / $HOOC$, H

(b) 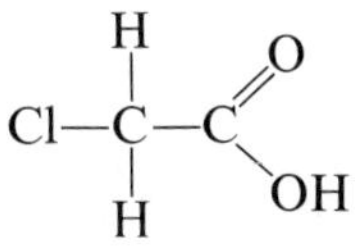 Cl—C—C with H, H on first C and $\overset{O}{\underset{OH}{}}$

(d) 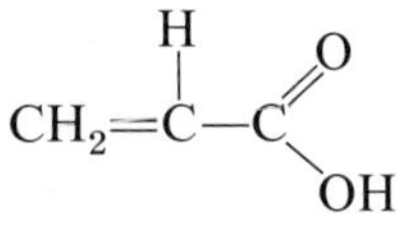 $CH_2=C$—C with H, and $\overset{O}{\underset{OH}{}}$

(e) 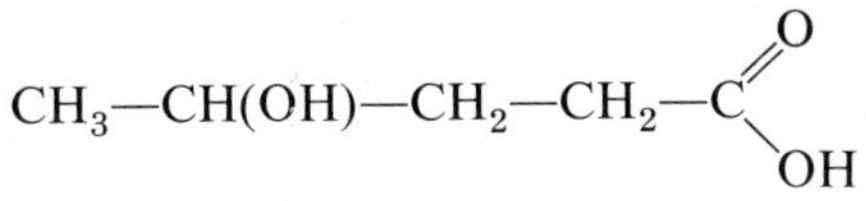 CH_3—CH(OH)—CH_2—CH_2—C$\overset{O}{\underset{OH}{}}$

(f) 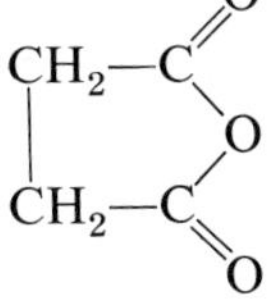 CH_2—C / CH_2—C ring with O and two =O

(g) H_2C—CH_2 / CH_3C—O ring, $C=O$, with H on CH_3C

(h) [cyclohexane ring]—COOH with OH

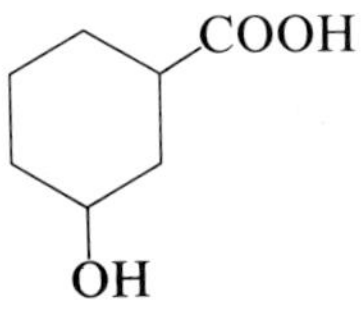

(*cis-*)

(i) 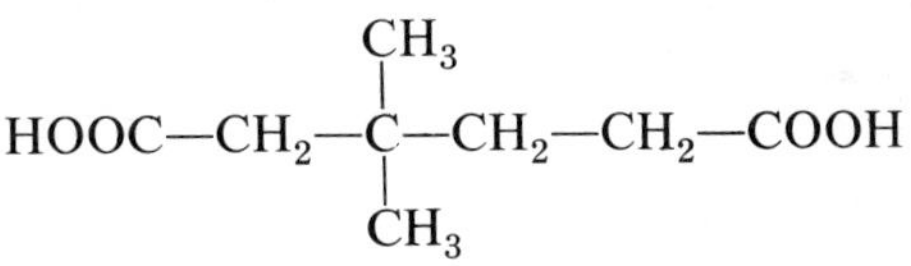 $HOOC$—CH_2—$\overset{CH_3}{\underset{CH_3}{C}}$—$CH_2$—$CH_2$—$COOH$

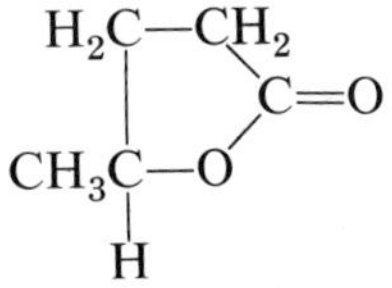

EXERCISE 12.7   Complete the following reaction sequences and identify each lettered product.

(a) $H_2C{=}CH_2 \xrightarrow{\text{HOCl}}$ (A) $\xrightarrow{[O]}$ (B) $\xrightarrow[\text{2. NaCN}]{\text{1. NaOH}}$ (C) $\xrightarrow{H_3O^+,\ \text{heat}}$ (D)

(b)

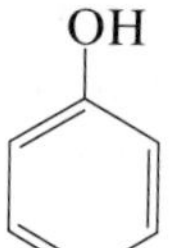

OH

$\xrightarrow{\text{NaOH}}$ (A) $\xrightarrow[\text{2. Heat, pressure}]{\text{1. CO}_2}$ (B) $\xrightarrow{H_3O^+}$ (C) $\xrightarrow[\text{H}_2\text{SO}_4]{(\text{CH}_3\overset{\text{O}}{\text{C}})_2\text{O}}$ (D)

(c) $CH_3{-}\overset{\overset{\text{H}}{|}}{C}{=}O \xrightarrow{\text{HCN}}$ (A) $\xrightarrow{H_3O^+,\ \text{reflux}}$ (B) $\xrightarrow{\text{heat}}$ (C)

(d) $CH_3CH_2CH_2COOH \xrightarrow[\text{2. H}_2\text{O}]{\text{1. Red P, Br}_2}$ (A) $\xrightarrow{\text{alcoholic KOH}}$ (B) $\xrightarrow{\text{HBr}}$ (C)

(e) $CH_2(COOC_2H_5)_2 \xrightarrow[\text{2. C}_6\text{H}_5\text{OCH}_2\text{CH}_2\text{Br}]{\text{1. C}_2\text{H}_5\text{O}^-\text{Na}^+}$ (A) $\xrightarrow[\text{heat}]{48\%\ \text{HBr,}}$ (B) + (C) + CO_2

(B) + $FeCl_3 \longrightarrow$ red color

(f) civetone $\xrightarrow[\text{2. H}_2\text{O + HCl}]{\text{1. KMnO}_4 + \text{H}_2\text{O, heat}}$ (A)

(g) maleic anhydride + 1,3-butadiene $\longrightarrow$ (A) $\xrightarrow[\text{Pt}]{H_2}$ (B)

EXERCISE 12.8   Arrange the following in an order of increasing acidity.
   (a) salicylic acid   (d) malonic acid
   (b) chloroacetic acid   (e) mandelic acid
   (c) oxalic acid

EXERCISE 12.9   Explain why the pK_a for the ionization of the first carboxyl group of maleic acid is *smaller* than that for fumaric acid and why the pK_a for the ionization of the second carboxyl group of maleic acid is *larger* than that for fumaric acid.

EXERCISE 12.10 Which of the following acids could exist as *cis-trans* isomers? Which as optical isomers?
   (a) acrylic   (d) 1,2-cyclopentanedicarboxylic
   (b) 2-butenoic   acid
   (c) malic   (e) ricinoleic acid

EXERCISE 12.11 Suggest a simple chemical test for distinguishing between
   (a) aspirin and phenyl salicylate.
   (b) acetyl chloride and chloroacetic acid.
   (c) maleic and malonic acid.

EXERCISE 12.12 Beginning with ethyl alcohol as your only organic starting material, and any other reagents you might require, show how you might prepare:

(a) acetic acid
(b) ethyl acetate
(c) ethyl acetoacetate
(d) β-hydroxybutyric acid

(e) 2-butenoic acid
(f) malonic acid
(g) 2-pentanone
(h) *n*-butyric acid

EXERCISE 12.13 A compound, $C_4H_8O_3$, responded to a series of tests as follows:

(a) Water → aqueous solution acid to litmus
(b) $Na_2Cr_2O_7 + H_2SO_4 + $ heat → blue-green coloration
(c) Strong heating → $C_4H_6O_2$
(d) Product of (c) + dilute $KMnO_4$ → decolorization
(e) $I_2 + NaOH$ → yellow solid
(f) Rotated plane polarized light

What was the compound? Write equations for reactions (a) through (e).

EXERCISE 12.14 An *optically active* acid containing no elements other than carbon, hydrogen, and oxygen has a molecular weight of 100. Suggest a structure for the acid.

EXERCISE 12.15 Show by means of equations how the following substances could be synthesized from either malonic ester or acetoacetic ester.

(a) glutaric acid
(b) cyclohexanecarboxylic acid
(c) 4-pentenoic acid

(d) 4-phenyl-2-butanone
(e) δ-ketocaproic acid
    (5-oxohexanoic acid)

EXERCISE 12.16 Suggest an explanation for the observation that the reaction of ethylene oxide with the anion of sodiomalonic ester gives the lactone shown.

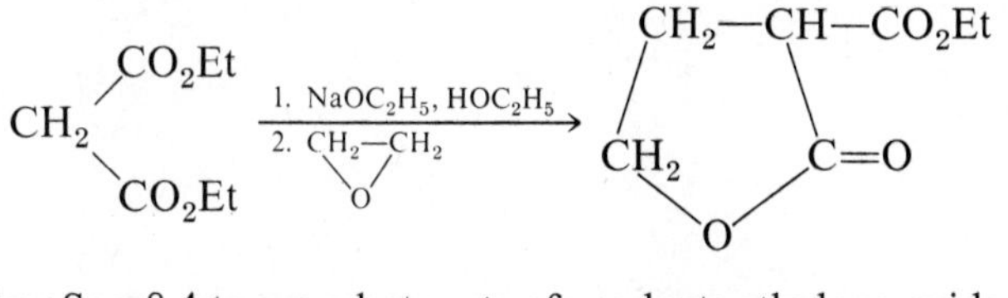

(*Hint:* Review Sec. 8.4 to see what sorts of products ethylene oxide forms with anionic reagents.)

13

Amines and Other Nitrogen Compounds; Sulfur Compounds

Introduction

The amines are our principal organic bases. Structurally, they are related to ammonia and may be considered as derivatives of this simple substance. Amines have the general formulas RNH_2, R_2NH, and R_3N in which one, two, or all three of the hydrogen atoms of the ammonia molecule have been replaced by alkyl or aryl groups. Many useful amines are synthetic, but some occur naturally in the decomposition products of nitrogenous substances such as the proteins. Some of our most valuable drugs obtained from plant extracts are characterized by the presence of one or more basic nitrogen atoms in their structures. On the basis of this finding they are assigned the general name "alkaloids." Of course, not all alkaloids are useful drugs.

Although the structures of many organic sulfur compounds parallel those of the oxygen compounds and can be considered to be derivatives of hydrogen sulfide, sulfur also can be found in organic compounds in higher oxidation states to provide us with many types of organic sulfur compounds. Our discussion will be limited to those of greatest importance in the chemistry of life processes: the thiols, sulfides, and disulfides.

13.1   Classification and Nomenclature

Amines are classified as primary, secondary, or tertiary according to the number of hydrogen atoms of ammonia that have been replaced.

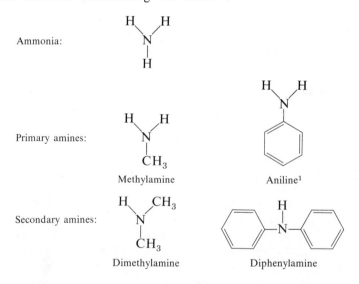

Aliphatic amines of low molecular weight and the aromatic amines are generally known by their common names. These are established simply by prefixing the names of the nitrogen-attached alkyl groups to the word **amine.** If the substituents are identical, the prefixes *di* and *tri* are employed. Examples of each class and their names are given below.

Ammonia:

Primary amines: Methylamine   Aniline[1]

Secondary amines: Dimethylamine   Diphenylamine

---

[1] Aniline, the most important aromatic amine, is named after an early Spanish name for indigo (añil) from which it first was obtained by distillation.

Tertiary amines:

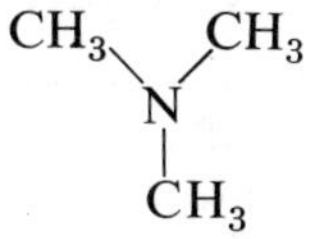

$$CH_3\diagdown N \diagup CH_3$$
$$|$$
$$CH_3$$

Trimethylamine

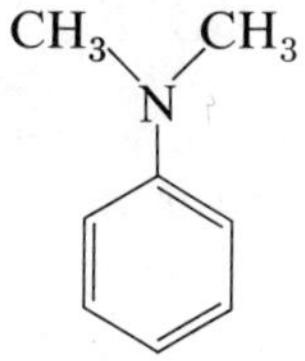

$$CH_3\diagdown N \diagup CH_3$$

Dimethylaniline

$$CH_3$$
$$|$$
$$\diagup N - CH_2 - $$

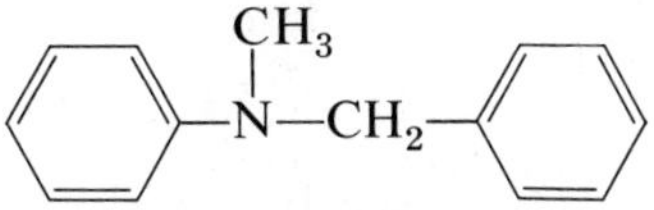

Methylphenylbenzylamine

The prefixes *sec* and *tert,* when part of the name of an amine, refer to the structure of an attached alkyl group, *not* to the class of the amine.

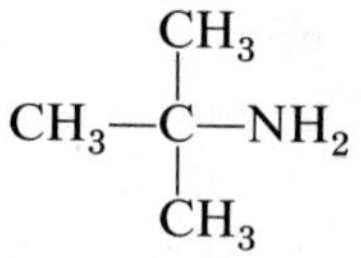

$$CH_3$$
$$|$$
$$CH_3 - C - NH_2$$
$$|$$
$$CH_3$$

tert-Butylamine (a primary amine)

$$H$$
$$|$$
$$CH_3 - CH_2 - C - NH_2$$
$$|$$
$$CH_3$$

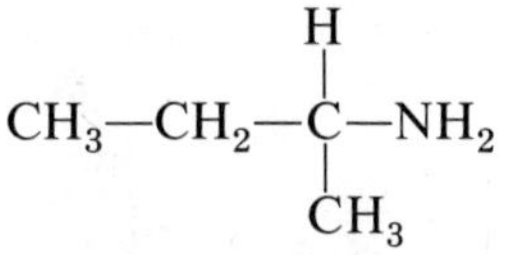

sec-Butylamine (a primary amine)

Frequently the amino group (—NH$_2$) is named simply as a substituent within another compound.

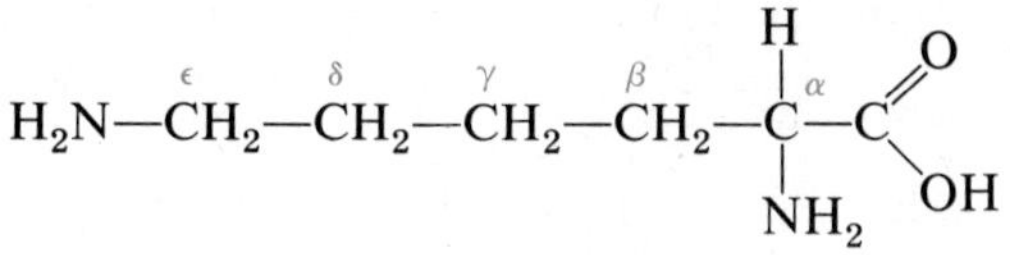

$$H_2N \overset{\epsilon}{-} CH_2 \overset{\delta}{-} CH_2 \overset{\gamma}{-} CH_2 \overset{\beta}{-} CH_2 \overset{\alpha}{-} C \overset{}{-} C \diagup \overset{O}{\diagdown OH}$$
$$|$$
$$NH_2$$

α,ε-Diaminocaproic acid (Lysine)
(2,6-Diaminohexanoic acid)

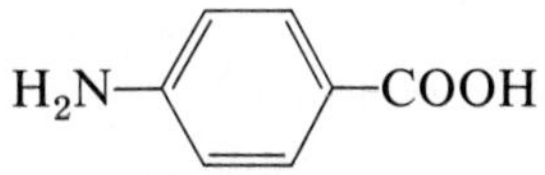

$$H_2N - \diagup \diagdown - COOH$$

p-Aminobenzoic acid

$$H_2N - CH_2 - CH_2 - CH_2 - CH_2 - CH_2 - CH_2 - NH_2$$

1,6-Diaminohexane
(Hexamethylenediamine)

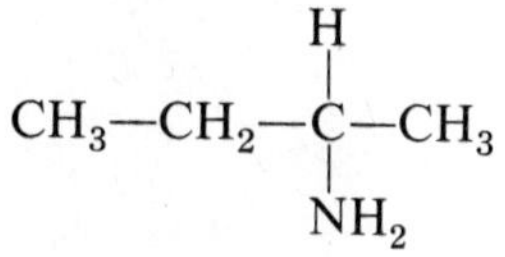

$$H$$
$$|$$
$$CH_3 - CH_2 - C - CH_3$$
$$|$$
$$NH_2$$

sec-Butylamine
(2-Aminobutane)

EXERCISE 13.1

How many different alcohols of each class are obtainable from $C_4H_{10}O$? How many different amines of each class are obtainable from $C_4H_{11}N$?

13.2 Properties of the Amines

The amines are basic compounds because only three of the five electrons in the valence shell of the nitrogen atom are used in covalent bonding. An amino nitrogen with two unshared electrons thus can function as an electron-pair donor (Lewis base). Structures deficient by an electron pair (Lewis acids) may combine with an amine and share in these two electrons to produce salts.

$$
\begin{array}{c} H \\ | \\ R-\overset{\displaystyle |}{\underset{\displaystyle |}{N}}: + H^+Cl^- \\ H \end{array} \longrightarrow \left[\begin{array}{c} H \\ | \\ R-\overset{\displaystyle |}{\underset{\displaystyle |}{N}}:H \\ H \end{array} \right]^+ Cl^-
$$

An alkyl
ammonium chloride
(a salt)

Amine salts, for the most part, are water-soluble solids. Their formation from mineral acids thus affords an easy method for the separation of amines from a mixture of water-insoluble compounds. Extraction of a mixture of organic substances with an aqueous acid solution removes the basic amines. Amines, when separated in this manner, may be liberated as free bases from the acid solution by making it strongly alkaline.

$$RNH_3^+Cl^- + Na^+OH^- \longrightarrow RNH_2 + NaCl + H_2O$$

The solution of ammonia in water produces ammonium hydroxide, i.e., ammonium and hydroxide ions. Amines, similarly, combine reversibly with water to produce basic, substituted ammonium hydroxides.

$$NH_3 + H_2O \rightleftharpoons NH_4^+ + OH^-$$

$$K_b = \frac{[NH_4^+][OH^-]}{[NH_3]} = 1.8 \times 10^{-5}$$

$$
\begin{array}{c} H \\ \diagup \\ R-N: \\ | \\ H \end{array} + \begin{array}{c} H-\overset{..}{\underset{..}{O}}: \\ | \\ H \end{array} \rightleftharpoons \left[\begin{array}{c} H \\ | \\ R-\overset{|}{\underset{|}{N}}:H \\ H \end{array} \right]^+ \ :\overset{..}{\underset{..}{O}}H^-
$$

$$CH_3NH_2 + H_2O \rightleftharpoons CH_3NH_3^+OH^-$$

Methylammonium
hydroxide

$$K_b = \frac{[CH_3NH_3^+][OH^-]}{[CH_3NH_2]} = 5 \times 10^{-4}$$

It is now common practice, particularly in biochemistry, to compare the *acid* strengths of the ammonium ions (the conjugate acids of the amines (Sec. 1.11)) rather than to compare the *base* strengths of the amines.

$$
\underset{\text{Ammonium ion}}{R-\overset{\overset{\displaystyle H}{|}}{\underset{\underset{\displaystyle H}{|}}{N}}\overset{+}{-}H} + :\overset{..}{\underset{\underset{\displaystyle H}{|}}{O}}-H \overset{K_a}{\rightleftharpoons} \underset{\text{Amine}}{R-\overset{\overset{\displaystyle H}{|}}{\underset{\underset{\displaystyle H}{|}}{N}}:} + H-\overset{..+}{\underset{\underset{\displaystyle H}{|}}{O}}-H
$$

$$
K_a = \frac{[RNH_2][H_3O^+]}{[RNH_3^+]}
$$

The K_b and pK_b of the amine RNH_2 and the K_a and pK_a of its conjugate acid RNH_3^+ are related as shown in the following equations.

$$
K_a = \frac{10^{-14}}{K_b} \quad \text{(at 25°C)}
$$

$$
pK_a = 14 - pK_b
$$

These equations lead to the following comparisons: the *stronger* the base, the *larger* the value of K_b, the smaller the value of pK_b, and the *larger* the value of pK_a (of the ammonium ion). (See Table 13.1.)

Do not confuse the pK_a of the ammonium ion with the pK_a of the amine itself. For example, ammonia is a very weak acid whose pK_a is about 35 for the dissociation to form amide ion, $NH_2:^-$. Amide ion is a powerful base, whose pK_b we can estimate from the equations to be about -21 (corresponding to a K_b of 10^{21}!). Because the amide ions derived from the amines are powerful bases, they are useful reagents in organic chemistry. One of the most useful is diisopropylamide anion (pK_b about -26), prepared by the reaction of the amine with the even more powerful base, *n*-butyllithium (pK_b about -36).

TABLE 13.1   *Physical Properties of Some Amines*

Name	Formula	B.P. (M.P.)°C	pK_a*	pK_b*
Ammonia	NH_3	-33.4	9.24	4.76
Methylamine	CH_3NH_2	-6.5	10.62	3.38
Dimethylamine	$(CH_3)_2NH$	7.4	10.77	3.23
Trimethylamine	$(CH_3)_3N$	3.5	9.80	4.20
Ethylamine	$C_2H_5NH_2$	16.6	10.63	3.37
n-Propylamine	$CH_3CH_2CH_2NH_2$	48.7		
Isopropylamine	$(CH_3)_2CHNH_2$	34	10.53	3.47
n-Butylamine	$CH_3CH_2CH_2CH_2NH_2$	77	10.61	3.39
tert-Butylamine	$(CH_3)_3CNH_2$	43.8	10.45	3.55
Aniline	$C_6H_5NH_2$	184	4.60	9.40
Dimethylaniline	$C_6H_5N(CH_3)_2$	193.5	5.21	8.79
p-Toluidine	$p\text{-}CH_3C_6H_4NH_2$	(43.7)	5.09	8.91
o-Nitroaniline	$o\text{-}O_2NC_6H_4NH_2$	(71.5)	0.99	13.01

* pK_a for conjugate acid of amine. All values at 25°C.

$$(CH_3)_2CH-\overset{\cdot\cdot}{\underset{\overset{|}{H}}{N}}-CH(CH_3)_2 + CH_3CH_2CH_2CH_2\overset{-}{\overset{\cdot\cdot}{:}}Li^+ \longrightarrow$$

Diisopropylamine   n-Butyllithium

$$(CH_3)_2CH-\overset{\overset{\cdot\cdot}{\cdot\cdot}\,-}{\underset{Li^+}{N}}-CH(CH_3)_2 + C_4H_{10}$$

Lithium diisopropylamide

A comparison of the pK_b values reveals that methylamine, CH_3NH_2, is a stronger base than ammonia, NH_3. This increased basicity is explained largely in terms of an inductive effect. An alkyl substituent will tend to release or donate electrons to the more electronegative nitrogen atom. This $+I$ inductive effect can be considered to make the unshared electron pair of the amine nitrogen more readily available for bonding to a proton, or it can be considered to stabilize the ammonium ion by somewhat reducing the net positive charge on the nitrogen atom. However, the inductive effect is not the only factor determining basicity, at least in aqueous solution. Hydrogen bonding to the water molecules may help to stabilize the ammonium ion. Thus, dimethylamine is a stronger base than methylamine, but trimethylamine is a weaker base than methylamine despite three alkyl substituents. This may be the result of the existence of only one hydrogen in the trimethylammonium ion to form hydrogen bonds to water, or the result of the methyl groups inhibiting solvation of the ion by keeping the solvent away from the positively charged nitrogen.

Aromatic amines tend to be weaker bases than the aliphatic amines or ammonia. One explanation of the difference is based on the comparison of the resonance forms for the amine and its ammonium ion, as shown for aniline.

Resonance structures of aniline

Resonance structures of anilinium ion

Aniline is highly stabilized by resonance interaction of the amino group with the ring; however, when the amino group is protonated, the resultant ammonium ion loses the extra resonance stabilization, making the formation of the ion less favorable. An alternative explanation would be that the resonance interaction shown for the aniline molecule makes the electron pair less available for donation to a base.

The lower molecular weight members of the amine family are water-soluble gases of an ammoniacal or fishlike odor. Those containing three to eleven carbons are liquids, and higher homologs are solids. Although the odors of most amines are unpleasant, their salts (called ammonium salts) are odorless.

13.3 *Preparation of Amines*

The amines may be prepared by a number of methods. The principal laboratory methods of preparation follow.

A. Alkylation of Ammonia. Alkyl groups may be introduced directly into the ammonia molecule by reaction with alkyl halides. The first product formed is an ammonium salt.

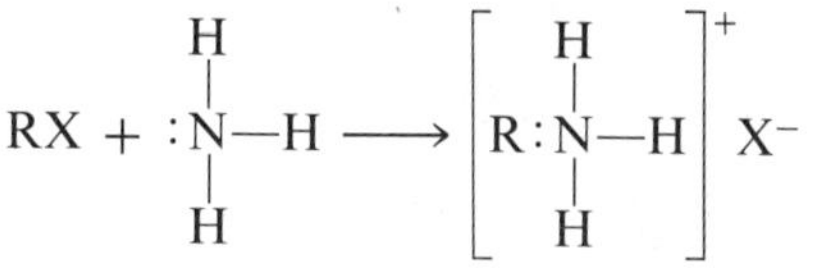

$$RX + :\overset{\overset{\displaystyle H}{|}}{\underset{\underset{\displaystyle H}{|}}{N}}-H \longrightarrow \left[R:\overset{\overset{\displaystyle H}{|}}{\underset{\underset{\displaystyle H}{|}}{N}}-H \right]^{+} X^{-}$$

Alkylammonium halide

Subsequent treatment of the ammonium salt with a stronger base (NaOH) liberates the free, primary amine.

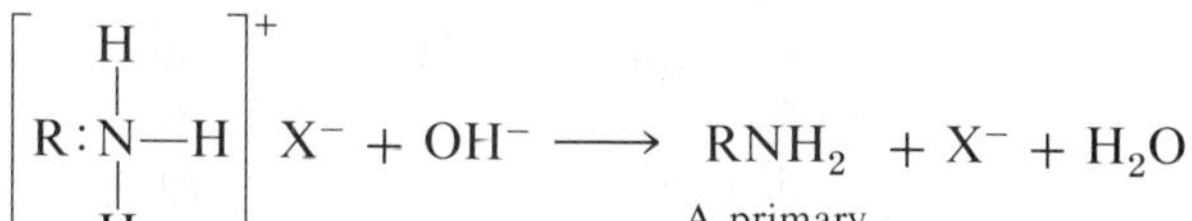

$$\left[R:\overset{\overset{\displaystyle H}{|}}{\underset{\underset{\displaystyle H}{|}}{N}}-H \right]^{+} X^{-} + OH^{-} \longrightarrow RNH_2 + X^{-} + H_2O$$

A primary
amine

The reaction, unfortunately, does not stop at the first stage as illustrated above, but continues until replacement of hydrogen by alkyl groups yields not only the primary amine but the secondary and the tertiary as well. The tertiary amine, with no hydrogen remaining, then may react with a fourth molecule of alkyl halide to produce a quaternary ammonium salt. The following series of reactions illustrates the progressive alkylation of ammonia to produce all of the above products.

$$RNH_2 + RX \longrightarrow R_2NH_2^+X^-$$
$$R_2NH_2^+X^- + NaOH \longrightarrow R_2NH + NaX + H_2O$$
$$R_2NH + RX \longrightarrow R_3NH^+X^-$$
$$R_3NH^+X^- + NaOH \longrightarrow R_3N + NaX + H_2O$$
$$R_3N + RX \longrightarrow R_4N^+X^-$$

A quaternary ammonium salt

B. Reduction of Unsaturated Nitrogen Compounds. Certain organic compounds that already contain nitrogen may be converted to primary amines by reduction methods. Most frequently used starting materials are oximes, nitriles, amides, and nitro compounds. The reduction can be accomplished either catalytically with hydrogen or directly with chemical reducing agents. The reactions below illustrate how the same primary amine may be obtained from four different starting materials.

1. $\underset{R}{\overset{H}{>}}C{=}N\overset{OH}{<}$ + 4 [H] $\xrightarrow{\text{Na, C}_2\text{H}_5\text{OH}}$ R—CH$_2$NH$_2$ + H$_2$O

    An aldoxime

2. R—C≡N + 4 [H] $\xrightarrow[\text{heat, pressure}]{\text{H}_2,\ \text{Ni,}}$ RCH$_2$NH$_2$

    A nitrile

3. $R{-}C\overset{O}{\underset{NH_2}{<}}$ + 4 [H] $\xrightarrow{\text{LiAlH}_4}$ R—CH$_2$NH$_2$ + H$_2$O

    An amide

4. R—CH$_2$—NO$_2$ + 6 [H] $\xrightarrow{\text{Fe, HCl}}$ R—CH$_2$NH$_2$

    A nitroalkane

Nitroalkanes may be prepared by the reaction of many primary and secondary alkyl bromides or iodides with sodium nitrite in dimethylformamide solution.

Lithiumaluminum hydride is the most versatile of the reducing agents cited and may be used to reduce any of the functional groups shown. Diborane (Sec. 8.4-A) may also be used to reduce all but the nitro compounds, which are inert to this reagent.

Aniline, by far the most important aromatic amine, is prepared by the reduction of nitrobenzene. Industrially, the reaction is carried out with iron and steam. In the laboratory, tin and hydrochloric acid or iron and acetic acid usually are used.

    C$_6$H$_5$—NO$_2$ + 2 Fe + 4 H$_2$O ⟶ C$_6$H$_5$—NH$_2$ + 2 Fe(OH)$_3$

Nitrobenzene   Aniline

2 C$_6$H$_5$—NO$_2$ + 3 Sn + 14 HCl ⟶ 2 C$_6$H$_5$—NH$_3^+$Cl$^-$ + 3 SnCl$_4$ + 4 H$_2$O

                                Anilinium chloride

The acid salt of aniline, when treated with sodium hydroxide, liberates the free amine.

$$\text{C}_6\text{H}_5{-}\text{NH}_3{}^+\text{Cl}^- + \text{Na}^+\text{OH}^- \longrightarrow \text{C}_6\text{H}_5{-}\text{NH}_2 + \text{Na}^+\text{Cl}^- + \text{H}_2\text{O}$$

C. Special Methods for the Preparation of Primary Amines.

The German chemist Hofmann[2] discovered that primary amines could be prepared by treating an amide with sodium hypobromite. The reaction involves a rearrangement in which the alkyl group (or aryl group) attached to the carbonyl carbon migrates to the nitrogen atom. Inasmuch as the carbonyl carbon is eliminated as carbon dioxide, the carbon chain of the amide is degraded to produce a primary amine with one carbon atom less than the original amide. The reaction is generally known as the **Hofmann Amide Hypohalite Degradation.** The reaction is believed to proceed through the following steps.

$$\underset{\substack{\text{An amide}}}{R-\overset{\displaystyle O}{\overset{\|}{C}}-NH_2} + \underset{\substack{\text{Sodium} \\ \text{hypobromite}}}{Na^+OBr^-} \longrightarrow \underset{\substack{\text{An} \\ \text{N-bromoamide}}}{R-\overset{\displaystyle O}{\overset{\|}{C}}\underset{\displaystyle H}{\overset{\displaystyle Br}{N}}}$$

$$R-\overset{\displaystyle O}{\overset{\|}{C}}-\underset{\displaystyle Br}{N}H + HO^-Na^+ \longrightarrow R-N{=}C{=}O + NaBr + H_2O$$

$$\underset{\substack{\text{An isocyanate}}}{R-N{=}C{=}O} + HOH \longrightarrow \left[R-N{=}\overset{\displaystyle O-H}{\underset{\displaystyle OH}{C}} \right] \longrightarrow \left[R-\overset{\displaystyle H}{N}-\overset{\displaystyle O}{\overset{\|}{C}}{-}OH \right]$$

$$\left[R-\overset{\displaystyle H}{N}-\overset{\displaystyle O}{\overset{\|}{C}}-OH \right] \xrightarrow{\text{heat}} R-NH_2 + CO_2$$

An alkyl-substituted   A primary amine
carbamic acid (unstable)

---

[2] August Wilhelm von Hofmann (1818–1895) was one of the most celebrated chemists of his time. He was noted particularly for his research in the chemistry of the amines. He was professor and the first director of the Royal College of Chemistry at London (1845–1864), professor and director of the laboratory, University of Berlin (1864–1895), and founder of the German Chemical Society (1868).

Another useful method for the preparation of primary amines, known as Gabriel's phthalimide synthesis, utilizes the reaction of an alkyl halide with the potassium salt of phthalimide.

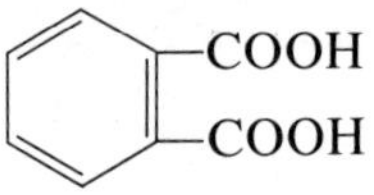

Phthalic acid 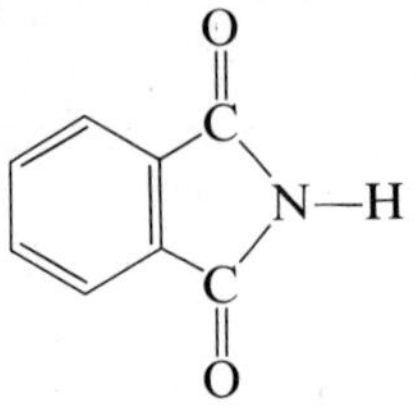   Phthalimide

The lone hydrogen atom bonded to the nitrogen of phthalimide is an acidic hydrogen due to the dual influence of the carbonyl groups. Salts can be formed from phthalimides and these salts are capable of reacting with alkyl halides.

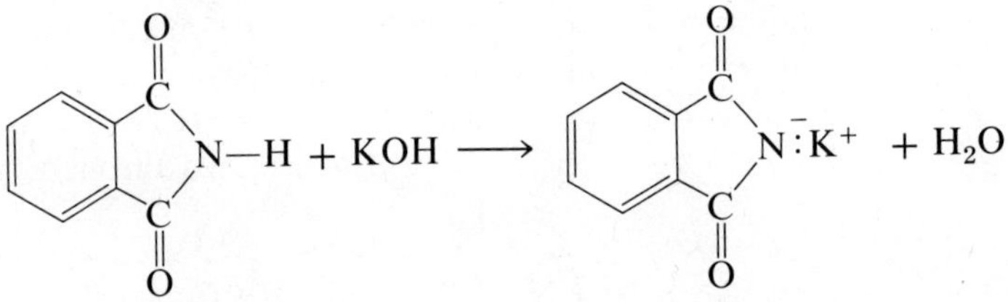

Phthalimide   N-Potassium phthalimide

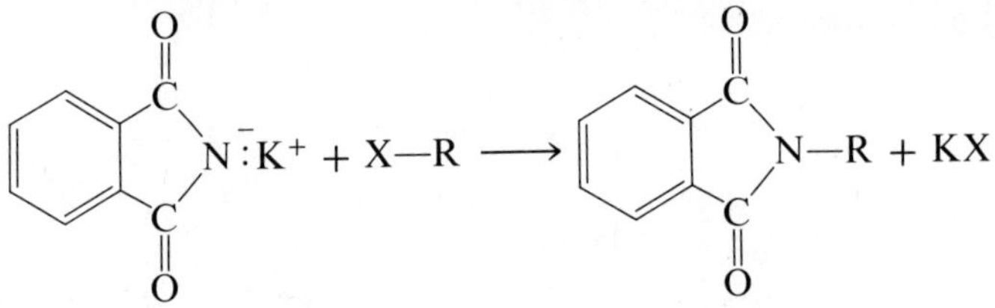

An N-alkyl
phthalimide

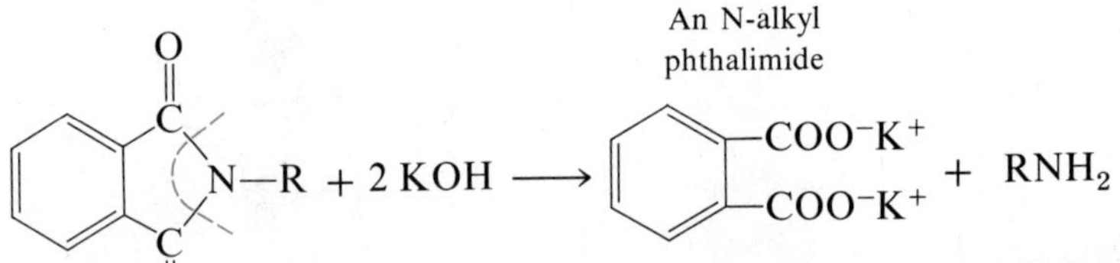

Potassium   A primary
phthalate   amine

EXERCISE 13.2

Which of the preparative methods outlined in Sec. 13.3 should be employed to accomplish the following conversions and to obtain the products indicated in a high state of purity, i.e., uncontaminated by traces of secondary and tertiary amines.

(a) $CH_2{=}CH{-}CH_2Cl \longrightarrow CH_2{=}CH{-}CH_2CH_2NH_2$
(b) $CH_3CH_2CH_2CH_2Br \longrightarrow CH_3CH_2CH_2CH_2NH_2$

What reaction mechanism does each of the above reactions illustrate?

13.4 *Reactions of Amines*

A. Salt Formation. As basic compounds, amines will react with acids to form water-soluble salts. This is a characteristic property of the amines the usefulness of which has already been pointed out (Sec. 13.2).

B. Alkylation. Primary amines, like ammonia, can be further alkylated with alkyl halides to give secondary and tertiary amines, and quaternary ammonium salts. Hofmann discovered that a primary amine, when exhaustively methylated with methyl iodide, produced a quaternary ammonium iodide. On treatment with silver oxide, the substituted ammonium iodide was converted to an ammonium hydroxide. The ammonium base, when heated, decomposed to yield trimethylamine and an olefin with a terminal double bond. The olefin thus had its origin in the alkyl group of the original amine. From the structures of the decomposition products it was possible to deduce the formula of the original amine. The steps in the reaction are illustrated using ethylamine.

$$C_2H_5NH_2 + 3\ CH_3I \longrightarrow \overset{\displaystyle CH_3}{\underset{\displaystyle CH_3}{C_2H_5\overset{+}{-}N-CH_3}}\ I^- + 2\ HI$$

Ethylamine   Trimethylethylammonium iodide

$$2\ C_2H_5\overset{+}{N}(CH_3)_3\ I^- + Ag_2O + H_2O \longrightarrow 2\ C_2H_5\overset{+}{N}(CH_3)_3\ OH^- + 2\ AgI$$

Trimethylethylammonium hydroxide

$$\underset{\displaystyle CH_3}{\overset{\displaystyle CH_3}{C_2H_5\overset{+}{-}N-CH_3}}\ OH^- \xrightarrow{\text{heat}} CH_2{=}CH_2 + (CH_3)_3N + H_2O$$

Ethylene   Trimethylamine

EXERCISE 13.3

Suggest a mechanism for the Hofmann elimination. (*Hint:* See Sec. 7.4.) Why does this elimination not take place with simple amines plus sodium hydroxide? (*Hint:* Compare the leaving groups and see Sec. 8.8-B.)

C. Acylation. The acyl group, $R-\overset{\displaystyle O}{\overset{\displaystyle \|}{C}}-$, may be substituted for hydrogen in both primary and secondary amines by reaction with acid anhydrides or acid chlorides. The use of either reagent produces an N-substituted amide. The reaction is illustrated below with ethylamine.

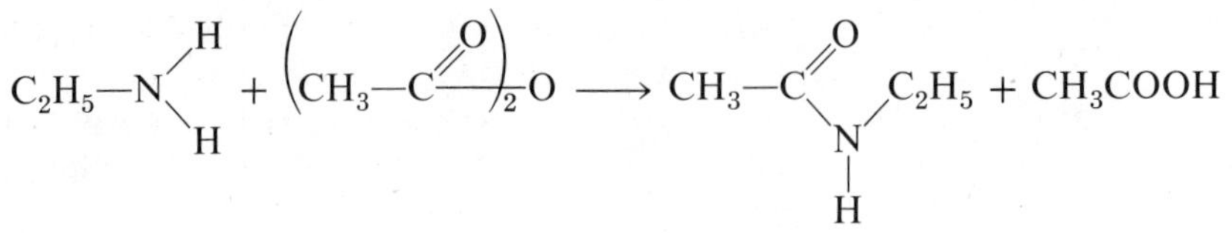

$$C_2H_5-N\overset{\displaystyle H}{\underset{\displaystyle H}{}} + \left(CH_3-C\overset{\displaystyle O}{\diagdown}\right)_2 O \longrightarrow CH_3-C\overset{\displaystyle O}{\underset{\displaystyle N}{\diagdown}} C_2H_5 + CH_3COOH$$

Ethylamine   Acetic anhydride   N-Ethylacetamide

The acylation of aniline produces an **anilide.**

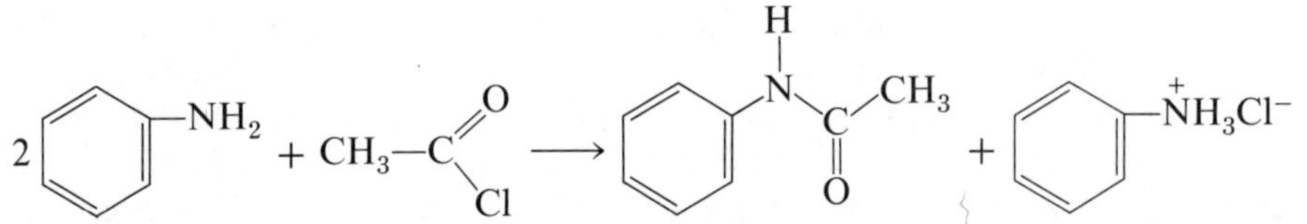

$$2 \underset{\text{Aniline}}{\boxed{}-NH_2} + \underset{\text{Acetyl chloride}}{CH_3-C\overset{O}{\underset{Cl}{\diagdown}}} \longrightarrow \underset{\substack{\text{Acetanilide}\\\text{(M.P. 114°C)}}}{\boxed{}-N\overset{H}{\underset{C=O}{}}-CH_3} + \underset{\substack{\text{Aniline}\\\text{hydrochloride}}}{\boxed{}-\overset{+}{N}H_3Cl^-}$$

Although acetanilide is an antipyretic it is also quite toxic. However, other acetanilide derivatives appear to be both safe and effective. A popular example is acetaminophen ("Tylenol"), which is used as an aspirin substitute.

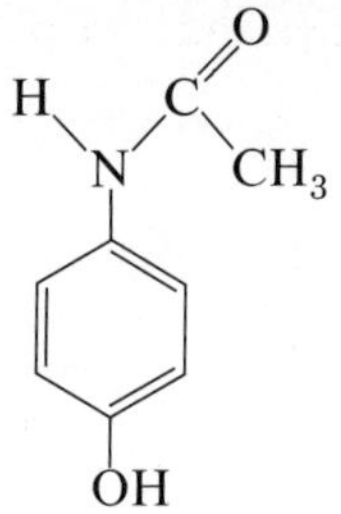

Acetaminophen

Ammonia and the amines have pyramidal structures, but the amides are planar. The planarity can be attributed to the effects of resonance stabilization of the amide group as shown in the following structures.

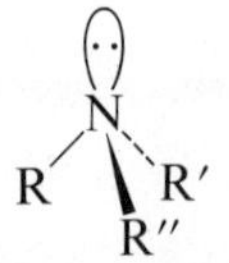

Pyramidal nitrogen   Planar nitrogen in amides
in amines

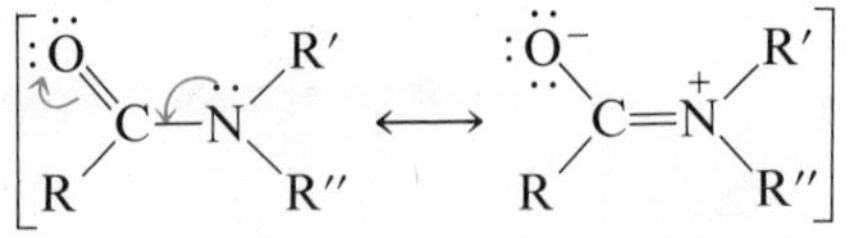

As a result of the unshared pair of electrons on nitrogen being delocalized by virtue of the resonance interaction, amides are much less basic than amines. Furthermore, there is some double bond character in the bond between the carbonyl carbon atom and the nitrogen atom. As a result rotation around the carbon-nitrogen bond of the amide function is somewhat restricted. This restriction of rotation is of considerable consequence in determining the three-dimensional conformations of protein molecules (Sec. 15.7).

Tertiary amines, lacking hydrogen atoms bonded to the nitrogen, cannot be converted to amides. The ability to form substituted amides thus provides a method for classifying an amine as primary, secondary, or tertiary. The **Hinsberg test** is one such method. In the Hinsberg test benzenesulfonyl chloride, $C_6H_5SO_2Cl$, is the acid chloride used. Reaction of a primary amine with benzenesulfonyl chloride yields an N-alkyl sulfonamide that is soluble in an alkaline solution. The hydrogen atom that remains attached to the nitrogen atom of the amide is acidic due to the electron-withdrawing power of the sulfonyl group. The amide, therefore, is soluble in an alkaline solution.

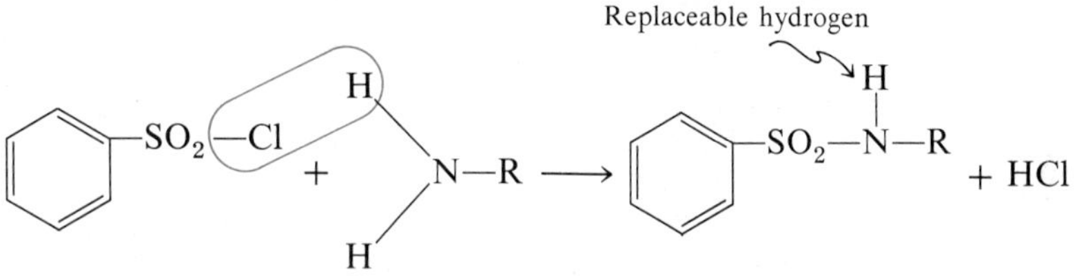

Replaceable hydrogen

Benzenesulfonyl
chloride

(Soluble in base)

A secondary amine, while reactive with benzenesulfonyl chloride, lacks the acidic hydrogen and forms an *alkali-insoluble* product.

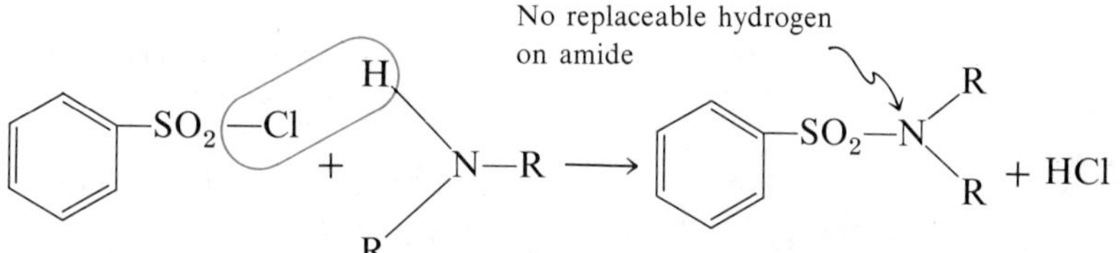

No replaceable hydrogen
on amide

(Insoluble in base)

Tertiary amines fail to react with benzenesulfonyl chloride for the same reason they failed to give acyl derivatives.

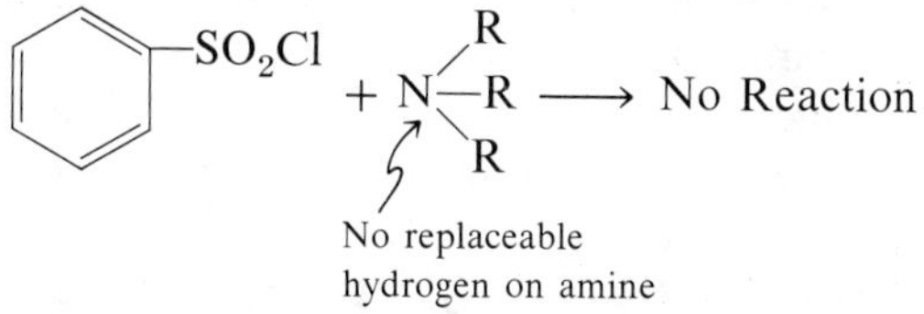

No replaceable
hydrogen on amine

5-Dimethylamino-1-naphthalenesulfonyl chloride (dansyl chloride) reacts with amino acids to give highly fluorescent dansyl derivatives and is used in determining the sequence of amino acids in polypeptides (Sec. 15.8).

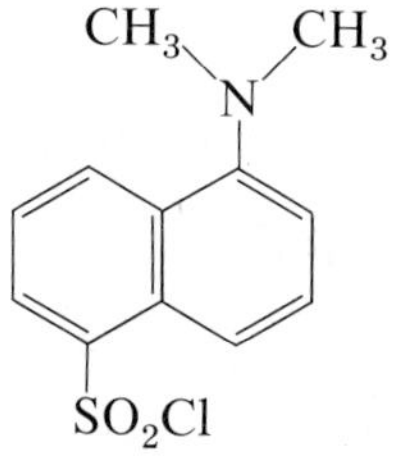

Dansyl chloride

D. Reaction with Nitrous Acid. Nitrous acid is an unstable substance prepared in solution only when needed, by reaction between a mineral acid and sodium nitrite.

$$Na^+NO_2^- + H^+Cl^- \longrightarrow Na^+Cl^- + HNO_2$$

Nitrous acid

Although the actual mechanism is somewhat more complex, in acid solution nitrous acid can be considered to be in equilibrium with nitrosonium ion, a weakly electrophilic reagent. Nitrosonium ion reacts with the nucleophilic amine to give a series of intermediates formed by successive proton transfers. The final product of the reaction depends upon the class of the amine.

$$H^+ + HO-N=O \rightleftharpoons H_2O + \quad {}^+N=O$$

Nitrosonium ion

With primary amines the reaction proceeds to the formation of the diazonium ion. Although such ions are moderately stable in cold, aqueous solution when they are derived from aromatic amines, diazonium ions formed from aliphatic amines are very unstable and decompose to form nitrogen and a carbonium ion. The carbonium ion then may react in several different ways. It may combine with a nucleophile, eliminate a proton to form an alkene, or rearrange to a more stable carbonium ion before following either of the first two pathways (Sec. 8.8-B). The routes leading to these different products are indicated by the following series of reactions.

$$CH_3CH_2CH_2CH_2-NH_2 \cdot HCl + NaNO_2 \longrightarrow$$

n-Butylamine hydrochloride

$$CH_3CH_2CH_2CH_2-\overset{\oplus}{N}\equiv N: + H_2O + OH^- + NaCl$$

A diazonium ion

$$CH_3CH_2CH_2CH_2\overset{\oplus}{N}\equiv N \longrightarrow CH_3CH_2CH_2-\overset{\overset{\displaystyle H}{|}}{\underset{\underset{\displaystyle H}{|}}{C}}{}^\oplus + N_2$$

A carbonium ion

$$CH_3-CH_2-\overset{\overset{\displaystyle H}{|}}{C}=CH_2$$

$-H^+$ ↑ 1-Butene
(26%)

n-Butyl chloride $\overset{\displaystyle Cl^-}{\longleftarrow}$ (5%) $H-\overset{\overset{\displaystyle H}{|}}{\underset{\underset{\displaystyle H}{|}}{C}}-\overset{\overset{\displaystyle H}{|}}{\underset{\underset{\displaystyle H}{|}}{C}}-\overset{\overset{\displaystyle \textcircled{H}}{|}}{\underset{\underset{\displaystyle H}{|}}{C}}-\overset{\overset{\displaystyle H}{|}}{\underset{\underset{\displaystyle H}{|}}{C}}\oplus$ $\overset{\displaystyle H_2O,\ -H^+}{\longrightarrow}$ n-Butyl alcohol (25%)

⇅

sec-Butyl chloride $\overset{\displaystyle Cl^-}{\longleftarrow}$ (3%) $CH_3-\overset{\overset{\displaystyle H}{|}}{\underset{\underset{\displaystyle \textcircled{H}}{|}}{C}}-\overset{\overset{\displaystyle H}{|}}{\underset{\underset{\displaystyle }{\oplus}}{C}}-CH_3$ $\overset{\displaystyle H_2O,\ -H^+}{\longrightarrow}$ sec-Butyl alcohol (13%)

$-H^+$ ↓

$$CH_3-CH=CH-CH_3$$
2-Butene
(10%)

Secondary amines react with nitrous acid to produce the neutral N-nitroso compounds. These are usually yellow oils. Many N-nitrosoamines are known to be carcinogenic.

$$\overset{\displaystyle R}{\underset{\displaystyle R}{>}}N-H + H-O-N=O \longrightarrow \overset{\displaystyle R}{\underset{\displaystyle R}{>}}N-N=O + H_2O$$

An N-nitroso
compound

The reaction of tertiary amines with nitrous acid is rather complex and of little practical importance. The initial product is the N-nitrosonium salt; however, attempts to isolate this product lead either to the recovery of the tertiary amine or to decomposition of the molecule.

$$R-\overset{\overset{\displaystyle R}{|}}{\underset{\underset{\displaystyle R}{|}}{N}}:\ +^+N=O \rightleftharpoons R-\overset{\overset{\displaystyle R}{|}}{\underset{\underset{\displaystyle R}{|}}{N}}\!\!\overset{+}{}-N=O$$

N-Nitrosonium ion
(unstable)

The reaction of aliphatic amines with nitrous acid has little value except as a diagnostic or as a separatory procedure. In either case, the primary amine is destroyed. On the other hand, the reaction of nitrous acid with aromatic primary amines produces intermediates known as diazonium salts.

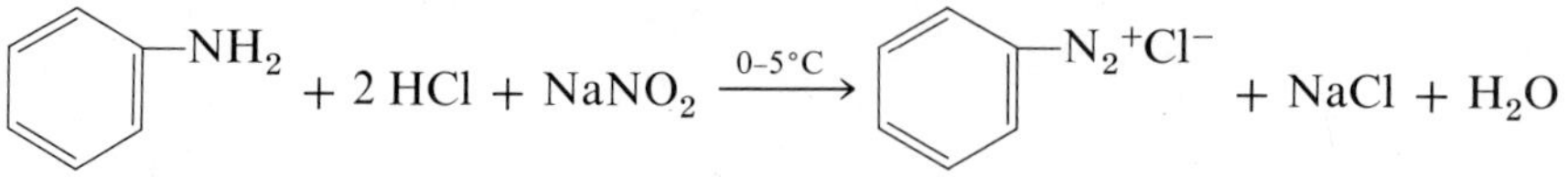

$$\text{C}_6\text{H}_5\text{-NH}_2 + 2\ \text{HCl} + \text{NaNO}_2 \xrightarrow{0\text{-}5°\text{C}} \text{C}_6\text{H}_5\text{-N}_2{}^+\text{Cl}^- + \text{NaCl} + \text{H}_2\text{O}$$

Benzenediazonium
chloride

A large number of very useful products, difficult, if not impossible to arrive at by any other route, are synthesized by way of a diazonium salt. We shall deal with the diazonium salts in a separate section.

EXERCISE 13.4

A primary aliphatic amine, when treated with nitrous acid, can form an olefin or an alcohol, whereas aniline under the same conditions yields only phenol. Explain.

E. Ring Substitution in Aromatic Amines.   The primary amino group of aniline, like the hydroxyl group of a phenol, is very sensitive to oxidation. In order to preserve the amino group in the presence of an oxidizing reagent it must be "protected" prior to reaction. The amino group usually is protected by acetylation. For example, the nitration of aniline to *p*-nitroaniline requires this preliminary step.

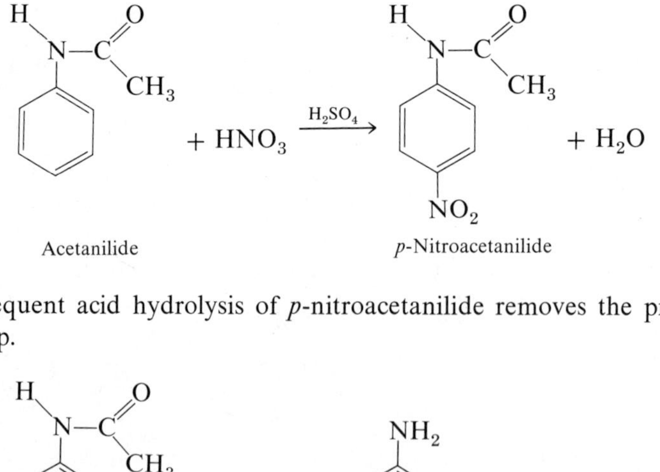

Acetanilide $+ \text{HNO}_3 \xrightarrow{\text{H}_2\text{SO}_4}$ *p*-Nitroacetanilide $+ \text{H}_2\text{O}$

Subsequent acid hydrolysis of *p*-nitroacetanilide removes the protecting acyl group.

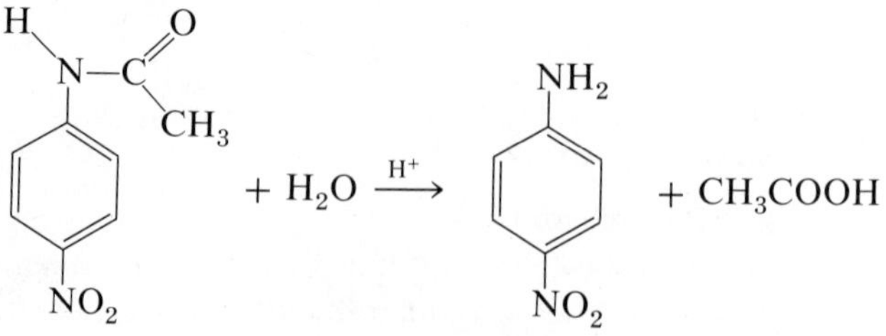

$+ \text{H}_2\text{O} \xrightarrow{\text{H}^+}$ $+ \text{CH}_3\text{COOH}$

p-Nitroaniline

EXERCISE 13.5

Explain why the nitration of "unprotected" aniline with a nitrating mixture of nitric and sulfuric acids gives *m*-nitroaniline. (*Clue: See Table 4.2.*)

In all reactions involving ring substitution the powerful *ortho-para* directive influence of the amino group is manifest. The tribromo derivative of aniline, for example, may be prepared simply by shaking aniline with a solution of bromine in water.

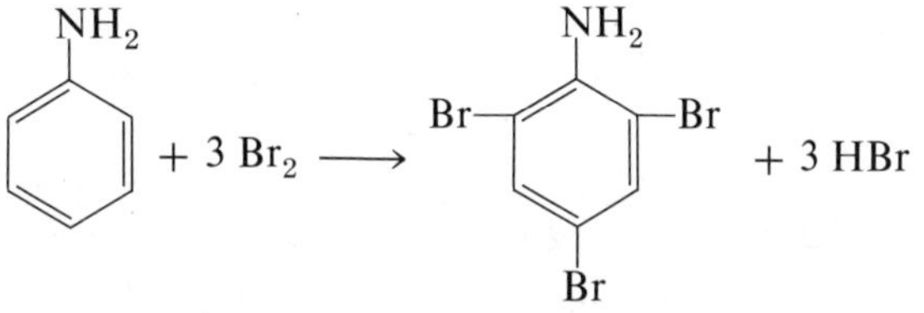

2,4,6-Tribromoaniline

Acetylation of aniline, besides protecting the amino group from oxidation, as explained under nitration, offers the additional advantage of diminishing the reactivity of aniline. For example, monobromoaniline is prepared by brominating acetanilide, followed by removal of the protecting group.

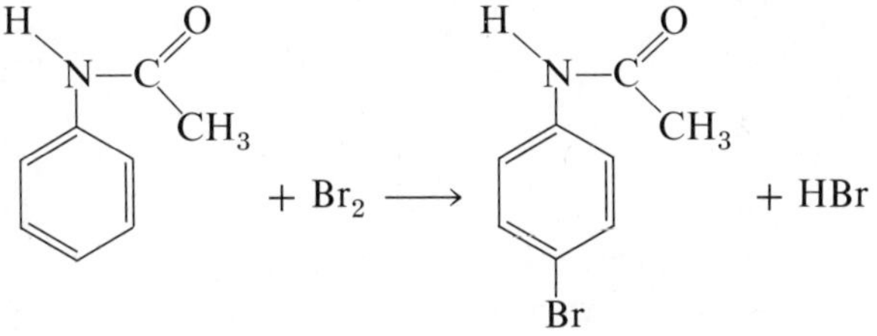

p-Bromoacetanilide

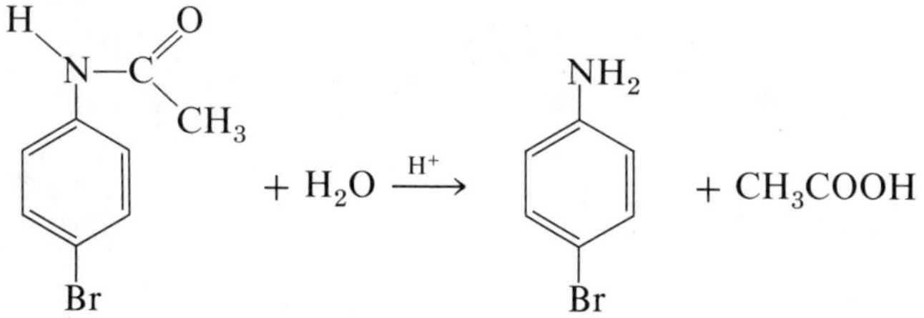

p-Bromoaniline

Sulfonation of aniline is a slow reaction in which the aniline hydrogen sulfate salt initially formed appears to undergo rearrangement upon further heating. The product is called sulfanilic acid.

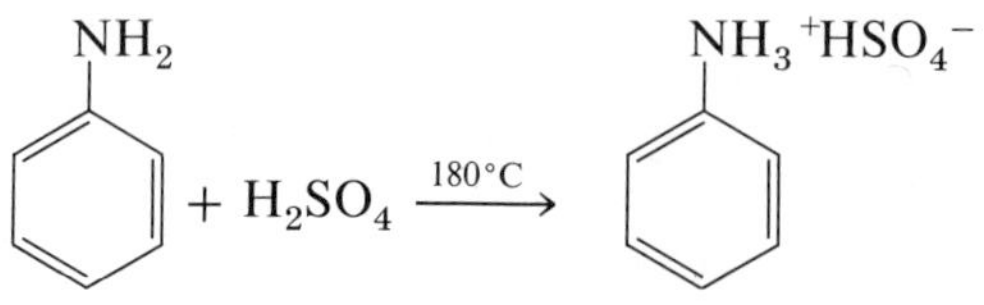

$$\text{C}_6\text{H}_5\text{-NH}_2 + \text{H}_2\text{SO}_4 \xrightarrow{180°C} \text{C}_6\text{H}_5\text{-NH}_3{}^+\text{HSO}_4{}^-$$

Aniline hydrogen sulfate

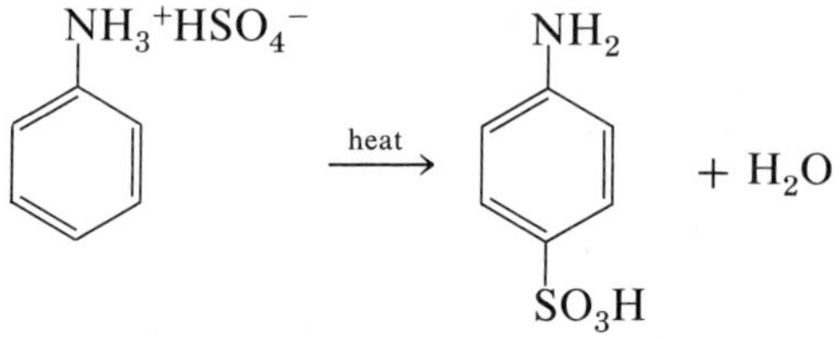

$$\text{C}_6\text{H}_5\text{-NH}_3{}^+\text{HSO}_4{}^- \xrightarrow{\text{heat}} \text{NH}_2\text{-C}_6\text{H}_4\text{-SO}_3\text{H} + \text{H}_2\text{O}$$

Sulfanilic acid

Sulfanilic acid appears to exist largely as an internal salt or a dipolar ion. Inner salts are possible when one group within a molecule is capable of acting as a proton donor (acid) to another within the same molecule capable of acting as a proton acceptor (base). The same phenomenon is shown by the amino acids (Sec. 15.4).

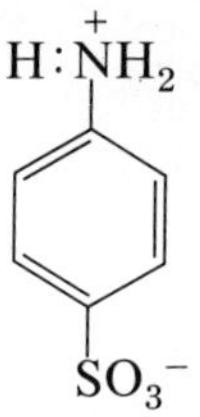

$$\overset{+}{\text{H}:\text{NH}_2}\text{-C}_6\text{H}_4\text{-SO}_3{}^-$$

Inner salt of sulfanilic acid

Sulfanilic acid is an important intermediate in the synthesis of certain dyes and in the preparation of the *sulfa drugs*. The sulfa drugs were widely used during World War II to prevent infection in wounds and still find some application in medicine. They now have been replaced to a large extent by the antibiotics (Sec. 17.24, 17.25). Sulfanilamide is effective internally against streptococci and staphylococci infections. Other sulfa drugs are derivatives of sulfanilamide. In these "sulfas" some other group appears as an N-substituent in the amide portion of the parent compound. The following sequence of reactions illustrates how sulfanilamide is synthesized.

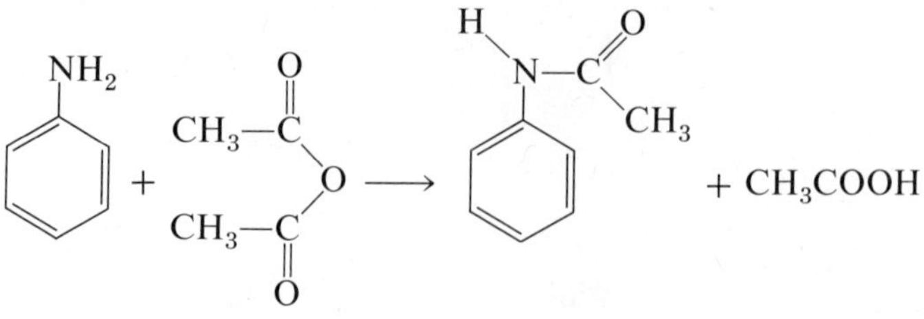

$$\text{Aniline} + (\text{CH}_3\text{CO})_2\text{O} \longrightarrow \text{Acetanilide} + \text{CH}_3\text{COOH}$$

Aniline   Acetic anhydride   Acetanilide

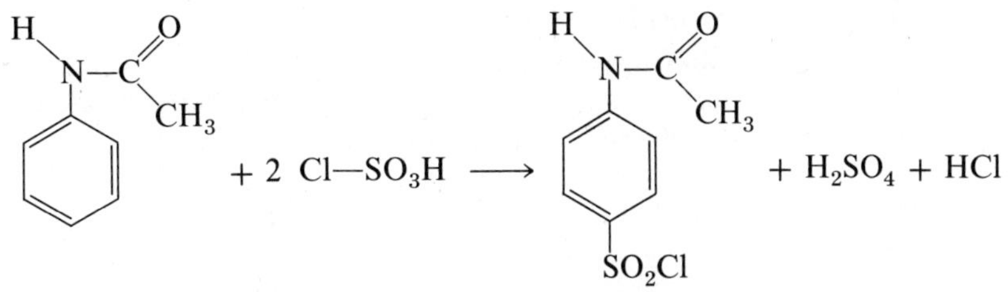

Chlorosulfonic
acid

p-Acetamido-
benzenesulfonyl
chloride

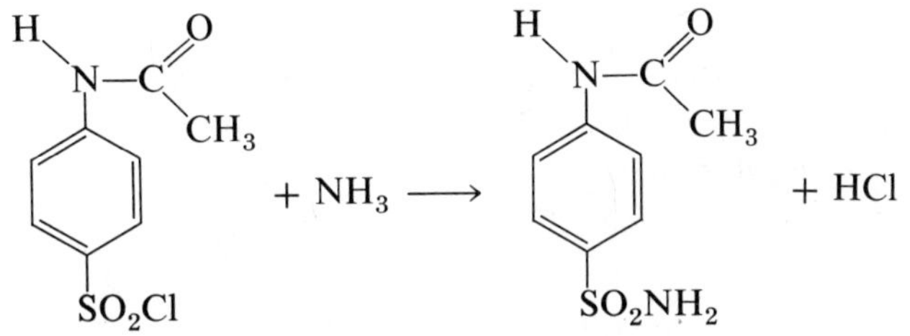

p-Acetamidobenzene-
sulfonamide

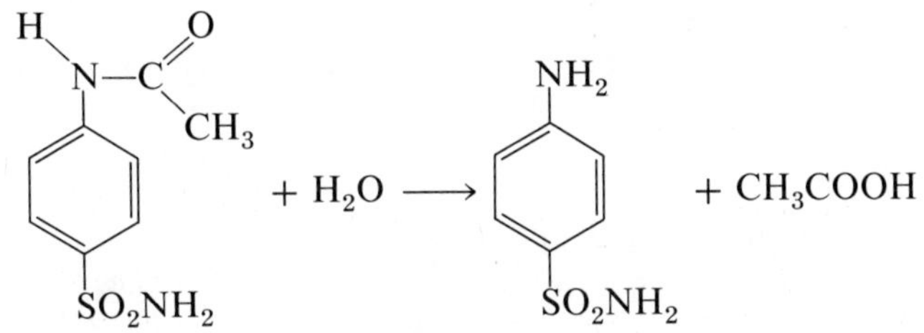

Sulfanilamide

13.5 *Diamines*

The diamines, compounds with two amino groups, can be prepared by methods similar to those employed for the preparation of the simple amines. The reaction of ammonia with dihalides, the reduction of dinitriles, and the Hofmann reaction of diamides all lead to the preparation of diamines. The primary requirement, in each case, is a bifunctional compound capable of being converted into one with two amino groups. Ethylene chloride, for example, reacts with ammonia to yield **ethylenediamine.**

$$ClCH_2CH_2Cl + 4\ NH_3 \xrightarrow{150°} H_2N—CH_2CH_2—NH_2 + 2\ NH_4Cl$$
Ethylenediamine

The ptomaines, tetramethylenediamine (putrescine) and pentamethylene-diamine (cadaverine) are natural, malodorous products formed by bacterial decomposition of protein material. The amino acids ornithine and lysine on breakdown yield these respective diamines. Ethylenediaminetetraacetic acid (EDTA), a useful chelating[3] agent for certain divalent metals, is made from ethylenediamine and sodium chloroacetate.

$$4\,Cl{-}CH_2{-}COO^-Na^+ + H_2N{-}CH_2CH_2{-}NH_2 + 4\,NaOH \longrightarrow$$

$$\begin{matrix} Na^{+\,-}OOC{-}CH_2 & & CH_2{-}COO^-Na^+ \\ & N{-}CH_2CH_2{-}N & \\ Na^{+\,-}OOC{-}CH_2 & & CH_2{-}COO^-Na^+ \end{matrix} + 4\,NaCl + 4\,H_2O$$

Sodium ethylenediaminetetraacetate (EDTA)

Sodium EDTA has the ability to sequester calcium and other metal ions to form soluble chelates and has been used with some success in the treatment of sclerosing diseases. The manner in which EDTA is able to chelate divalent metals is shown.

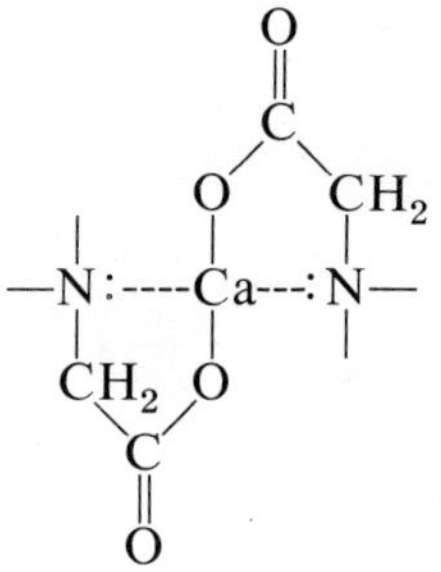

Hexamethylenediamine, used in the preparation of nylon, can be prepared from furfural (Sec. 17.2) by the following sequence.

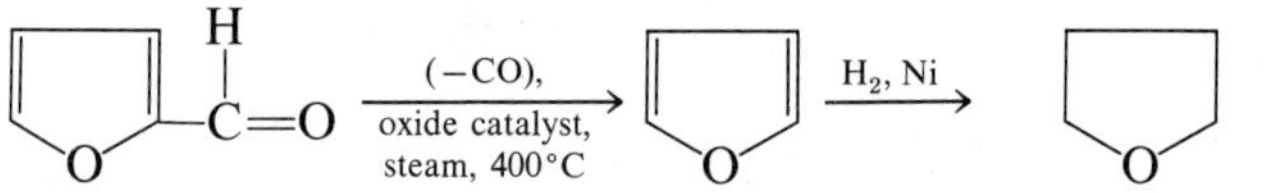

Furfural Furan Tetrahydrofuran

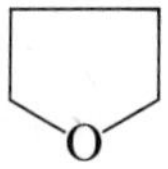 $+ HCl \longrightarrow HOCH_2CH_2CH_2CH_2Cl$

Tetramethylene chlorohydrin

$$HO(CH_2)_4{-}Cl + HCl \longrightarrow Cl(CH_2)_4Cl + H_2O$$

Tetramethylene chloride

$$Cl{-}(CH_2)_4{-}Cl + 2\,NaCN \longrightarrow NC{-}(CH_2)_4{-}CN + 2\,NaCl$$

Adiponitrile

---

[3] A chelating agent (Gr., *chele,* claw) is a chemical compound capable of grasping a metallic ion in a clawlike manner. A ring closure results through a coordination of unshared pairs of electrons on the chelating agent with the metallic ion.

The reduction of adiponitrile produces hexamethylenediamine.

$$NC-(CH_2)_4-CN + 4 H_2 \xrightarrow{\text{Ni}} H_2N-(CH_2)_6-NH_2$$

Hexamethylenediamine

Adiponitrile on hydrolysis yields adipic acid. The condensation reaction between adipic acid and hexamethylenediamine produces the polyamide known as Nylon 6–6.

The aromatic diamines may be prepared by the reduction of the appropriate nitro compounds.

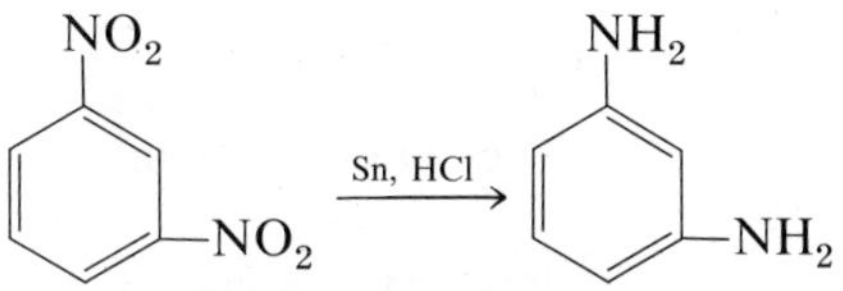

Benzidine, or p,p′-diaminobiphenyl, an important intermediate in the synthesis of dyes, is prepared *via* the rearrangement of hydrazobenzene. This reaction is known as the **benzidine rearrangement.** Unfortunately, benzidine is now thought to be carcinogenic, and replacements for it must be sought.

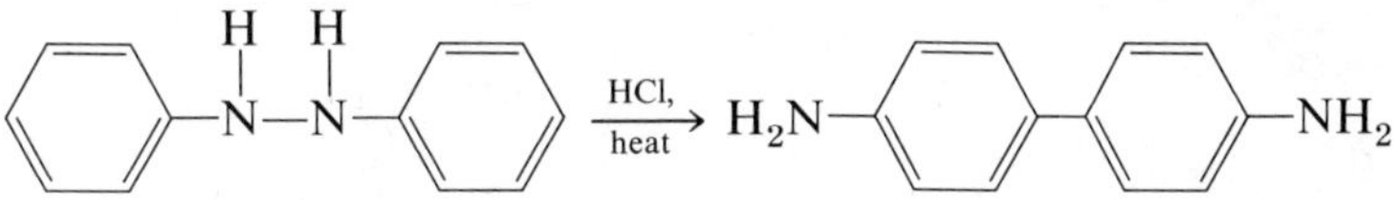

13.6 *Nitrogen Compounds Used as Dietary Sugar Substitutes*

Certain individuals (diabetics) are incapable of producing sufficient insulin for the complete utilization of glucose (Sec. 14.2). Others, through an abnormal hormonal activity, overproduce glucose in the body. In either case, an abnormal amount of this sugar accumulates in the blood and results in a condition known as *hyperglycemia*. For these persons and for those who are overweight and wish to lower their caloric intake, a number of artificial sweeteners are available. These substances are nonnutritive agents with a sweetness many times that of sucrose. *Sucaryl,* the calcium salt of cyclohexylsulfamic acid,

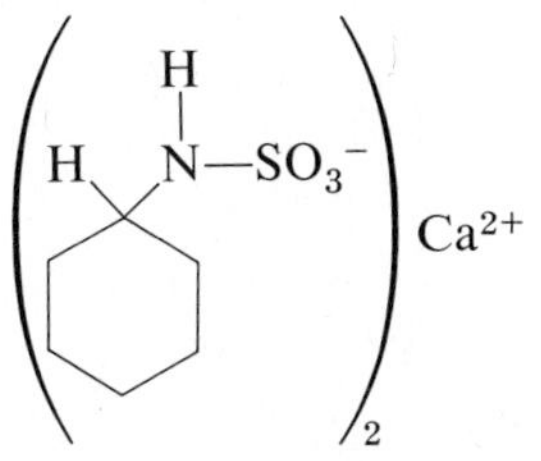

Sucaryl calcium
(Calcium cyclohexylsulfamate or "Cyclamate")

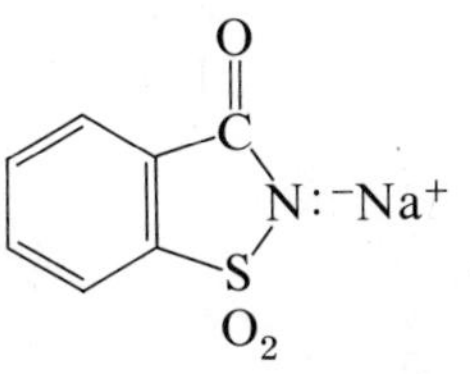

Saccharin sodium
(Sodium benzosulfimide)

and *saccharin,* the imide of *o*-sulfobenzoic acid, are two sweetening agents that have been widely used. However, sucaryl was banned as a food additive in 1969 by the federal government because of the possibility that it might be carcinogenic. The banning of saccharin in 1977 for the same reason led to consumer protests, and the ultimate fate of saccharin had not been resolved at this writing. Saccharin has a sweetness of approximately 500 times that of sucrose, which is assigned a sweetness of 1.0. The sweetness of sucaryl is approximately 30 when measured on the same scale.

13.7   *Amino Alcohols and Some Related Physiologically Active Compounds*

The aminoalcohols appear in a number of compounds of potent physiological activity.

Choline, $(CH_3)_3\overset{+}{N}CH_2CH_2OHOH^-$, trimethyl-$\beta$-hydroxyethyl ammonium hydroxide, forms part of the structure of lecithin (Sec. 11.6). Choline appears in lecithin with glycerol as a mixed ester of phosphoric acid and fatty acids. Acetylated choline, or *acetylcholine,* plays a vital role in the generation and conduction of nerve impulses in the body.

$$\left[(CH_3)_3 \overset{+}{N}-CH_2CH_2-O-\overset{\displaystyle O}{\overset{\|}{C}}-CH_3 \right] OH^-$$

Acetylcholine

The ability of certain local anesthetics to mitigate pain appears to be due (in part at least) to their structural similarity to choline and an ability to replace acetylcholine for a short duration. A number of local anesthetizing agents have in their molecular architecture an aminoalcohol grouping in the form of either an ester or an ether unit. The local anesthetic *procaine,* to which most of us have been introduced while in a dentist's chair, is a synthetic replacement for the natural drug *cocaine* (Sec. 17.7).

$$\begin{matrix} C_2H_5 \\ \\ C_2H_5 \end{matrix} \overset{+}{\underset{H}{N}}-CH_2CH_2-O-\overset{\displaystyle O}{\overset{\|}{C}}-\!\!\!\!\bigcirc\!\!\!\!-NH_2 \; Cl^-$$

Procaine hydrochloride
(Novocaine)

A large number of amines and amino alcohols, both naturally occurring and synthetic, have an effect upon the sympathetic nervous system and are known as **sympathomimetic agents.** These agents are powerful stimulants and dangerous if used promiscuously. Some may result in habituation when used for prolonged periods. The structures for a few of these are given on the next page.

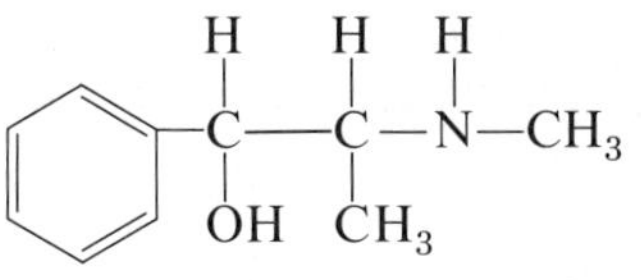

Ephedrine

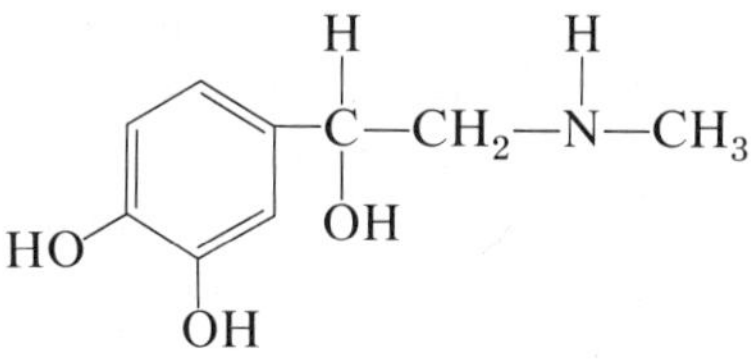

Epinephrine (Adrenalin)
(1-(3,4-Dihydroxyphenyl)-
2-methylaminoethanol))

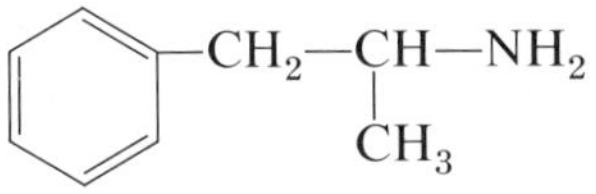

Benzedrine
(β-Aminopropylbenzene)

13.8 *Diazonium Salts*

As we have learned, the action of nitrous acid on primary aliphatic amines is of little importance as a preparative reaction. The action of nitrous acid on primary aromatic amines, on the other hand, has a number of applications, which lead to many useful products. The acid salt of an aromatic primary amine, when treated with a cold aqueous solution of sodium nitrite, does not liberate nitrogen as would an aliphatic primary amine. Instead, a water-soluble **diazonium salt** is produced. This reaction is called **diazotization** and is an easy one to carry out.

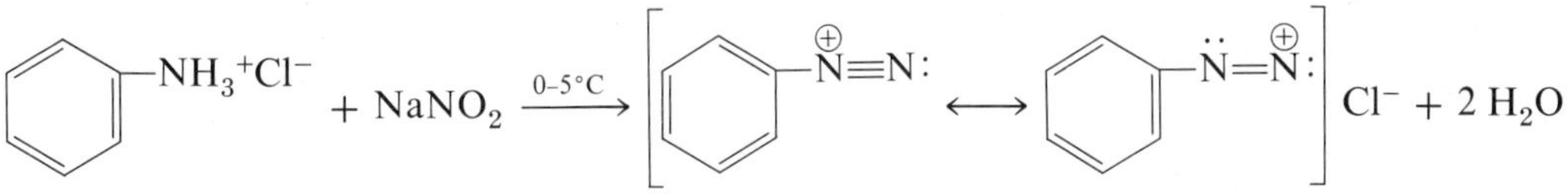

Aniline
hydrochloride

Benzenediazonium chloride

The benzenediazonium ion usually is written as

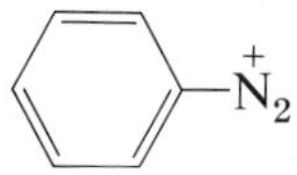

Although we have illustrated its formation from the hydrochloric acid salt of aniline, the sulfuric acid salt serves equally well, but produces the diazonium salt as the sulfate. The diazonium salt, once formed, is not isolated but instead is treated with the reagent required to give the product desired. The reactions of the diazonium salts may be divided into three principal types. In one type of reaction a displacement of the diazonium group by some other group takes place with an accompanying loss of nitrogen. In the second type, a coupling reaction takes place in which the nitrogen atoms are retained as an azo grouping, —N=N—, to become part of a new molecular structure. In the third type of reaction, a reduction, either partial or complete, of the diazonium group occurs. Reactions of each type are illustrated.

A. Displacement Reactions of Diazonium Salts. The following examples illustrate the usefulness of diazonium salts as reaction intermediates, where the diazonium group is replaced by some other group.

(1) 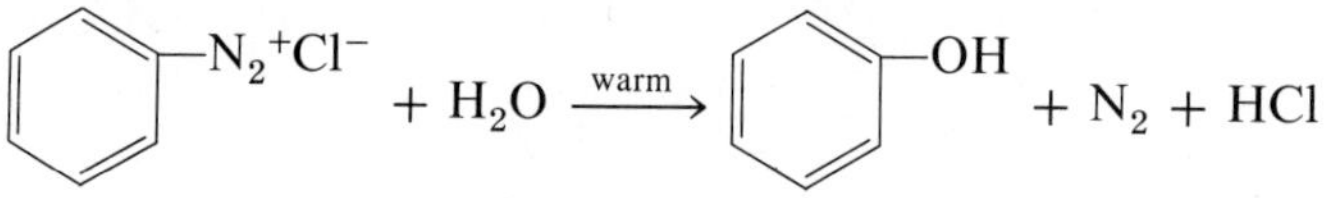 benzene—$N_2^+Cl^-$ + H_2O $\xrightarrow{\text{warm}}$ benzene—OH + N_2 + HCl

Phenol

(2)
(3) benzene—$N_2^+Cl^-$ — $\begin{pmatrix} CuCl \\ CuBr \\ CuCN \end{pmatrix}$ → benzene—Cl or (—Br) or (—CN) + N_2
(4)

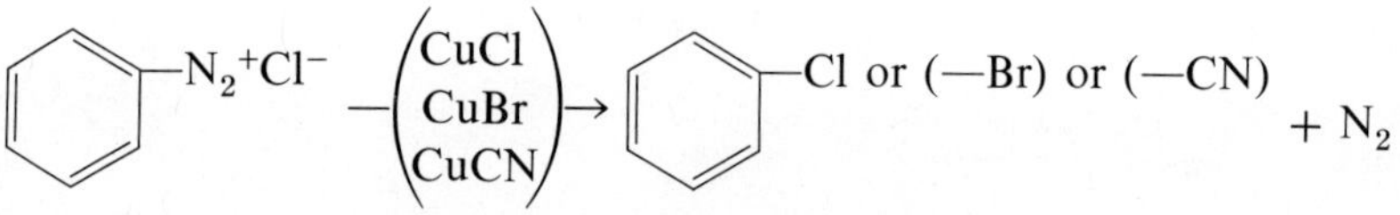

Chlorobenzene

(5) 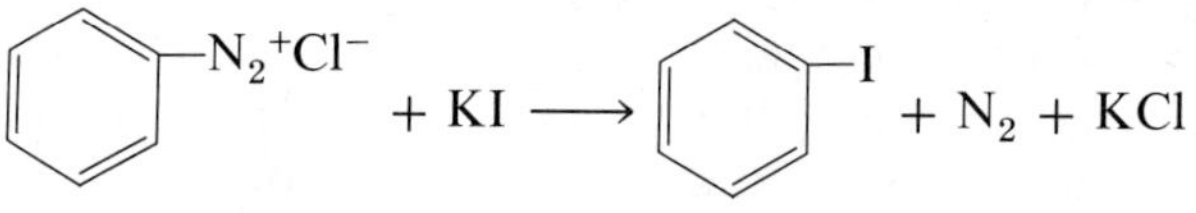 benzene—$N_2^+Cl^-$ + KI $\longrightarrow$ benzene—I + N_2 + KCl

Iodobenzene

(6) 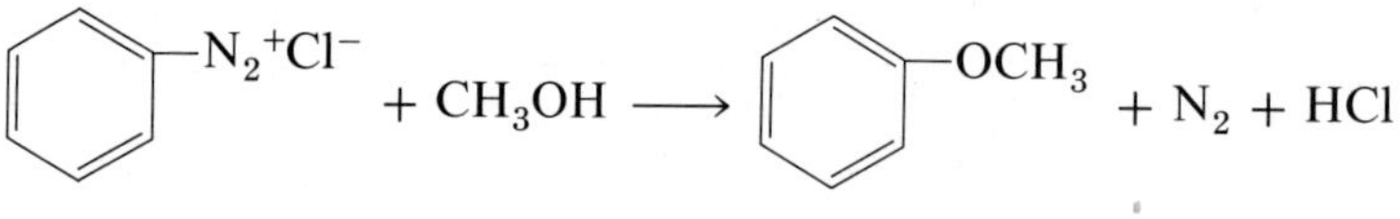 benzene—$N_2^+Cl^-$ + CH_3OH $\longrightarrow$ benzene—OCH_3 + N_2 + HCl

Methyl phenyl ether
(Anisole)

(7) 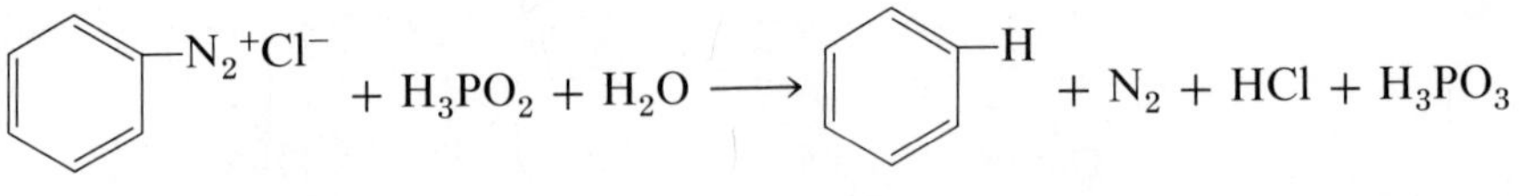 benzene—$N_2^+Cl^-$ + H_3PO_2 + H_2O $\longrightarrow$ benzene—H + N_2 + HCl + H_3PO_3

Benzene

The reactions of a benzenediazonium compound with cuprous salts, reactions (2), (3), and (4) above, usually are referred to as **Sandmeyer reactions.**

B. Coupling Reactions of Diazonium Salts. Diazonium salts undergo coupling reactions readily with phenols and aromatic amines. Since the coupling involves an attack of the positive diazonium ion on a center of high electron density in the benzene ring, the presence of substituents that increase the electron density of the ring enhances the coupling reaction. Conversely, electron-withdrawing substituents on the ring inhibit the coupling reaction or prevent it entirely. Coupling takes place preferably in the position *para* to the activating group. Should the *para* position be occupied, coupling then occurs at one of the *ortho* positions. The coupling reaction is carried out in a neutral, an alkaline, or a weakly acidic solution.

Aniline   Diazoaminobenzene

The diazoaminobenzene initially formed when aniline couples with the benzenediazonium ion rearranges on heating to *p*-aminoazobenzene.

p-Aminoazobenzene

Coupling will not take place with aniline in a strongly acid solution because in a strong acid environment the primary amino group of aniline becomes a positive anilinium ion which deactivates the ring.

Ring active   Ring deactivated
(couples   (does not
readily)   couple)

Coupling with a phenol is inhibited in a strongly acid solution because the strong activating influence of the phenoxide ion is lacking. Phenols, you will recall, are acidic (Sec. 8.3) and for this reason would couple with the benzenediazonium ion most readily in a solution slightly alkaline.

Weakly active   Strongly active
(couples   (couples
slowly)   readily)

When there is a choice of ring positions open to the benzenediazonium ion, the group on the ring with the stronger directive influence prevails.

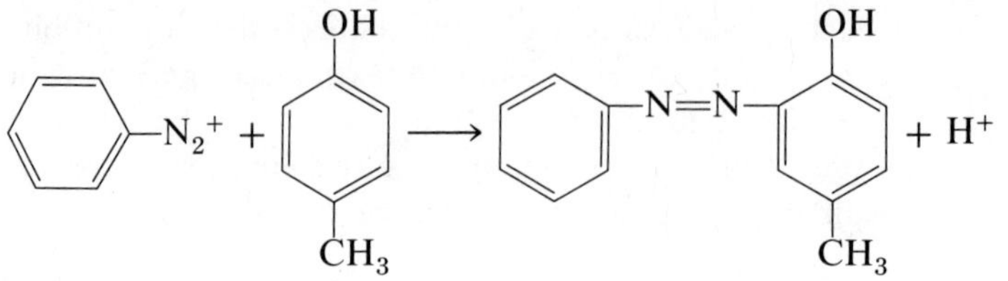

p-Cresol   Benzeneazo-*p*-cresol

The azo compounds are highly colored substances. The azo grouping, for this reason, is the structural feature in a wide variety of dyes called **azo dyes** (Sec. 19.3-B).

EXERCISE 13.6

Anisole, ⟨ ⟩—OCH$_3$, is not sufficiently active to couple with benzenediazo-nium chloride but will couple with 2,4-dinitrobenzenediazonium chloride. Explain.

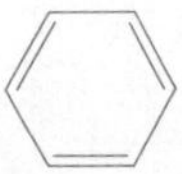

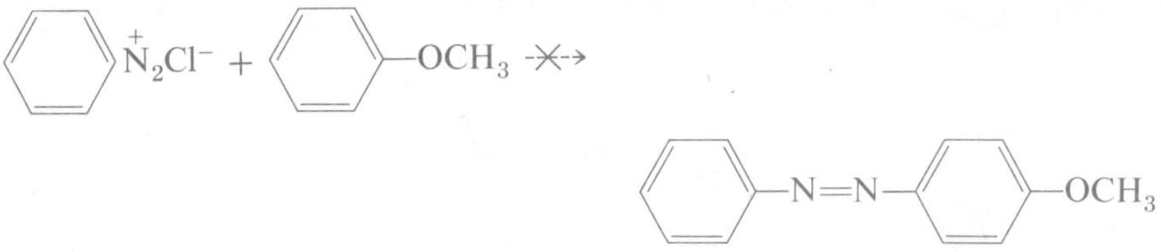

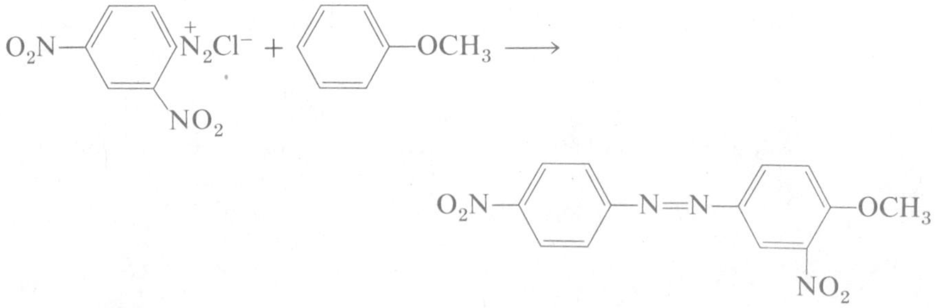

C. Reduction of Diazonium Compounds. The partial reduction of a diazo-nium salt can be accomplished without the loss of nitrogen by the use of sodium sulfite as the reducing agent. The product of such a reduction is phenylhydra-zine, a reagent useful in the identification of sugars (Sec. 14.8) and carbonyl compounds (Sec. 9.7).

$$\text{(benzene with } N_2^+Cl^-) + 2\,Na_2SO_3 + 2\,H_2O \longrightarrow \text{(benzene with } N{-}NH_2 \cdot HCl, \; H\text{ on N)} + 2\,Na_2SO_4$$

Phenylhydrazine
hydrochloride

A complete reduction of a diazonium salt that results in its replacement by hydrogen is called **deamination.** Hypophosphorous acid, H_3PO_2, is an excellent reagent for this reaction (Sec. 13.8, Eq. A-7).

13.9 *Organic Sulfur Compounds*

The divalent sulfur compounds can be regarded as the sulfur analogs of the corresponding oxygen compounds. This relationship is shown in the parallelism of their nomenclature and, to a certain extent, their chemistry. The rules of nomenclature for the thiols or mercaptans (R—SH), sulfides or thioethers (R—S—R), and disulfides (R—S—S—R) will be apparent from the examples given. Note that the prefix *thio-* indicates that sulfur has replaced oxygen in an organic compound and that *alkylthio-* is the sulfur equivalent of *alkoxy-*. The —SH group is often called the *sulfhydryl* group in biochemistry.

$CH_3{-}CH_2{-}SH$	$HS{-}CH_2{-}CH_2{-}SH$	$CH_3{-}CH{=}CH{-}CH_2{-}SH$
Ethanethiol (Ethyl mercaptan)	1,2-Ethanedithiol	2-Butene-1-thiol

$CH_3CH_2{-}S{-}CH_2CH_3$	(cyclohexyl)—S—CH$_3$	(benzene ring)—SH
Diethyl sulfide ((Ethylthio)ethane)	Cyclohexyl methyl sulfide ((Methylthio)cyclohexane)	Thiophenol

$CH_3CH_2{-}S{-}S{-}CH_2CH_3$	$HS{-}CH_2{-}CH_2{-}OH$
Diethyl disulfide ((Ethyldithio)ethane)	2-Mercaptoethanol (β-Mercaptoethyl alcohol)

The thiols are not satisfactorily prepared by the S_N2 reaction of sodium hydrosulfide (NaSH) with an alkyl halide but may be prepared by the alkylation of the highly nucleophilic thiourea followed by basic hydrolysis.

$$\underset{\text{Thiourea}}{\overset{H_2N}{\underset{H_2N}{\diagup}}\!\!C{=}S} + R{-}Br \longrightarrow \underset{\substack{\text{S-Alkylisothiouronium} \\ \text{bromide}}}{\overset{H_2N^+}{\underset{H_2N}{\diagup}}\!\!C{-}S{-}R\ Br^-} \xrightarrow[\text{2. } H^+]{\text{1. } H_2O + NaOH} \underset{\text{Thiol}}{R{-}SH} + \underset{\text{Polymer}}{(NH_2CN)_x}$$

An alternative route is the treatment of an alkyl or aryl Grignard reagent with sulfur.

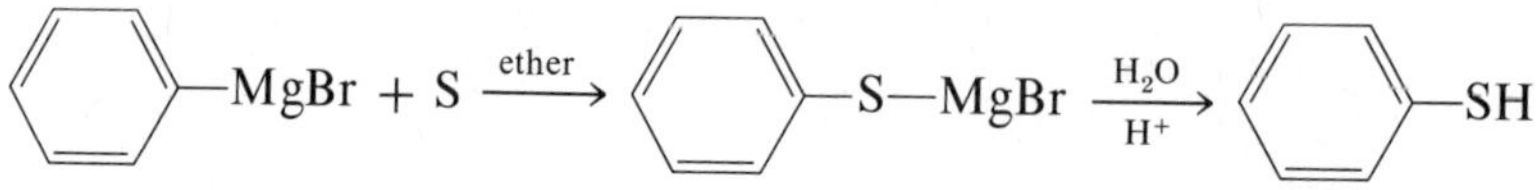

Phenylmagnesium bromide   Thiophenol

Sulfides are prepared by a variation of the Williamson synthesis.

$$CH_3CH_2\!-\!SH + NaOH \longrightarrow CH_3CH_2\!-\!S^-Na^+ + H_2O$$
Ethanethiol

$$CH_3CH_2\!-\!S^-Na^+ + CH_3\!-\!Br \longrightarrow CH_3CH_2\!-\!S\!-\!CH_3 + NaBr$$
Ethyl methyl sulfide

The greater acidity of ethanethiol (pK_a 10.5) relative to that of ethanol (pK_a 15.9) permits the preparation of the alkylthio (mercaptide) anion by treatment with aqueous sodium hydroxide rather than with metallic sodium.

The thiols and the disulfides form a redox system that is extremely important in biological systems. In the laboratory, thiols are readily oxidized to disulfides with *mild* oxidizing agents such as iodine, and disulfides are reduced to thiols with reagents such as lithiumaluminum hydride or lithium metal in ammonia.

$$2\,R\!-\!SH \underset{LiAlH_4}{\overset{I_2}{\rightleftarrows}} R\!-\!S\!-\!S\!-\!R$$
Thiol   Disulfide

In nature the disulfide bond serves to bind together different parts of the same protein chain or two (or more) different protein chains (Sec. 15.7) and is involved in the functions of the coenzyme, lipoic acid.

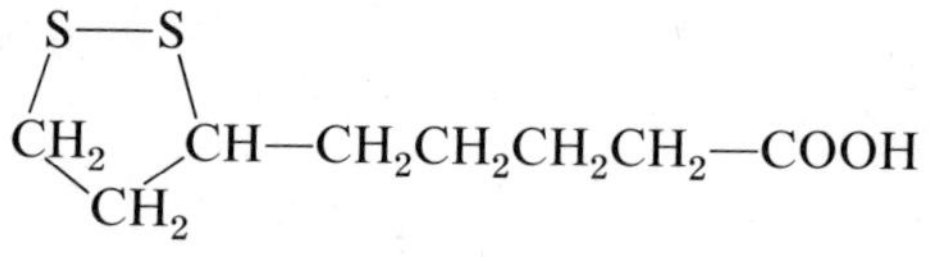

Lipoic acid

Summary

1. **Structure**

   Amines may be considered as ammonia derivatives. Amines are classified as primary, secondary or tertiary; also, as aliphatic or aromatic.

2. **Nomenclature**

   The amines usually are named by naming the alkyl or aryl groups attached to the nitrogen of the amino group, followed by the word "amine."

3. **Physical Properties**

   Amines are weakly basic compounds. The low molecular weight members (C_1-C_2) are gases and low boiling liquids. They are soluble in acids. Nearly all are foul-smelling.

4. **Preparation**

   The amines may be prepared by

      (a) the direct alkylation of ammonia

      (b) the reduction of nitriles ($-C\equiv N$), amides $\left(-\overset{\displaystyle O}{\underset{}{C}}-NH_2\right)$, oximes ($-C=NOH$), and nitro ($-NO_2$) compounds

      (c) Hofmann's hypobromite degradation of amides

      (d) the Gabriel phthalimide synthesis

5. Amines react with each of the following reagents to give the product indicated.

Amine	*Reactant*	*Product*
(a) RNH_2	+ acids	$\longrightarrow$ salts
(b) RNH_2	+ acyl halides or (acid anhydrides)	$\longrightarrow$ amides
(c) RNH_2	 + ⬡—SO_2Cl	$\longrightarrow$ alkali-soluble benzenesulfonamide
(d) R_2NH	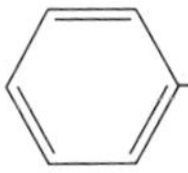 + ⬡—SO_2Cl	$\longrightarrow$ alkali-insoluble benzenesulfonamide
(e) RNH_2	+ HNO_2	$\longrightarrow$ N_2 + alcohols + olefins
(f) R_2NH	+ HNO_2	$\longrightarrow$ N-nitroso compounds
 (g) ⬡—NH_2	+ HNO_2	$\xrightarrow[0\text{-}5°]{HCl}$ diazonium salts
 (h) ⬡—NH_2	+ Br_2	$\longrightarrow$ *ortho-para* substitution products
(i) $R-CH_2CH_2NH_2 + CH_3I$		$\longrightarrow R-CH_2CH_2\overset{+}{N}(CH_3)_3X^-$
		$\xrightarrow[\text{heat}]{Ag_2O} R-CH=CH_2 + (CH_3)_3N$

6. Diamines may be prepared by the same general methods used to prepare the simple amines.

7. Amino ethanols can be prepared from ethylene oxide and ammonia. Physiologically active compounds frequently contain an aminoalcohol unit.

8. Diazonium salts formed from aromatic primary amines are useful intermediates in organic synthesis. The diazonium group may be
   (a) replaced by: halogen, cyanide, hydroxyl, and alkoxyl
   (b) coupled to: other aromatic rings
   (c) reduced to: phenylhydrazine, benzene (deamination)

9. Divalent sulfur compounds may be considered to be sulfur analogs of the alcohols, ethers, and peroxides. IUPAC nomenclature uses the suffix thiol to name the —SH group of thio alcohols (mercaptans). Sulfide and disulfide nomenclature parallels that of the ethers, the prefix alkylthio corresponding to alkoxy.

10. Preparation of sulfur compounds
    (a) Thiols are prepared from alkyl halides by successive treatment with thiourea and aqueous base or by treatment of a Grignard reagent with sulfur.
    (b) Sulfides are prepared by a Williamson synthesis using an alkyl halide and an alkali metal mercaptide.
    (c) Disulfides are prepared by oxidation of thiols with iodine.

11. The thiols and disulfides form a redox system that is important in biological systems.

New Terms

acetylcholine	Gabriel phthalimide synthesis
chelating agent	Hinsberg test
deamination	Hofmann amide degradation
diazonium salt	inner salts
diazotization	sympathomimetic agents

Supplementary Exercises

EXERCISE 13.7 Name the following compounds. If amines, classify each as primary, secondary, or tertiary.

(a) $CH_3-\underset{\underset{CH_3}{|}}{\overset{\overset{CH_3}{|}}{C}}-NH_2$

(b) $CH_3-\underset{\overset{|}{N}}{\overset{\overset{CH_3}{|}}{}}-CH_3$

(c) [naphthalene with NH_2 substituent]

(d) $HOCH_2-CH_2-N\underset{CH_3}{\overset{CH_3}{<}}$

(e) [benzene ring with $H_3C\underset{N}{\diagdown}\ \diagup CH_3$]

(f) [cyclohexane ring with H and NH_2]

(g) $H_2N-CH_2-CH_2-CH_2-CH_2-CH_2-NH_2$

(h) [benzene ring with $-\overset{\overset{H}{|}}{N}-NH_2$]

(i) [benzene ring with $-N=C=O$]

(j) $CH_3-CH_2-\underset{\underset{SH}{|}}{CH}-CH_3$

(k) $(CH_2=CH-CH_2)_2S$

EXERCISE 13.8 Write structures for the following.
    (a) aniline
    (b) benzenediazonium sulfate
    (c) p-aminobenzenesulfonic acid (sulfanilic acid)
    (d) tetramethylammonium chloride
    (e) acetanilide
    (f) β-diethylaminoethyl-p-aminobenzoate hydrochloride (Procaine)
    (g) p-nitrosodimethylaniline
    (h) N-ethylphthalimide
    (i) ethyldimethylamine

EXERCISE 13.9 Without consulting a table of ionization constants, arrange the following compounds in an order of diminishing basic strength.
    (a) aniline (b) ammonia (c) dimethylamine (d) ethylamine (e) benzamide

EXERCISE 13.10 Taking advantage of acidic, basic, and other properties, outline a procedure for separating a mixture that includes (a) aniline, (b) benzoic acid, (c) acetophenone, and (d) bromobenzene.

EXERCISE 13.11 What reagents and conditions are required to accomplish each of the following conversions? (Some require more than one step.)

(a) $\underset{CH_3}{\overset{CH_3}{\diagdown}} C=O \longrightarrow \underset{CH_3}{\overset{CH_3}{\diagdown}} \underset{}{\overset{H}{\diagup}} C-NH_2$

(b) $CH_3-CH_2-I \longrightarrow CH_3-CH_2-CH_2-NH_2$

(c) $CH_3-\overset{O}{\overset{\|}{C}}-NH_2 \longrightarrow CH_3-NH_2$

(d) $CH_3-CH_2-CH_2-CH_2-NH_2 \longrightarrow (CH_3)_3N + CH_3-CH_2-CH=CH_2$

(e) ⟨benzene⟩$-NO_2 \longrightarrow$ ⟨benzene⟩$-NH_2$

(f) ⟨benzene⟩$-NH_2 \longrightarrow$ ⟨benzene⟩$-\overset{H}{\overset{|}{N}}-\overset{O}{\overset{\|}{C}}-CH_3$

(g) ⟨benzene⟩$-NH_2 \longrightarrow$ ⟨benzene⟩$-N_2{}^+Cl^-$

(h) ⟨benzene⟩$-N_2{}^+Cl^- \longrightarrow$ ⟨benzene⟩$-CN$

(i) ⟨benzene⟩$-CH_3 \longrightarrow H_2N-$⟨benzene⟩$\overset{NH_2}{\underset{-CH_3}{|}}$

(j) ⟨benzene⟩$-NH_2 \longrightarrow$ ⟨benzene⟩$-N=N-$⟨benzene⟩$-NH_2$

(k) $CH_3-\underset{CH_3}{\overset{CH_3}{\overset{|}{\underset{|}{C}}}}-Br \longrightarrow CH_3-\underset{CH_3}{\overset{CH_3}{\overset{|}{\underset{|}{C}}}}-SH$

EXERCISE 13.12 What simple test tube reactions would serve to distinguish between:
  (a) *n*-butylamine and triethylamine
  (b) aniline hydrochloride and benzamide
  (c) *sec*-butyl alcohol and *sec*-butylamine
  (d) aniline and benzylamine
  (e) 1-butanol and 1-butanethiol

EXERCISE 13.13 When 1.83 g of an unknown amine was treated with nitrous acid, the evolved nitrogen, corrected to standard temperature and pressure, measured 560 ml. The alcohol isolated from the reaction mixture gave a positive iodoform reaction. What is the structural formula of the unknown amine?

EXERCISE 13.14 Give the structures and names of the principal products expected from each of the following reactions.

(a) chlorobenzene + 15% NaOH $\xrightarrow[\text{200 atm.}]{\text{300°}}$

(b) *p*-chloroaniline + HCl, then NaNO$_2$ $\xrightarrow{\text{0-5°}}$

(c) product of (a) + product of (b) $\longrightarrow$

(d) 2,4-dinitrochlorobenzene + NH$_3$ $\xrightarrow{\text{warm}}$

(e) α-phenylpropionamide + NaOH + Br$_2$ $\longrightarrow$

EXERCISE 13.15 Suggest a method for converting *p*-toluic acid (*p*-methylbenzoic acid) into *p*-toluidine (*p*-methylaniline).

EXERCISE 13.16 Suggest a reaction mechanism that would account for the following transformation.

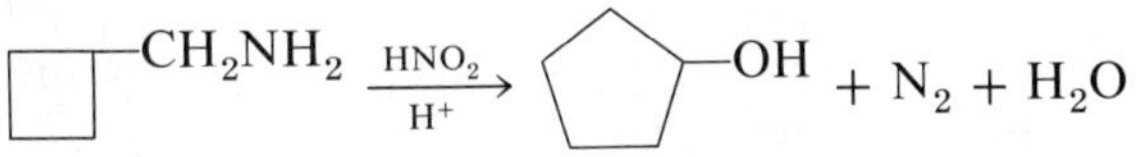

$$\square\text{--CH}_2\text{NH}_2 \xrightarrow[\text{H}^+]{\text{HNO}_2} \pentagon\text{--OH} + \text{N}_2 + \text{H}_2\text{O}$$

EXERCISE 13.17 Calculate the ratio [CH$_3$CH$_2$NH$_2$]/[CH$_3$CH$_2$NH$_3{}^+$] in water at pH 6, 8, 10, and 12. (*Hint:* Review footnote 2, Sec. 10.2.)

EXERCISE 13.18 Draw reasonable structures for the isomeric compounds A, B, and C, C$_{10}$H$_{13}$NO$_2$, if:

A {
(a) it is a white solid that is insoluble in cold, dilute acid or base;
(b) refluxing with an alkaline solution produces a gas with a strong ammoniacal odor;
(c) acidification of the hydrolysis mixture liberates an acid with a N.E. of 152 ± 1.
}

B {
(a) it is a white solid insoluble in cold, dilute acid or base;
(b) refluxing with sodium hydroxide solution produces an oily layer. The oily layer, when separated and purified, can be diazotized.
(c) acidification of the alkaline hydrolysis mixture, followed by extraction with ether, isolated a compound with a neutralization equivalent of 74 ± 1.
}

C {
(a) it is an oil insoluble in water, acid, or base;
(b) oxidation with sodium dichromate and concentrated sulfuric acid produced a solid compound, C$_8$H$_5$NO$_6$, that was soluble in hot water and in cold, dilute base;
(c) strong heating of the compound caused it to melt only.
}

EXERCISE 13.19 A white solid compound, $C_{13}H_{11}NO$, was insoluble in water but appeared to go into solution when heated with sodium hydroxide. Continued heating produced an oily layer which was acid-soluble. The mixture was cooled, and acidified with hydrochloric acid. This caused the precipitation of a white acid whose neutralization equivalent was found to be 122 ± 1. Draw and name the original compound.

EXERCISE 13.20 The reaction of sulfides with periodic acid (HIO_4) yields sulfoxides:

$$R—S—R' \xrightarrow{HIO_4} R—\overset{:\ddot{O}:^-}{\underset{+}{S}}—R'$$

Sulfide   Sulfoxide

If the two alkyl substituents on the sulfur atom are different, the resultant sulfoxide is found to be resolvable into a pair of enantiomeric sulfoxides (mirror image isomers). What kind of structure must the sulfoxides have to allow the existence of optical isomerism?

EXERCISE 13.21 Compound A, $C_8H_{19}N$, gave an nmr spectrum consisting of a 9-proton singlet at δ 1.00, a 6-proton singlet at δ 1.17, a 2-proton singlet at δ 1.28, and a 2-proton singlet at δ 1.42. When compound A was treated with D_2O, the peak at δ 1.42 was no longer present. Suggest a structure for compound A.

EXERCISE 13.22 Compound B, $C_6H_{15}N$, gave an nmr spectrum consisting of a 1-proton, rather broad singlet at δ 0.67, a 12-proton doublet at δ 1.00, and a 2-proton septet at δ 2.88. Suggest a structure for compound B.

EXERCISE 13.23 Compound C contained 63.01% carbon, 12.24% hydrogen, and 24.65% nitrogen (according to the report from the analyst). Its nmr spectrum consisted of two singlets at δ 2.29 and 2.43 with an integral ratio of 3:4 (the downfield peak being the larger). Compound C reacted with two equivalents of an alkyl halide to form a *bis*-quarternary salt (actually two moles of 2-bromoethanol were used, but the specific halide is not important in the solution of the problem). The *bis*-quarternary salt could be separated into two isomers, which were shown by X-ray crystallography to be *cis-trans* isomers. Suggest a structure for compound C and for its two *bis*-quarternary salts. You may use R—Br to represent the alkyl halide. (*Note:* When two or more identical complex functions (such as chloromethyl, or trimethylammonium) are present, the prefixes *bis-*, *tris-*, *tetrakis-*, etc. are used instead of *di-*, *tri*, *tetra*, etc.)

EXERCISE 13.24 The nmr spectrum of dimethylformamide (N,N-dimethylformamide) in $CDCl_3$ consists of a 3-proton singlet at δ 2.88, a 3-proton singlet at δ 2.97, and a 1-proton singlet at δ 8.02. The distance between the two 3-proton singlets increases by up to 1.7 ppm on replacement of the $CDCl_3$ by benzene. At temperatures above about 115° the two 3-proton singlets coalesce into a single, sharp 6-proton peak. Suggest an explanation for these observations. (*Hint:* Review Sec. 13.4-C.)

14

Carbohydrates

Introduction

The name **carbohydrate** originated from the French "hydrate de carbon." Early analyses of a number of naturally occurring compounds of this class gave empirical formulas in which the ratios of carbon, hydrogen, and oxygen appeared to indicate hydrates of carbon of the type $C_xH_{2y}O_y$ or $C_x(H_2O)_y$ where x and y may be the same or different. For example, the simple sugar **glucose** (grape sugar) has a molecular formula of $C_6H_{12}O_6$, and **sucrose** (cane sugar) the formula $C_{12}H_{22}O_{11}$, but such definite ratios of water to carbon are not to be found in every carbohydrate. Therefore the name carbohydrate, as often is the case in chemical nomenclature, is not descriptive but has persisted and probably will be with us for a long time.

The carbohydrates comprise a great class of natural substances which includes the sugars, starches, cellulose and related products. Carbohydrates may be described as polyhydroxy aldehydes or polyhydroxy ketones, or substances which, when hydrolyzed, give polyhydroxy aldehydes or polyhydroxy ketones.

14.1　*Classification and Nomenclature*

The carbohydrates may be subdivided conveniently into three principal classes: monosaccharides, oligosaccharides (Gr., *oligos,* a few), and polysaccharides.

Monosaccharides include all sugars that contain a single carbohydrate unit—that is, one incapable of producing a simpler carbohydrate on further hydrolysis. Most of the monosaccharides are five- and six-carbon structures. The five-carbon monosaccharides are called **pentoses** and those of six-carbons are called **hexoses.** The suffix *-ose* is a generic designation of any sugar. The monosaccharides frequently are referred to as simple sugars and are either polyhydroxy aldehydes or polyhydroxy ketones. If an aldehyde, the name **aldose** is applicable; if a ketone, the name **ketose** frequently is used to describe the sugar.

Oligosaccharides consist of two or more (but a relatively small number) monosaccharide units joined by acetal linkages between the aldehyde or ketone group of one simple sugar and a hydroxy group of another. This kind of coupling in sugar chemistry gives rise to what is called a **glycosidic** linkage. Hydrolysis of an oligosaccharide yields the simple sugar components. **Disaccharides** are composed of two simple sugar units, **trisaccharides** of three, etc.

Polysaccharides consist of hundreds or even thousands of monosaccharide units joined together through glycosidic linkages to form macro molecules, or polymers.

14.2　*Glucose: A Typical Monosaccharide*

Glucose (frequently called dextrose because of its dextrorotation) is the most important of the monosaccharides. Not only is it the most widely occurring sugar, but in free or combined form it is perhaps the most abundant of organic compounds. Glucose is the end product of the hydrolysis of starch and cellulose and is closely associated with the metabolic processes. Glucose is a main source of energy for all living organisms. It comprises 0.08–0.1% of the blood content of all normal mammals and is one of the few organic compounds which may be injected as a food directly into the blood stream. Our best introduction to carbohydrate chemistry is offered by a review of the chemical and physical properties of glucose.

14.3 *The Structure of Glucose*

The experimental evidence which led to the elucidation of the glucose structure provides one of the most fascinating chapters in organic chemistry. Combustion analysis and a molecular weight determination established the molecular formula of glucose as $C_6H_{12}O_6$. Structural evidence was supplied by the following: (a) reaction with acetic anhydride produced a crystalline pentacetate and suggested the presence of five hydroxyl groups; (b) reaction with hydroxylamine (Sec. 9.7) produced an oxime and suggested the presence of a carbonyl group; (c) mild oxidation with bromine in an aqueous solution yielded gluconic acid ($C_6H_{11}O_5COOH$), an **aldonic acid,** and indicated an aldehyde function at one end of the molecule. This observation was reënforced when the addition of HCN yielded a cyanohydrin; (d) hydrolysis of the cyanohydrin, followed by reduction, produced heptanoic acid.

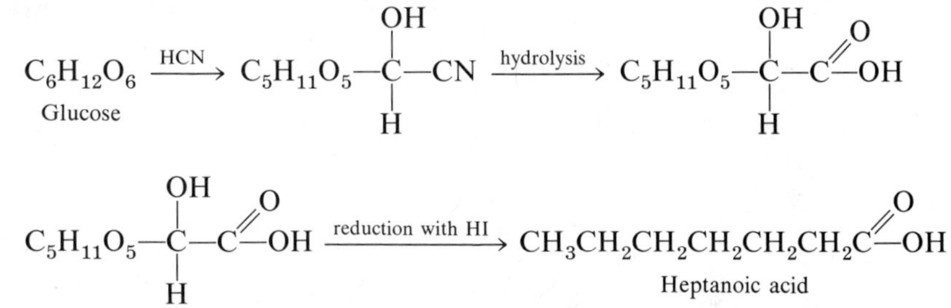

$$C_6H_{12}O_6 \xrightarrow{\text{HCN}} C_5H_{11}O_5\text{—}\underset{\underset{H}{|}}{\overset{\overset{OH}{|}}{C}}\text{—CN} \xrightarrow{\text{hydrolysis}} C_5H_{11}O_5\text{—}\underset{\underset{H}{|}}{\overset{\overset{OH}{|}}{C}}\text{—}C\overset{O}{\diagup}\text{—OH}$$

Glucose

$$C_5H_{11}O_5\text{—}\underset{\underset{H}{|}}{\overset{\overset{OH}{|}}{C}}\text{—}C\overset{O}{\diagup}\text{—OH} \xrightarrow{\text{reduction with HI}} CH_3CH_2CH_2CH_2CH_2CH_2C\overset{O}{\diagup}\text{—OH}$$

Heptanoic acid

The above reactions indicate a six-carbon chain with a terminal aldehyde group. A six-carbon aldehyde with five hydroxyl groups can have a stable structure only if each hydroxyl group is attached to a different carbon atom. The structure of glucose thus was proposed as the following:

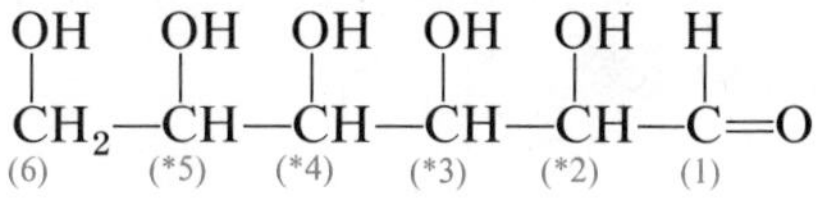

$$\underset{(6)}{\overset{\overset{OH}{|}}{CH_2}}\text{—}\underset{(*5)}{\overset{\overset{OH}{|}}{CH}}\text{—}\underset{(*4)}{\overset{\overset{OH}{|}}{CH}}\text{—}\underset{(*3)}{\overset{\overset{OH}{|}}{CH}}\text{—}\underset{(*2)}{\overset{\overset{OH}{|}}{CH}}\text{—}\underset{(1)}{\overset{\overset{H}{|}}{C}}\text{=O}$$

Glucose

14.4 *The Configuration of Glucose*

Inspection of the above formula reveals the presence of four different asymmetric carbon atoms. No less than sixteen ($2^4 = 16$) optical isomers with the above structure are possible. Only two other aldohexoses (mannose and galactose) occur in nature. Which of the sixteen possible structures is glucose? Which is mannose? Which is galactose? The answers to these perplexing questions were obtained through a brilliant series of syntheses and degradation studies performed by a research group headed by Emil Fischer, 1891–1896.

Most of the remaining thirteen aldohexoses were synthesized and the configurations of all known isomers elucidated. To review all the chemistry which finally led to this great accomplishment is beyond the scope of this text but a simplified scheme may be written to show how all eight D-forms of the aldohexoses were obtained.

The reference standard chosen for relating configurations of optically active compounds is D-glyceraldehyde (Sec. 5.3). The carbon chain of D-glyceraldehyde can be lengthened into a sugar molecule through repeated cyanohydrin formation. A lengthening of the carbon chain in this manner is known as the **Kiliani synthesis.**

$$
\begin{array}{ccc}
\text{H} & & \text{CN} \\
| & & | \\
\text{C}=\text{O} & & \text{H}-\text{C}-\text{OH} \\
| & \xrightarrow{\text{HCN}} & | \\
\text{H}-\text{C}-\text{OH} & & \text{H}-\text{C}-\text{OH} \\
| & & | \\
\text{CH}_2\text{OH} & & \text{CH}_2\text{OH}
\end{array}
\qquad + \qquad
\begin{array}{c}
\text{CN} \\
| \\
\text{HO}-\text{C}-\text{H} \\
| \\
\text{H}-\text{C}-\text{OH} \\
| \\
\text{CH}_2\text{OH}
\end{array}
$$

D (+) Glyceraldehyde

(A mixture of optically active diastereoisomers. Both retain the D configuration.)

$$
\left[
\begin{array}{c}
\text{CN} \\
| \\
\text{H}-\text{C}-\text{OH} \\
| \\
\text{H}-\text{C}-\text{OH} \\
| \\
\text{CH}_2\text{OH} \\
\\
\text{CN} \\
| \\
\text{HO}-\text{C}-\text{H} \\
| \\
\text{H}-\text{C}-\text{OH} \\
| \\
\text{CH}_2\text{OH}
\end{array}
\right]
\xrightarrow{\text{hydrolysis}}
\left[
\begin{array}{c}
\text{COOH} \\
| \\
\text{H}-\text{C}-\text{OH} \\
| \\
\text{H}-\text{C}-\text{OH} \\
| \\
\text{CH}_2\text{OH} \\
\\
\text{COOH} \\
| \\
\text{HO}-\text{C}-\text{H} \\
| \\
\text{H}-\text{C}-\text{OH} \\
| \\
\text{CH}_2\text{OH}
\end{array}
\right]
\xrightarrow{\text{reduction*}}
\begin{array}{c}
\text{H} \\
| \\
\text{C}=\text{O} \\
| \\
\text{H}-\text{C}-\text{OH} \\
| \\
\text{H}-\text{C}-\text{OH} \\
| \\
\text{CH}_2\text{OH} \\
\text{D-Erythrose} \\
\\
\text{H} \\
| \\
\text{C}=\text{O} \\
| \\
\text{HO}-\text{C}-\text{H} \\
| \\
\text{H}-\text{C}-\text{OH} \\
| \\
\text{CH}_2\text{OH} \\
\text{D-Threose}
\end{array}
$$

* Conversion to lactone and reduction with sodium borohydride at pH 3-5.

The schematic structure on the opposite page shows how the configurations of the pentoses are related to those of their parent structures, the tetroses. Similarly, the configurations of the hexoses may be related to those of their antecedents, the pentoses. The hydroxyl groups in each structure are denoted by a short horizontal spur and the carbon chain as a vertical line. The dotted line

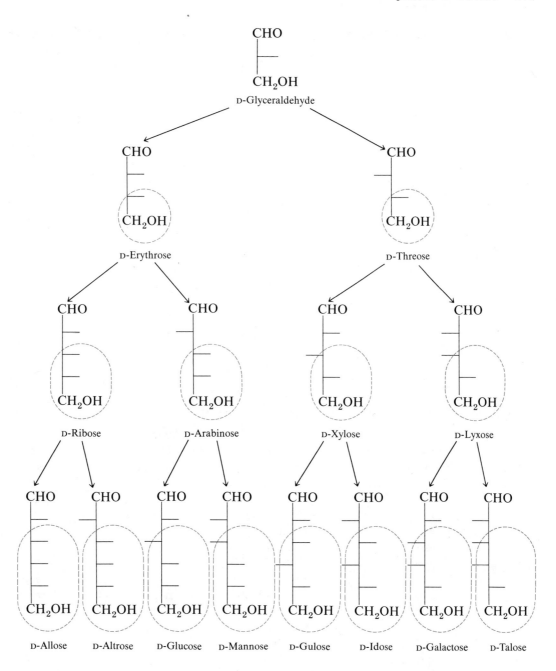

encloses only that portion of each structure that has a configuration identical with that of its parent. Mirror images of all structures shown would have resulted had we used L-glyceraldehyde as our starting material.

EXERCISE 14.1

The oxidation of glucose by strong nitric acid converts both the aldehyde and the primary alcohol group into carboxyl groups. The resulting dicarboxylic acid, called **glucaric acid,** was found to be optically active. Why did this result eliminate the structure shown for galactose as being that of glucose?

The Fischer projection formula for glucose usually is written as shown below and assigned a D-configuration (Sec. 5.3). The "tail" or lower extremity of the formula is $HOCH_2-$, that is, the hydroxymethyl group.

14.5   *The Cyclic Structure of Sugars. Mutarotation*

Several properties of glucose are not explainable by the open-chain formula. For example, when D-glucose is heated with methanol in the presence of hydrogen chloride, the expected dimethyl acetal is not obtained. Instead, two optically active isomeric compounds result, each containing but *one* methoxyl group. Ethers such as those shown in structures II and III are named **glycosides.** If glycosides are formed from a sugar and a nonsugar hydroxy compound, the nonsugar component is called an **aglycone. Glucoside** is a name specifically assigned to a glycoside produced from glucose. The easily oxidized aldehyde group no longer exists in a glucoside. The methyl glucosides (structures II and III), therefore, are incapable of reducing Fehling's solution or Tollens' reagent. A second observation, unexplainable by the straight-chain formula, was that two crystalline forms of D-glucose could be isolated. One form, designated the α-form, crystallized from a concentrated aqueous solution at 30°C, decomposed at its melting point (147°C) and, in a freshly prepared solution, showed a specific rotation of +113°. The other form, designated the β-form, crystallized from a hot, glacial acetic acid solution, melted at 148–150°C, and, in a freshly prepared solution, showed a specific rotation of +19°. The rotation of an aqueous solution of either the α- or the β-form of glucose, when allowed to stand, was found to change. The rotation of each solution reached the equilib-

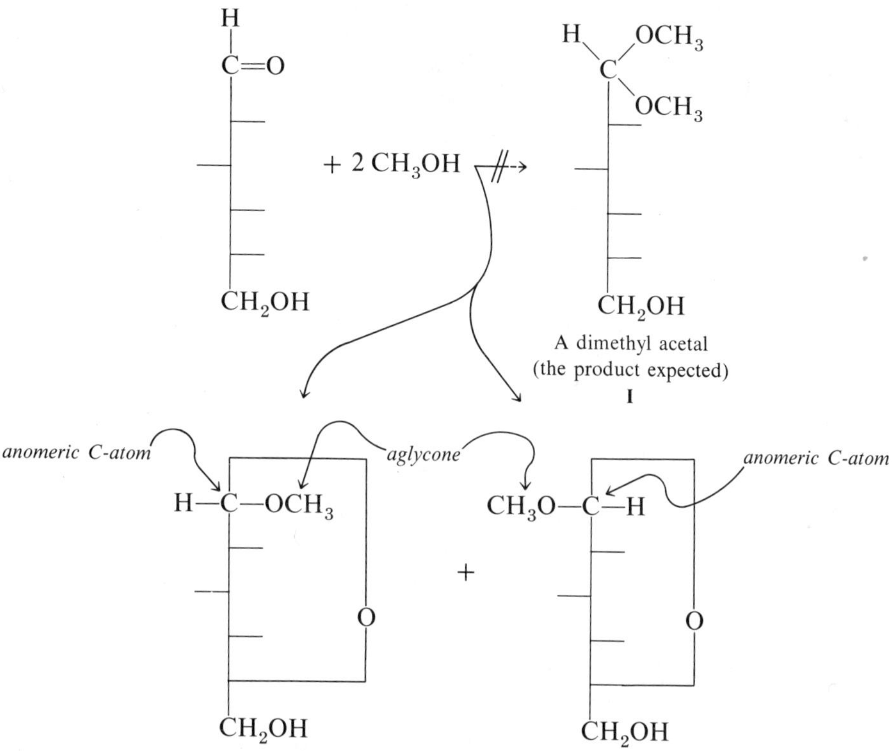

$$\text{+ 2 CH}_3\text{OH} \quad \xrightarrow{\;\;\not\;\;}$$

A dimethyl acetal
(the product expected)

I

anomeric C-atom   *aglycone*   *anomeric C-atom*

$$H-\overset{|}{\underset{|}{C}}-OCH_3 \qquad + \qquad CH_3O-\overset{|}{\underset{|}{C}}-H$$

α-Methylglucoside
$[\alpha]_D^{20} = +159°$
II

β-Methylglucoside
$[\alpha]_D^{20} = -34°$
III

rium value of +52°. Such a change in rotation on standing is called **mutarotation.** Mutarotation is caused by a change in configuration of one asymmetric center in each isomer. The explanation for this behavior for glucose rests in the fact that the glucose molecule exists largely in one of two cyclic hemiacetal forms. Ring closure can result easily through intramolecular acetal formation when the hydroxyl group of carbon (5) is brought into close proximity to the carbonyl group. Zigzag carbon chains, as illustrated in structures IV and V, provide this condition. You will note that two different modes of addition are possible.

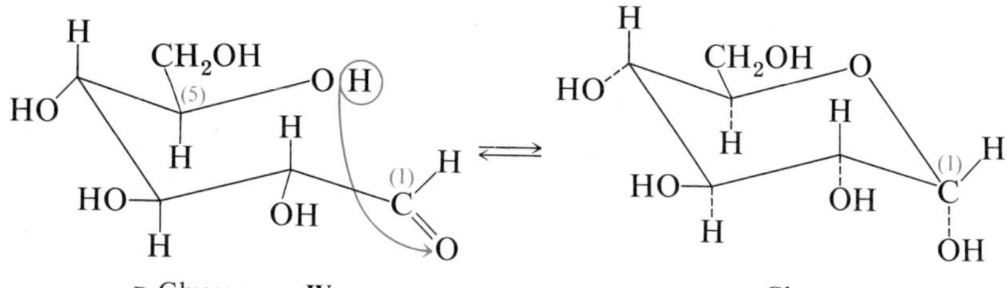

D-Glucose   **IV**   α-D-Glucose

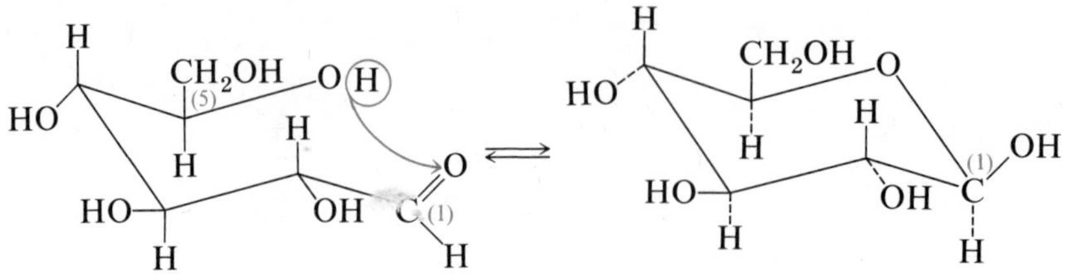

D-Glucose   **V**   β-D-Glucose

Carbon (1) in each of the cyclic hemiacetal structures shown represents a new asymmetric center, and two diastereoisomers thus are possible: α- and β-D-glucose. Such stereoisomers are called **anomers** and the hemiacetal carbon (or acetal carbon in the glycosides) is called the **anomeric** carbon atom. Enough of the open chain (about 1%) form of glucose is present in the equilibrium mixture to give some of the reactions typical of the aldehyde group, but glucose exists preferentially in the cyclic form. The structures below represent typical ways of formulating the glucose molecule.

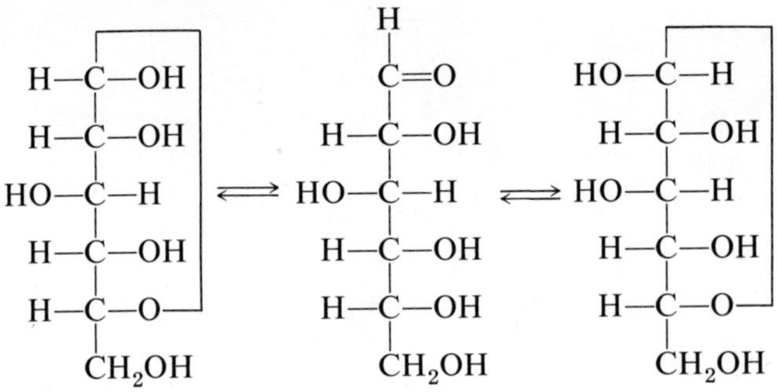

α-D-Glucose   D-Glucose   β-D-Glucose

The Haworth[1] formulas for the cyclic forms of glucose are drawn as planar, hexagonal slabs with darkened edges toward the viewer. Hydroxyl groups and hydrogen atoms are shown either as above (solid bonds) or below (dotted bonds) the plane of the hexagon.

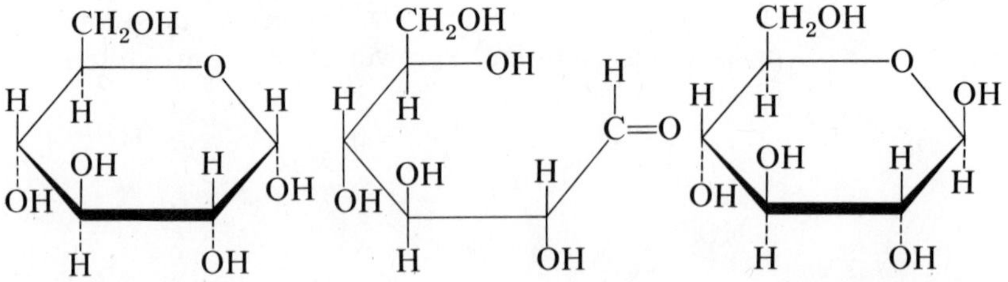

α-D-Glucose   D-Glucose   β-D-Glucose
(α-D-Glucopyranose)   (β-D-Glucopyranose)

---

[1] Walter Haworth (1883–1950), Professor of Chemistry, University of Birmingham. Winner of the Nobel Prize in Chemistry in 1937.

The *pyranose* designation for the cyclic forms of glucose indicates a structural similarity to the six-membered, heterocyclic **pyran** ring, ⬡O. A *furanose* designation for a sugar indicates a five-membered ring with a structural relationship to the heterocycle, **furan**, ⬠O (Sec. 17.2). The Haworth structures are easy to draw, but one must not forget that the six-membered ring is very much like that of cyclohexane and can pucker (Sec. 2.11). Correct cyclic structures for glucose are indicated by the chair representations that follow. Of the two structures illustrated below for α-D-glucopyranose, the conformation in which most of the bulkier groups are bonded equatorially (structure VII) appears to be the more stable one. Similar conformations for β-D-glucopyranose can be drawn by interchanging the positions of the hydrogen atom and the hydroxyl group on carbon (1).

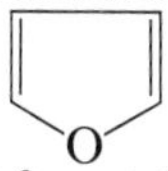

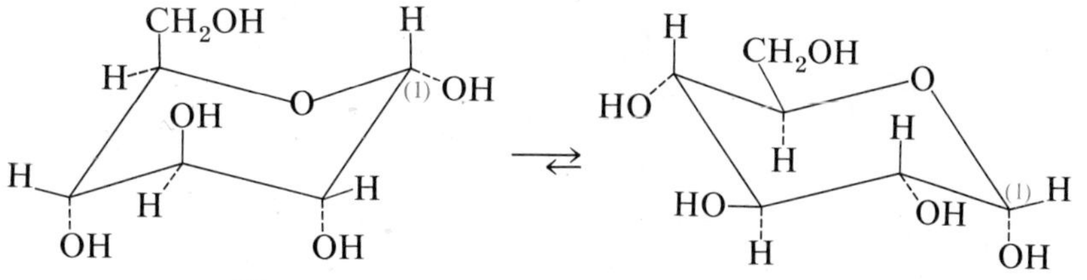

VI α-D-Glucopyranose   VII

14.6 *Fructose*

Fructose, $C_6H_{12}O_6$, also called *levulose* because it is levorotatory, is the most widely distributed ketose. Fructose is found along with glucose in the juices of ripe fruits and in honey. It also occurs with glucose as a component of the disaccharide *sucrose*.

A comparison of the structures of glucose and fructose shows that fructose has only three asymmetric carbon atoms, and the configurations about these are identical to those of corresponding carbon atoms in the glucose chain.

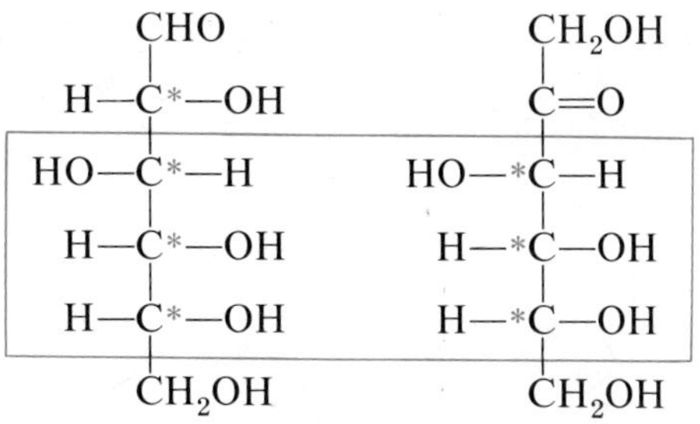

D(+)Glucose   D(−)Fructose

Fructose, unlike glucose, cannot be oxidized by aqueous bromine and, as a keto sugar, one would not expect fructose to reduce Fehling's solution.

However, the alkalinity of Fehling's solution is sufficient to cause a rearrangement of fructose to glucose or mannose, both easily oxidizable sugars.

The cyclic form of fructose may be either that of a pyranose or a furanose. In more complicated structures of which fructose is a part, it is found in the furanose form.

● = *anomeric C-atom*

β-D-Fructopyranose   *β*-D-Fructofuranose

14.7 *Reactions of the Hexoses*

A number of reactions involving both the carbonyl and hydroxylic groups of the monosaccharides were reviewed in previous sections. Oxidation to aldonic acids by bromine, Tollens' reagent, and Fehling's solution, the addition of hydrogen cyanide, the formation of acetals and acetates, all were reactions helpful in elucidating the structure of glucose. In addition to the preceding reactions, the following also are important in sugar chemistry.

14.8 *Reaction with Phenylhydrazine. Osazone Formation*

The carbonyl function in a sugar molecule will, if free to react, condense with phenylhydrazine to produce a phenylhydrazone. The phenylhydrazones of sugars, unlike the phenylhydrazones of simple compounds containing a carbonyl function, usually are difficult to isolate. On the other hand, when an excess of phenylhydrazine is used it forms with sugars a diphenylhydrazone derivative called an **osazone.** The osazones of sugars are easily crystallized and can be used for identification purposes. When excess phenylhydrazine is used the carbinol adjacent to the carbonyl, that is, carbon No. 2 in aldoses but carbon No. 1 in ketoses, becomes oxidized to a second carbonyl group. The carbonyl formed by this oxidation, along with the one originally present, condenses with phenylhydrazine to produce the osazone. The mechanism of the oxidation step is not entirely clear but it appears that for osazone formation to occur, the ratio of phenylhydrazine to sugar in the reaction mixture must be at least 3 : 1 because one molecule of phenylhydrazine is changed to aniline and ammonia. Inasmuch as only carbons (1) and (2) are involved in osazone formation, D-glucose, D-mannose, and D-fructose all form identical osazones.

The configurations of carbon atoms (3), (4), and (5) in each of these three sugars thus are revealed to be identical. A pair of aldoses that differ only in the configuration about the No. 2 carbon atom such as glucose and mannose are called **epimers.**

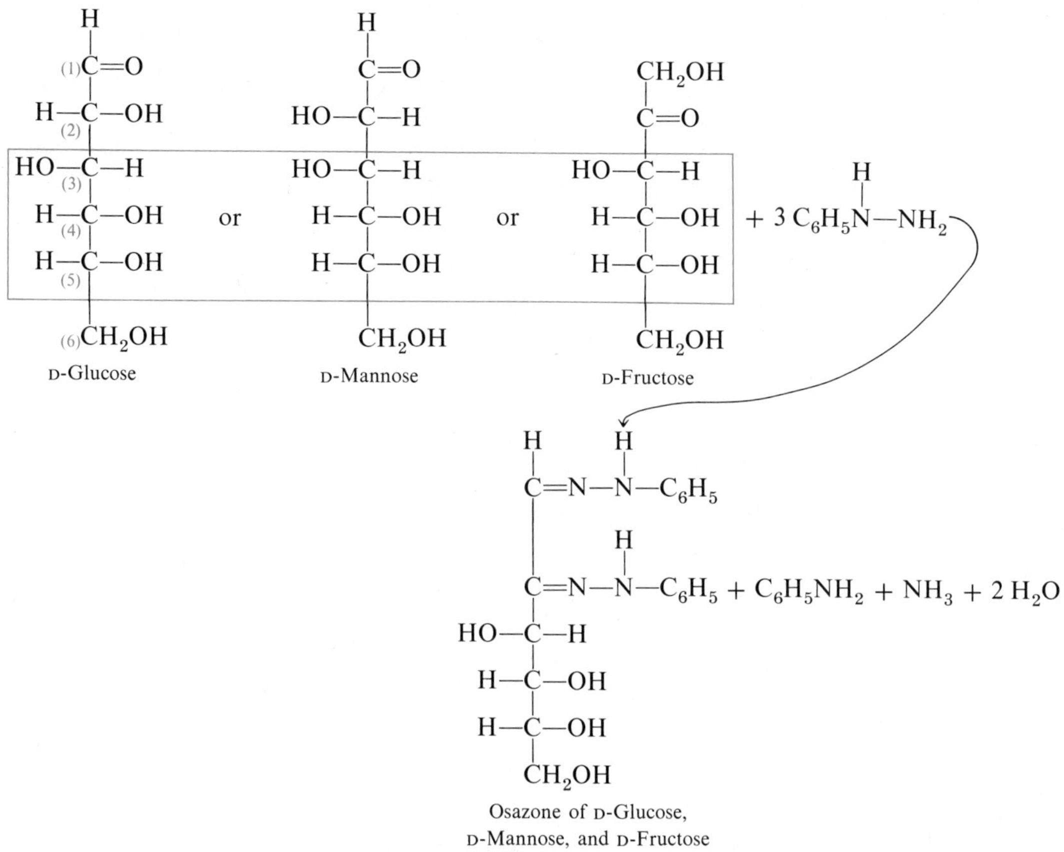

Osazone of D-Glucose,
D-Mannose, and D-Fructose

EXERCISE 14.2

Criticize the following statement: "Two hexoses that react with phenylhydrazine and yield identical osazones are epimers."

14.9 *O-Acylation and O-Alkylation*

The free hydroxyl groups of monosaccharides (and polysaccharides) show many of the normal reactions of alcohols. Thus, treatment of α-D-glucose with excess acetic anhydride gives penta-O-acetyl-α-D-glucose.

α-D-Glucose

Penta-O-acetyl-α-D-glucose

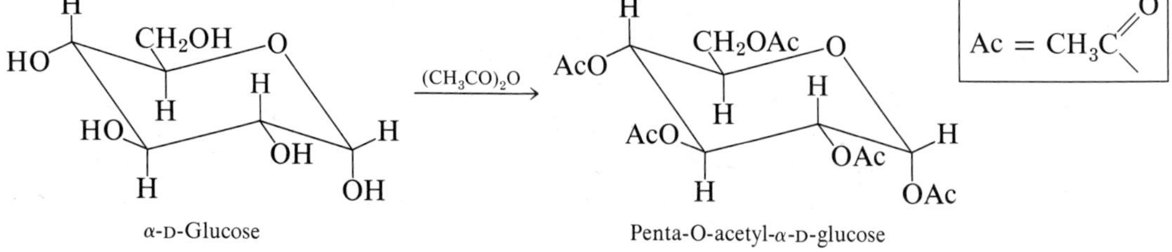

Methylation of the hydroxyl group on the anomeric carbon atom of a mono-saccharide is accomplished with the use of methanol and hydrogen chloride (Sec. 14.5). Methylation of the remaining free hydroxyl groups requires the use of a modified Williamson synthesis (Sec. 8.13), that is, by treatment with methyl iodide and silver oxide or with dimethyl sulfate and 30% aqueous sodium hydroxide. Thus, methylation of methyl α-D-glucopyranoside yields methyl 2,3,4,6-tetra-O-methyl-D-glucopyranoside. Mild acid hydrolysis may be used to cleave the glycosidic linkage (which is an acetal) without affecting the other O-alkyl groups (which are ethers). Methylation is commonly used to establish the position of substituents on the carbohydrate molecule or to determine whether a glycoside has the furanose and pyranose ring structure.

Methyl α-D-glucopyranoside

Methyl 2,3,4,6-tetra-O-methyl-
α-D-glucopyranoside

2,3,4,6-Tetra-O-methyl-
D-glucopyranose

14.10   *The Fermentation of Sugars*

Emil Fischer found that of the sixteen aldohexoses only those found in nature (D-glucose, D-mannose, and D-galactose) could be fermented by yeast. The three simple sugars just named, along with D-fructose, are acted upon by the enzyme **zymase** to yield the same decomposition products—ethyl alcohol and carbon dioxide. The reactions by which alcohol is produced from a sugar are complex but the net result may be indicated by the equation below.

$$C_6H_{12}O_6 \xrightarrow{\text{zymase}} 2\,C_2H_5OH + 2\,CO_2$$

The fermentation of sugar, the oldest chemical reaction known, provided the ancients with wine and the leavening action for making bread.

14.11 *Oligosaccharides*

Disaccharides, the most important of the oligosaccharides, may be regarded as glycosides. However, unlike the simple methylglucosides (Sec. 14.5), the hydroxylic compound that is coupled through the glycosidic link is a second monosaccharide unit—not an aglycone. A molecule of water is split out from two monosaccharide units to form a disaccharide. Restoration of the water by hydrolysis of the disaccharide reforms the two monosaccharide components. Hydrolysis may be effected by dilute acids or by enzymes. The most important of the disaccharides and the only ones we shall consider are *sucrose, maltose, cellobiose,* and *lactose.*

14.12 *Sucrose*

Sucrose, ordinary table sugar, is obtained from the juices extracted from sugar cane and sugar beets. The juice from either source contains about 14–25% of sucrose. To a small extent, sucrose is obtained from the sap of certain species of maple trees. Maple sugar is marketed largely as a syrup. Sucrose is dextro-rotatory but on hydrolysis produces equimolar quantities of D-glucose (dextrose), and D-fructose (levulose). During the hydrolysis of sucrose the sign of the specific rotation changes from positive to negative, or is said to **invert.** The hydrolysis mixture is called *invert sugar.* The enzyme *invertase,* carried by the bee, is able to accomplish the same result in the production of honey.

$$C_{12}H_{22}O_{11} \xrightarrow[\text{(or \textit{invertase})}]{\text{acid hydrolysis}} \text{D-Glucose} + \text{D-Fructose}$$

Sucrose

$[\alpha]_D^{20} = +66.5°$

$\underbrace{[\alpha]_D^{20} = +52° \quad [\alpha]_D^{20} = -92°}_{\text{Invert sugar}}$

$$\frac{+52 - 92}{2} = [\alpha]_D^{20} = -20°$$

The sweetness of honey is largely due to the presence of fructose which is approximately three times as sweet as glucose. Invert sugar shows less tendency to crystallize than does sucrose and for this reason is used to a large extent in the manufacture of candy. Sucrose will not reduce Fehling's solution and therefore is classified as a *nonreducing* disaccharide. The α-D-glucopyranose ring, as shown in the sucrose structure below, is joined to the β-D-fructofuranose ring through a glycosidic link. The anomeric carbon atoms of glucose and of fructose are involved in this union. A "head-to-head" arrangement of this type leaves no potential aldehyde function to be oxidized and explains why sucrose is nonreducing. A junction at these ring positions further explains why sucrose fails to form an osazone with phenylhydrazine and why it fails to show muta-rotation.

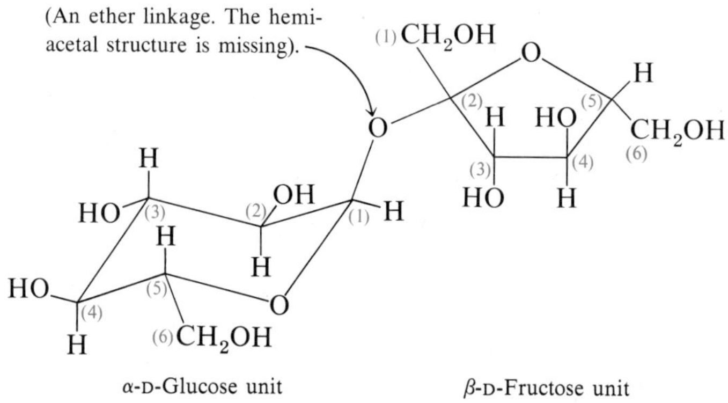

(An ether linkage. The hemiacetal structure is missing).

α-D-Glucose unit　　　　　　　β-D-Fructose unit

Sucrose
(α-D-Glucopyranosyl β-D-fructofuranoside)

14.13　Maltose

Maltose, or malt sugar, is a disaccharide produced when starch is hydrolyzed by malt *diastase,* an enzyme found in sprouting barley. Another enzyme,
maltase, selectively splits the alpha-glycosidic link and completely hydrolyzes
maltose to yield two D-glucose units. Maltose is a *reducing sugar.* This chemical
evidence suggests the presence of an aldehyde group either uncombined or in
equilibrium with the hemiacetal form. In agreement with this, maltose forms an
osazone and also exhibits mutarotation. Maltose is dextrorotatory and gives at
equilibrium a specific rotation value of +136°. The structure of maltose is that
of two glucose units in a "head-to-tail" arrangement joined through an α-
linkage from carbon (1) of one glucose unit to carbon (4) of a second glucose
unit.

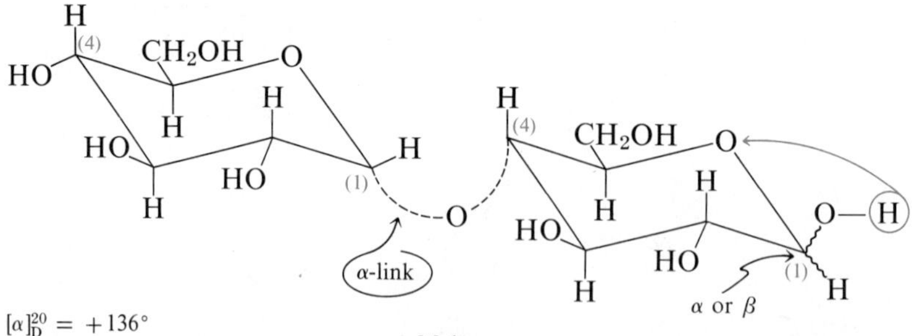

$[\alpha]_D^{20} = +136°$

Maltose

(Potential aldehyde group
in hemiacetal structure)

14.14　Cellobiose

Cellobiose is a disaccharide which is obtained by the hydrolysis of cellulose.
It is a reducing sugar consisting of two glucose units joined as in maltose, *but
through a β-linkage.* The enzyme maltase is incapable of hydrolyzing cellobiose.

In all other respects the behavior of cellobiose is identical to that described for maltose.

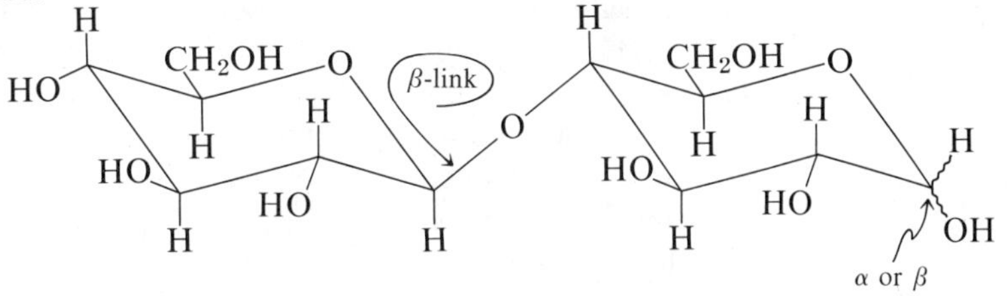

Cellobiose

14.15 *Lactose*

Lactose is known as milk sugar because it is present in the milk of mammals. It is present in cows' milk to the extent of about 5% and in human milk to about 7%. It is produced commercially as a by-product in the manufacture of cheese. Lactose is a reducing sugar, forms an osazone, and exhibits mutarotation. Lactose is dextrorotatory and, at equilibrium, gives a specific rotation of +55°. Lactose, when hydrolyzed by mineral acids or by the action of the enzyme *lactase*, produces equimolar quantities of D-glucose and D-galactose. Lactose is a β-galactoside in which the anomeric (1) carbon of galactose is joined through a β-linkage to the number (4) carbon of glucose in a "head-to-tail" arrangement. This is shown below.

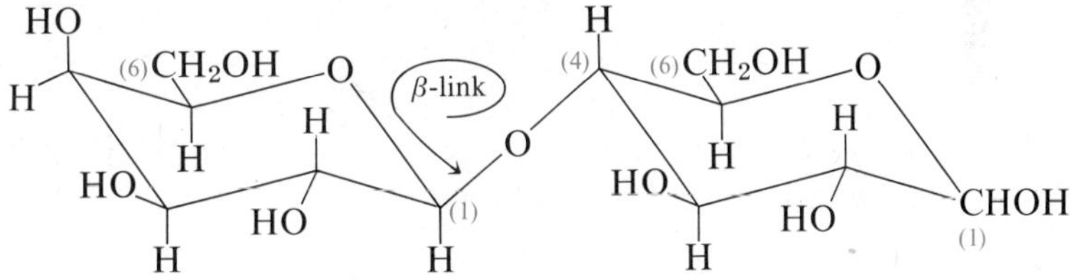

D(+)-Galactose unit   D(+)-Glucose unit

Lactose

14.16 *Polysaccharides*

The polysaccharides are high molecular weight (25,000–15,000,000) natural polymers in which hundreds (or even thousands) of pentose or hexose units have been joined through glycosidic linkages. The most important polysaccharides are starch, glycogen, inulin, and cellulose.

14.17 *Starch, Glycogen, and Inulin*

Starch is the reserve carbohydrate of most plants. It comprises the major part of all cereal grains and most plant tubers, where it is stored. Starch is used as a principal food source throughout the world. Glycogen is the reserve carbohydrate of animals and a relatively small amount is stored in the liver and muscles. Structurally, starch and glycogen are similar. Both have the empirical formula $(C_6H_{10}O_5)_n$ and both, when completely hydrolyzed, yield glucose. Starch and glycogen are only partially hydrolyzed to maltose by the enzyme amylase. When a paste made of starch and water is heated, two fractions may be separated. One, called the **amylose fraction,** is water soluble and has a molecular weight range of 20,000–225,000. A second fraction, called **amylopectin,** is water insoluble and has a molecular weight range of 200,000–1,000,000.

Amylose can be hydrolyzed almost completely to maltose by the enzyme β-amylase. This hydrolysis indicates that the maltose units arise from glucose units joined through alpha 1-4 linkages. Amylopectin, on the other hand, is hydrolyzed to maltose to a much lesser degree by the same enzyme. Chemical evidence indicates that considerable branching occurs in the amylopectin fraction of the starch molecule. Such branches are formed from glucose units linked through carbon atoms (1) and (6). The glycosidic link at these positions resists hydrolysis by β-amylase. A segment of a branched starch molecule is shown.

A segment of the starch molecule

Inulin is a polysaccharide comparable to starch. It is found in the underground stems (tubers) of the Jerusalem artichoke and in dahlia roots. Inulin, on hydrolysis, yields fructose as its end product.

14.18 Cellulose

Cellulose comprises the skeletal material of plants and is the most abundant organic substance found in nature. It is the chief constituent of wood and cotton.

Cotton, almost pure cellulose, is the principal source of cellulose used as fiber for fabrics. The cellulose content of wood is approximately 50 per cent. The separation of cellulose from other plant components is an important commercial process upon which the textile, paper, and plastics industries are largely dependent.

The general formula for cellulose, like that for starch, may be written $(C_6H_{10}O_5)_n$, but the numerical value of n in the formula for cellulose is much larger than that for starch. Methylation studies indicate that the structure of cellulose, unlike that of starch, is largely unbranched.

Complete hydrolysis of cellulose produces D-glucose. Partial hydrolysis produces cellobiose (Sec. 14.14), cellotriose, and higher oligosaccharides. The presence of β-glycosidic linkages establishes the structure of cellulose as

Structure of cellulose unit

Conformation of cellulose

Man is incapable of utilizing cellulose for food because his digestive juices lack enzymes capable of hydrolyzing the β-glycosidic linkage. Ruminants (cud-chewing animals) are able to digest cellulose because certain microorganisms present in their compartmented stomachs cause a preliminary hydrolysis of cellulose before it reaches the intestine. Certain lower orders of animals (snails, termites) having similar assistance also are able to feed upon cellulose.

14.19   *Derivatives of Cellulose*

The derivatives of cellulose, like those of any polyhydric alcohol, are principally esters and ethers. Each $(C_6H_{10}O_5)$ unit of cellulose has three hydroxyl groups available. The conversion of some or all of these groups to ethers or to ester functions alters the properties of cellulose remarkably.

Cellulose nitrates are formed when cellulose is nitrated in the presence of sulfuric acid. A product known as *pyroxylin* is formed when cellulose is nitrated under conditions which esterifies approximately two-thirds of the available hydroxyl groups.

$$[C_6H_7O_2(OH)_3]_n \xrightarrow{2nHNO_3, H_2SO_4} [C_6H_7O_2(OH)(ONO_2)_2]_n$$
Cellulose   Pyroxylin

Pyroxylin is used in plastics and in lacquers. Unlike cellulose, pyroxylin is very soluble in a number of highly volatile organic solvents and can be used in such solutions as a quick-drying finish. Evaporation of the solvent, after application of such finishes, leaves a thin, hard, glossy film of cellulose nitrate deposited upon a surface. A product known as "patent leather" is produced when solutions of pyroxylin are applied to cloth. *Celluloid* (*cellu*lose + *oid*), one of the earliest plastics, is made by mixing pyroxylin and camphor with alcohol. The gelatinous mass which results hardens after evaporation of the solvent.

Guncotton is a product obtained when cellulose is almost completely nitrated under conditions carefully controlled to prevent degradation of the cellulose molecule. Guncotton contains about 12–13% of nitrogen, is explosive, and is used in the manufacture of smokeless powder.

Cellulose acetate is obtained by the reaction of cellulose with acetic anhydride. Like the nitrates, cellulose acetate also is an ester but is not explosive. A viscous solution of cellulose acetate when extruded through the fine openings of a die, called a "spinneret," produces thin fibers. From such fibers the popular "acetate" fabrics are woven. Photographic film also is prepared from cellulose acetate. Cellulose from cotton linters or wood pulp, when treated with sodium hydroxide and carbon disulfide, can be converted into a product known as a **xanthate.** Xanthates with dilute alkali form a heavy, viscous solution called "viscose." Cellulose in the form of *rayon* is regenerated when viscose is extruded through a spinneret into a bath of dilute mineral acid. Extrusion of viscose through a long, thin slit produces *cellophane.* The chemical steps involved in the formation of viscose rayon may be illustrated by applying the same reactions to a simple alcohol.

$$ROH + NaOH + CS_2 \longrightarrow RO\!-\!\overset{\displaystyle S}{\underset{}{C}}\!-\!S^-Na^+ + H_2O$$
A xanthate

$$RO\overset{\displaystyle S}{C}\!-\!S^-Na^+ + H_2SO_4 \longrightarrow ROH + CS_2 + NaHSO_4$$

Ethyl cellulose is a cellulose ether prepared by a modified Williamson synthesis (Sec. 14.9). Alkali-cellulose (produced as in the viscose process) is treated with ethyl chloride to produce the ether.

Two or three hydroxyl groups per six-carbon unit are converted into ethoxy ethers. The cellulose ethers are widely used in the plastics and textile industries.

14.20 *Mercerized Cotton*

Mercerized cotton is cotton fiber modified by treatment with sodium hydroxide while the yarn is under tension. Subsequent washing and drying produces a smooth fiber with a lustrous finish and greatly improved tensile strength. The process, called **mercerization,** was discovered by an English chemist, John Mercer, in 1814.

14.21 *Paper*

Most paper is made from wood pulp. The cellulose of wood is separated from a noncellulosic substance called lignin. The latter is converted into an alkali-soluble substance by treatment with calcium hydrogen sulfite, $Ca(HSO_3)_2$, and removed from the insoluble cellulose fibers. The cellulose fibers, after removal of lignin, are washed and removed from the mixture as a matting on a large, flat filter. Compression of the matting, followed by drying, produces paper. Additional treatment produces paper products for every purpose—stationery, newsprint, hardboard, tissue, and wax paper. The paper industry, which depends upon cellulose as its raw material, is one of the country's largest consumers of forest products.

14.22 *Digestion and Metabolism of Carbohydrates*

The digestion of carbohydrates involves the enzymatic hydrolysis of carbohydrates to produce the simple sugars: glucose, fructose, and galactose. These simple sugars are absorbed into the blood stream and are transported first to the liver and eventually to the muscles. At these sites, glycogen can be synthesized and temporarily stored. These carbohydrate stores, when called upon to supply energy, are hydrolyzed to glucose. A series of enzymatic reactions cleaves the 6-carbon glucose chain into two 3-carbon pyruvic acid molecules,

$$CH_3-\overset{\overset{\displaystyle O}{\|}}{C}-COOH.$$ Oxidative decarboxylation of pyruvic acid, in association with CoA (Sec. 11.10), forms acetyl CoA, which then enters the tricarboxylic acid cycle as shown in Fig. 14.1 (page 434). All carbohydrates, in meeting the energy requirements of the body, ultimately are oxidized to carbon dioxide and water.

FIGURE 14.1 *The Tricarboxylic Acid Cycle*

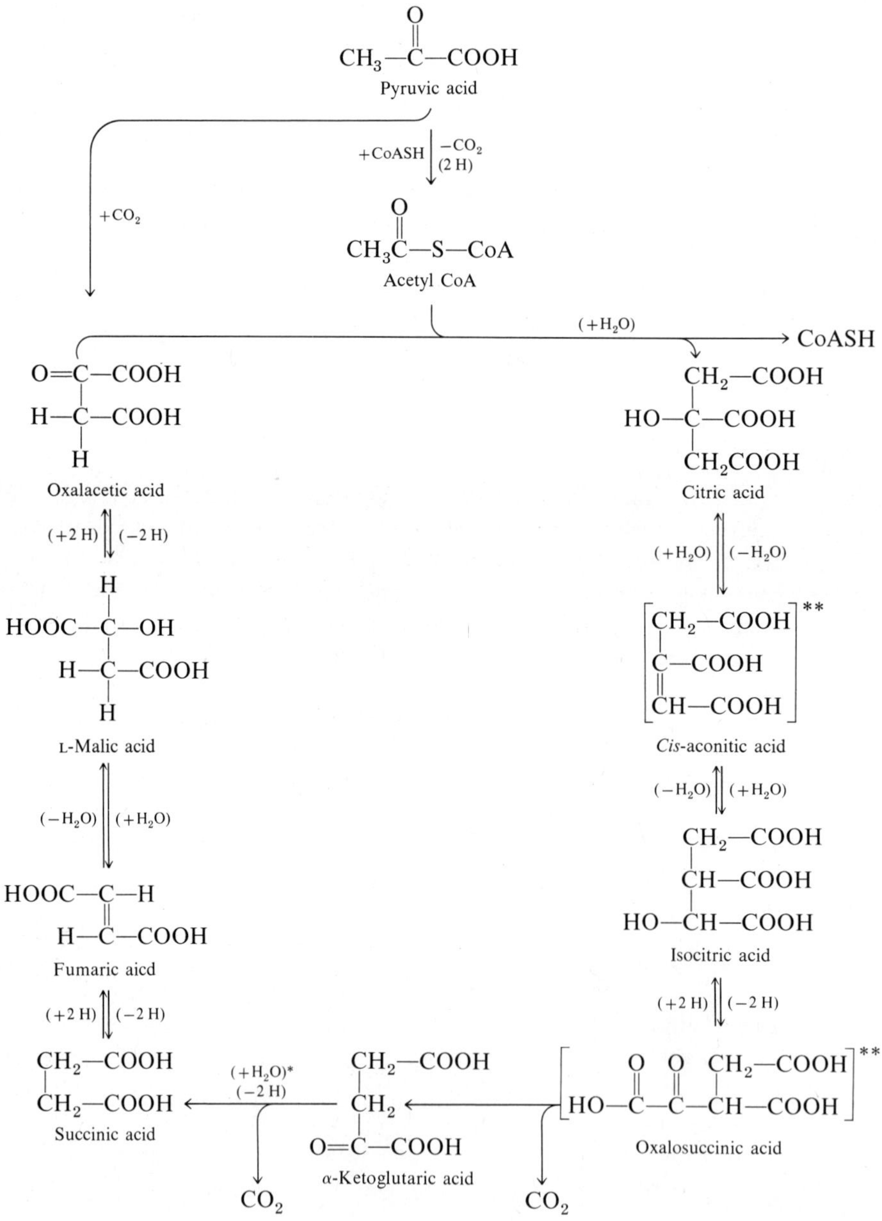

$$O$$
$$\text{CH}_3-\overset{\|}{\text{C}}-\text{COOH}$$
Pyruvic acid

$+\text{CoASH}$ | $\overset{-\text{CO}_2}{(2\text{ H})}$

$+\text{CO}_2$

$$O$$
$$\text{CH}_3\overset{\|}{\text{C}}-\text{S}-\text{CoA}$$
Acetyl CoA

$(+\text{H}_2\text{O})$ → CoASH

$$\text{O}=\text{C}-\text{COOH}$$
$$\text{H}-\text{C}-\text{COOH}$$
$$\text{H}$$
Oxalacetic acid

$(+2\text{ H})$ ‖ (-2 H)

$$\text{H}$$
$$\text{HOOC}-\text{C}-\text{OH}$$
$$\text{H}-\text{C}-\text{COOH}$$
$$\text{H}$$
L-Malic acid

$(-\text{H}_2\text{O})$ ‖ $(+\text{H}_2\text{O})$

$$\text{HOOC}-\text{C}-\text{H}$$
$$\text{H}-\text{C}-\text{COOH}$$
Fumaric aicd

$(+2\text{ H})$ ‖ (-2 H)

$$\text{CH}_2-\text{COOH}$$
$$\text{HO}-\text{C}-\text{COOH}$$
$$\text{CH}_2\text{COOH}$$
Citric acid

$(+\text{H}_2\text{O})$ ‖ $(-\text{H}_2\text{O})$

$$\left[\begin{array}{l}\text{CH}_2-\text{COOH} \\ \text{C}-\text{COOH} \\ \text{CH}-\text{COOH}\end{array}\right]^{**}$$
Cis-aconitic acid

$(-\text{H}_2\text{O})$ ‖ $(+\text{H}_2\text{O})$

$$\text{CH}_2-\text{COOH}$$
$$\text{CH}-\text{COOH}$$
$$\text{HO}-\text{CH}-\text{COOH}$$
Isocitric acid

$(+2\text{ H})$ ‖ (-2 H)

$$\text{CH}_2-\text{COOH}$$
$$\text{CH}_2-\text{COOH}$$
Succinic acid

$\overset{(+\text{H}_2\text{O})*}{(-2\text{ H})}$

$$\text{CH}_2-\text{COOH}$$
$$\text{CH}_2$$
$$\text{O}=\text{C}-\text{COOH}$$
α-Ketoglutaric acid

$$\left[\begin{array}{l}\text{O}\quad\text{O}\quad\text{CH}_2-\text{COOH} \\ \text{HO}-\text{C}-\text{C}-\text{CH}-\text{COOH}\end{array}\right]^{**}$$
Oxalosuccinic acid

CO_2 CO_2

* Succinoyl-CoA is an intermediate in this step.

** Enzyme-bound intermediates.

Summary

1. Carbohydrates are polyhydroxy aldehydes or polyhydroxy ketones.

2. Carbohydrates, in general, may be classified as: mono-, oligo-, and poly-saccharides. Monosaccharides may be classified functionally, as aldoses or ketoses; according to chain length, as pentoses or hexoses.

3. Glucose is the most important monosaccharide. It is an aldohexose but exists largely in a cyclic hemiacetal structure. It exhibits *mutarotation* and has an equilibrium rotation value of $+52°$.

4. Important reactions of glucose include

   (a) Reduction of Fehling's solution.
   (b) Formation of a hemiacetal (glucoside) with an alcohol.
   (c) Formation of an osazone with phenylhydrazine.
   (d) Formation of a pentacetate with acetic anhydride.
   (e) Addition of HCN to form a cyanohydrin.
   (f) Fermentation to ethyl alcohol and carbon dioxide.
   (g) Formation of ethers with dimethyl sulfate and a base.

5. The disaccharides sucrose, maltose, cellobiose, and lactose are the most important oligosaccharides. Sucrose and maltose are α-glucosides. Lactose is a β-galactoside. Maltose and lactose reduce Fehling's solution, form osazones, and exhibit mutarotation; sucrose does none of these. The hydrolysis of sucrose "inverts," or changes, its rotation value from $(+)$ to $(-)$. Invert sugar is hydrolyzed sucrose.

6. Polysaccharides are high molecular weight polymers in which thousands of basic pentose or hexose units are combined. The most important polysaccharides are starch, glycogen, and cellulose.

7. Starch (cereal grains, plant roots and tubers) is composed of thousands of glucose units joined predominantly through α-linkages at carbons (1) and (4). It is the reserve carbohydrate of plants.

8. Glycogen is the reserve carbohydrate of animals. It is stored in the liver and muscle.

9. As with starch, cellulose is composed of glucose units, but the connection between glucose molecules in cellulose are β-linkages. Moreover, cellulose appears to be an unbranched structure.

10. Cellulose derivatives (esters and ethers) are useful plastics. Among these are:
    (a) Cellulose nitrates, which are used in smokeless powder, in lacquers, and other surface coatings;
    (b) Cellulose acetate, which is used in textiles and photographic film;
    (c) Ethyl cellulose, which is used in a variety of plastic articles.

11. Cellulose which has been regenerated from viscose (a cellulose xanthate) may be formed into *rayon* and *cellophane*.

12. Mercerized cotton is cotton yarn treated with sodium hydroxide to make the fiber smooth and stronger.

13. The manufacture of paper products is one of the principal uses for cellulose.

14. The digestion of carbohydrates results in the production of simple sugars.

15. The metabolism of simple sugars involves a series of degradation reactions *via* the tricarboxylic acid cycle.

New Terms

aglycone	ketose
aldose	mercerization
anomeric carbon	mutarotation
anomers	nonreducing sugar
epimer	oligosaccharide
glucopyranose	polysaccharide
glucopyranoside	pyroxylin
glycoside	reducing sugar
guncotton	xanthate
invert sugar	

Supplementary Exercises

EXERCISE 14.3 Write Fischer open-chain projection formulas for D-glucose and D-fructose. Draw cyclic structures for each and show the manner in which they are joined to form sucrose, maltose, and lactose.

EXERCISE 14.4 Write balanced equations to show the products formed when glucose is caused to react with each of the following.

(a) phenylhydrazine
(b) Fehling's solution
(c) CH_3OH, anhydrous HCl
(d) HNO_3

(e) acetic anhydride
(f) zymase in yeast
(g) Br_2 (aqueous)
(h) hydroxylamine

EXERCISE 14.5 Explain why both maltose and lactose are reducing sugars but sucrose is not.

EXERCISE 14.6 Explain the apparent contradiction in the following statement: "Fructose is not a reducing sugar but may be oxidized by Fehling's solution."

EXERCISE 14.7 Draw structures for each of the following.

(a) methyl α-D-glucoside
(b) sucrose

(c) a structural unit of starch
(d) a structural unit of cellulose

EXERCISE 14.8 Using R—OH to represent a cellulose unit, write equations illustrating the preparation of

(a) cellulose acetate
(b) smokeless powder

(c) viscose rayon
(d) ethyl cellulose

EXERCISE 14.9 Which of the following tests would serve to distinguish between the sugars in each of the pairs listed below: (a) bromine (aqueous), (b) Fehling's solution, (c) phenylhydrazine, (d) a polarimetric measurement.

(A) glucose and fructose
(B) glucose and mannose
(C) maltose and lactose
(D) sucrose and maltose

EXERCISE 14.10 An optically active hexose, (A), $C_6H_{12}O_6$, was degraded to an optically active pentose (B), $C_5H_{10}O_5$, by the following series of reactions:

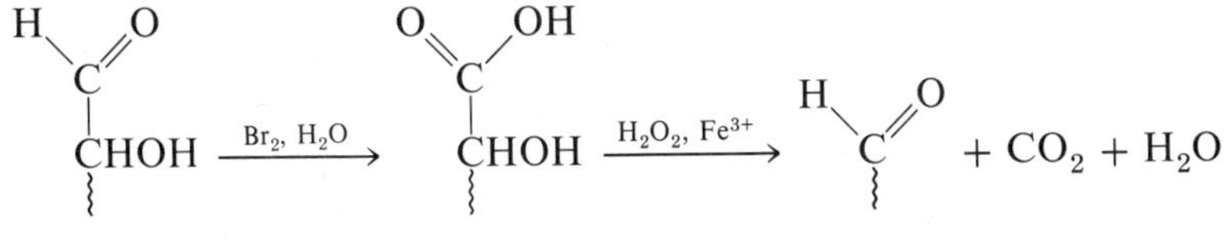

$$(B) + HNO_3 \longrightarrow C_5H_8O_7 \text{ (optically active)}$$

Compound (B) was degraded to an optically active tetrose (C), $C_4H_8O_4$, through the same series of reactions as shown above.

$$(C) + HNO_3 \longrightarrow \textit{meso} \text{ tartaric acid}$$

Give possible structures for (A) and (B).

EXERCISE 14.11 Which would be more palatable and nutritious if used on your breakfast cereal, 4-O-β-glucopyranosyl D-glucopyranose or α-D-glucopyranosyl β-D-fructofuranoside? Why?

EXERCISE 14.12 To which carbohydrates or their derivatives do the following phrases apply: (a) highly explosive, (b) stored in the liver, (c) a polyfructofuranoside, (d) on hydrolysis yields equimolar amounts of galactose and glucose, (e) regenerated cellulose, (f) a polyglucoglucoside.

EXERCISE 14.13 When naturally occurring aldotetrose A is reduced with sodium borohydride, *meso*-1,2,3,4-butanetetrol is formed. What are the name and structure of A?

EXERCISE 14.14 The α- and β-anomers of D-mannose have the optical rotations $[\alpha]_D = +29.3°$ and $[\alpha]_D = -16.3°$, respectively, in aqueous solution. On standing, solutions of either anomer mutarotate to an equilibrium value of $[\alpha]_D = +14.5°$. Calculate the relative amounts of each anomer present at equilibrium.

EXERCISE 14.15 For the aldohexose, D-gulose, draw the perspective (Haworth) formulas for the following:

(a) α-D-gulose

(b) α-L-gulose

(c) Anomer of (a)

(d) Epimer of (a)

(e) A reducing disaccharide formed from D-glucose and D-gulose in which the C-4 hydroxyl of D-glucose is one point of attachment.

EXERCISE 14.16 A pentose of the D-family gave an optically active glycaric acid (a dicarboxylic acid (Exercise 14.1)) on oxidation with nitric acid. The pentose was converted by the Kiliani reaction into a pair of diastereoisomeric hexoses, one of which gave an optically active glycaric acid when oxidized with nitric acid and the other gave an optically inactive glycaric acid. What are the names and configurations of the pentose and the two hexoses?

EXERCISE 14.17 Two aldoses A and B give the same osazone. Oxidation of A gives an optically inactive glycaric acid, and oxidation of B gives an optically active glycaric acid. Treatment of D-glucose with hydrogen cyanide followed by hydrolysis gives a mixture of the glycaric acids obtained from A and B. From these observations deduce the structures of A and B.

EXERCISE 14.18 Oxidative degradation of a glycoside with periodic acid (Exercise 9.26) may be used to determine whether the glycoside is a furanoside or pyranoside. Thus, treatment of methyl α-D-glucopyranoside with periodic acid gives a dialdehyde and formic acid, while similar treatment of methyl α-D-arabinofuranoside gives the same dialdehyde but no formic acid. Write equations for these oxidative degradations and explain the formation of the products.

EXERCISE 14.19 On treatment with periodic acid, a 0.243-g sample of cellulose gave 0.00100 mmole (1.00×10^{-6} mole) of formic acid. What is the approximate number of glucose units in each molecule of the cellulose? (*Hint:* there will be ($2n + 2$) glucose units based on the structure given in Sec. 14.18.)

EXERCISE 14.20 A disaccharide, $C_{11}H_{20}O_{10}$, may be hydrolyzed by α-glucosidase (an enzyme that is a specific catalyst for the hydrolysis of α-glucosides) to yield a hexose and a pentose. The disaccharide does not reduce Fehling's solution. Methylation of the disaccharide with methyl bromide and silver oxide, followed by acid hydrolysis, yields 2,3,4,6-tetra-O-methyl-D-glucose and a tri-O-methyl pentose. Oxidation of the latter with bromine water gives 2,3,5-tri-O-methyl-D-xylonic acid. From these data deduce the structure of the disaccharide and write equations for the transformations described.

15

Amino Acids, Peptides, and Proteins

Introduction

The name protein has its origin in the Greek word *proteios,* meaning "of first importance." The name is well chosen for proteins are the basis of protoplasm and comprise the underlying structure of all living organisms. Proteins, in the form of muscle, skin, hair, and other tissue, make up the bulk of the body's nonbony structure. In addition to providing such structural material, proteins have many other functions. Proteins, as enzymes, catalyze biochemical reactions; as hormones, they regulate metabolic processes; as antibodies, they resist and nullify the effects of toxic substances. Such specialized functions illustrate the great importance of proteins.

Proteins are high molecular weight, long-chain polymers made up largely of various amino acids linked together. The constituent amino acids are obtained when a protein is hydrolyzed by dilute acids, by dilute alkalis, or by protein-digesting enzymes. It would be difficult to consider the properties of molecules as complex as the proteins without first examining the properties of the α-amino acids from which they are constructed.

Amino Acids

15.1 Structure of Amino Acids

Nearly all amino acids obtained from plant and animal proteins have an amino group on the carbon atom alpha (α) to the carboxyl function. An α-amino acid has the following general formula.

$$R-\underset{\underset{NH_2}{|}}{\overset{\overset{H}{|}}{C}}-\overset{\overset{O}{\parallel}}{C}-OH$$

The R in the general formula for an α-amino acid may be a hydrogen, a straight- or branched-chain aliphatic group, an aromatic ring, or a heterocyclic ring (Chap. 17, Intro.). Most amino acids have one amino group and one carboxyl group and usually are classified as neutral amino acids. A few have a second amino group joined to other carbon atoms in the molecule and show basic properties. Others contain a second carboxyl group and behave as acids. With the exception of glycine (aminoacetic acid) all amino acids contain at least one center of asymmetry and are optically active. Amino acids of protein origin all possess the L-configuration. With few exceptions, the body is able to utilize completely only the L-isomers of those amino acids which it is not able to synthesize itself.

15.2 Nomenclature and Classification by Structure

Of the known α-amino acids, about twenty have been found to be constituents of the more common plant and animal proteins. The structures for these, together with their names and abbreviations, are given in Table 15.1.

15.3 Nutritive Classification of Amino Acids

A nutritive classification of the α-amino acids has resulted from nutritional experiments carried out on laboratory animals. It is not a strict classification but varies with the requirements of the different species of animals tested. The

TABLE 15.1 *Amino Acids Derived from Proteins*

A.1. *Neutral Amino Acids with Non-polar R Group*

Name (Abbreviation)	Formula

1. **Glycine (Gly)***
   (Aminoacetic acid)

$$\underset{\underset{NH_2}{|}}{\overset{\overset{H}{|}}{H-C}}-COOH$$

2. **Alanine (Ala)**
   (α-Aminopropionic acid)

$$\underset{\underset{NH_2}{|}}{\overset{\overset{H}{|}}{CH_3-C}}-COOH$$

3. **Valine (Val)**
   (α-Aminoisovaleric acid)

$$CH_3-\underset{\underset{H}{|}}{\overset{\overset{CH_3}{|}}{C}}-\underset{\underset{NH_2}{|}}{\overset{\overset{H}{|}}{C}}-COOH$$

4. **Leucine (Leu)**
   (α-Aminoisocaproic acid)

$$CH_3-\underset{\underset{H}{|}}{\overset{\overset{CH_3}{|}}{C}}-\underset{\underset{H}{|}}{\overset{\overset{H}{|}}{C}}-\underset{\underset{NH_2}{|}}{\overset{\overset{H}{|}}{C}}-COOH$$

5. **Isoleucine (Ile)**
   (α-Amino-β-methylvaleric acid)

$$CH_3-CH_2-\underset{\underset{H}{|}}{\overset{\overset{CH_3}{|}}{C}}-\underset{\underset{NH_2}{|}}{\overset{\overset{H}{|}}{C}}-COOH$$

6. **Phenylalanine (Phe)**
   (α-Amino-β-phenylpropionic acid)

7. **Tryptophan (Trp)**
   (α-Amino-β-(3-indolyl) propionic acid)

8. **Proline (Pro)†**
   (2-Pyrrolidine carboxylic acid)

9. **Methionine (Met)**
   (α-Amino-γ-methylthiobutyric acid)

$$CH_3-S-CH_2-CH_2-\underset{\underset{NH_2}{|}}{\overset{\overset{H}{|}}{C}}-COOH$$

*Because the R group on glycine is hydrogen it has little effect on the rest of the molecule and glycine is often classified with the neutral amino acids having a polar R group.

†Proline and hydroxyproline are *imino acids*. The nitrogen atom, although joined to the α-carbon, is part of a ring. An imino nitrogen bears only one hydrogen atom but can still take part in the formation of proteins.

TABLE 15.1 *Continued*

A.2. Neutral Amino Acids with Polar R Group

10. **Serine (Ser)**
    (α-Amino-β-hydroxypropionic acid)

$$HOC\!-\!C\!-\!COOH$$

(with H, H above and H, NH$_2$ below)

11. **Threonine (Thr)**
    (α-Amino-β-hydroxybutyric acid)

$$CH_3\!-\!C\!-\!-\!C\!-\!COOH$$

(with H, H above and OH, NH$_2$ below)

12. **Cysteine (Cys)**
    (α-Amino-β-mercaptopropionic acid)

$$HS\!-\!CH_2\!-\!C\!-\!COOH$$

(with H above and NH$_2$ below)

13. **Tyrosine (Tyr)**
    (α-Amino-p-hydroxyhydrocinnamic
    acid)

$$HO\!-\!\bigcirc\!-\!C\!-\!C\!-\!COOH$$

(with H, H above and H, NH$_2$ below)

14. **Asparagine (Asn)**
    (β-Carbamoyl-α-
    aminopropionic acid)

$$\underset{H_2N}{\overset{O}{\diagup}}C\!-\!CH_2\!-\!C\!-\!COOH$$

(with H above and NH$_2$ below)

15. **Glutamine (Gln)**
    (γ-Carbamoyl-α-
    aminobutyric acid)

$$\underset{H_2N}{\overset{O}{\diagup}}C\!-\!CH_2\!-\!CH_2\!-\!C\!-\!COOH$$

(with H above and NH$_2$ below)

omission of certain amino acids in the diet of some animals prevents their normal growth and development. The animal organism either is incapable of synthesizing certain amino acids or is incapable of synthesizing them in sufficient quantity to maintain a normal state of good health. The effects of such malnutrition disappear when the missing amino acids are supplied in the diet. On the basis of such studies, α-amino acids are divided into two categories as **essential** or **nonessential.** Table 15.2 lists the α-amino acids in these two general classes.

15.4   *Properties of the Amino Acids. Isoelectric Point*

The amino acids are colorless crystalline solids and have melting points (with decomposition) in excess of 200°C. Most amino acids are soluble in water but sparingly soluble in organic solvents. These properties are not characteristic of most simple organic acids or simple amines but are more like those of salts.

TABLE 15.1 *Continued*

B. *Basic Amino Acids*

16. **Histidine (His)**
    (α-Amino-
    β-4-imidazolylpropionic acid)

$$\underset{\displaystyle \underset{\displaystyle H-C=\underset{\displaystyle NH}{C}-CH_2-\underset{\displaystyle \underset{\displaystyle NH_2}{|}}{\overset{\displaystyle \overset{\displaystyle H}{|}}{C}}-COOH}{\overset{\displaystyle N\diagup \diagdown NH}{}}}{\overset{\displaystyle \overset{\displaystyle H}{\overset{|}{C}}}{}}$$

17. **Lysine (Lys)**
    (α,ε-Diaminocaproic acid)

$$CH_2-CH_2-CH_2-CH_2-\overset{\displaystyle \overset{H}{|}}{\underset{\displaystyle \underset{NH_2}{|}}{C}}-COOH$$
$$\underset{NH_2}{|}$$

18. **Arginine (Arg)**
    (α-Amino-δ-guanidinovaleric acid)

$$H_2N-\overset{\displaystyle \overset{N}{\|}}{C}-\overset{\displaystyle \overset{H}{|}}{N}-CH_2CH_2CH_2-\overset{\displaystyle \overset{H}{|}}{\underset{\displaystyle \underset{NH_2}{|}}{C}}-COOH$$

C. *Acidic Amino Acids*

19. **Aspartic acid (Asp)**
    (Aminosuccinic acid)

$$\overset{\displaystyle \overset{O}{\|}}{HOC}-CH_2-\overset{\displaystyle \overset{H}{|}}{\underset{\displaystyle \underset{NH_2}{|}}{C}}-COOH$$

20. **Glutamic acid (Glu)**
    (α-Aminoglutaric acid)

$$\overset{\displaystyle \overset{O}{\|}}{HOC}-CH_2-CH_2-\overset{\displaystyle \overset{H}{|}}{\underset{\displaystyle \underset{NH_2}{|}}{C}}-COOH$$

TABLE 15.2 *The Amino Acids According to Nutritional Requirements*

Essential (indispensable)	**Nonessential** (dispensable)
Arginine*	Alanine
Glycine†	Asparagine
Histidine‡	Aspartic acid
Isoleucine§	Cysteine
Leucine§	Glutamic acid
Lysine§	Glutamine
Methionine§	Proline
Phenylalanine§	Serine
Threonine§	Tyrosine
Tryptophan§	
Valine§	

*Required for optimum growth in the rat and chick.
†Required for optimum growth in the chick.
‡Required by all subhuman species tested and by infants, but not by adult man.
§Required by all species tested, including adult man.

A neutral amino acid possesses an amino group and a carboxyl group and thus can behave either as a base or an acid. An amino acid in an alkaline solution reacts like an acid to produce a *metal salt*.

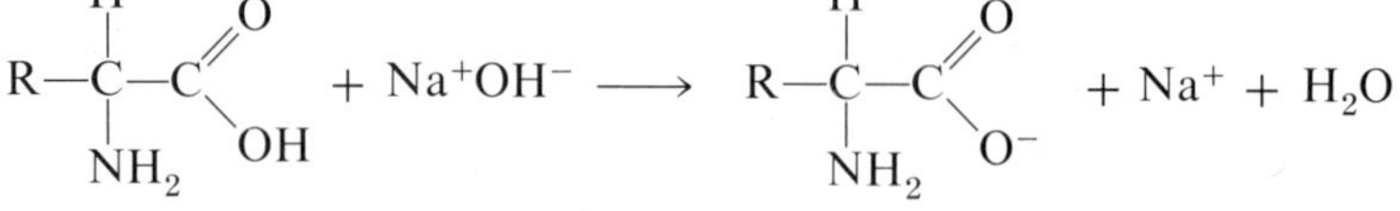

An amino acid anion

If an alkaline solution of an amino acid is electrolyzed, the anion of the amino acid salt migrates toward the anode or positive electrode.

An amino acid in an acidic solution behaves like a base to form an *amine salt*.

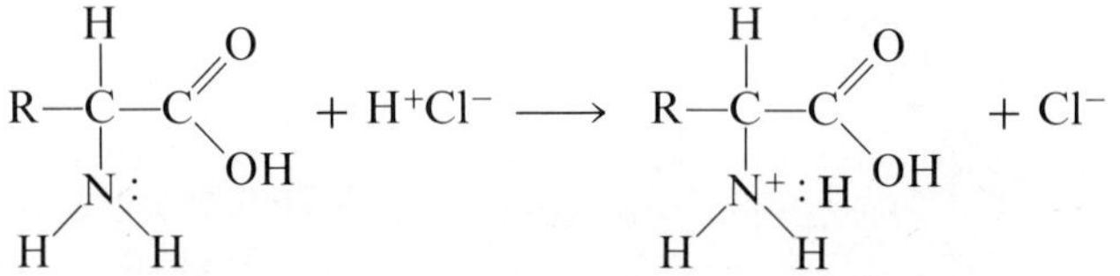

An amino acid cation

If an acidic solution of an amino acid is electrolyzed, the cation of the amine salt migrates to the cathode or negative electrode.

The hydrogen ion concentration at which the amino acid shows no net migration to either electrode is called the **isoelectric point.** Isoelectric points are given in pH values and vary from low values for acidic amino acids (pH 3 for aspartic acid) to high values for basic amino acids (pH 10.8 for arginine). Neutral amino acids do not have isoelectric points at the neutral figure (pH 7.0), as might be expected, but slightly on the acid side. At its isoelectric point, a "neutral" amino acid is completely ionized, with the proton shifting from the carboxyl group to the amino group to produce an inner salt or **dipolar ion.**

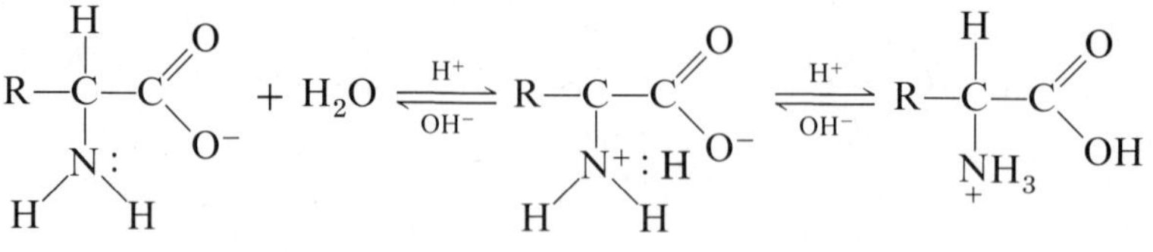

Dipolar ion

EXERCISE 15.1

The amino acid cation (second equation Sec. 15.4) has two acidic groups. What are they? Which group is more acidic—that is, will give up a proton more readily when base is added to a solution of the amino acid hydrochloride? (*Hint: Observe the changes in the dipolar ion (same section) with a change in pH.*)

Dipolar ions, as the preceding equation illustrates, are *amphoteric* and can act either as acids or bases.

The pK_a of the carboxylic acid group can be determined by adding one-half equivalent of aqueous sodium hydroxide (enough to half-neutralize the carboxyl group) to a solution of the amino acid cation (III) and measuring the pH (Sec. 10.2). At this point the solution will contain approximately equal amounts of III and the dipolar ion (II). Addition of a full of equivalent of base (total) will convert most of the III into II and bring the pH of the solution to (approximately) the isoelectric point. Addition of a further half-equivalent of base will half-neutralize the ammonium ion (conjugate acid of the amino group), at which point the pH will equal (approximately) the pK_a of the ammonium ion. The solution will contain roughly equal amounts of II and the carboxylate salt (I). In general, the isoelectric pH will be the average of the two pK_a values (that is, one-half their sum). Some pK_a data are given in Table 15.3.

TABLE 15.3 *Ionization Constants of the Amino Acids in Water at 25°C**

Amino Acid	pK_1	pK_2	pK_3
Alanine	2.29	9.74	
Arginine	2.01	9.04	12.48
Asparagine	2.02	8.80	
Aspartic acid	2.10	3.86	9.82
Cysteine	2.05	8.00	10.25
Glutamic acid	2.10	4.07	9.47
Glutamine	2.19	9.13	
Glycine	2.35	9.78	
Histidine	1.77	6.10	9.18
Isoleucine	2.32	9.76	
Leucine	2.33	9.74	
Lysine	2.18	8.95	10.53
Methionine	2.28	9.21	
Phenylalanine	2.58	9.24	
Proline	2.00	10.60	
Serine	2.21	9.15	
Threonine	2.09	9.10	
Tryptophan	2.38	9.39	
Tyrosine	2.20	9.11	10.07
Valine	2.29	9.72	

* pK_1 is for the COOH. pK_2 is for the second COOH, if present; otherwise, it is for the ammonium ion. pK_3 is for an ammonium ion or other weak acid function.

15.5   Synthesis of Amino Acids

Interest in obtaining the individual amino acids in pure form for use in nutrition experiments has led to a number of synthetic procedures for the preparation of amino acids. Some of these are adaptations of methods previously reviewed for the preparation of simple primary amines (Sec. 13.3). Of the many methods that have been employed for the preparation of amino acids, we shall consider only three. These three methods are general methods and are satisfactory for the preparation of *some* amino acids. No single method has been developed for the synthesis of *all* amino acids.

A. Direct Amination[1] of α-Halogen Acids. In this method the halogen atom of an α-halo acid is replaced by an amino group directly by treating the acid with ammonia. In practice, a large excess of ammonia is used to prevent the formation of a disubstituted ammonia derivative.

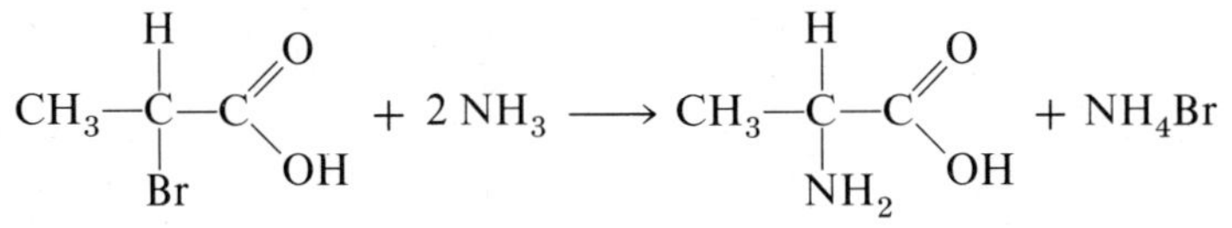

α-Bromopropionic acid   D,L-Alanine

B. Indirect Amination of α-Halogen Acids. The halogen of α-halo acids may be replaced by the —NH_2 group indirectly by adaptation of Gabriel's primary amine synthesis (Sec. 13.3-C). In this synthesis, an α-halo ester is condensed with the potassium salt of phthalimide.

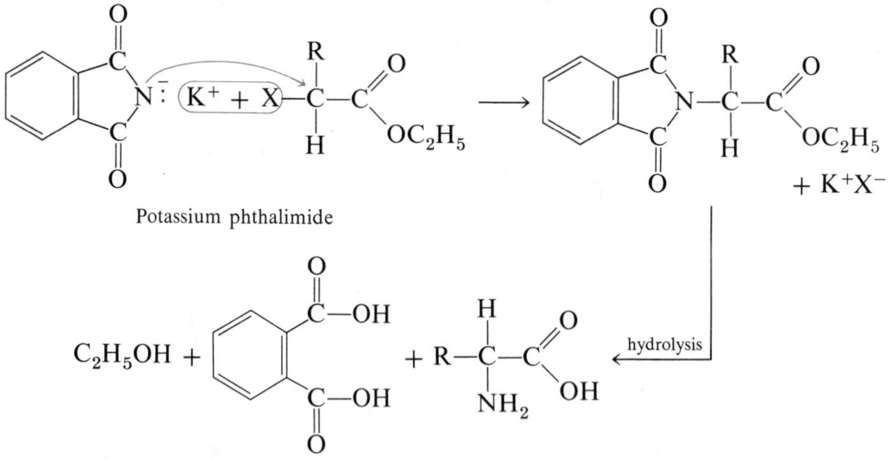

Potassium phthalimide

Phthalic acid

C. The Hydrolysis of α-Amino Nitriles (Strecker Synthesis). In the **Strecker synthesis** of an amino acid, an aldehyde is treated with ammonium cyanide (ammonia and hydrogen cyanide). The aminonitrile that results from this reaction then is hydrolyzed to the amino acid. The method lends itself very well to the preparation of simple neutral amino acids. The preparation of D,L-alanine by the Strecker synthesis begins with acetaldehyde.

---

[1] A reaction that introduces the amino group into the molecule is called *amination*.

$$CH_3-\overset{\overset{\displaystyle H}{|}}{C}=O + HCN + NH_3 \longrightarrow CH_3-\overset{\overset{\displaystyle H}{|}}{\underset{\underset{\displaystyle NH_2}{|}}{C}}-CN + H_2O$$

Acetaldehyde   Aminonitrile

$$CH_3-\overset{\overset{\displaystyle H}{|}}{\underset{\underset{\displaystyle NH_2}{|}}{C}}-CN + 2\,H_2O \longrightarrow CH_3-\overset{\overset{\displaystyle H}{|}}{\underset{\underset{\displaystyle NH_2}{|}}{C}}-C\overset{\displaystyle O}{\underset{\displaystyle OH}{\diagup}} + NH_3$$

D,L-Alanine

Any synthesis of amino acids (except glycine) leads to a racemic mixture and must be followed by resolution if the natural form of the amino acid is required.

EXERCISE 15.2

Using acrylic acid, $CH_2{=}CHCOOH$, and any other reagents you might require, outline a synthesis leading to aspartic acid.

15.6 *Reactions of the Amino Acids*

The reactions of the amino acids are, in general, reactions characteristic of both carboxylic acids and primary amines.

A. Esterification. All amino acids can be esterified. Emil Fischer utilized this reaction as early as 1901 as a technique for the separation of the constituent amino acids obtained from a protein hydrolysate. The protein was hydrolyzed to its constituent amino acids, the mixture was esterified, and the liquid amino acid esters separated by fractional distillation. Fischer's method is illustrated by the following equations.

$$CH_3-\overset{\overset{\displaystyle }{|}}{\underset{\underset{\displaystyle NH_2}{|}}{CH}}-C\overset{\displaystyle O}{\underset{\displaystyle OH}{\diagup}} + HCl + C_2H_5OH \longrightarrow CH_3-\overset{\overset{\displaystyle H}{|}}{\underset{\underset{\displaystyle \overset{+}{N}H_3}{|}}{C}}-C\overset{\displaystyle O}{\underset{\displaystyle OC_2H_5}{\diagup}} + H_2O + Cl^-$$

$$CH_3-\overset{\overset{\displaystyle H}{|}}{\underset{\underset{\displaystyle \overset{+}{N}H_3}{|}}{C}}-C\overset{\displaystyle O}{\underset{\displaystyle OC_2H_5}{\diagup}} + OH^- \longrightarrow CH_3-\overset{\overset{\displaystyle H}{|}}{\underset{\underset{\displaystyle NH_2}{|}}{C}}-C\overset{\displaystyle O}{\underset{\displaystyle OC_2H_5}{\diagup}} + H_2O$$

Ethyl ester of alanine

B. Reaction with Nitrous Acid. The amino acids, with the exception of proline, react with nitrous acid to liberate nitrogen gas. This reaction is the basis for the **Van Slyke method** for determining "free" amino groups (uncombined α-amino groups) in protein material. The reaction gives an index to the number of uncombined —NH_2 groups such as would be provided by certain basic amino acids.

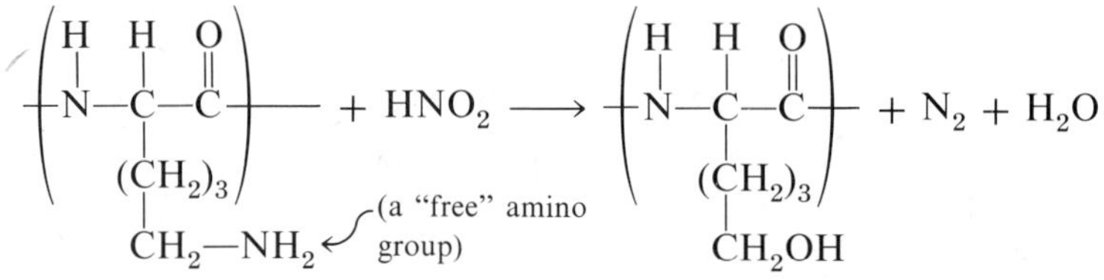

Lysine unit

C. Reaction with Acid Halides or Acid Anhydrides. The amino group of amino acids is readily converted to an amide by reaction with an acid halide or anhydride. Thus, glycine reacts with benzoyl chloride to produce hippuric acid. Benzoates, when ingested, are rendered water-soluble and voided from the body in the urine as hippuric acid.

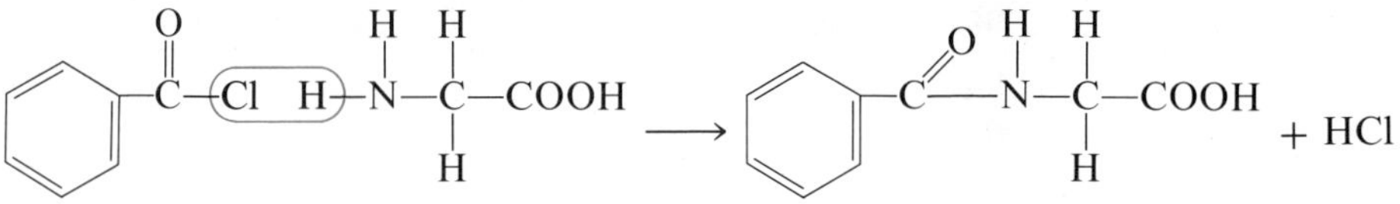

Benzoyl chloride   Glycine   Hippuric acid

D. Reaction with Ninhydrin. Amino acids react with a ninhydrin solution (triketohydrindene hydrate) to produce purple compounds. The reaction is of value in the assay of protein material and can be used for the quantitative determination of amino acids. The following sequence of reactions illustrates how ninhydrin converts an amino acid into an aldehyde and carbon dioxide.

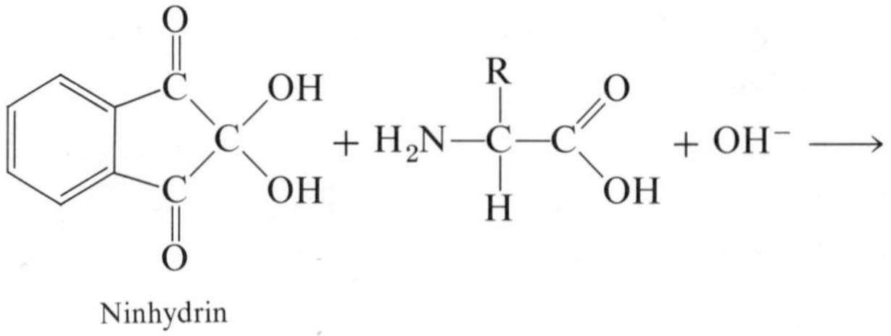

Ninhydrin

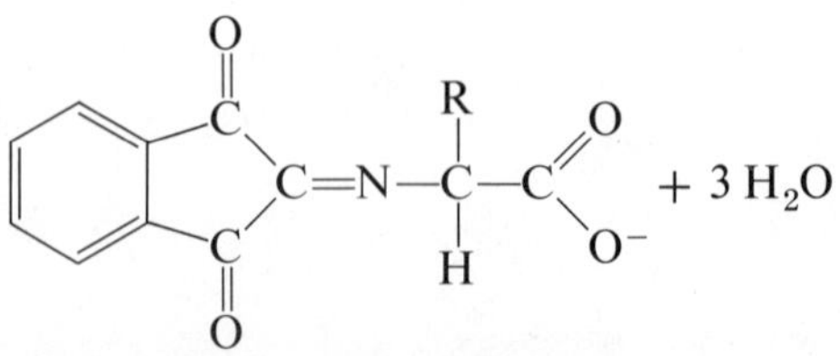

(Colored product)

E. Peptide Formation. Amino acids can be condensed with each other to form peptides (Sec. 15.7). However, the sequence of the individual amino acids in a peptide cannot be controlled unless steps are taken to block reaction at an amino group or at a carboxyl function. Without such protective measures, random condensation can occur. In practice, when amino acids are to be joined into peptides, the amino acid to appear first in the peptide chain (counting from left to right) must have its amino group protected. The protecting group is removed after condensation is completed. The method is illustrated in the preparation of the dipeptide, alanylglycine (Ala-Gly), with both amino and carboxyl group protection.

Step 1.
Amino group
of alanine
protected.

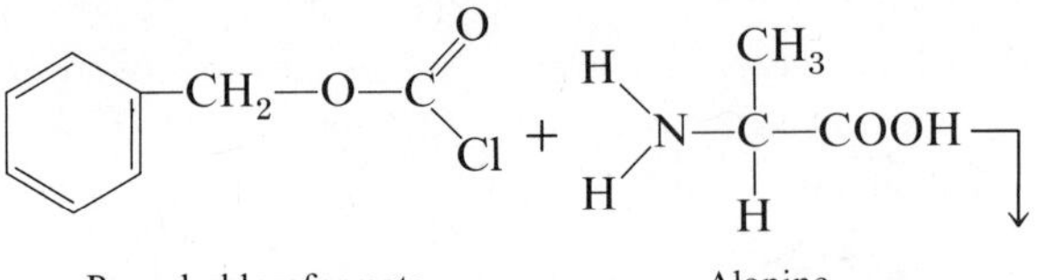

Benzyl chloroformate
(Benzyl chlorocarbonate
or carbobenzoxy chloride)

Alanine

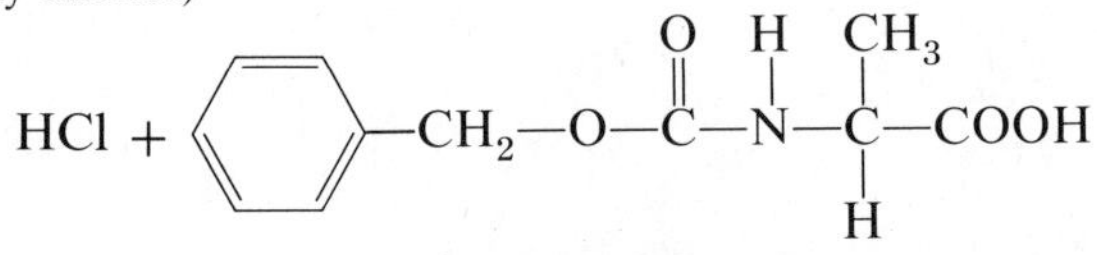

HCl +

Benzyloxycarbonylalanine
(Carbobenzoxyalanine)
Cbz · Ala

Step 2.
Carboxyl group
of glycine
protected.

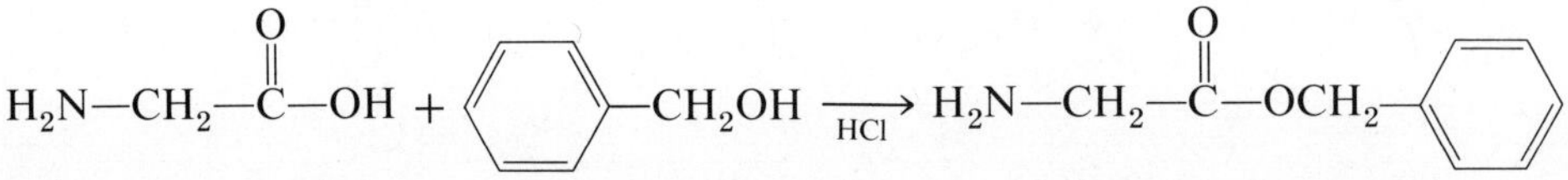

$$H_2N-CH_2-\overset{\overset{\displaystyle O}{\|}}{C}-OH \ + \ \text{—CH}_2\text{OH} \ \xrightarrow{HCl} \ H_2N-CH_2-\overset{\overset{\displaystyle O}{\|}}{C}-OCH_2-$$

Glycine

Benzyl alcohol

Glycine benzyl ester

Next the two protected amino acids may be condensed using the generally
useful coupling reagent, dicyclohexylcarbodiimide (DCC), which is a specific
catalyst for the condensation of carboxylic acids with alcohols and amines.

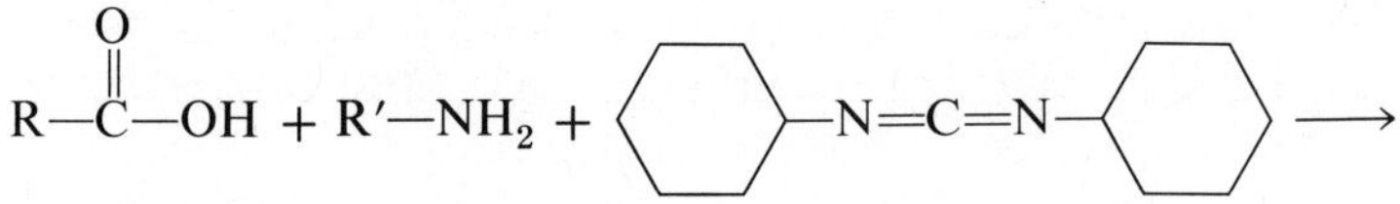

$$R-\overset{\overset{\displaystyle O}{\|}}{C}-OH \ + \ R'-NH_2 \ + \ \text{—N}=C=N\text{—} \ \longrightarrow$$

Carboxylic
acid

Amine

Dicyclohexylcarbo-
diimide

$$R-\overset{\overset{\displaystyle O}{\|}}{C}-NH-R' \ + \ \text{—NH}-\overset{\overset{\displaystyle O}{\|}}{C}-NH\text{—}$$

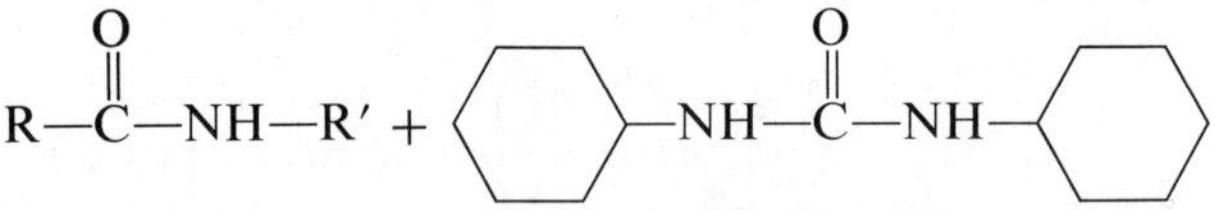

Amide

Dicyclohexylurea
(*very insoluble*)

Step 3.
Condensation.

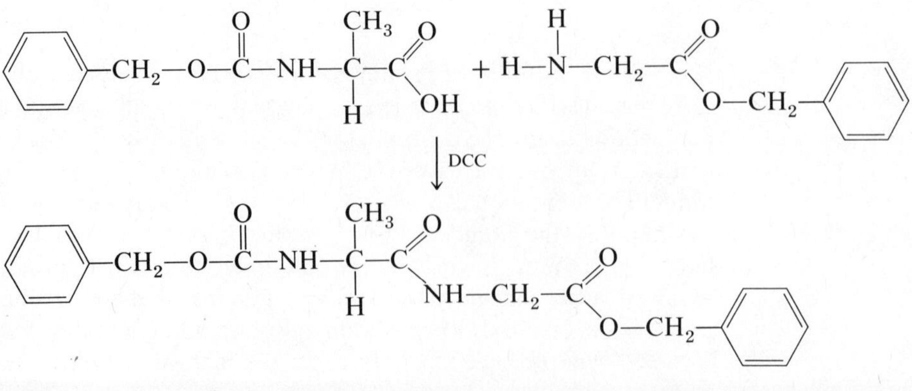

DCC

Finally, both protecting groups are removed by catalytic **hydrogenolysis,** a procedure that is specific for the reduction of benzyl esters to the acid plus toluene.

Step 4.
Hydrogenolysis.

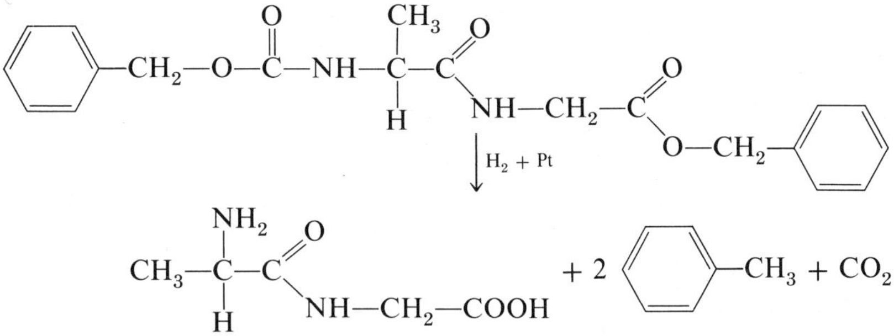

Alanylglycine (a dipeptide)   Toluene

Peptides and Proteins

15.7 *Structure and Nomenclature*

In a protein molecule the α-amino acids are joined together through amide linkages formed between the amino group of one acid molecule and the carboxyl group of another. Such amide linkages are called peptide links and serve to unite hundreds of amino acid residues in a protein molecule. When only two α-amino acids are joined, as in alanylglycine (Sec. 15.6-E), the product is a dipeptide. A dipeptide, with free amino and carboxyl groups on opposite ends of the chain, can unite at either end with a third α-amino acid to form a tripeptide. A tripeptide can form a tetrapeptide, and so on, until finally a long-chain polypeptide results. If n is the number of different amino acids in a polypeptide, the number of possible sequences is *n factorial,* $(n!)$. Thus, three different amino acids could combine in one of six different sequences, $(3 \times 2 \times 1 = 6)$. One such sequence is illustrated below.

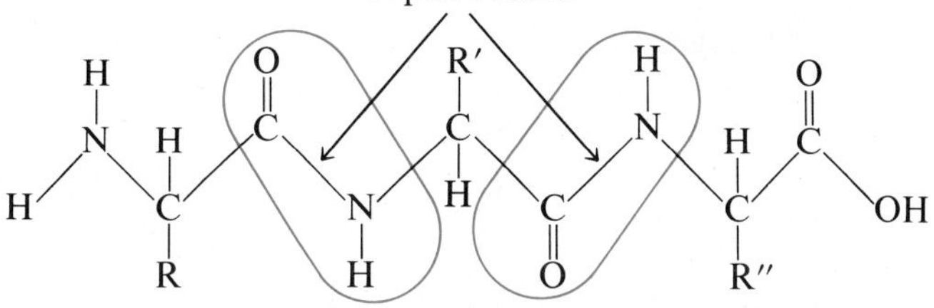

Peptide links

A variation in sequence, as well as a variation in the number and kind of amino acids joined, makes possible an almost infinite number of arrangements. It must not be thought, however, that the manner in which amino acids are joined in nature to form the polypeptide links in proteins is a haphazard one. On the contrary, the amino acids which comprise a protein are joined uniformly to give specificity to certain types of tissue within each organism.

Peptides are named as derivatives of the C-terminal amino acid which still has a free carboxyl group. The C-terminal amino acid is the last one, counting from left to right. The method of naming is illustrated by the following tetrapeptide.

Glycyl-alanyl-lysyl-tyrosine

15.8 *The Sequence of Amino Acids in Peptides*

Determination of the amino acid sequence of a natural peptide is a tedious and laborious process but has been accomplished in a few cases. In practice, the N-terminal group of a peptide is "tagged" by reaction with Sanger's reagent, 2,4-dinitrofluorobenzene, or with dansyl chloride (5-*d*imethylamino-1-*n*aphthalene*s*ulfon*yl* chloride (Sec. 13.4-C) before the peptide is hydrolyzed.

The tagged amino acid is liberated by hydrolysis. Since it is colored, it is easily distinguished from the other amino acids that comprise the peptide structure. The advantage of the dansyl chloride procedure, which is chemically similar to the Sanger method, is that it yields easily detected fluorescent products and, thus, is about 100 times as sensitive as the Sanger method.

EXERCISE 15.3

Write equations similar to those given for the Sanger method, substituting dansyl chloride for Sanger's reagent.

A third procedure, called the **Edman degradation,** may be used, not only for the identification of N-terminal groups but also for the establishment of the sequence of amino acids in the polypeptide chain by a *stepwise* cleavage of the amino acid from the N-terminus. The reagent used is phenylisothiocyanate, which gives a product that can be split by treatment with anhydrous acid to form the phenylthiohydantoin derivative of the N-terminal amino acid.

Phenylisothiocyanate   Polypeptide

Phenylthiohydantoin derivative
of N-terminal amino acid

The process can be repeated to remove the new N-terminal amino acid and has been automated to permit the rapid determination of sequences of up to 20 amino acids by the sequential removal and identification of amino acids. Other methods are available for the sequential degradation of polypeptides from the C-terminal end. To illustrate the general procedure let us review the work that led to the elucidation of the *glutathione* structure.

Glutathione is a tripeptide isolated from yeast. A complete enzymatic or acid hydrolysis of the peptide revealed that it was composed of only three α-amino acids: L-glutamic acid, L-cysteine, and glycine. If each of these α-amino acids had but one amino group and one carboxyl group, the number of possible sequences would be six (Sec. 15.7). However, glutamic acid (see Table 15.1) has two carboxyl groups. This increases the total number of possible sequences to twelve, for we do not know whether the α-carboxyl or the γ-carboxyl group of glutamic acid is involved in a peptide link. Mild hydrolysis of glutathione gave two dipeptides. One of these, on further treatment, yielded cysteine and glutamic acid, the other gave cysteine and glycine. This much is now known about glutathione: the cysteine structure is linked to both glycine and glutamic acid. But how? A little pencil work now will show that only the three sequences I, II, and III could possibly fit that of glutathione.

Cys-Glu

I

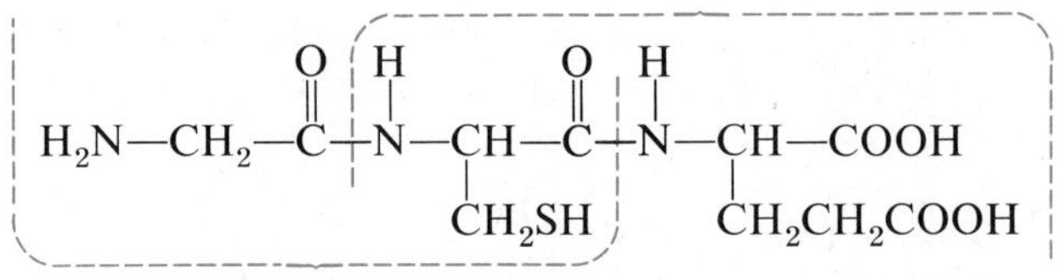

Gly-Cys

Cys-Gly

II

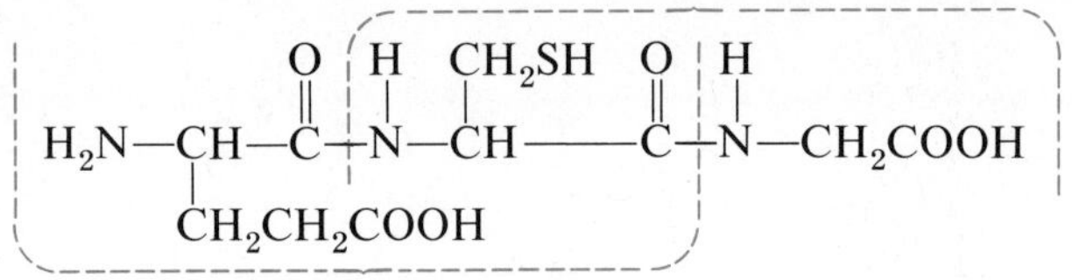

α-(COOH) Glu-Cys

Cys-Gly

III

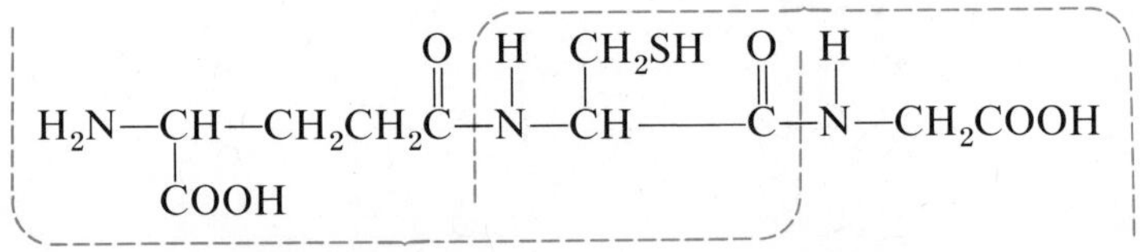

γ-(COOH) Glu-Cys

The tagging technique identified glutamic acid as the N-terminal amino acid. This finding eliminates structure I as a possibility. The synthesis of the remaining two dipeptides is now in order. In one of these the α-carboxyl of glutamic acid must be linked to cysteine, and in the other the γ-carboxyl must be joined. The latter structure proved to be identical with that of the dipeptide obtained by hydrolysis. The correct structure for glutathione is thus established as that shown by the tripeptide III.

Once the proper sequence of amino acids in a peptide is known, a synthesis for it usually follows. The historic accomplishment of duVigneaud[2] and his co-workers in determining the oxytocin structure provides such an outstanding example.

---

[2] Vincent duVigneaud (1901–   ), Cornell University College of Medicine. Winner of the Nobel Prize in Chemistry, 1955.

Oxytocin is a peptide hormone produced by the posterior lobe of the pituitary gland. It has the function of causing a contraction of smooth muscle, particularly uterine muscle, and finds application in obstetrics. In addition to this function, the hormone also promotes the flow of milk from the mammary glands.

Oxytocin, on hydrolysis, produced one molecule each of leucine, isoleucine, proline, glutamic acid, tyrosine, aspartic acid, cystine, glycine, and three equivalents of ammonia. The sequence of the α-amino acids in oxytocin was determined and the structure shown in Fig. 15.1 was assigned to it. The research team subsequently synthesized an octapeptide of the same sequence. This outstanding achievement represents the first synthesis of a peptide hormone.

FIGURE 15.1 *Oxytocin**

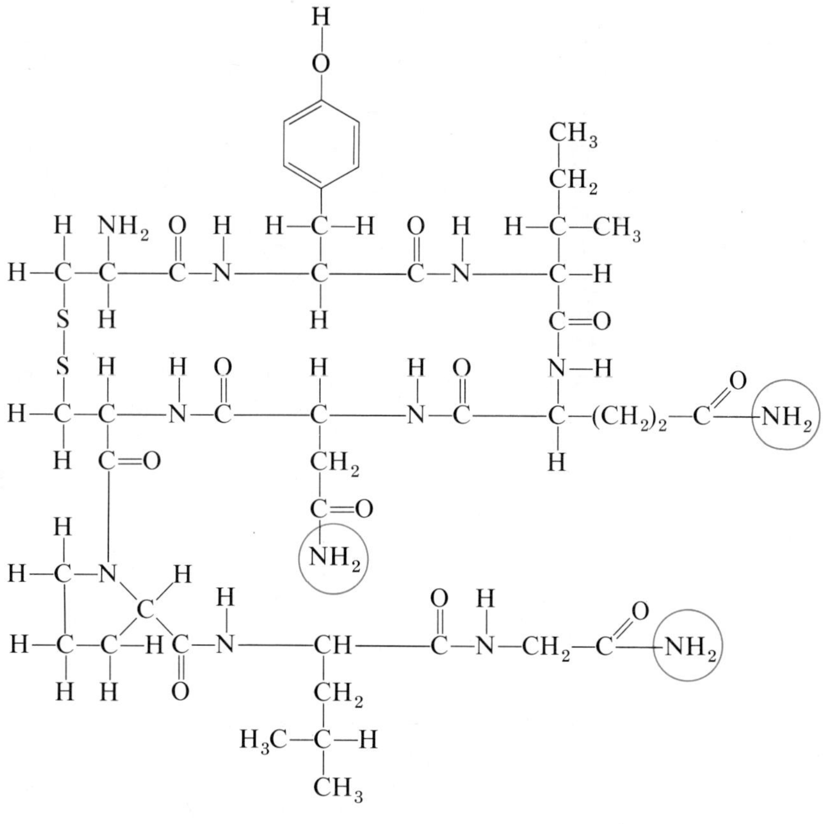

* Areas circled in color indicate the free amino groups of the amides of glycine, glutamic acid, and aspartic acids.

Another brilliant example of determining the sequence of amino acids in a natural polymer was that performed by Sanger[3] of England. He and his associates, after years of diligent investigation and by the use of ingenious techniques, were able to elucidate the amino acid sequence of the hormone insulin. Until about 1955 the amino acid sequence in any polypeptide within the protein range was unknown. The molecular weight of beef insulin was determined to be 5,734 and its composition that of forty-eight amino acid residues of sixteen different kinds!

EXERCISE 15.4

Why could not bromobenzene be used for tagging the N-terminal α-amino acid of a peptide? (*Hint: See Section 7.5.*)

15.9 *The Composition and Structure of Proteins*

Proteins differ from carbohydrates and fats in elementary chemical composition. All proteins contain, in addition to carbon, hydrogen, and oxygen, other elements in the approximate percentages as follows: nitrogen (15%), sulfur (1.0%), and phosphorus (0.5%). The molecular weights of proteins are unbelievably high, ranging from 10,000 to 10,000,000.

Acceptance of the peptide hypothesis of protein structure, first proposed in 1902 by Fischer and Hofmeister, has led to many other questions regarding protein structures. How are peptide chains held together? How are they arranged in space? X-ray analysis and the persistent efforts of numerous investigators have revealed the answer to some of these questions.

The long polypeptide chains which comprise the fibrous proteins of hair and wool (the α-keratins) are arranged in right-handed helical coils with 3.6 amino acid units per turn. The coils are held together by hydrogen bonding between the amide hydrogen atom in one peptide link and the carbonyl oxygen atom of another peptide link three amino acid units beyond the first. Such a coil is called an α-helix. All of the polypeptide chains run in the same direction (from N-terminal to C-terminal), and the coiled chains are held together by the formation of disulfide bonds, —S—S— (Sec. 13.9), between adjacent polypeptide chains. This arrangement is shown graphically on the next page, where *intra*chain hydrogen bonds are shown by dotted lines.

In silk fibroin (a β-keratin) the polypeptide chains are grouped together in side-by-side chains joined by hydrogen bonds *between* chains. Adjacent polypeptides run in opposite directions (antiparallel). The overall appearance of the

---

[3] Frederick Sanger (1918–   ), Cambridge University. Winner of the Nobel Prize in Chemistry, 1958.

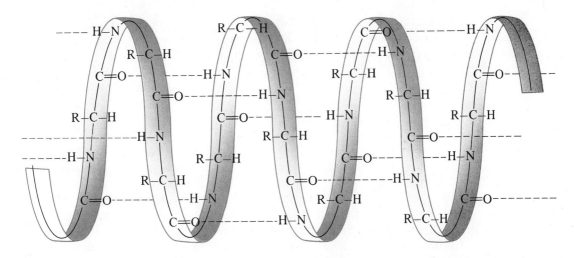

β-conformation is of a *pleated sheet,* which you may visualize by studying the schematic structure below in which the *inter*chain hydrogen bonds are shown by dotted lines. Note that the R groups of the amino acids project out from the pleated surface, while the smaller hydrogen atoms on the α-carbon atoms of the amino acid units are located in the folds of pleats. No disulfides cross-links are found in β-keratins.

The collagen of connective tissue consists of three kinked, left-handed helices held together by hydrogen bonds. The globular proteins such as the catalytic proteins (enzymes) and transport proteins (myoglobin, hemoglobin, cyctochrome C, etc.) are more complex structures in which the chains are folded into three-dimensional globular shapes, as implied by the name. In the olio-gomeric globular proteins two or more chains (four in hemoglobin) are inter-twined. Segments of these folded chains may have the α-helix structure or the β-conformation. Generally, the folds are so arranged as to have polar groups on the exterior (to promote water solubility) and non-polar (hydrophobic) groups inside. The catalytic or transport function of a protein requires the presence of a unique structural feature called the **active site.** Thus, for example, in an enzyme the substrate is bound to the active site while being acted upon. How enzymes and the transport proteins function and the nature of the active sites are among the major objectives of contemporary research in biochemistry.

15.10 *Classification of Proteins*

Proteins may be classified as simple proteins or as conjugated proteins. Simple proteins are those which yield, on hydrolysis, only α-amino acids. Albumin in eggs, gluten in wheat, keratin in hair, and collagen in connective tissue are examples of simple proteins. Conjugated proteins are those which, on hydrolysis, yield other compounds in addition to α-amino acids. Such nonprotein materials are called **prosthetic** groups. Hemoglobin is an example of a conjugated protein. The prosthetic group, in this case, is the iron-containing porphyrin structure called heme (Sec. 17.2).

15.11 *Properties of the Proteins*

Proteins when heated coagulate, or precipitate. The protein albumin, found in egg white, is a common example of a protein easily coagulated by heat. The salts of certain heavy metals (silver, mercury, lead) also cause proteins to precipitate. The immediate ingestion of protein material such as egg white as an antidote for accidental heavy metal poisonings is based on this behavior. The cauterizing action of silver nitrate is another application. Proteins also are precipitated by certain acids. Precipitation of protein by any agent is sometimes an irreversible change and the precipitated protein is said to be **denatured.**

A number of chemical tests on proteins produce color reactions. One of these, the **biuret test,** produces a pink or purple color when an alkaline solution of protein material is treated with a very dilute cupric sulfate solution. The test is specific for multiple peptide links and is not given by α-amino acids. It is a convenient test and often is used on a protein hydrolysate to determine the completeness of hydrolysis.

The **xanthoproteic test** is a yellow color reaction produced when a protein is treated with concentrated nitric acid. The test is simply a nitration of the aromatic ring of certain amino acids (tyrosine, phenylalanine, and tryptophan) and is one recognized by every student as the familiar nitric acid stain. Another color test for proteins is the **Millon test.** The reagent used for this test is a mixture of mercuric and mercurous nitrates. A protein when treated with Millon's reagent and heated produces a red color. Any phenolic compound will give the reaction, and the Millon test, like others which give colored reactions, is dependent upon the presence of certain individual amino acids in the protein molecule.

15.12 Nutritional Importance of Proteins

Proteins provide one of the major nutrients for the body, but their utilization differs from that of the fats and carbohydrates. Whereas fats and carbohydrates are used primarily to supply heat and energy, the proteins are used mainly to repair and replace wornout tissue. Such repairs and replacements are made of protein material which the animal organism has synthesized from other ingested proteins. Unlike certain of the lower plants, animals are not capable of "fixing"[4] atmospheric nitrogen, or converting ammonium or nitrate salts into proteins. Animals obtain their protein by eating plants that have synthesized protein material, or by eating other animals that have eaten such plants. The digestion of proteins to α-amino acids by the body supplies the required building material from which the animal's own protein can be formed. So far as is known, and in contrast to carbohydrates and fat depots, there are no body depots of proteins which serve as stores and have no other function.

15.13 Metabolism of Proteins

The digestion of proteins leads to mixtures of simple amino acids and polypeptides of varying lengths. Digestion destroys the specificity of a protein and frees the constituent amino acids for the synthesis of new proteins that suit the requirements of the individual. Amino acids not required for such syntheses are converted to other necessary foods such as carbohydrates and fat and ultimately are oxidized to yield energy. The metabolic changes involved in these conversions are very complex.

The amino acids produced by the digestion of a protein are absorbed through the intestinal wall into the blood and are transported to the liver. Certain of the amino acids then proceed from the liver to other tissues. Amino acids are required by the cells for the synthesis of proteins, enzymes, certain hormones, and other nitrogen-containing substances. Body tissues are capable of synthesizing some amino acids (nonessential) by removing the required

---

[4] The conversion of atmospheric nitrogen into nitrates or other nitrogenous compounds. Certain free-living soil bacteria and others that live in nodules on the roots of leguminous plants (peas, beans, clover) are able to "fix" and store nitrogen in the form of nitrates.

amino groups from other amino acids. The amino group taken from one acid is transferred to an α-keto analogue of the amino acid to be synthesized. This transfer is called **transamination** and is illustrated below.

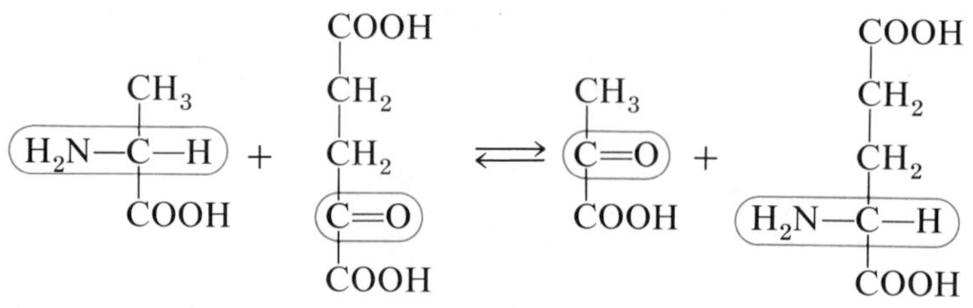

$$
\begin{array}{c}
\text{CH}_3 \\
\text{H}_2\text{N}-\overset{|}{\text{C}}-\text{H} \\
\text{COOH}
\end{array}
+
\begin{array}{c}
\text{COOH} \\
\text{CH}_2 \\
\text{CH}_2 \\
\text{C}=\text{O} \\
\text{COOH}
\end{array}
\rightleftharpoons
\begin{array}{c}
\text{CH}_3 \\
\text{C}=\text{O} \\
\text{COOH}
\end{array}
+
\begin{array}{c}
\text{COOH} \\
\text{CH}_2 \\
\text{CH}_2 \\
\text{H}_2\text{N}-\overset{|}{\text{C}}-\text{H} \\
\text{COOH}
\end{array}
$$

Alanine α-Ketoglutaric acid   Pyruvic acid   Glutamic acid

The amino group of α-amino acids, if not required for the synthesis of new amino acids, can be oxidatively removed. This process is called **deamination.**

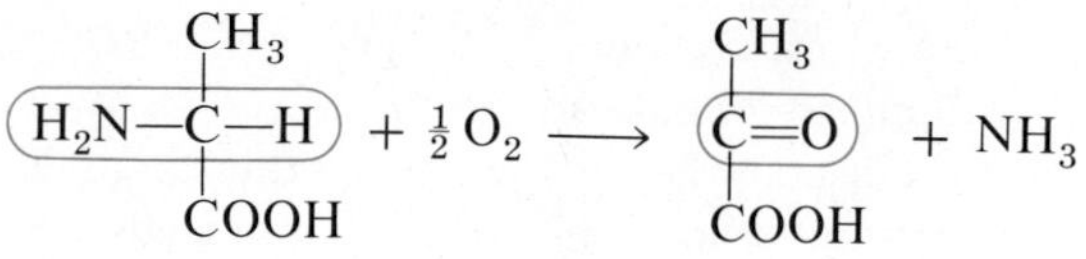

$$
\begin{array}{c}
\text{CH}_3 \\
\text{H}_2\text{N}-\overset{|}{\text{C}}-\text{H} \\
\text{COOH}
\end{array}
+ \tfrac{1}{2}\text{O}_2 \longrightarrow
\begin{array}{c}
\text{CH}_3 \\
\text{C}=\text{O} \\
\text{COOH}
\end{array}
+ \text{NH}_3
$$

Alanine   Pyruvic acid

The ammonia formed by deamination combines with carbon dioxide through a series of enzymatic reactions to produce urea.

$$
2\,\text{NH}_3 + \text{CO}_2 \longrightarrow \text{H}_2\text{N}-\overset{\overset{\displaystyle \text{O}}{\|}}{\text{C}}-\text{NH}_2 + \text{H}_2\text{O}
$$

Urea

Urea is eliminated from the body by way of the urine and is a major end-product of protein metabolism.

The carbon skeleton of an amino acid, after the amino group is removed by either of the processes described, may be used in the synthesis of other amino acids or it may enter the pathways of carbohydrate and fatty acid metabolism. The latter course proceeds by way of the tricarboxylic acid cycle (Sec. 14.22). Thus the deaminated amino acid, like a fat or a carbohydrate, may ultimately be converted to carbon dioxide, water, and energy.

15.14   *Improper Metabolism of Proteins. Allergies*

The inability of some persons to accomplish complete hydrolysis of certain protein material may result in the absorption of minute amounts of unchanged protein from the intestinal tract. Such metabolic failures cause the individual to become extremely sensitive or allergic to certain foods. When present even in minute amounts and eaten unknowingly as ingredients in other foods, the effects of such allergens can be very distressing. Sneezing, hives, eczema, and general discomfort result. The old phrase "one man's food is another's poison" has some basis in fact. Proteins injected as serums or as antibiotics sometimes, when incompatible with the individual, produce even more serious results. Incompatibilities of the kind described appear to be genetically related and part of the heredity of the individual. The complex chemistry that controls heredity rests in the area of nucleoproteins (Chapter 16) and is currently the area of greatest excitement to biochemists.

Summary

1. Proteins are high molecular weight natural polymers made up largely by combination of various α-amino acids.

2. Approximately twenty amino acids comprise the bulk of plant and animal protein. Names and structures for the common amino acids are listed in Table 15.1.

3. Amino acids may be classified according to properties as (a) neutral (one amino group and one carboxyl group), (b) acidic (more than one carboxyl group per amino group), and (c) basic (more than one amino group per carboxyl function).

4. The α-amino acids with the exception of glycine are optically active.

5. The α-amino acids obtained from plant or animal protein have the L-configuration.

6. Amino acids may be grouped under a nutritive classification into two categories: (a) **essential** (required in the diet because the animal organism is incapable of synthesizing it), and (b) **nonessential** (not required in the diet).

7. Amino acids have both the properties of carboxylic acids and primary amines. Amino acids form inner salts or dipolar ions.

8. Amino acids have an isoelectric point.

9. Individual amino acids can be separated from protein hydrolysates. Amino acids can be synthesized by (a) amination of α-halogen acids with ammonia, (b) an adaptation of the Gabriel primary amine synthesis, and (c) the Strecker synthesis.

10. Amino acids generally show reactions characteristic of both the primary amines and the carboxylic acids. The principal reactions of amino acids are:

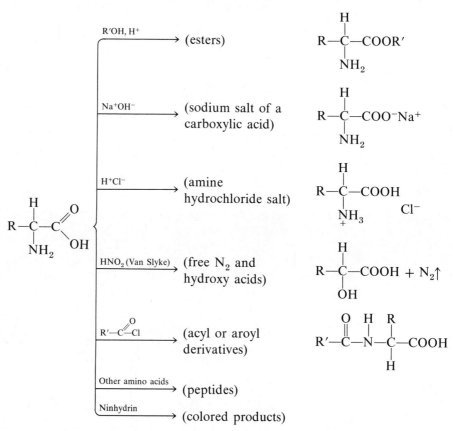

11. Two or more amino acids joined through amide linkages form peptides. A peptide is named as a derivative of the amino acid with a C-terminal carboxyl function.

12. Proteins, when hydrolyzed, yield amino acids and polypeptides.

13. Proteins are classified as simple proteins or conjugated proteins. Simple proteins on complete hydrolysis give only amino acids. Conjugated proteins on hydrolysis give, in addition to amino acids, nonprotein *prosthetic* groups.

14. Proteins are easily precipitated, or coagulated, by heat, by acids, and by certain heavy metals. Such coagulation is sometimes irreversible and is called *denaturation*.

15. Proteins are used in the body mainly to repair and replace worn-out tissue.

16. The digestion and metabolism of protein material begins with its hydrolysis to α-amino acids. The α-amino acids, by **transamination** and/or **deamination,** are converted into other α-amino acids or become oxidized to carbon dioxide and water.

New Terms

active site

conjugated protein

deamination

denaturation

dipolar ion

essential amino acid

α-helix

isoelectric point

peptide

pleated sheet (β-conformation)

prosthetic group

simple protein

transamination

Supplementary Exercises

EXERCISE 15.6 Draw the structures and name two sulfur-containing α-amino acids. Which of these can form disulfide bonds?

EXERCISE 15.7 Draw the structures of three amino acids which would give a yellow color with concentrated nitric acid.

EXERCISE 15.8 Beginning with a 3-carbon alcohol outline all necessary steps in the preparation of D,L-alanine. Include as one step an adaptation of Gabriel's method for the preparation of primary amines.

EXERCISE 15.9 Complete the following equations.

(a) $CH_3-\underset{\underset{NH_2}{|}}{\overset{\overset{H}{|}}{C}}-COOH + NaNO_2 + HCl \longrightarrow$

(b) $C_6H_5-\overset{\overset{O}{\parallel}}{C}\diagdown_{Cl} + H_2NCH_2COOH \longrightarrow$

(c) $HS-CH_2-\underset{\underset{NH_2}{|}}{\overset{\overset{H}{|}}{C}}-COOH + I_2 \longrightarrow$

$$\text{(d) } CH_3-\underset{\underset{NH_2}{|}}{\overset{\overset{H}{|}}{C}}-COOH + NaOH \longrightarrow$$

(e) [benzene]—$CH_2-O-\overset{\overset{O}{\|}}{C}\diagdown_{Cl}$ + $CH_3-\underset{\underset{NH_2}{|}}{\overset{\overset{H}{|}}{C}}-\overset{\overset{O}{\|}}{C}\diagdown_{OH}$ $\longrightarrow$

(f) Product of (e) + $SOCl_2$ $\longrightarrow$

(g) Product of (f) + alanine $\longrightarrow$

EXERCISE 15.10 Draw all tripeptides that could possibly be synthesized from glycine, alanine, and phenylalanine.

EXERCISE 15.11 A kilogram of human hemoglobin contains approximately 3.33 grams of iron. If each molecule of hemoglobin contains four heme structures (the iron-containing prosthetic group), what is the minimum molecular weight of this protein?

EXERCISE 15.12 A Van Slyke nitrogen determination made on a solution of 8.74 mg of an unknown amino acid liberated 2.50 ml of N_2 at 740 mm and 25°. What is the minimum molecular weight of the amino acid? If the isoelectric point of this unknown amino acid occurs at a pH value of 10.8, what amino acid could it be?

EXERCISE 15.13 A peptide on complete hydrolysis yielded five different amino acids in the following amounts: one unit each of serine and glycine, two units each of arginine and phenylalanine, and three units of proline. The N-terminal amino acid was identified as arginine. Partial hydrolysis liberated di- and tripeptides of the following compositions:

Ser-Pro-Phe   Gly-Phe-Ser   Pro-Gly

Phe-Ser-Pro   Pro-Phe-Arg   Arg-Pro

Deduce and write out the structure of this nonapeptide.

EXERCISE 15.14 Calculate the isoelectric pH values for the following amino acids.
(a) glycine
(b) alanine
(c) glutamine
(d) serine

EXERCISE 15.15 A solution turns pink litmus blue and the indicator phenolphthalein slightly pink. The solution also contains the amino acid, glycine. If the solution was electrolyzed, would the glycine migrate to the cathode or anode?

EXERCISE 15.16 What is the probable net charge (such as $+\frac{1}{2}$) on the following amino acids in solution of (1) pH 1.0, (2) pH 2.2, (3) pH 4.0, and (4) pH 9.8.
(a) alanine
(b) aspartic acid
(c) asparagine

EXERCISE 15.17 Draw the structures of the following amino acids as they would appear in solutions of (1) pH 3.0 and (b) pH 10.

    (a) lysine   (c) alanine

    (b) glutamic acid   (d) glycine

EXERCISE 15.18 The dipolar (zwitterionic) form of arginine has the following structure. Suggest an explanation for the protonation of the δ-guanadino group (which is responsible for the high pK_3) rather than the α-amino group.

$$\overset{\overset{\textstyle +NH_2}{\|}}{H_2N-C}-NH-CH_2CH_2CH_2-\overset{\overset{\textstyle NH_2}{|}}{CH}-CO_2^-$$

EXERCISE 15.19 The use of amino acids labeled with radioactive isotopes is a valuable technique in biochemistry. Show how the following substances could be prepared from commercially available sources of ^{14}C: $Ba^{14}CO_3$ and $Na^{14}CN$.

    (a) alanine labeled at the C-1 position (carboxyl group)

    (b) alanine labeled at the C-2 position (α-carbon atom)

    (c) aspartic acid labeled at the C-4 position (second carboxyl group)

    (d) phenylalanine labeled at the C-3 position (β-carbon atom)

EXERCISE 15.20 Draw the structures of the products obtained in the following reactions

    (a) Ala-Leu-Cy-Gly-Ser-Leu-His-Cy-Phe $\xrightarrow{\text{I}_2}$

                 | |

              SH   SH

    (b) Ala-His-Gly-Arg $\xrightarrow[\text{2. HCl}]{\text{1. } C_6H_5N=C=S}$

    (c) Cbz-Phe + Ala$-OCH_2C_6H_5$ $\xrightarrow{\text{DCC}}$

16

The Nucleic Acids

Introduction

The nucleic acids are high molecular weight natural polymers found in the nuclei and cytoplasm of all living cells, where they play a highly specialized role in life chemistry. The nucleic acids not only direct the synthesis of specialized cellular material in each individual according to a predetermined code, but also control the genetic continuity of the species.

Your background knowledge of sugars, amino acids, and peptides has prepared you to understand and, to a degree, participate in the excitement which the study of nucleic acids has brought to the field of biochemistry. In no other part of your semester's study will you find organic chemistry more closely related to biology than in the area of nucleic acids, for it is here that the creative capabilities of the living laboratory are so wondrously manifest.

16.1   *The Structure of Nucleic Acids*

The structure of a nucleic acid has been determined, as have the structures of many other natural substances, through a series of degradative studies in which the whole structure is broken down into its component parts. Once the parts of an unknown have been identified, a determination of the complete structure usually follows. There are two families of nucleic acids, but the hydrolysis of either type results in the production of three different chemical substances—orthophosphoric acid, a 5-carbon sugar, and four different heterocyclic basic compounds. The sugar, D-ribose (Sec. 14.4), is obtained from **ribonucleic acid** (RNA), the nucleic acid found largely in the cytoplasm. The sugar 2-deoxy-D-ribose is obtained from **deoxyribonucleic acid** (DNA), the constituent of the genetic material in the cell nucleus. The heterocyclic bases found as parts of the nucleic acid structure are derivatives of either pyrimidine or purine (Sec. 17.1).

The nucleic acid polymer actually is a polyester structure in which phosphoric acid-sugar units are repeated over and over. This polyphosphate chain in the nucleic acid polymer is sometimes described as the "backbone" structure. To each sugar molecule in the ester unit there is bonded *via* an N-glycosidic link one of five different heterocyclic bases (see Sec. 17.1). The following schematic drawing shows the fundamental nucleic acid structure.

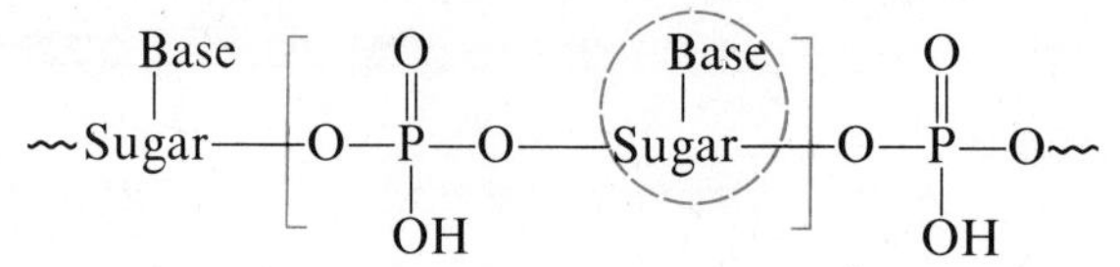

A section of the polynucleotide chain

The portion of the polynucleotide enclosed by brackets in the structure above shows all three components of a nucleic acid—base, sugar, and phosphoric acid unit. This portion is called a *nucleotide.* A nucleotide must include the phosphoric acid structure, otherwise it would not be a nucleic acid. The N-glycosidic portion shown enclosed in a broken circle—the sugar with attached base but lacking the phosphoric unit—is called a *nucleoside.* Examples of both nucleotides and nucleosides will be given in following sections.

Let us now examine the component parts of a nucleic acid, and then we will be able to understand better how the parts combine to form DNA and RNA and also how the nucleic acids can perform their vital roles.

16.2   *The Sugar-Phosphate Ester*

The sugar components of DNA and RNA differ only at the number 2 carbon atom of the furanose ring (page 471). In ribose the number 2 carbon atom has bonded to it both hydrogen and a hydroxyl group. The hydroxyl group is lacking at this position in deoxyribose as is indicated by the prefix *2-deoxy.*

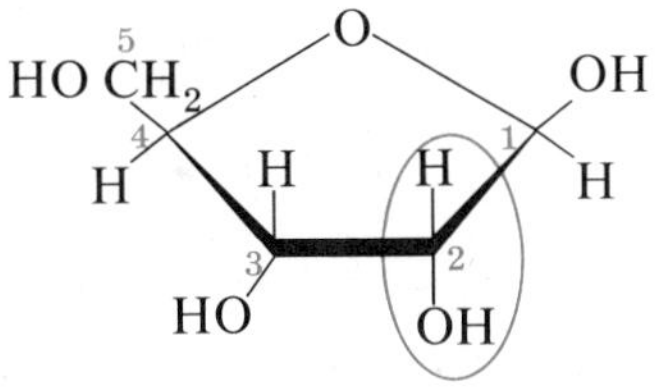

Ribose
(β-D-Ribofuranose)

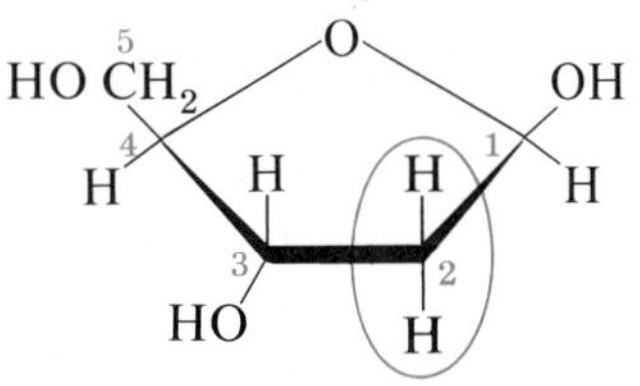

Deoxyribose
(2-Deoxy-β-D-ribofuranose)

Both ribose and deoxyribose have the β-configuration (Sec. 14.5) at the number 1 carbon atom, and in both RNA and DNA the nitrogen bases are attached to the sugar molecule at this position. Also, in both nucleic acids each phosphoric acid molecule forms a diester by joining carbon 5 of one sugar molecule to carbon 3 of the next. The mode of esterification in DNA is illustrated in Figure 16.1.

FIGURE 16.1   *A Segment of the DNA Polynucleotide "Backbone"*

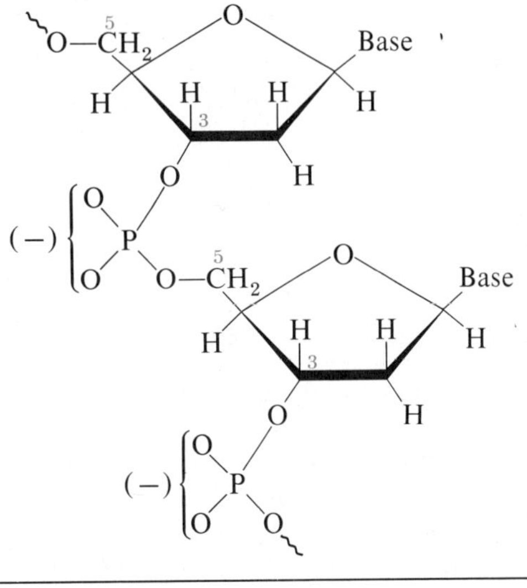

16.3   *Purines and Pyrimidines*

The basic portions of nucleic acids are derivatives of two heterocyclic structures, purine and pyrimidine, and are the parts of nucleic acids responsible for the storage and transmission of genetic information. Although there are a number of minor, but important, bases, five constitute the major bases found in nucleotides. Two of these, **adenine** and **guanine,** are purines and are found in both RNA and DNA. The other three, **thymine, cytosine** and **uracil,** are pyrimidines. DNA contains thymine and RNA contains uracil. Cytosine is common to both nucleotides.

The structures and numbering systems of the two parent heterocycles are shown along with the derivatives of each that occur in RNA and DNA.

Pyrimidine   Purine

Adenine   Guanine

Thymine   Cytosine   Uracil

16.4   *Nucleotides and Nucleosides*

The structural units in both RNA and DNA are called nucleotides and, as has been pointed out earlier, are the phosphate esters of a substituted pentose, the substituent being either a purine or a pyrimidine derivative attached at carbon atom number 1′ of the sugar ring.[1] The point of attachment in the heterocyclic base is at position 1 in the pyrimidine, at position 9 in the purine. A nucleoside is only the substituted sugar portion of a nucleotide and structurally is analogous to a glycoside (Sec. 14.5), except that here we have a nitrogen atom instead of an oxygen atom bonded at carbon 1′. Examples of both nucleoside and nucleotide formation are illustrated on the opposite page.

The name of a nucleotide incorporates that of its nitrogen base using the suffix *ylic* and followed by *acid* as a separate word. The name of a nucleoside is similar to that given a nucleotide with the same sugar and base except that the name ends with the suffix *idine* if the base portion is a pyrim*idine,* with *osine* if the base portion is a purine. Of course the word acid does not apply. Names of structural units found in RNA and DNA are given in Table 16.1.

---

[1] In order to distinguish between two different ring systems present within a nucleic acid, primed numbers will henceforth be used to designate positions in the sugar, unprimed numbers to designate ring positions in the heterocyclic base.

SUGAR + HETEROCYCLIC BASE ⟶ NUCLEOSIDE

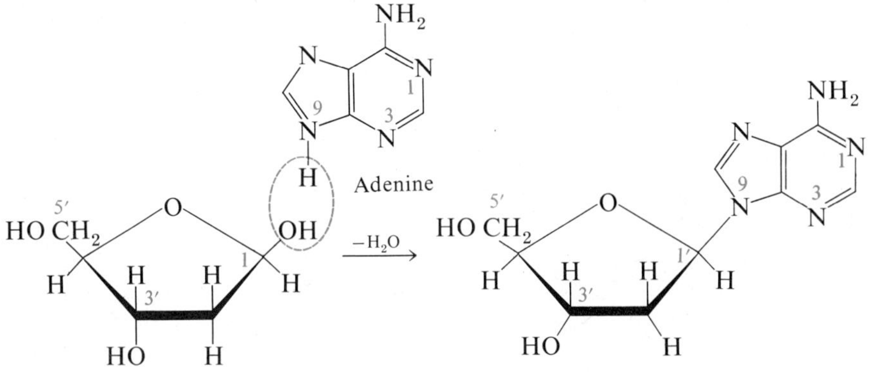

Adenine

Deoxyribose

Deoxyadenosine
(A nucleoside found in DNA)

PHOSPHORIC ACID + NUCLEOSIDE ⟶ NUCLEOTIDE

Sites for further
esterification or acid
anhydride formation

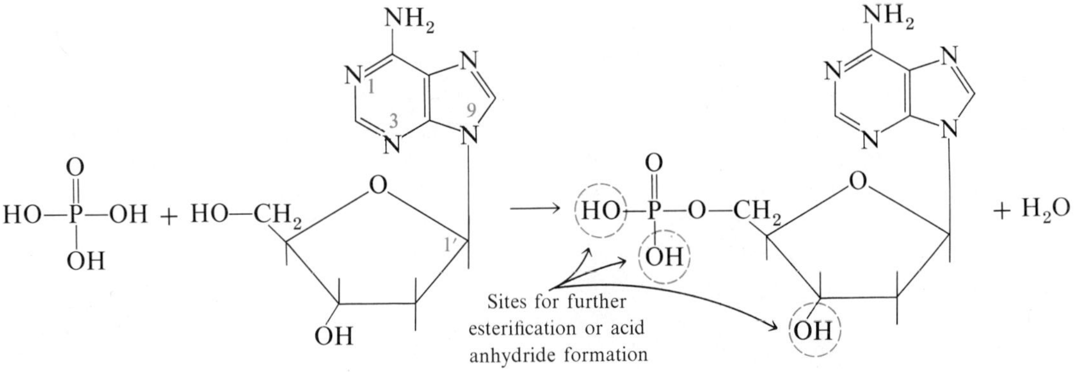

Orthophosphoric
acid

Deoxyadenosine

Deoxyadenylic acid
(A nucleotide found in DNA)

TABLE 16.1 *Some Nucleosides and Nucleotides*

	Base	Pentose	Nucleoside	Nucleotide
Purines	Adenine	Ribose	Adenosine	Adenylic acid
	Guanine	Deoxyribose	Deoxyguanosine	Deoxyguanylic acid
Pyrimidines	Uracil	Ribose	Uridine	Uridylic acid*
	Thymine	Deoxyribose	Deoxythymidine	Deoxythymidylic acid‡
	Cytosine	Deoxyribose	Deoxycytidine	Deoxycytidylic acid

* Found only in RNA.
‡ Found only in DNA.

16.5   The DNA Polymer

One of the most significant observations that led to a deduction of the DNA structure was the finding that the ratio of adenine to thymine and of guanine to cytosine in the polynucleotide was almost 1:1. Such equimolar quantities of pyrimidines and purines suggested that the bases must be present in the DNA structure in pairs, but how could such equivalence be accomplished? The answer to this question was supplied by Watson and Crick,[2] who built a model which appeared to embody all facts known about DNA. The Watson-Crick model was based upon measurements and calculations made from X-ray diffraction data obtained by M. Wilkins from crystallographic studies on a prepared sample of DNA polymer. To account for the molecular ratio of unity between pyrimidine and purine derivatives, Watson and Crick proposed that the DNA structure is not that of a simple helix as are many other protein structures, but has instead the form of a *double* helix. A double helix structure would allow the nitrogen bases in each DNA strand to be projected toward the center. Uniform intermolecular (hydrogen) bonding would then be possible between adenine in one strand of the helix, with thymine in the other strand and also between cytosine in one strand and guanine in the other. Such hydrogen bonding between complementary strands would result in connecting both strands somewhat like the treads connect the two rising parts of a spiral staircase. A structure of this kind not only would require base pairing, but also would provide the stability that holds the two strands together. The hydrogen bonding capabilities of complementary base pairs are shown in Figure 16.2.

Watson and Crick also postulated that the two polynucleotide strands entwine about a common axis and are right-handed spirals. The 3',5'-inter-

FIGURE 16.2   *Hydrogen Bonding Between Base Pairs in DNA*

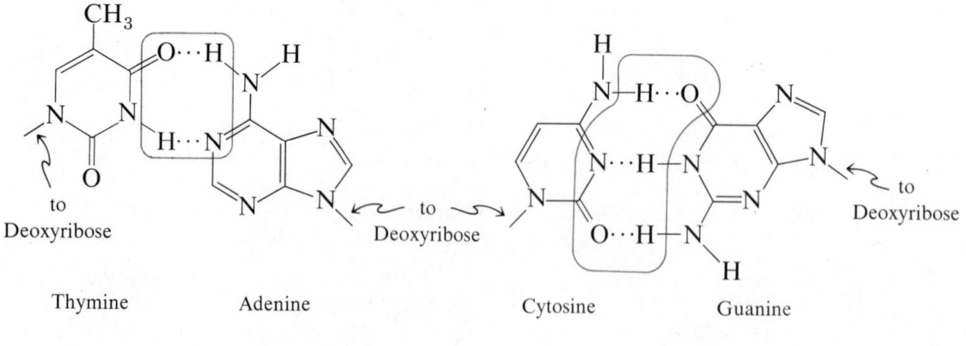

Thymine   Adenine   Cytosine   Guanine

---

[2] James D. Watson (1928–   ), Professor of Biology, Harvard University, and Francis H. Crick (1916–   ), British Medical Research Council, shared with M. H. F. Wilkins (1916–   ), Kings College, University of London, the 1962 Nobel Prize for Medicine and Physiology.

FIGURE 16.3   *A Segment of the DNA Double Helix*

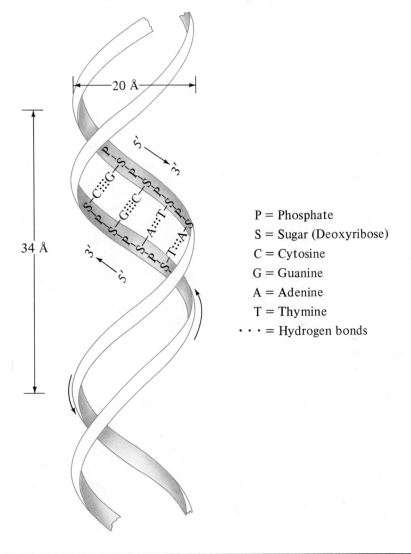

P = Phosphate
S = Sugar (Deoxyribose)
C = Cytosine
G = Guanine
A = Adenine
T = Thymine
· · · = Hydrogen bonds

nucleotide phosphodiester bridge (Fig. 16.1) runs in one direction for one strand and in the opposite direction for the other. The approximate diameter of the double helix was revealed to be about 20 Å and consisted of ten nucleic acid residues per turn—each turn measuring a distance of 34 Å. A diagrammatic segment of the double helix with base pairings between strands is shown in Figure 16.3.

16.6 Replication of DNA

The concept of guanine-cytosine and adenine-thymine pairing between two coiled polynucleotide chains not only offered a plausible structure for DNA but also suggested the mechanism whereby genetic information is transmitted from one generation to the next.

The biosynthesis of new strands of DNA containing the same sequence of base pairs as the parent can result if the parent DNA double helix uncoils and each single strand functions as a template on which a new complementary strand of DNA forms. In this way two new daughter strands result with the same base-pair sequence (and hereditary information) as was present in the parent. As each new strand is formed, hydrogen bonding between complementary bases leads to an identical replica of the parent. A schematic mechanism of DNA replication is shown in Figure 16.4.

FIGURE 16.4 *The Schematic Representation of DNA Replication*

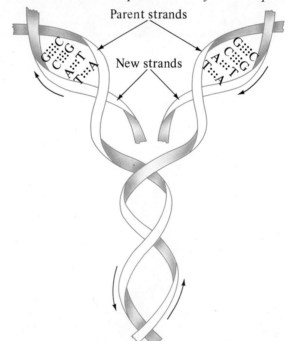

16.7 The Genetic Code

Protein material designed to perform a particular function—hair, enzymes, muscle tissue, etc.—must be formed from amino acids selected and combined in a certain sequence. The genetic information which must be programmed for the

TABLE 16.2   *m-RNA Codons Proposed for Amino Acids Used in Protein Synthesis*

Amino acid	Codons	Amino acid	Codons
Alanine	GCA, GCC, GCG, GCU	Lysine	AAA, AAG
		Methionine	AUG
Arginine	AGA, AGG, CGA, CGC, CGG, CGU	Phenylalanine	UUC, UUU
		Proline	CCA, CCC, CCG, CCU
Asparagine	AAC, AAU		
Aspartic acid	GAC, GAU	Serine	UCA, UCC, UCG, UCU, AGC, AGU
Cysteine	UGC, UGU		
Glutamic acid	GAA, GAG	Threonine	ACA, ACC, ACG, ACU
Glutamine	CAA, CAG		
Glycine	GGA, GGC, GGG, GGU	Tryptophan	UGA, UGG
		Tyrosine	UAC, UAU
Histidine	CAC, CAU	Valine	GUA, GUC, GUG, GUU
Isoleucine	AUA, AUC, AUU		
Leucine	CUA, CUC, CUG, CUU, UUA, UUG		

synthesis of such specialized material is embodied in the *sequence* of the four bases found in the DNA strand and is referred to as the *genetic code*. How is such coding accomplished for the proper selection of the twenty different amino acids from which proteins are synthesized? Obviously, one nucleotide could not code for only one particular amino acid for, in this case, we could use only four different amino acids. The coding information, therefore, must lie in *combinations* of nucleotides. Let us abbreviate with capital letters the names of the four bases found in DNA and consider combinations of bases in groups of two, e.g., AT, TG, CG, AA. By so doing we could possibly form 4^2, or sixteen, combinations—not quite enough. However, if we use combinations of three nucleotides in different sequences we could form 4^3, or sixty-four, coding combinations—more than enough to accommodate all twenty amino acids. In fact, we would have enough combinations to allow more than one coding triplet for some amino acids and additional combinations that do not appear to code for any. Coding units made up of such triplet combinations are called *codons* and may be thought of as 3-lettered words from which an entire sentence may be constructed—the sentence, in this case, being the polypeptide chain. Codons for the twenty amino acids are given in Table 16.2. The code is usually expressed in terms of codons for messenger RNA.

16.8   *RNA*

The primary structure of RNA differs from that of DNA mainly in the nature of the nucleosides esterified. RNA incorporates the base uracil instead of thymine. Except for lacking the 5-methyl substituent, uracil is structurally similar to thymine and its hydrogen bonding characteristics allow it also to pair with adenine. Thus, uracil appears in the RNA strand instead of thymine.

Thymine-Adenine pairing in DNA　　　　　　Uracil-Adenine pairing in RNA

Another difference between DNA and RNA lies in their secondary structures. RNA molecules exist mainly as single strands, rather than in the form of double helices, and therefore unlike DNA may have a pyrimidine-purine ratio other than 1:1.

Three different types of RNA molecules appear to be involved in the vital business of protein synthesis. One of these, **ribosomal RNA** (*r*-RNA), is of the greatest molecular weight and is associated with the ribosomes, or nucleoprotein particles, in the cytoplasm. The ribosomes provide the site for protein synthesis. A second type of RNA, called **messenger RNA** (*m*-RNA), is a polynucleotide of lower molecular weight than *r*-RNA and carries the transcribed genetic code from the DNA in the nucleus of the cell to the ribosomes at the protein-building site. A third type of RNA is called **transfer RNA** (*t*-RNA). Transfer RNA has a molecular weight of approximately 30,000 and is the smallest of the polynucleotides. Because it is soluble, it also is referred to as **soluble RNA** (*s*-RNA). The function of *t*-RNA is to select and to carry the amino acids required for a specific protein to the correct *m*-RNA-ribosome site. The amino acid selected by a *t*-RNA molecule is the one called for by the codon in the *m*-RNA. The triplet of complementary bases in *t*-RNA that corresponds to this "order from above" is called an anticodon. All *t*-RNA molecules, while possibly of a different base composition individually, have as a common feature the terminal base triplet C-C-A at the 3'-end. In Figure 16.5 the unfolded two-dimensional structure of yeast alanine *t*-RNA, as deduced by Holley,[3] is shown. All *t*-RNAs have similar cloverleaf structures, sometimes with an extra arm. In its active, three-dimensional conformation, the cloverleaf is twisted into a more compact and specific form. The role that each type of RNA molecule plays in the synthesis of a protein is discussed in Sec. 16.9.

The code is read from the 5' to the 3' direction. The coding triplets for that strand of DNA responsible for transcribing the genetic message to *m*-RNA are complementary to those in *m*-RNA, and T replaces U in DNA. In the anti-

[3] Robert W. Holley, (1922–　　), Cornell University, Ithaca, New York (now at the Salk Institute, La Jolla, Ca.), Nobel Prize in Medicine and Physiology 1968.

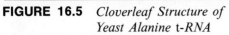

FIGURE 16.5 *Cloverleaf Structure of Yeast Alanine t-RNA*

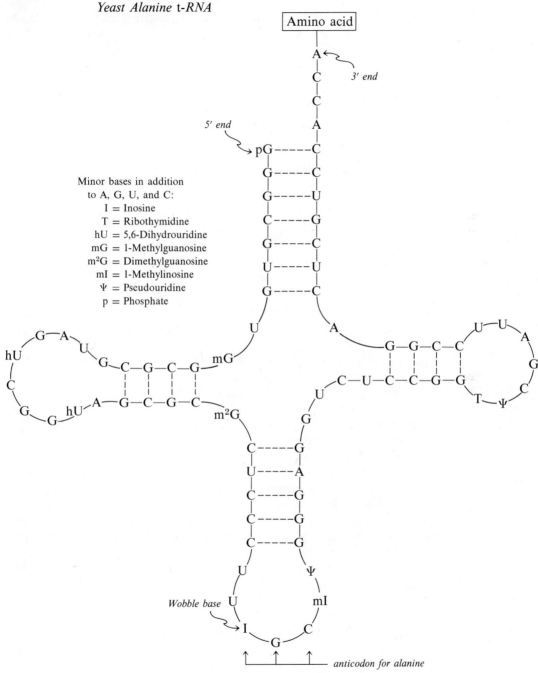

codons of *t*-RNA the coding triplets are also complementary to those in *m*-RNA; however, the first base (called the "wobble" base) of the anticodon may not be the exact complement of the third base in the codon and may be one of the minor bases (see Fig. 16.5). The various interrelationships are shown in the following diagram and in Figures 16.4 and 16.5 and on pages 482 and 483.

DNA

5'-end   3'-end

AAC—AGA—AGT—ACG—TCC—GGC—AAA—TCC—GTA

m-RNA

3'-end   5'-end

UUG—UCU—UCA—UGC—AGG—CCG—UUU—AGG—CAU

t-RNA

5'-end   3'-end

AAC⁄ ⁄IGA⁄ ⁄AGU⁄ ⁄ICG⁄ ⁄UCC⁄ ⁄IGC⁄ ⁄AAA⁄ ⁄UCC⁄ ⁄GΨA

      *   *   *   *

Amino acid

Carboxyl end   Amino end

Val—Ser—Thr—Arg—Gly—Ala—Phe—Gly—Tyr

*I and Ψ are minor bases.

16.9 *Protein Synthesis*

The first step in protein synthesis is the enzyme-initiated reaction of an amino acid with adenosine triphosphate (ATP) to form a phosphoric-amino acid anhydride complex bound to the active site of the enzyme. This enzyme-bound amino acid complex then reacts with a *t*-RNA molecule to produce an activated amino acid ester at the 3′ position of the terminal adenylic acid in the C-C-A "tail." The following series of chemical changes on page 482 shows the attachment of valine to *t*-RNA.

The attached valine ester is then transferred by the *t*-RNA to the ribosomes where it is combined according to a predetermined code with another amino acid previously added to the polypeptide chain. After the *t*-RNA has delivered and deposited an amino acid it returns for others. The exact details of this transfer of amino acids and their incorporation into an ever growing polypeptide structure are yet unknown. It appears, however, that *m*-RNA is synthesized at the site of a portion of unravelled DNA and has had transcribed to it the genetic information necessary for the proper sequential alignment of certain amino acids at the protein-building site. See page 483.

Experiments carried out by Nirenberg, Khorana, and Ochoa[4] provided some insight into the role played by *m*-RNA in protein synthesis. Nirenberg

---

[4] Marshall W. Nirenberg (1927–   ), National Institutes of Health, Bethesda, Md., Nobel Prize in Medicine and Physiology 1968; Servero Ochoa (1905–   ), New York University School of Medicine, New York, N.Y. (now at La Roche Research Institute, Nutley, N.J.), Nobel Prize in 1959; Har Gobind Khorana (1922–   ), University of Wisconsin, Madison, Wisconsin (now at the Massachusetts Institute of Technology), Nobel Prize in Medicine and Physiology 1968.

combined a synthetic RNA molecule composed entirely of polyuridylic acid (UUUUUUUUU····) with the twenty amino acids used in protein synthesis and found that only the peptide, polyphenylalanine, was formed. Similar experiments were carried out by Ochoa. Khorana prepared synthetic polynucleotides with repeating dinucleotides, such as UGUGUGUGUG·····. Polynucleotides of this type contain two alternating triplets, UGU and GUG, along the chain. This synthetic messenger gave no homopolypeptide but led only to the formation of the alternating polypeptide containing valine and cysteine, Val-Cys-Val-Cys-Val-Cys·····. Experiments of the latter type definitely established the existence of the triplet code, and the deciphering of the code followed quickly thereafter.

The mechanism of protein synthesis and the roles played by DNA and the different RNA molecules as shown in the illustrations in this chapter certainly are an oversimplification of the extremely complex and highly specialized function of the nucleic acids in protein synthesis. Much research remains to be done before more of the details of this intricate chemistry can be discovered. Some of the finest scientific minds of our time are dedicated to the attainment of this goal.

16.10   *Miscoding and Mutations*

One would imagine that any biosynthesis involving hundreds of amino acid residues of twenty different kinds all combined in a great variety of different sequences would also involve the possibility of errors. Such indeed is the case, and errors do occur. Correct genetic information rests in the proper sequence of bases in the DNA molecule, and when one base in the sequence appears in an altered form then the genetic message also is altered by a triplet that miscodes. Such miscoding in DNA results in an erroneous transcription to *m*-RNA and finally in the wrong amino acid sequence for the protein under construction. If such genetic errors or mutations should prove beneficial, a better species could result. On the other hand, if such genetic defects leave the organism handicapped or weakened to the extent that it will not reproduce, the species will die.

It is believed that the bases in the DNA sequence might be altered structurally by exposure to high energy radiation. Certain of the bases present in DNA can exist in tautomeric forms (pairs of equilibrium isomers, Sec. 9.5) in which one isomeric structure is much more stable than the other. Conceivably, the absorption of high energy radiation such as that from X rays, γ-rays, or cosmic rays could transform the structure of a base into that of its higher energy tautomer. Should this occur, then the altered structure of the normal base in DNA no longer will be able to hydrogen-bond with the proper complementary base in *m*-RNA, and miscoding results. An example of such a tautomeric shift and resultant miscoding is illustrated for adenine on page 484.

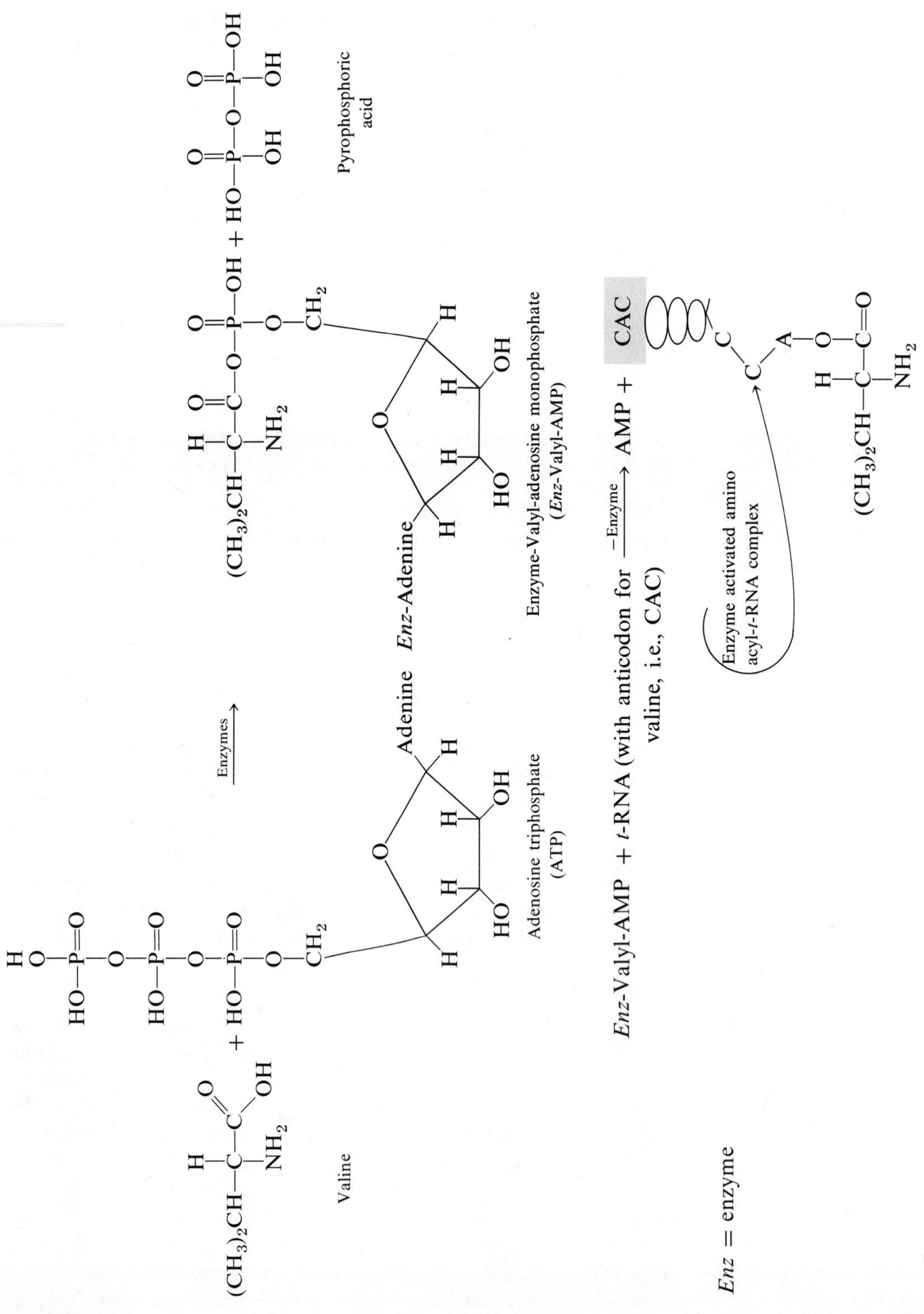
Valine
Adenosine triphosphate (ATP)
Enzymes
Adenine
Enz-Adenine
Enzyme-Valyl-adenosine monophosphate (Enz-Valyl-AMP)
Pyrophosphoric acid
Enz-Valyl-AMP + t-RNA (with anticodon for valine, i.e., CAC) —Enzyme→ AMP +
CAC
Enzyme activated amino acyl-t-RNA complex
Enz = enzyme

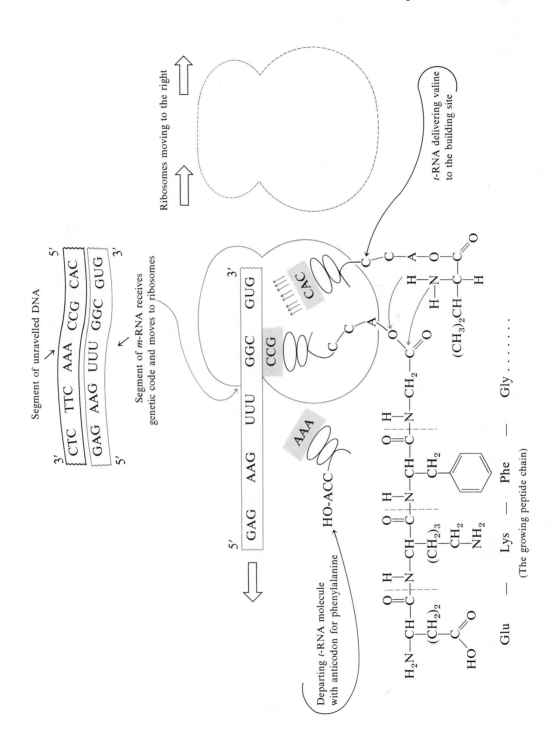

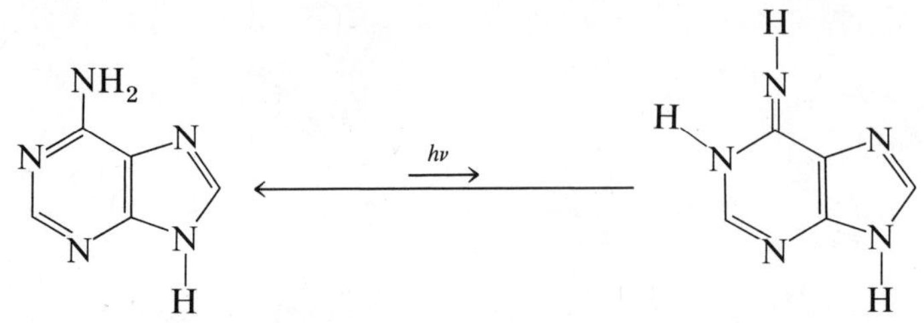

(Stable amino form)   (Unstable imino form)

Adenine

$$\text{Adenine} \begin{cases} \text{(amino form)} + \text{Thymine} \rightarrow \text{normal pairing} \\ \text{(imino form)} + \text{Cytosine} \rightarrow \text{abnormal pairing} \end{cases}$$

Mutations are also thought to be the result of chemical changes. You will recall from our study of amines that nitrous acid may be used to convert a primary amine to an alcohol according to the following reaction:

$$R\text{—}NH_2 + \{NaNO_2 + HCl\} \rightarrow R\text{—}OH + H_2O + NaCl + N_2$$

In the same manner nitrous acid could also possibly convert adenine to hypoxanthine and cytosine to uracil, since both contain primary amino groups.

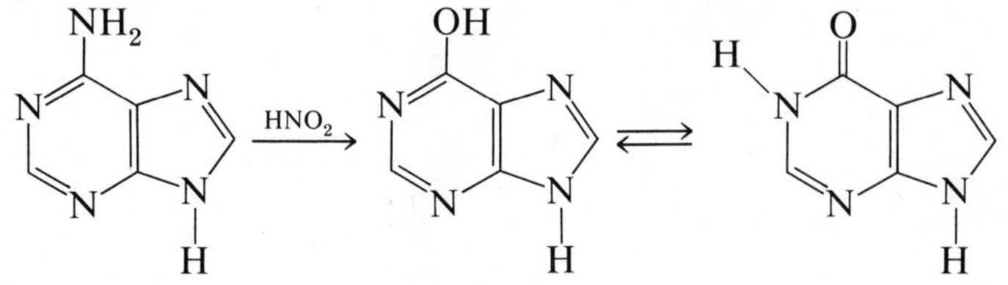

Adenine   Enol form   Keto form

Hypoxanthine

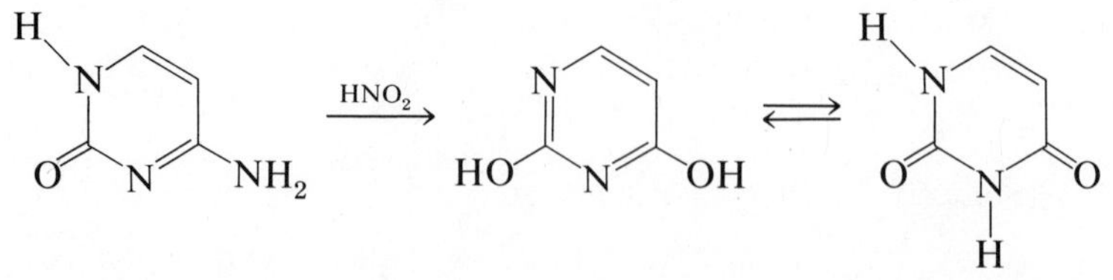

Cytosine   Enol form   Keto form

Uracil

Many of our foods processed from meat products have sodium nitrite added as a preservative, and since the environment of the stomach is acidic the proper conditions for the formation of nitrous acid exist in this part of the digestive tract. While it is doubtful that nitrous acid from this source exists beyond the limits of the animal stomach, mutations have been effected in viral proteins through the use of nitrous acid, and therefore the possibility of such chemically induced miscoding does indeed exist.

Research in this area of chemistry is urgently needed, in order to provide answers to many of the questions that arise regarding the possibility of chemical mutagenesis from the use of other food additives such as artificial sweeteners and certain condiments. We might also hold suspect certain insecticides and herbicides, which unwittingly get into our food supply.

Summary

1. The nucleic acids are base-substituted sugar esters of phosphoric acid.

2. The two principal types of nucleic acids are *ribose nucleic acid* (RNA), and *deoxyribose nucleic acid* (DNA).

3. Both ribose and deoxyribose appear in nucleic acids as N-glycosides of five different heterocyclic bases.

4. The base portion of the nucleic acid molecule may be either a pyrimidine derivative or a purine derivative.

5. A nucleoside is an N-substituted glycoside; a nucleotide is an N-substituted glyosidic ester of phosphoric acid.

6. The DNA polymer has the form of a double helix with the pyrimidine and purine base structures oriented toward the axis of the helix.

7. Hydrogen bonding between complementary bases makes possible the double helix DNA structure, as well as replication of the DNA molecule.

8. The genetic code is embodied in DNA but transcribed to *m*-RNA.

9. A sequence of three bases in the DNA polymer may code for only one of the amino acids used in protein synthesis.

10. Amino acids are carried to the protein building site by *t*-RNA.

11. Alteration of the base structures in DNA alter their hydrogen bonding characteristics. Such changes can result in an erroneous sequence that leads to miscoding.

New Terms

anticodon

codons

deoxyribose

DNA

nucleic acids

nucleosides

nucleotides

RNA

m-RNA

r-RNA

t-RNA

replication

ribosomes

Supplementary Exercises

EXERCISE 16.1 Distinguish between the following:
   (a) a pyrimidine ring system and a purine ring system
   (b) a nucleoside and a nucleotide
   (c) ribose and deoxyribose
   (d) a codon and an anticodon

EXERCISE 16.2 Draw the structures for
   (a) deoxyadenosine
   (b) adenylic acid
   (c) cytidine
   (d) cytidylic acid
   (e) adenosine triphosphate
   (f) a nucleotide found only in RNA
   (g) a nucleotide found only in DNA

EXERCISE 16.3 If ATP is *completely* hydrolyzed what chemical compounds result?

EXERCISE 16.4 The ratio of cytosine to guanine in the DNA double helix is $1:1$, but the ratio of cytosine to adenine need not be unity. Explain.

EXERCISE 16.5 Why do neither adenine and cytosine nor guanine and thymine form complementary base pairs in DNA?

EXERCISE 16.6 How does *m*-RNA differ from *t*-RNA?
   (a) in base composition
   (b) in function

EXERCISE 16.7 If only cytosine and guanine are used to form 3-base sequences and if each sequence could code for a different amino acid, theoretically for how many amino acids could codes be formed? How many amino acids are actually coded for with only these bases in sequence?

EXERCISE 16.8 Draw the imine form of adenine and show how it could then pair with cytosine rather than with thymine.

EXERCISE 16.9 If the unstable tautomer of adenine (identified by an asterisk) were to appear in DNA in the sequence C-Å-G, which amino acid would be erroneously built into a peptide chain?

EXERCISE 16.10 Which amino acid sequences will be found in peptides formed by ribosomes in response to the synthetic *m*-RNA's given below. Assume that the polypeptide chain will begin with the codon on the left.

        (a) GGUGCAAAGUCCUGACACAUA
        (b) UGCGUAAUACUUUUCCCUAAU
        (c) AAACCCUUUUUCCCUAGGGAA

EXERCISE 16.11 One segment of a strand of DNA consists of the following sequence going in the 5′ to 3′ direction (the code is read in this direction):

AAACACTTCTCGACGATCGGCGGCTAC

        (a) Write the sequence of bases in the other strand of DNA.
        (b) Write the sequence of bases in the *m*-RNA transcribed from the first strand of DNA.
        (c) Write the sequence of amino acids called for by the message in the *m*-RNA.

EXERCISE 16.12 Draw the anticodon sequence of the *t*-RNA and the codon sequence of the *m*-RNA that would be required for the synthesis of oxytocin (Fig. 15.1).

17

Heterocyclic Compounds—
Natural Products

Introduction

The cyclic compounds studied up to this point were largely ring systems in which only carbon atoms were joined together. Such ring systems are spoken of as **carbocyclic.** Cyclic compounds which contain, in addition to carbon, one or more other atoms in the ring are called **heterocyclic** (Gr., *heteros, other*). The hetero atoms which occur most frequently in heterocyclic rings are nitrogen, sulfur, and oxygen. A number of cyclic structures containing nitrogen and oxygen as ring atoms have been discussed or at least mentioned in earlier sections—e.g., ethylene oxide, cyclic acid anhydrides, lactones, imides, and the hemiacetal carbohydrate structures. Strictly speaking, the preceding examples are heterocycles, but each is converted rather easily into a noncyclic form. The heterocycles which we refer to in this section are much more stable. The five- and six-membered heterocycles are nearly free of strain and possess a stability which is characteristic of the benzene nucleus. In fact, some heterocyclic compounds are even more stable than benzene. Heterocyclic compounds are very abundant in nature, and a study of plant and animal products would be difficult without at least a brief survey of the common heterocyclic ring systems.

17.1 *Heterocyclic Ring Systems*

Natural products that contain one or more heterocyclic rings in their molecular structures are so numerous and their properties, in many cases, are so complex that we can consider only a very few of them in the present chapter. The names and structures of the five- and six-membered heterocyclic ring systems most frequently encountered in natural products are given in Table 17.1. You will note that numbering the ring members begins with the hetero atom. Other ring members (in rings with but one hetero atom) often are designated in nonsystematic nomenclature as α, β and γ. The α atom is the ring member adjacent to the hetero atom. For example, β-picoline (Sec. 17.12-b) is 3-methylpyridine. There are three isomeric picolines. The other two, of course, are α- and γ-picoline.

17.2 *Five-Membered Heterocycles*

The simplest five-membered heterocycles are those with only one hetero atom. We shall consider only three: furan, pyrrole, and thiophene.

| Furan | Pyrrole | Thiophene |

Furan, C_4H_4O, appears as part of the skeletal structure of a number of natural products but is most readily available as the α-aldehyde, *furfural.* Furfural is obtained from corn cobs, oat hulls, bran, and straw. These agricultural wastes are a rich source of polymeric pentoses known as *pentosans.* Hydrolysis of the latter to pentoses, followed by treatment with hot, dilute sulfuric acid, yields furfural.

TABLE 17.1 *Common Heterocyclic Ring Systems*

Structure	Name	Occurs in These Examples
(A) Five-membered Rings		

| | (a) Furan | Furfural |
| | (b) Tetrahydrofuran | Morphine |

| | Thiophene | |

	(a) Pyrrole	Vitamin B_{12}
		Chlorophyll-a
		Heme
	(b) Pyrroline	Tryptophan
	(c) Pyrrolidine	Proline
		Nicotine

| | Thiazole | Penicillins |
| | | Vitamin B_1 |

	Indole	Tryptophan
		Strychnine
		Reserpine

| **(B) Six-membered Rings** | | |

	(a) Pyridine	NAD, Nicotine,
	(b) Piperidine	Vitamin B_{12}
	(c) Tetrahydropyran	Quinine
		Morphine
		Cocaine, Coniine
		Reserpine
		Marijuana
	Pyrimidine	Vitamin B_1
		Barbiturates
		Nucleic acids

	Purine	CoA enzyme
		NAD enzyme
		Nucleic acids

| | (a) Quinoline | Curare |
| | (b) Isoquinoline | |

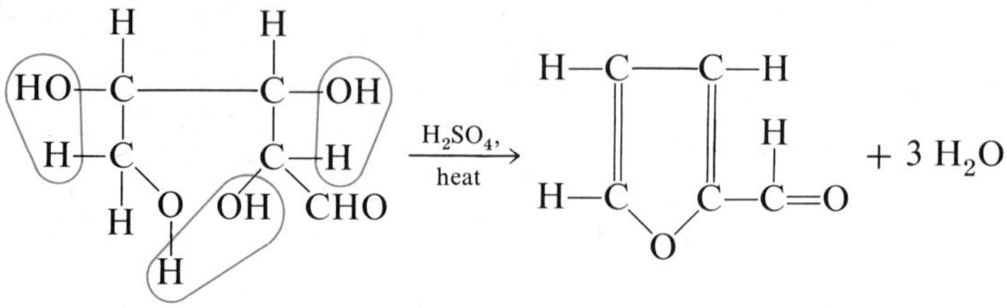

A pentose → Furfural (B.P. 162°C) + 3 H$_2$O

The ready availability of large quantities of furfural from low-cost raw materials had led to a number of industrial applications. One of these is the manufacture of nylon.

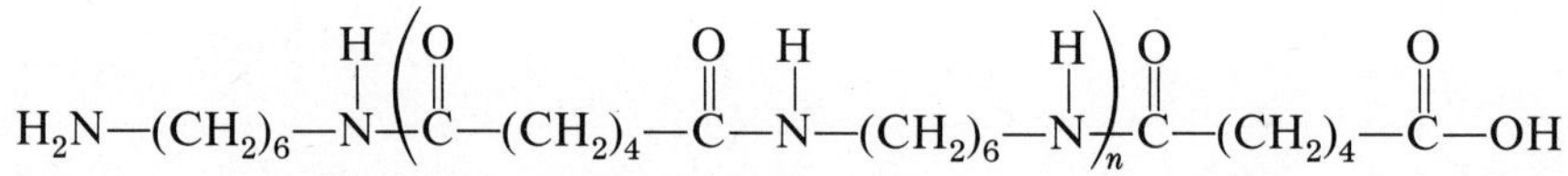

Nylon 66 (n = 450–500)

Pyrrole, (Gr., *pyrros,* fiery + L. *oleum,* oil), C_4H_5N, is found in coal tar and in bone oil. The latter is obtained by the dry distillation, or pyrolysis, of animal by-products such as horns, hooves, and bones.

The pyrrole structure, as well as its partial and completely reduced forms, *pyrroline,* C_4H_7N, and *pyrrolidine,* C_4H_9N, are part of many natural products.

A union of four pyrrole rings makes up the **porphyrin structure** which appears in both the heme of blood and in chlorophyll-a. You will note that both of these natural products are chelates (Sec. 13.5). The biologically important redox enzymes, the cytochromes, are classified as hemoproteins and contain iron-porphyrin groups.

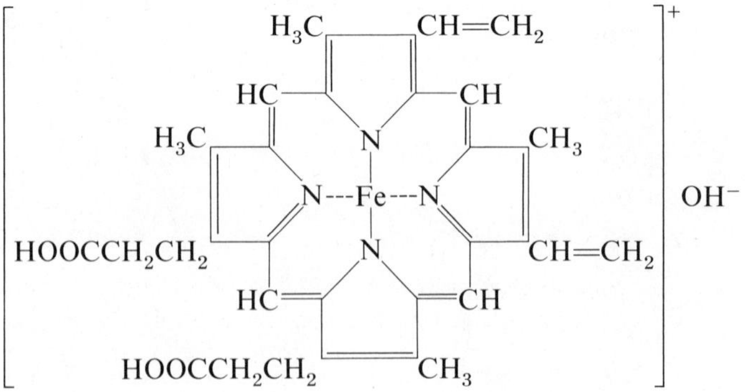

Heme

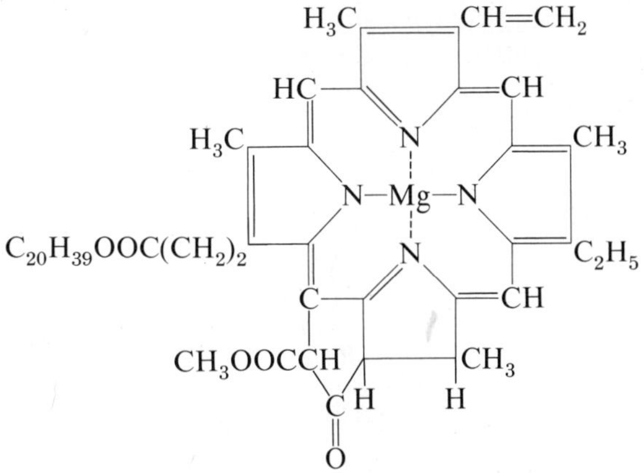

Chlorophyll-a

Indole, C_8H_7N, also a coal tar constituent, is found as part of the total structure of a number of alkaloids (Sec. 17.6, 17.7, 17.8). Indole, or 2-benzopyrrole,[1] and its 3-methyl derivative, skatole, are degradation products of the α-amino acid tryptophan. These two compounds are largely responsible for the odor of feces.

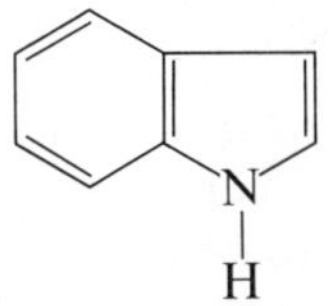

Indole	Tryptophan	Skatole
(2-Benzopyrrole)	[2-Amino-3-(3-indolyl)-propanoic acid]	(3-Methylindole)

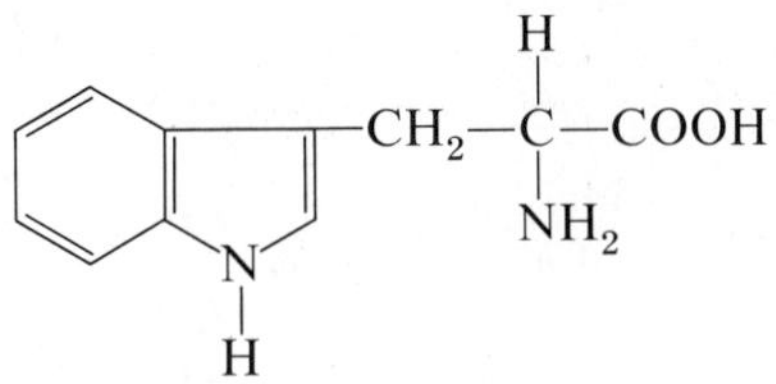

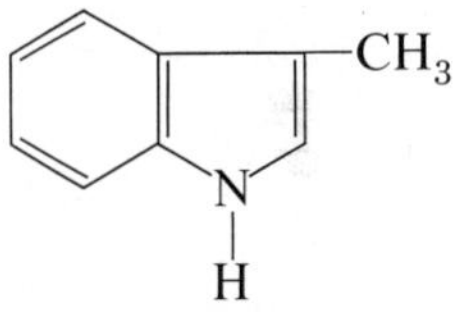

Thiophene, C_4H_4S, like many of the nitrogen-containing heterocycles, also is found in coal tar. Because its boiling point (87°C) is so near that of benzene (80°C), thiophene invariably appears as an impurity in any benzene obtained from coal tar.

17.3 Properties of the Five-Membered Heterocycles

The chemical properties of furan, pyrrole, and thiophene are somewhat similar to those of benzene. All three five-membered heterocyclic ring systems have an electron cloud—i.e., the aromatic sextet (Sec. 4.2), above and below the plane of the ring and engage in reactions characteristic of the benzene nucleus.

---

[1] A cyclic structure which has part of its ring in common with a benzene ring sometimes is called a *benzo* compound.

Furan, pyrrole, and thiophene undergo nitration, sulfonation, halogenation, and the Friedel-Crafts reaction. Substitution on the ring in each case takes place preferentially at the 2,5-positions.

17.4 *Six-Membered Heterocycles*

The six-membered heterocycle containing an oxygen atom and two double bonds is named *pyran* (Sec. 14.5). No simple pyran ring structure is known, but numerous natural products incorporate the reduced form, either dihydro- or tetrahydropyran, in their structures. One of the most notorious of these is Δ^1-3,4-*trans*-tetrahydrocannabinol,[2] the physiologically active constituent found in marijuana, which is the dried leaves and flowering tops of the hemp plant, *Cannabis sativa.*

Δ^1-3,4-*trans*-Tetrahydrocannabinol

The most widely occurring six-membered heterocycle is *pyridine,* C_5H_5N. Pyridine occurs along with pyrrole in bone oil but commercially is obtained from coal tar. Pyridine is a malodorous liquid, soluble in water and in most organic solvents. Chemically, it is less reactive than benzene and undergoes substitution with difficulty. The pyridine ring, its reduced form, *piperidine,* $C_5H_{11}N$, or one of the derivatives of pyridine can be found in a great many plant products. A number of these are described in the section devoted to the alkaloids (Sec. 17.5). Pyridine, along with the purine and pyrimidine structures, also is found in nucleic acids—the prosthetic groups of nucleoproteins. The nucleoproteins are of particular interest to the biochemist because they are intimately related to the life processes in both plants and animals. For example, an important enzyme necessary for a large number of metabolic reactions to occur is *nicotinamide-adenine-dinucleotide* (NAD), or sometimes referred to as diphosphopyridine nucleotide (DPN).

The nicotinamide residue (upper left-hand part of structure) is the amide of nicotinic acid. Nicotinic acid, commonly called *niacin,* is the anti-pellagra vitamin (Sec. 17.12). The fused ring heterocycle adenine (upper right-hand part of structure) may be recognized as the purine part of the coenzyme A molecule (Sec. 11.10).

---

[2] The Greek capital delta (Δ) is sometimes used in complex structures to denote the location of a double bond.

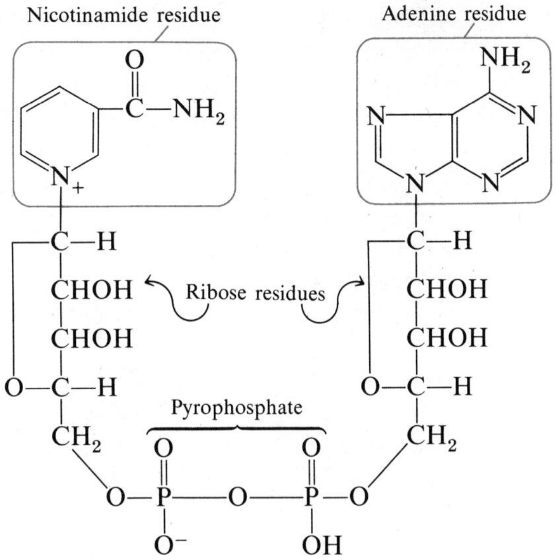

NAD

The Alkaloids

17.5 Introduction

The alkaloids are naturally occurring nitrogenous substances of plant origin that possess marked physiological properties. The term alkaloid means "alkali-like" and originates in the fact that nearly all alkaloids are nitrogen heterocycles with basic properties. Alkaloid chemistry represents an extremely complex study and one to which volumes have been devoted. In this book we can examine only a few members of each general class.

17.6 Alkaloids with Isolated Five- and Six-Membered Heterocyclic Systems

Coniine, 2-*n*-propylpiperidine, has a relatively simple structure when compared to those of other alkaloids.

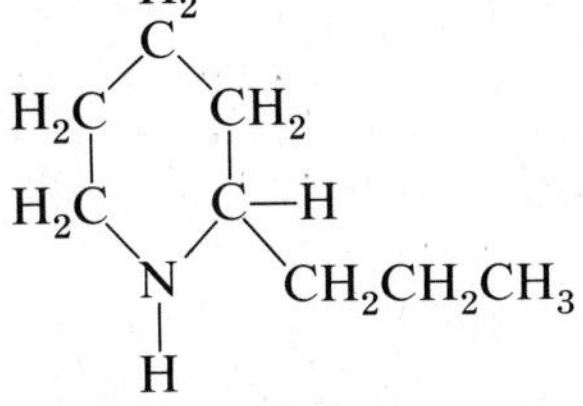

Coniine

Coniine is the toxic substance in *Conium maculatum* and certain other closely related herbs. Its common name is "hemlock," but, of course, this is not the hemlock tree of our forests. Students of history will remember reading of hemlock as the death potion Socrates was forced to drink.

Meperidine, a synthetic phenylpiperidine sometimes called "Demerol," has a physiological action similar to, but less marked than, that of morphine (Sec. 17.8). The analgesic property of meperidine was accidentally discovered when an injection into a rat caused the animal to manifest the same symptoms as those produced by morphine.

Meperidine

Nicotine

Nicotine, one of the principal alkaloids found in the tobacco plant, is a β-pyridine derivative in which N-methylpyrrolidine is joined at its 2-position to the pyridine ring.

Nicotine is extremely toxic to animals when ingested. It kills a number of insects on contact and is used in sprays against a number of leaf-sucking pests.

17.7 Alkaloids Containing Bridged Heterocyclic Systems

Cocaine and atropine belong to a class of alkaloids known as the **tropane alkaloids.** Cocaine is obtained from coca leaves and is a powerful anesthetic but has the disadvantages of being very toxic and habit forming. A synthetic substitute, *procaine* (sometimes called novocaine), incorporates the beneficial properties of cocaine without the undesirable side effects. The part of the cocaine structure enclosed in the formula (solid line) appeared to be responsible for its anesthetic property and served as a pattern in early attempts to synthesize local anesthetics.

However, in the evolution of hundreds of compounds sought for their anesthetic properties, it was discovered that a benzoic acid ester is not necessarily a requisite for anesthetic action. *Xylocaine,* and more recently *mepivacaine,* are amides that retain certain structural similarities to cocaine.

Atropine occurs in the dried roots, leaves, and tops of the belladonna plant, *Atropa belladonna.* It is used in the form of its sulfate salt as a mydriatic for the dilation of the pupil of the eye. The action of the drug was known to early Europeans, and the belladonna plant, supposedly, was named "beautiful lady" in reference to the cosmetic effects of enlarged pupils. Structurally, atropine is very similar to cocaine.

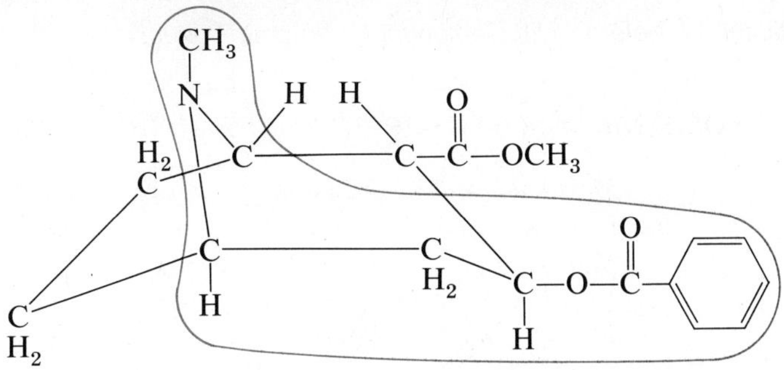

Cocaine

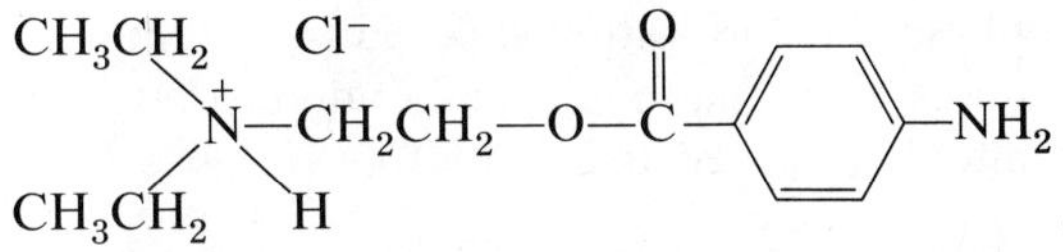

β-Diethylaminoethyl-*p*-aminobenzoate hydrochloride
(Procaine)

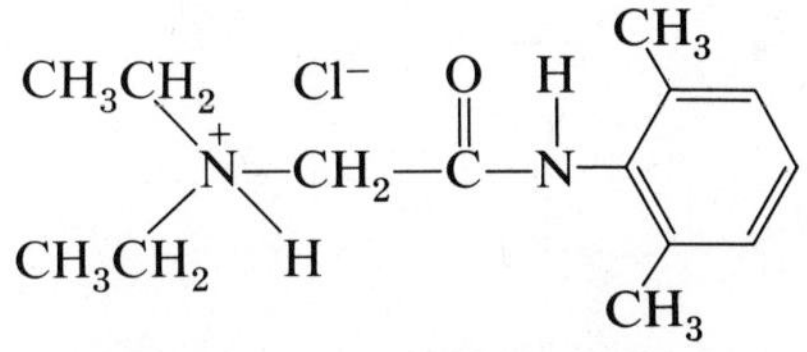

Diethylaminoacet-(2,6-dimethylanilide) hydrochloride
(Xylocaine)

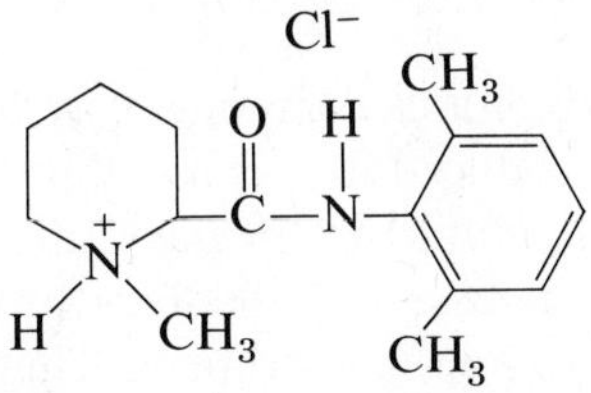

($\pm$)-1-Methyl-2′,6′-pipecoloxylidide hydrochloride
(Mepivacaine)

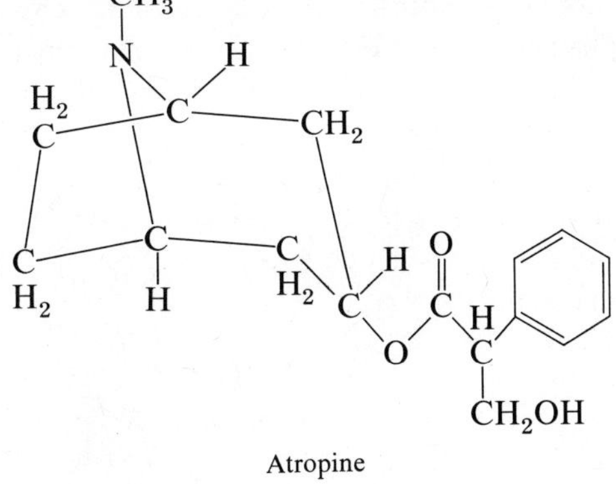

Atropine

17.8 *Alkaloids with Fused Heterocyclic Systems*

Morphine, one of the most useful of drugs, occurs along with twenty-four other alkaloids in the dried latex of the opium poppy, *Papaver somniferum.* Morphine is a powerful analgesic, but its use, like that of other narcotics, leads to addiction. A synthetic compound called *methadone* was developed by the Germans during World War II as a substitute for morphine and found to be an even more effective analgesic than the natural product. While the use of methadone also leads to addiction, it does not produce the same mental and physical deterioration suffered by heroin addicts. By treatment with daily oral doses of methadone, heroin addicts have been socially rehabilitated but not always returned to a drug-free status.

Morphine

Methadone

Codeine is an ether derivative of morphine in which the phenolic hydroxyl has been replaced by a methoxyl group. Codeine is widely used as a cough depressant.

Heroin is not a component of opium but is the diacetyl derivative of morphine produced by acetylating both phenolic and alcoholic hydroxyl groups. Its manufacture in the United States is forbidden.

Ergot alkaloids are alkaloids produced by the fungus *ergot,* a parasitic growth on rye and other cereals. They are amides of lysergic acid, a structure which includes the indole nucleus. Perhaps the most notorious of the ergot alkaloids is the synthetic diethyl amide of lysergic acid known as LSD. The oral administration of as little as 50 μ g (0.000,050 g) of this amide produces psychotic symptoms in man resembling schizophrenia. It is believed by psychopharmacologists that the drug is antagonistic to serotonin, a substance naturally present in brain tissue.

Lysergic acid

Serotonin

Reserpine is a complex polycyclic system which, like the ergot alkaloids, also includes the indole nucleus. Reserpine is one of the alkaloids obtained from the extracts of a species of Indian snake root called *Rauwolfia serpentina*. Such extracts have been used for centuries by the natives of southern Asia to treat a variety of disorders including snake bite, dysentery, insanity, and epilepsy. Reserpine has remarkable tranquilizing powers and is called a behavioral drug. Its synthesis was accomplished by Woodward[3] in 1956.

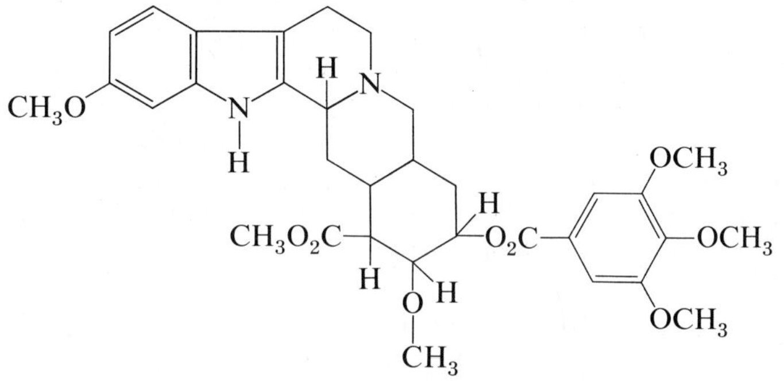

Reserpine

Tubocurarine is an alkaloid which contains two isoquinoline residues. It is the active principle of curare, a paralyzing arrow and dart poison used by South American aborigines. Tubocurarine and other curariform drugs act as skeletal muscle relaxants and have been used as valuable aids during surgery. A lethal dose of the drug results in respiratory failure. The tubocurarine structure includes two quaternary ammonium ions and in this respect bears a functional similarity to choline (Sec. 13.7).

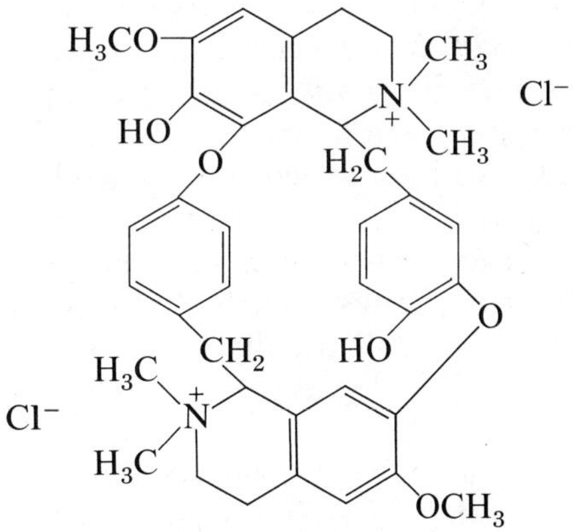

Tubocurarine chloride

[3] Robert B. Woodward (1917–   ), Professor of Chemistry, Harvard University. Winner of the Nobel Prize in Chemistry (1965).

Vitamins

17.9 *Introduction*

The major nutrients (fats, carbohydrates, proteins) comprise the bulk of the animal diet, but certain other nutrients in minute amounts also are necessary constituents of the diet. The lack of these trace nutrients in the animal diet results in deficiency diseases manifested by improper growth, metabolism, and behavior. Among these essential dietetic constituents are the vitamins. Many enzymes require the assistance of relatively small, non-protein molecules called coenzymes in order to carry out their catalytic functions. Some of the water-soluble vitamins serve as components of a larger coenzyme molecule. The fat-soluble vitamins do not appear to serve in coenzyme roles but have other important functions. At present the full scope of the functions of some vitamins are not known with any certainty. Vitamins in this category include vitamin A, C, D, and E; thus, for our purposes the vitamins may be defined simply as abundant, potent, naturally-produced substances that the animal organism requires, but usually is incapable of synthesizing. The need for a certain vitamin varies with the species, and a vitamin required by one animal may not always be required by another. Vitamin research is one of several areas in which organic chemistry and nutrition are interrelated. It is a vast and complex field of study and our discussion will be limited to only a brief review of the principal vitamins.

17.10 *Historical*

One of the major hardships encountered by the early explorers—Columbus, Vasco da Gama, Jacques Cartier, Henry Hudson and others—on their expeditions into the new world was the disease *scurvy*. Scurvy, caused by a lack of vitamin C, was especially common among sailors on long voyages without fresh food. It was recognized as early as 1750 to be a deficiency disease, and British naval surgeons, in their attempts to treat it, observed the beneficial effects of fruit juices—especially, citrus fruits. They recommended fruit as a regular part of a sailor's ration, but the recommendation went unheeded by the British Admiralty for almost fifty years. Finally, the inclusion of limes in ships provisions eliminated scurvy as one of a British seaman's hazards. The name "limey" in reference to a British seaman is still heard today.

The first serious inquiry into the cause and effects of a vitamin deficiency began with the observations of Dr. Eijkman, a Dutch physician working in Java. In 1897, he observed that a diet of polished rice caused beriberi among the natives. This disease is characterized by a polyneuritis, muscular atrophy, and a general debilitation. The protective principle against this disease, now known as thiamine, appeared to have been removed in the rice polishings. The symptoms of beriberi disappeared when the rice polishings were restored to the diet. Funk, in 1911, succeeded in isolating from rice polishings the vital substance. On analysis, it was found to contain nitrogen, and Funk concluded that the chemical structures of these essential agents were those of amino compounds. He therefore named them **vitamines** (vital amines). The name since has been shortened by dropping the final "e." As used today the term designates a number of substances some of which lack nitrogen entirely.

17.11 *The Fat-Soluble Vitamins*

The fat-soluble vitamins are those soluble in fats and in fat solvents. Included in this classification are vitamins A, D, E, and K. The role of vitamins as food accessories usually is considered in relation to nutritional deficiencies. It should be pointed out that doses of the fat-soluble vitamins, when given far in excess of normal requirements, also can have toxic effects. Vitamin poisoning occurred in a number of arctic explorers who became seriously ill after eating polar bear liver. There also have been numerous cases of vitamin poisoning in infants. Young mothers, eager to fulfill all vitamin requirements of their first offspring, sometimes give overdoses of fat-soluble vitamins. Poisoning by water-soluble vitamins is not possible because any amounts not required are voided from the body in urine.

Each of the fat-soluble vitamins is discussed briefly in the following sections.

(a) **Vitamin A** may be obtained from the coloring matter of many green and yellow vegetables. Vitamin A, as such, is not found in plants, only β-carotene, its precursor or *provitamin*. The β-carotene molecule (Sec. 3.12), when cleaved in the center of the linear chain and converted at each end to a carbinol, yields two molecules of vitamin A. Other sources of vitamin A[4] are fish-liver oil, the livers of other animals, eggs, butter, and cheese.

A deficiency of vitamin A causes night blindness—an inability to see in dim light or to adjust to decreased intensity of light. Another disease of the eye known as *xerophthalmia,* in which the tear glands cease to function, results from a lack of vitamin A.

Vitamin A

---

[4] There are two vitamins A. One is known as vitamin A_1 or retinol, and the other as vitamin A_2, or 3-dehydroretinol. Vitamin A_2 is found in the liver oils of fresh water fish and differs structurally from vitamin A_1, found in the livers of cod and other salt-water fish, by having a second double bond between carbons 3 and 4. Physiologically, the two vitamins have the same activity and both are called vitamin A.

(b) **Vitamin D** is sometimes referred to as the "antirachitic" vitamin. It is related to the proper deposition of calcium phosphate and controls the normal development of the teeth and bones. There are ten or more compounds which have antirachitic properties and are designated D_1, D_2, D_3, etc. Vitamin D from fish oils is D_3, while that produced by irradiation of the skin with ultraviolet or sunlight is D_2. Vitamin D_2 is known as *calciferol* and is derived from ergosterol, a plant sterol (Sec. 17.21). Milk, by irradiation, is fortified with vitamin D. Vitamin D is produced in or on the skin by irradiation of the sterols present there. Sunlight supplies *not* the vitamin, only the necessary radiation. Vitamin D_2 has the following structure.

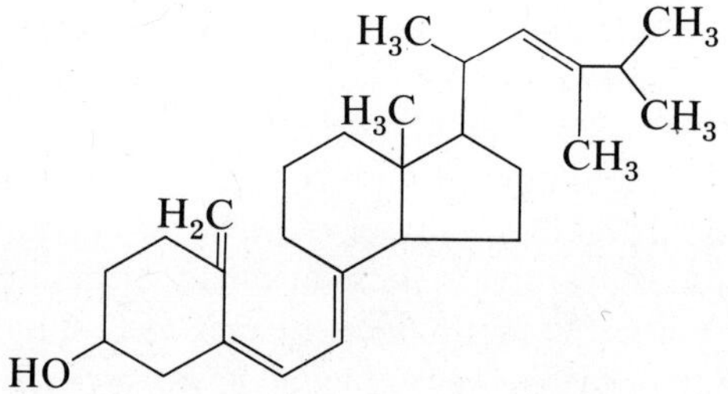

Calciferol (vitamin D_2)

(c) **Vitamin E,** sometimes called the fertility factor, is related to the proper functioning of the reproductive system. Vitamin E is found in the nonsaponifiable fraction of vegetable oils such as corn-germ oil, cottonseed oil, wheat-germ oil, and peanut oil. It also occurs in green leafy vegetables. As in the case of the A and D vitamins, there also is more than one form of vitamin E. Four different structures called *tocopherols* have vitamin E activity. These are designated α-, β-, γ-, and δ-tocopherols. The structure of α-tocopherol, the most potent, is shown.

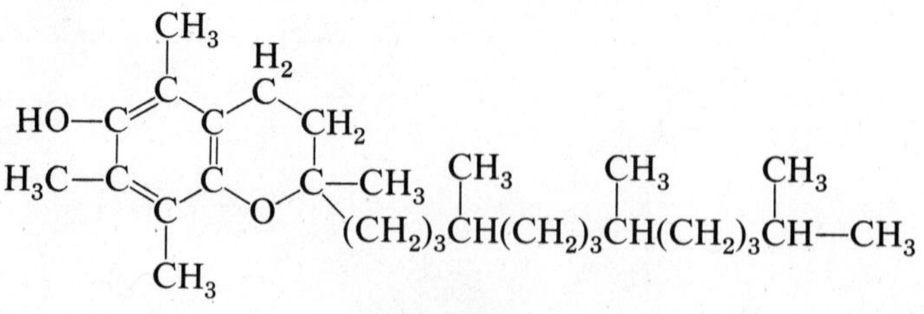

α-Tocopherol

Vitamin E, as was pointed out earlier (Sec. 11.6), is also used as an antioxidant for the prevention of oxidative rancidity in vegetable oils.

(d) **Vitamin K** is the antihemorrhagic factor related to the bloodclotting mechanism. This vitamin is important especially from a surgical standpoint.

There are at least two K vitamins. Vitamin K_1 is obtained from the alfalfa leaf; vitamin K_2 is produced by bacterial action in the intestinal canal. The structure of vitamin K_1 is shown.

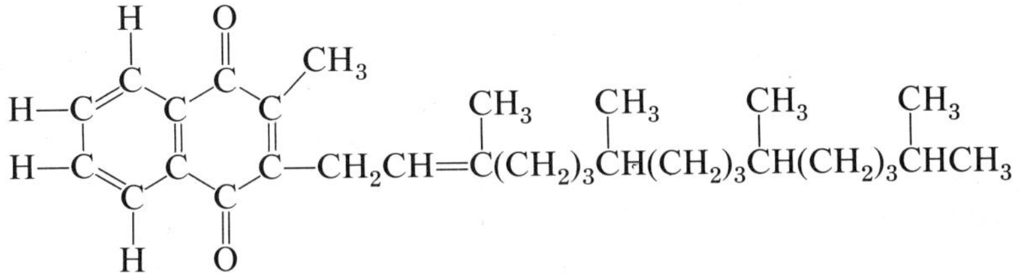

Vitamin K_1

17.12 *The Water-Soluble Vitamins*

The water-soluble vitamins include vitamin C and all other vitamins designated B. The latter are collectively referred to as the vitamin B complex. The water-soluble vitamins bear less resemblance to each other than do the fat-soluble vitamins. Whereas the fat-soluble vitamins are isoprenoid structures (Sec. 17.13) and largely of hydrocarbon composition, the water-soluble vitamins all possess polar groupings to render them water-soluble.

(a) **Vitamin C,** now called **ascorbic acid,** is the vitamin which prevents scurvy. A person with scurvy suffers pains in the joints and hemorrhages from the mucous membranes of the mouth. The gums especially are affected and become red, ulcerated, and even gangrenous. Ascorbic acid is abundantly found in citrus fruits, tomatoes, green peppers, and parsley. It should be obtained from fresh sources for it loses its potency when heated or when exposed to air for any length of time. Ascorbic acid is a hexose derivative with the following structure.

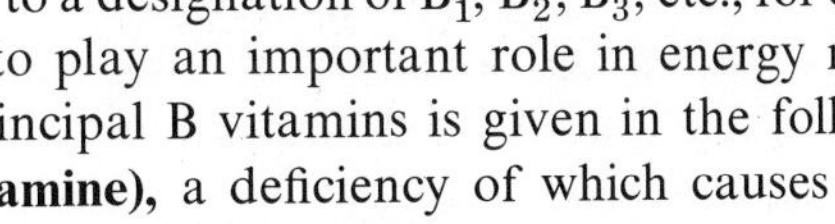

Ascorbic acid

(b) **The B Vitamin Complex.** Vitamins designated as B vitamins were at one time all thought to be the same. The isolation of each new factor from vitamin B preparations has led to a designation of B_1, B_2, B_3, etc., for each new factor. The B vitamins appear to play an important role in energy metabolism. A brief discussion of the principal B vitamins is given in the following sections.

Vitamin B_1 (thiamine), a deficiency of which causes beriberi in man, is present in whole cereal grains, legumes, lean meat, nuts, and yeast. The vitamin has been found to contain both the pyrimidine and thiazole heterocyclic systems. The structure of thiamine has been determined and the vitamin has been prepared. Thiamine usually is prepared in the form of an acid salt.

$$CH_3-C \underset{N}{\overset{N}{\cdots}} C-NH_3{}^+Cl^- \quad CH \overset{S}{\cdots} C-CH_2CH_2OH$$

Thiamine hydrochloride
(Vitamin B₁)

Thiamine occurs in nature either as the pyrophosphoric acid ester or as the free vitamin.

Thiamine hydrochloride pyrophosphate

Vitamin B₂ (riboflavin) is an orange-yellow, crystalline compound widely distributed in nature. It is found in milk, lean meats, liver, fish, eggs, and leafy vegetables. A lack of riboflavin in the diet causes an inflammation of the lips, dermatitis, and a dryness and burning of the eyes, accompanied by a sensitivity to light.

The structure of riboflavin is shown.

Riboflavin

Niacin (nicotinic acid), another of the B vitamins, is the antipellagra factor. Pellagra is a disease characterized by dermatitis, a pigmentation and thickening of the skin, and soreness and inflammation of the tongue and mouth. Niacin may be found in most of the same foods that supply riboflavin. Especially rich sources of niacin are lean meats and liver. Although present in whole cereal grains, niacin is lost in the milling process. The niacin structure is part of the enzyme NAD (Sec. 17.4). The structural formula of niacin, shown below, is relatively simple when compared to those of other vitamins. The oxidation of nicotine (Sec. 17.6) or β-picoline (3-methylpyridine) produces nicotinic acid, or niacin.

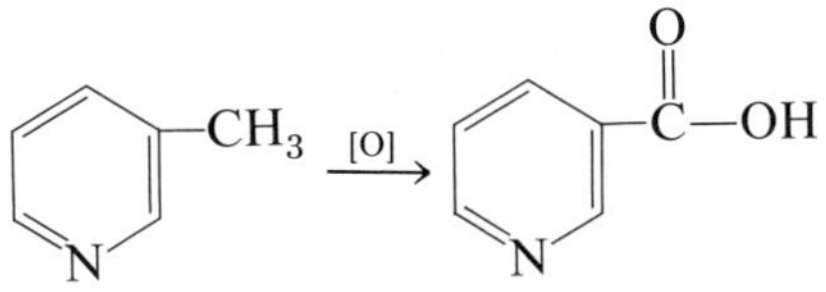

β-Picoline   Niacin

Vitamin B₆ (pyridoxine) is a vitamin whose function appears to be intimately related to the proper metabolism of fats and amino acids. Meat, fish, egg yolk, and whole cereal grains are rich sources of vitamin B₆. The exact requirements of vitamin B₆ for adult man has not been established, but a lack of pyridoxine in the diet of experimental animals leads to dermatitis, anemia, and epileptic seizures. The structures of pyridoxine and two of its derivatives which also show vitamin B₆ activity are given below.

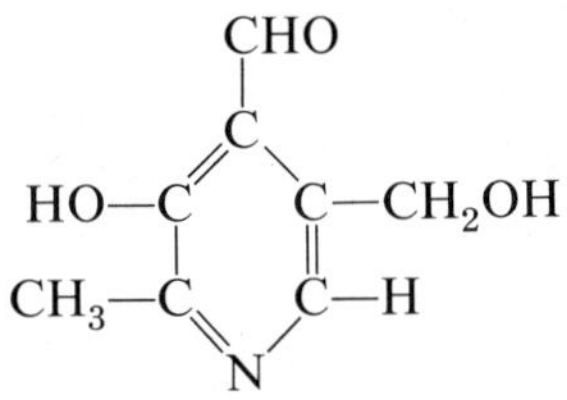

Pyridoxal 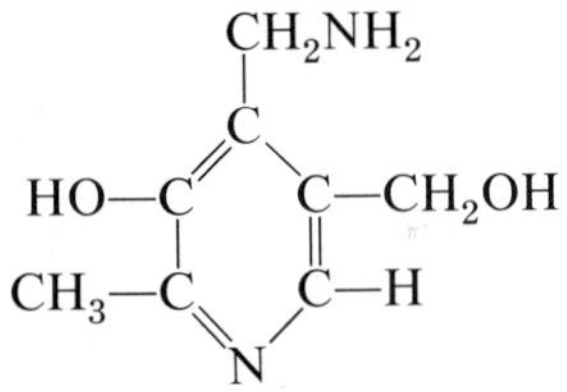   Pyridoxamine

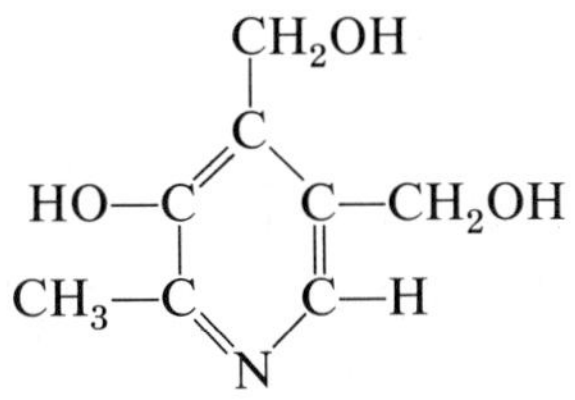

Pyridoxine

Vitamin B₁₂ is the vitamin which prevents pernicious anemia. It is a dark red crystalline compound with a complex structure. Like heme and chlorophyll-a, it also contains a porphyrin nucleus, but a unique feature of its structure is the presence of trivalent cobalt. Vitamin B₁₂ was the first cobalt-containing organic compound to be found. It sometimes is referred to as *cobalamin.* It is found most abundantly in liver but also occurs in meat, eggs, and sea foods. The structure of vitamin B₁₂ is shown on page 506.

Vitamin B₁₂
(Cyanocobalamin)

Terpenes

17.13 Introduction

. The name **terpene** (Gr., *terebinthos,* turpentine tree) is used to describe, in a *broad* sense, a class of natural products whose carbon skeletons are multiples of the C_5 isoprene unit. The name **isoprenoid** is sometimes used to describe these compounds. Although isoprene itself does not occur naturally, it is apparent that in nature a large number of useful products are built from isoprene units. The isoprene unit appears in most natural substances in a regular head-to-tail sequence, although in some cases a head-to-head or a tail-to-tail arrangement is found. Recognition of the head-to-tail arrangement as a recurring architectural feature in a number of natural products led to what has come to be known as the **isoprene rule.** This rule simply states that the structure of a terpene most likely to be correct is one that allows its carbon skeleton to be divisible into iso-C_5 units. The rule has been helpful in deriving the structures not only of the terpenes, but also those of a number of other natural products.

$$\text{head (h)} \longrightarrow CH_2{=}\underset{\underset{\displaystyle CH_3}{|}}{C}{-}CH{=}CH_2 \longleftarrow \text{tail (t)}$$

Isoprene

Many of the terpenes are hydrocarbons while others are alcohols, ethers, aldehydes, ketones, or acids. Terpenes, for the most part, are fragrant compounds and usually can be separated from other plant materials by gentle heating or by steam distillation. A number are classified as "essential oils"[5] and are used in perfumes, in flavoring agents, and in medicinals.

17.14 *Monoterpenes (C_{10})*

Terpenes, in a *strict* sense, include those unsaturated hydrocarbons or their derivatives that contain only *two* isoprene units. These two C_5 units may appear in either a cyclic or in an open chain structure. Two very common terpenes are *limonene,* found in the rinds of lemons and oranges, and *α-pinene,* the chief constituent of turpentine. Turpentine is obtained as the steam volatile component of pine tree sap. Pine stumps are another source of this important commercial product. Turpentine is widely used as a solvent and paint thinner. Turpentine, natural resins, and pitch, all essential to the maintenance of early wooden sailing ships, are still referred to as "naval stores." The structures for limonene and α-pinene, and for two other common terpenes, are shown. The junction of separate isoprene units in each is indicated by a dotted dividing line.

Limonene   Citronellol   α-Pinene   Camphor

[5] Essential oils are "oils of essence"—that is, volatile and pleasantly scented.

17.15 Sesquiterpenes (C_{15})

Sesquiterpenes (three isoprene units) appear in a number of cyclic and acyclic structures. Many are bicyclic and represent fused ring systems which can be converted into naphthalene derivatives. The structures of *selinene, cadinene,* and *farnesol* are shown.

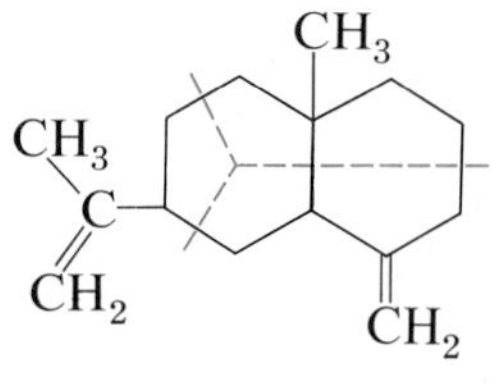

β-Selinene
(oil of celery)

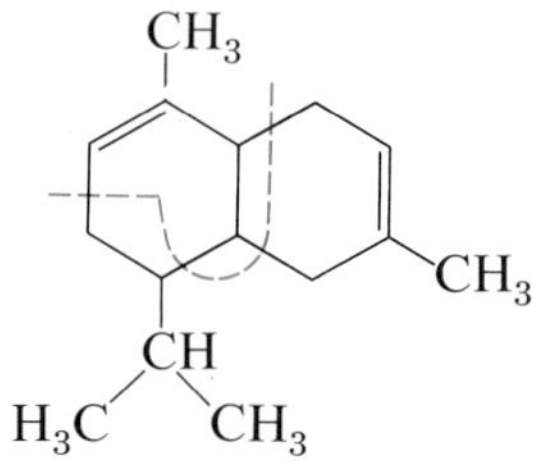

Cadinene
(oil of cade in juniper
and cedar oils)

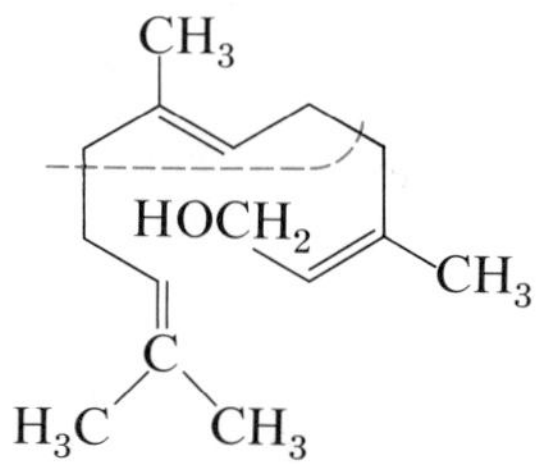

Farnesol
(lily of the valley)

17.16 Diterpenes (C_{20})

Diterpenes (four isoprene units), like the sesquiterpenes, have either cyclic or acyclic structures, or a combination of both. A diterpene alcohol already discussed in some detail is vitamin A (Sec. 17.11-a). The precursor of vitamin A is the *tetraterpene, β-carotene* (Sec. 3.12). The resin acid, *abietic acid,* is a tricyclic diterpene that remains as *rosin* in the residue after turpentine is removed from pine sap. Rosin is an important commercial raw material widely used in the manufacture of varnish, sizing for paper and textiles, and in soaps. Abietic acid is one of the most abundant and cheapest of organic acids. Its structure is shown.

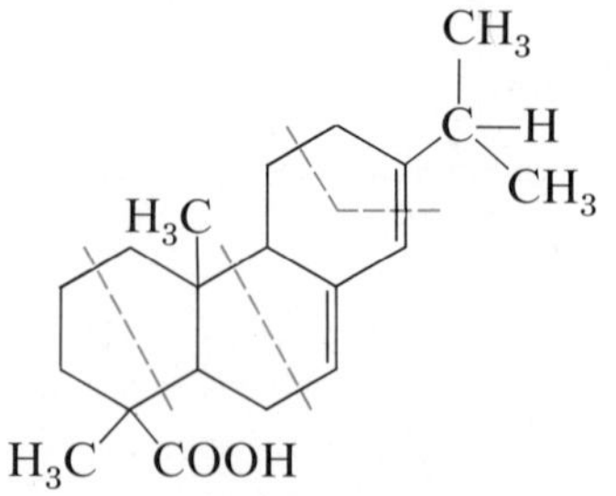

Abietic acid

17.17 Triterpenes (C_{30})

Triterpenes (six isoprene units) are very abundant natural substances. Of particular interest is *squalene,* a triterpene found in shark-liver oil. Squalene is an acyclic isoprenoid in which two regular C_{15} units are joined tail-to-tail at the center of the chain.

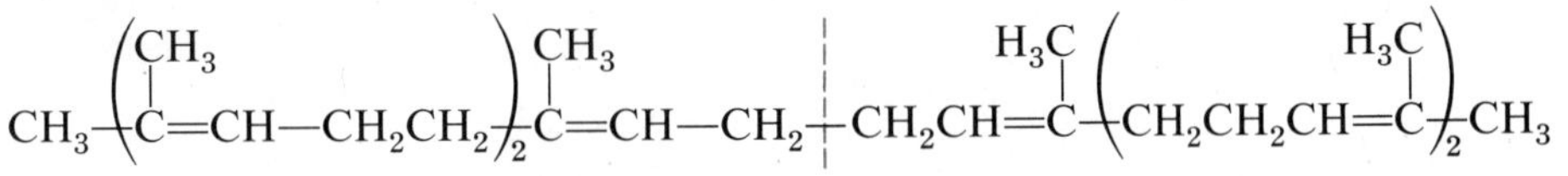

$$CH_3 \left(\overset{CH_3}{\underset{|}{C}} = CH - CH_2CH_2 \right)_2 \overset{CH_3}{\underset{|}{C}} = CH - CH_2 \;\Big|\; CH_2CH = \overset{H_3C}{\underset{|}{C}} \left(CH_2CH_2CH = \overset{H_3C}{\underset{|}{C}} \right)_2 CH_3$$

Recent evidence has shown squalene to be a precursor in the biosynthesis of cholesterol (Sec. 17.20).

17.18 *Polyterpenes*

Natural rubber, our best example of a polyterpene, is found in the latex of the rubber plant (*Hevea braziliensis*). Approximately 2,000 isoprene units are combined in a rubber molecule to produce a long linear structure. Cross-linking of adjacent chains at unsaturated sites by heating with sulfur (vulcanization) improves the properties of rubber. Natural rubber, when subjected to distillation, produces isoprene.

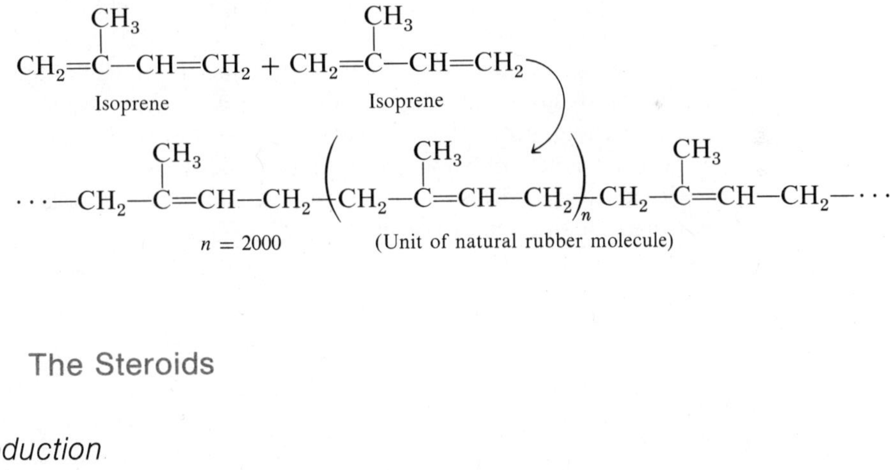

$$CH_2 = \overset{CH_3}{\underset{|}{C}} - CH = CH_2 + CH_2 = \overset{CH_3}{\underset{|}{C}} - CH = CH_2$$

Isoprene   Isoprene

$$\cdots - CH_2 - \overset{CH_3}{\underset{|}{C}} = CH - CH_2 \left(CH_2 - \overset{CH_3}{\underset{|}{C}} = CH - CH_2 \right)_n CH_2 - \overset{CH_3}{\underset{|}{C}} = CH - CH_2 - \cdots$$

$n = 2000$ (Unit of natural rubber molecule)

The Steroids

17.19 *Introduction*

The steroids are a family of compounds widely distributed in plants and animals. Common to the structure of all compounds of this class is a tetracyclic framework composed of the phenanthrene nucleus (Sec. 4.5) to which is fused at the 1,2-positions a cyclopentane ring.

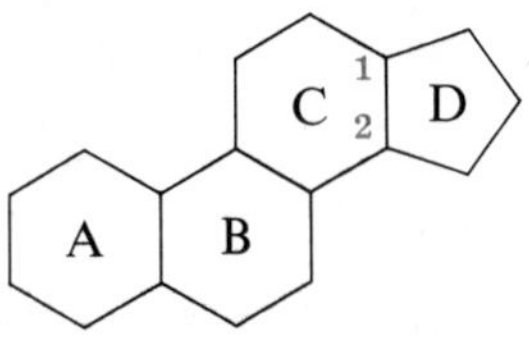

The steroid ring system

The rings in the steroid molecule usually are not aromatic but often contain one or more isolated double bonds. The total structure of one steroid differs from that of another, usually by a variation in the side chain or by a variation in the number and type of functional groups. To the family of steroids with this common ring system belong the sterols, the sex hormones, the bile acids, and other biologically important materials.

For purposes of nomenclature the steroid ring skeleton is numbered as shown.

17.20 *Cholesterol*

The **sterols** are solid alcohols which possess a hydroxyl group at position 3, a double bond between carbons 5 and 6, a side chain on carbon 17, and methyl groups joined to ring carbons numbered 10 and 13. *Cholesterol,* $C_{27}H_{46}O$, one of the most widely distributed sterols, is found in almost all animal tissue but is particularly abundant in the brain, the spinal cord, and in gallstones. Deposition of cholesterol or its derivatives in the arteries (hardening of the arteries) restricts the flow of blood, causes high blood pressure, and leads to some forms of cardio-vascular disease. The structure of cholesterol is shown with asymmetric carbon atoms indicated by asterisks.

Structure and numbering system of cholesterol

Stereochemical configuration of cholesterol

17.21 *Ergosterol*

Although cholesterol is found only in animals, a large number of closely related compounds known as **phytosterols** are found in plants. One of these, *ergosterol*, $C_{28}H_{44}O$, is produced by yeast. Ergosterol is of particular interest because, when irradiated, it yields calciferol, vitamin D_2.

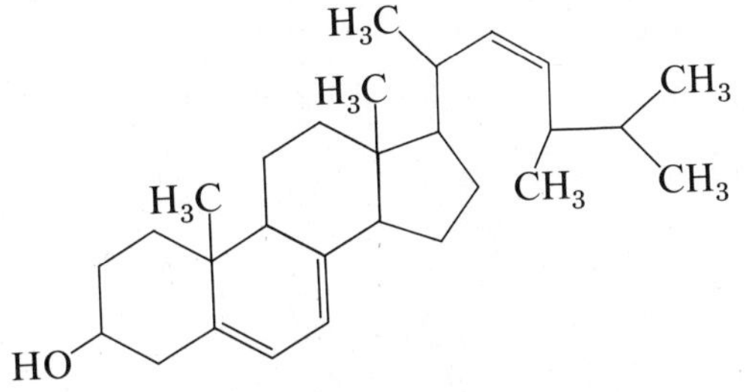

Ergosterol

17.22 *Sex Hormones*

The male and female sex hormones are structurally related steroids responsible for the development of sex characteristics and sexual processes in animals. Sex hormones are produced in the gonads (ovaries and testes) when the latter are stimulated by other gonadotropic hormones. The female sex hormones are involved in the menstrual cycle, the changes in the uterus, and in the preparation for and maintenance of pregnancy. The structures and names of the principal female sex hormones are shown below.

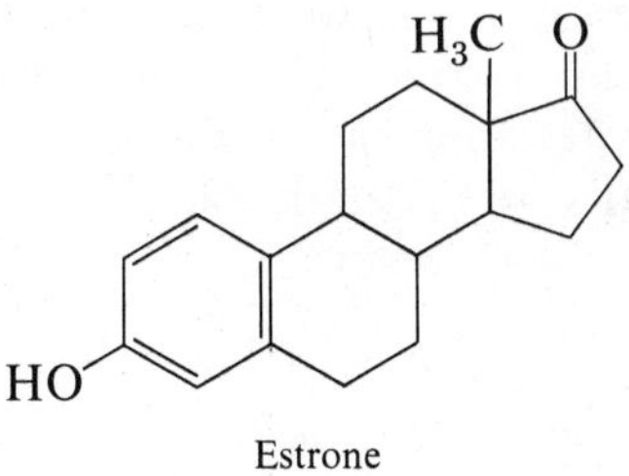

Estrone

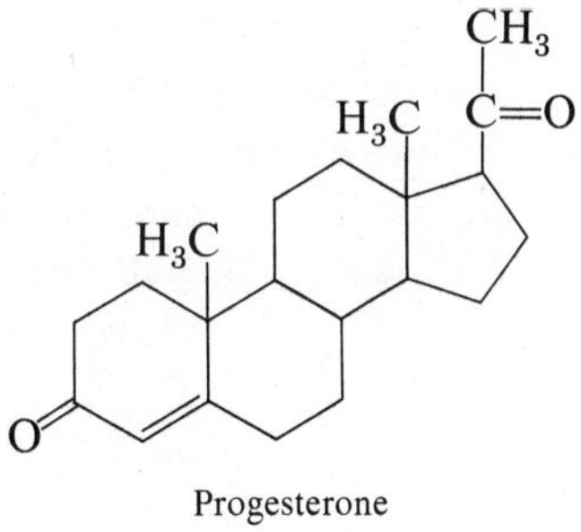

Progesterone

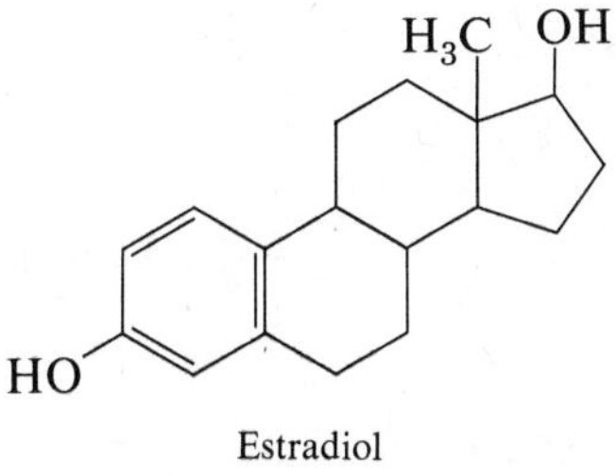

Estradiol

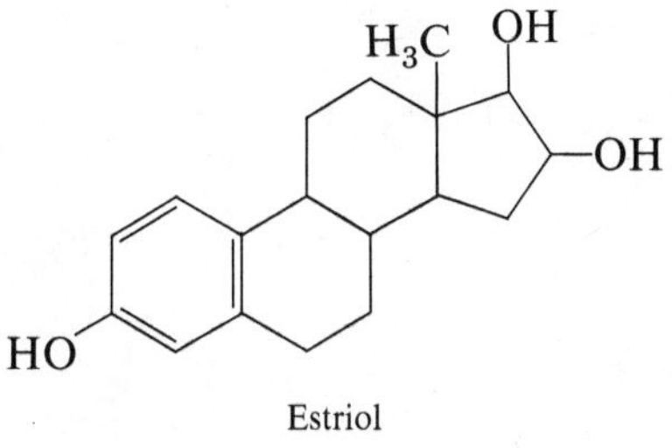

Estriol

Oral contraceptives have been developed that contain synthetic compounds structurally similar to progesterone and estriol but modified chemically to permit easier assimilation into the bloodstream. These synthetic agents, when taken orally, suppress ovulation and mimic pregnancy. The structures of two synthetic progesterones and a synthetic estrogen[6] are shown. Both types of synthetic hormones are used either in combined form or sequentially.

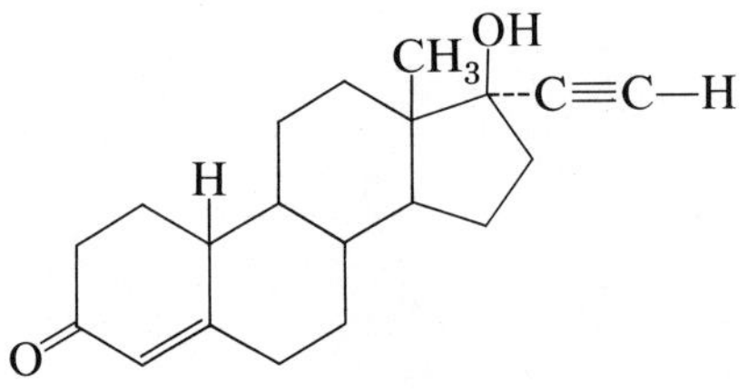

(a)

Norethindrone

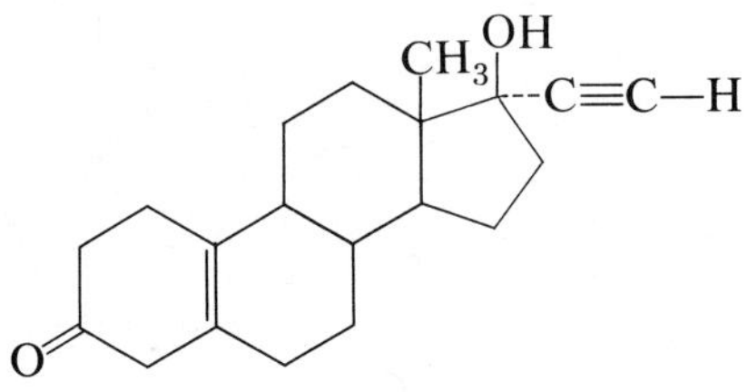

(b)

Norethynodrel

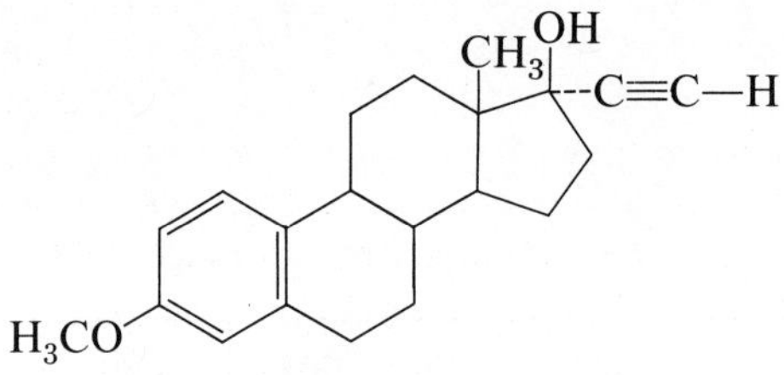

(c)

Mestranol

Orthonovum = (a) + (c)
Enovid　　 = (b) + (c)

The male sex hormones, called **androgens,** except for the absence of an aromatic ring, are very similar in structure to the female hormones. The male sex hormones control the development of the male genital tract and the secondary male characteristics—e.g., beard, voice. The structures of the principal male sex hormones are shown below.

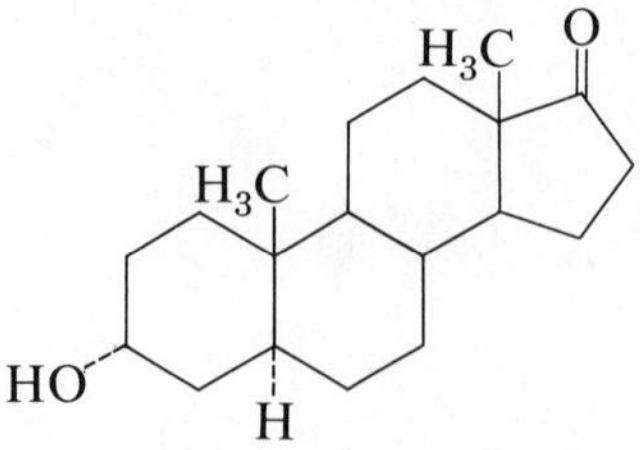

Androsterone

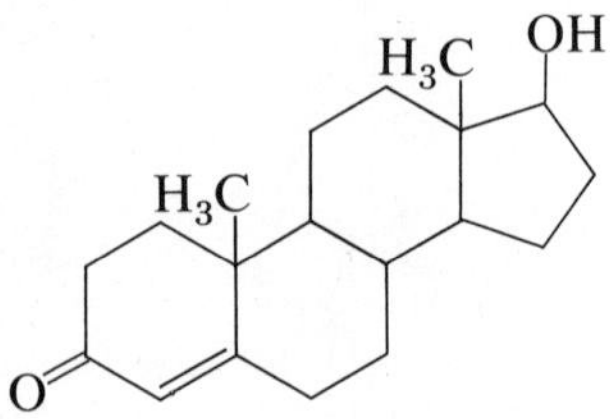

Testosterone

[6] Estrogen is a generic term for a substance that induces estrus—the cyclic phenomenon of the female reproductive system.

17.23 *Adrenal Steroids*

Cortisone, and its derivative, *17-hydroxycorticosterone,* are two steroids produced by the adrenal cortex. These compounds have been used with beneficial results in the treatment of inflammatory and allergic diseases. The structures of cortisone and its reduced form, 17-hydroxycorticosterone, are shown.

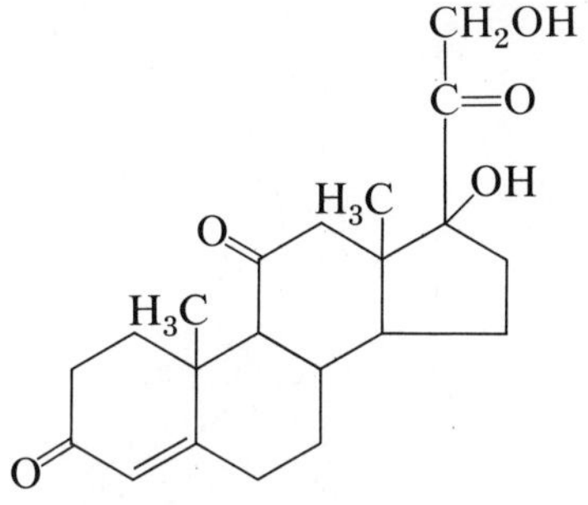

Cortisone 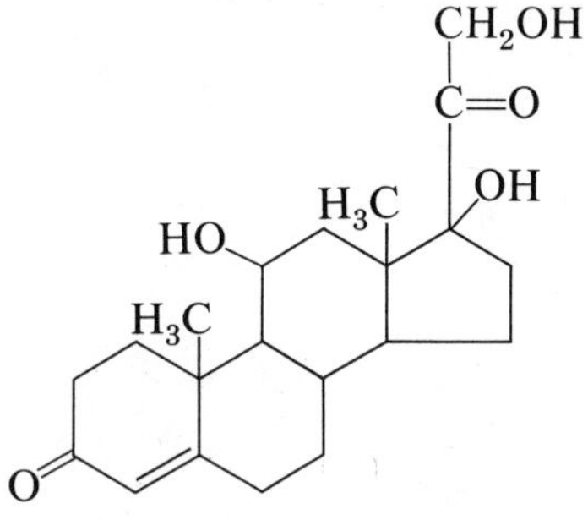   17-Hydroxycorticosterone

Antibiotics

Our discussion of natural products would not be complete without a brief treatment of the antibiotics. These chemotherapeutic agents are potent antibacterials which possess the power to inhibit the growth of, or destroy, microorganisms.

17.24 *Penicillins*

Penicillin, the first antibiotic to be isolated from a mold, was introduced into clinical practice in 1941 with remarkable results. It was found to be active against a number of *Gram-positive*[7] microorganisms of the cocci type and against spirochetes. A number of penicillins have been prepared. Their skeletal structures are the same, but they differ from one another in the character of the side chain R (see formula for penicillin). Commercial preparations of penicillin are mainly penicillin G (R = $C_6H_5CH_2-$, benzyl).

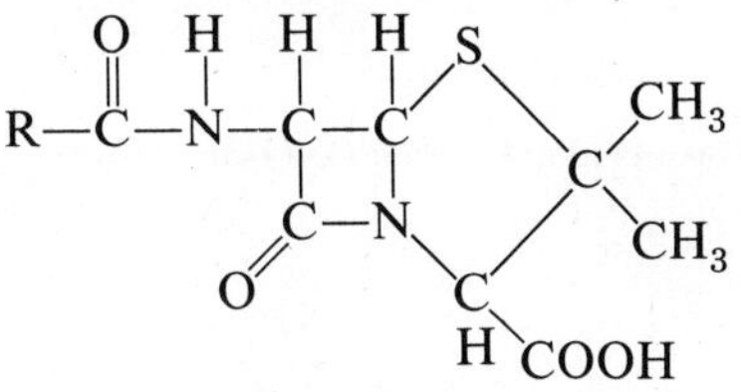

Penicillin

---

[7] *Gram-positive* refers to organisms which stain "positive" when treated with Gram stain; *Gram-negative* refers to those which stain "negative" to the same reagent. A *broad spectrum antibiotic* is one which may be effective against both Gram-negative and Gram-positive organisms.

17.25 *Streptomycin*

A more recent antibiotic is **streptomycin** whose structure is that of a trisaccharide. Streptomycin is active against *Gram-negative* bacteria and is used in tuberculosis therapy. The structure of streptomycin is indicated below.

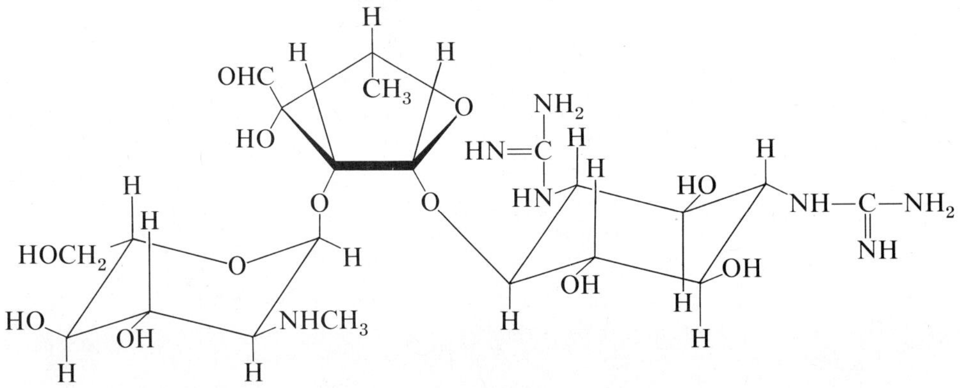

Streptomycin

17.26 *The Tetracyclines*

The **tetracyclines** comprise a family of compounds, each member of which has a four fused-ring structure. Each member differs from the other only in minor detail. Tetracycline, its 7-chloro derivative, *aureomycin,* and its 5-hydroxy derivative, *terramycin,* are *broad spectrum* antibiotics widely used against a number of bacterial and viral diseases.

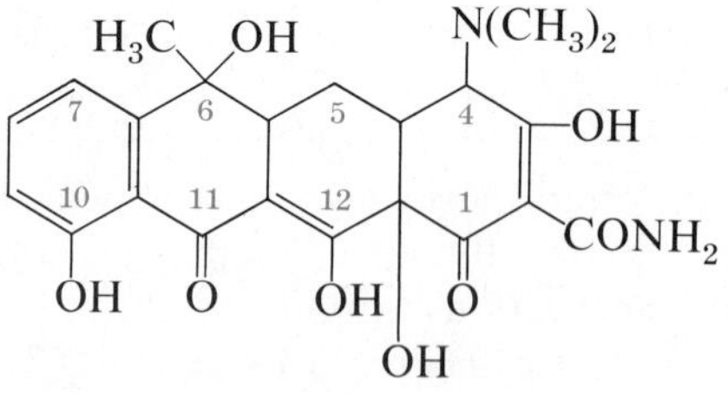

Tetracycline

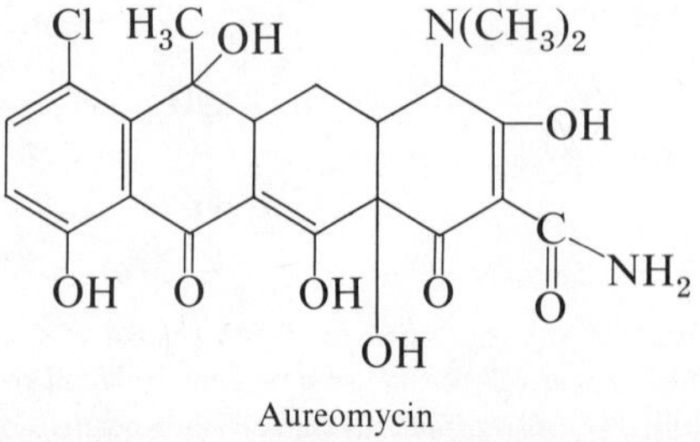

Aureomycin

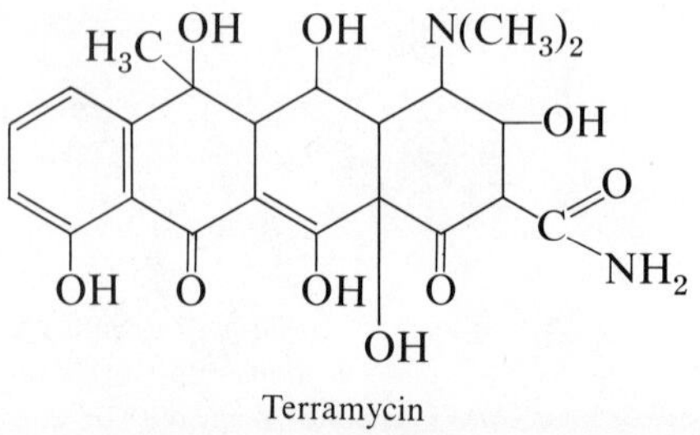

Terramycin

Summary

1. The heterocyclic compounds are ring compounds containing atoms other than carbon. The most common *hetero* atoms are **nitrogen, oxygen,** and **sulfur.** The hetero atoms appear most frequently in five- or six-membered rings.

2. The heterocycles possess *aromatic* properties similar to those of benzene.

3. The heterocyclic systems frequently make up all or part of a great number of polycyclic natural products.

Alkaloids
4. The alkaloids are nitrogen-containing plant products that possess marked physiological properties.

5. Nearly all of the alkaloids are made up of more than one heterocyclic ring. Such rings usually are fused rings—that is, two rings have one or more bonds in common.

6. Many of our most valuable drugs are obtained from alkaloids.

7. The promiscuous use of alkaloids is dangerous and usually leads to addiction.

Vitamins
8. The vitamins are trace nutrients required in the diet of animals for proper growth, development, and good health.

9. The absence of sufficient amounts of vitamins in the diet results in deficiency diseases.

10. Vitamins are generally classified as fat-soluble or water-soluble.

11. The fat-soluble vitamins are vitamins A, D, E, and K; the water-soluble vitamins are vitamin C and those usually designated B vitamins.

Terpenes
12. The name terpene is a general term used to describe a large class of natural products with carbon skeletons that are multiples of the C_5 isoprene unit.

13. Many of the terpenes are unsaturated hydrocarbons. Others possess, in addition to double bonds, the functional groups of the alcohols, carbonyls, and acids. Many are "essential oils" used in perfumery and in flavoring agents.

14. Terpenes may be branched, open-chain structures; they may be cyclic; or they may have structures that contain both features.

15. The terpenes are classified according to the number of C_{10} units present. Monoterpenes are C_{10} compounds, diterpenes, C_{20} compounds, etc.

Steroids

16. The steroids are natural products that have in common the 4-cycle, skeletal structure of 1,2-pentanophenanthrene.

17. Steroids with one or more hydroxyl groups may be called *sterols*.

18. The most abundant steroids of animal origin are cholesterol, ergosterol, the sex hormones, and cortisone.

Antibiotics

19. The antibiotics are antibacterials used in chemotherapeutic agents to inhibit the growth of, or to destroy, microorganisms. A number are of plant (mold) origin.

20. Commonly used antibiotics are the penicillins, streptomycin, and those of the tetracycline family—aureomycin and terramycin.

New Terms

alkaloid	estrogen	sterol
androgen	heterocycle	terpene
antibiotic	isoprene rule	vitamin
essential oil	steroid	

Supplementary Exercises

EXERCISE 17.1   2-Methylpyridine (α-picoline) and 4-methylpyridine (γ-picoline) both enter into base-catalyzed condensation reactions with benzaldehyde. Suggest a reason why the methyl hydrogen atoms in these two compounds appear to be weakly acidic like those in malonic ester, acetoacetic ester, and other dicarbonyls.

EXERCISE 17.2   Refer to the structure of farnesol (Sec. 17.15) and draw it as a horizontal chain. Number all carbon atoms in the longest chain and name the compound according to IUPAC rules.

EXERCISE 17.3 To the "stretched out" farnesol structure of Exercise 17.2 add a second mole-cule of farnesol but set the hydroxyl groups together. If the hydroxyl groups were now removed via reduction and the two carbon atoms to which they were bonded joined, which compound would result?

EXERCISE 17.4 Draw the structures of β-carotene and abietic acid (Sec. 17.16). Are all C_5 units in a head-to-tail arrangement? Does either compound contain asymmetric centers?

EXERCISE 17.5 Natural rubber (Sec. 17.18, also Exercise 5.6) has a *cis* configuration. Is it likely that β-carotene also has isomers? Could these be geometric isomers, structural isomers, or both types of isomers?

EXERCISE 17.6 Cholesterol (Sec. 17.20) is shown both as a planar structure and in its natural conformation. We have been told why cyclohexane rings prefer a chair confor-mation, but why is the D ring not a planar pentagonal structure?

EXERCISE 17.7 What natural structural unit is evident in the fat-soluble vitamins?

EXERCISE 17.8 What simple chemical test would serve to distinguish between the following terpenes? Which spectroscopic method would most easily distinguish between them?

    (a) citronellol   (c) α-pinene

    (b) camphor   (d) abietic acid

EXERCISE 17.9 In the biosynthesis of cholesterol from squalene, the triterpene lanosterol is an important intermediate. Apparently the immediate precursor of lanosterol is squalene oxide. Suggest a mechanism for the formation of lanosterol from squalene oxide. The reaction is acid-catalyzed (H^+) and involves rearrange-ments of the type discussed in Sec. 8.8-B.

Squalene oxide
(Squalene 2,3-epoxide)

Lanosterol

EXERCISE 17.10 Ozonolysis of natural rubber followed by oxidative work-up with hydrogen peroxide (oxidizes —CHO to —COOH) yields an acid, $C_5H_8O_3$, as the major product. What is the structure of the acid?

18

Natural Gas, Petroleum, and Petrochemicals

Introduction

The drilling of the first oil well at Titusville, Pennsylvania, in 1859 gave birth not only to a mammoth new industry but also to a new way of life. The discovery of a vast new source of fuels and lubricants ushered in the machine age. However, the production of fuels and lubricants from crude oils no longer remains the only function of the petroleum industry. Petroleum provides us with much more. Today, not only the high octane fuel which powers our automobiles, but the upholstery, the dozens of plastic components built into the car, much of the rubber in the tires on the wheels—indeed, the asphalt roadway beneath them, all have this common origin. Although the petroleum industry is still the principal supporting industry for all modes of transportation, every facet of our economy is in some measure dependent upon this valuable natural resource.

The demand for ever increasing quantities of petroleum products is presently so great that our current rate of consumption can not be maintained unless vast new reservoirs of petroleum deposits are discovered and developed. Either that or other sources of energy must be utilized.

18.1   *Natural Gases*

According to one theory, petroleum resulted from the decomposition of plant and animal matter of marine origin. This appears to be borne out by the high percentages of nitrogen and carbon dioxide found in gas samples from certain oil fields. Natural gas, other than these noncombustible components, is largely methane with lesser amounts of lower alkanes. The composition of a natural gas varies according to the source from which it is taken. Some natural gases have a carbon dioxide content as high as 30% or more. Other natural gases contain high percentages of nitrogen, hydrogen sulfide, and some helium. Obviously, any gas sold as fuel cannot contain such high percentages of nonfuel components. The latter must be removed by chemical means or by fractionation. Analysis of a representative gas sample supplied as fuel to the consumer has an approximate percentage composition shown in Table 18.1.

TABLE 18.1   *Composition of Natural Gas Sold as Fuel**

Component	Percent
Methane	78–80
Nitrogen	10–12
Ethane	5.9
Propane	2.9
n-Butane	0.71
Isobutane	0.26
C_5-C_7 hydrocarbons	0.13
Carbon dioxide	0.07

* Courtesy of Northern Natural Gas Company, Omaha, Nebraska

18.2   *Liquefied Petroleum (L. P.) Gases*

Certain of the low molecular weight gaseous hydrocarbons are found dissolved in crude oil and also occur as by-products of gasoline manufacture. These low-boiling hydrocarbons (largely propane), after removal by distillation, are compressed until liquefied and then "bottled" in steel cylinders. The liquefied gas reverts to gaseous form when the pressure is released. It is a common sight to see such cylinders of L. P. gas in rural areas, on house trailers, at lakeside cabins, and wherever the convenience of piped natural gas is desired but not available.

18.3 Composition of Crude Oil

Crude oil, with the exception of the very volatile and easily removed gaseous fractions, is an extremely complex mixture made up of hundreds of compounds. A sample of mid-continent crude oil taken from an oil field in Oklahoma and analyzed over a forty-year period (1926–1967) yielded 295 different hydrocarbons. These, altogether, make up but approximately 50% of the volume of the petroleum sample analyzed. The 295 hydrocarbons isolated and identified thus far have been the "easy" ones leaving half of the sample still to be resolved and identified. Table 18.2 lists the classes of hydrocarbons and the number of compounds within each class that have been found in the sample.

TABLE 18.2 *Hydrocarbons Found in Crude Oil**

Class of Hydrocarbon	Type of Compounds within Each Class	Number of Compounds within Each Class
Alkanes	Saturated, straight-chain hydrocarbons	33
	Saturated, branched-chain hydrocarbons	52
Cycloalkanes	Cyclopentane derivatives	27
	Cyclohexane derivatives	25
	Other cycloparaffins	31
Aromatic	Benzene derivatives	40
	Aromatic cycloparaffins	12
	Naphthalene and polynuclear aromatic hydrocarbons	67
	Oxygen-containing compounds	4
	Sulfur-containing compounds	4

* Source: American Petroleum Institute

The number of hydrocarbons within any particular class will, of course, vary according to the geographical location of the producing area. Table 18.3 shows the principal hydrocarbon fractions into which petroleum is separated, the approximate boiling range of each fraction, and their principal uses.

TABLE 18.3 *Petroleum Fractions*

Fractions	Approx. Composition	Approx. Boiling Range, °C	Principal Uses
Gas	C_1–C_4	0–20	fuel
Gasoline	C_6–C_9	69–150	motor fuel, solvent naphtha
Kerosine	C_{10}–C_{16}	175–300	jet fuel, fuel oil, diesel fuel
Gas-oil	C_{16}–C_{18}	300 up	diesel fuel, fuel oil, cracking stock*
Wax-oil	C_{18}–C_{20}	—	lubricants, mineral oils, cracking stock
Paraffin wax	C_{21}–C_{40}	—	packaging (wax paper), candles
Residuum	—	—	roofing, waterproofing, road building materials

* See Sec. 18.4 for a discussion of cracking.

18.4   *The Cracking, or Pyrolysis, of Petroleum*

The appearance of the automobile created an immediate demand for a high energy, volatile, liquid fuel. The supply of **straight-run gasoline** (the C_6–C_9 fraction obtained by simple fractional distillation of petroleum) soon became inadequate for the number of automobiles. Moreover, straight-run gasoline is a low quality motor fuel unsuitable for use in the present-day, high compression automobile engine.

In order to increase the supply of high octane gasoline, it was found necessary to break down, or "crack," high molecular weight hydrocarbons not suitable for motor fuel into smaller molecules. A number of the lower molecular weight members that result from this treatment fall within the gasoline range. Cracking of petroleum usually is accomplished by heating hydrocarbons to high temperatures in the presence of certain catalysts. This treatment, called **catalytic cracking,** yields not only lower alkanes, but alkenes and hydrogen as well. These various fragmentary products then are combined, rearranged, or in some other manner used in the synthesis of high grade gasoline. Certain of the methods used to make gasoline are illustrated in Section 18.6. Figure 18.1 shows schematically the separation and refinement of petroleum fractions.

FIGURE 18.1 *A Schematic Flow Diagram of an Oil Refinery's Principal Processing Units (Courtesy of Texaco, Inc.)*

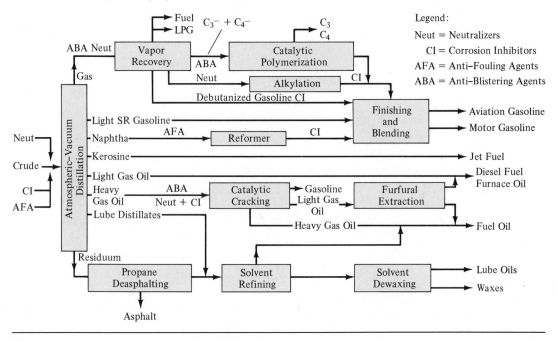

18.5 Gasoline. Octane Number

The ability of a gasoline to perform well in an internal combustion engine is given by its **octane number.** In order to standardize the performance of a gasoline, 2,2,4-trimethylpentane, $(CH_3)_3C$—$CH_2CH(CH_3)_2$, a very fine motor fuel with no tendency to premature explosion under high compression, is assigned an octane number of 100. Normal heptane, $CH_3CH_2CH_2CH_2$-$CH_2CH_2CH_3$, an extremely bad "knocker," is assigned a value of zero. The octane number of any fuel is determined from a comparison of its performance to that of a blend of 2,2,4-trimethylpentane and *n*-heptane. Both the fuel in question and the prepared blend are tested in a specially instrumented engine. The percent of 2,2,4-trimethylpentane in the blend that gives the same performance as the fuel in question establishes the octane number of the latter. Motor fuels can be prepared that have octane numbers well in excess of 100, that is, they will perform even better as motor fuels than 2,2,4-trimethylpentane.

Octane ratings of motor fuels have been improved further by the use of certain additives. These agents improve the performance of a gasoline largely

TABLE 18.4 *Octane Ratings (Number) of Some Hydrocarbons**

Hydrocarbon	Formula	ml TEL**/gallon	
		0.0	3.0
Methane	CH_4	>120	—
Ethane	C_2H_6	118.5	—
Propane	C_3H_8	112.5	—
n-Butane	C_4H_{10}	93.6	101.6
n-Pentane	C_5H_{12}	61.7	88.7
2-Methylbutane (Isopentane)	C_5H_{12}	92.6	102.0
n-Hexane	C_6H_{14}	24.8	65.3
Methylcyclopentane	C_6H_{12}	91.3	105.3
n-Heptane	C_7H_{16}	**0.0**	43.5
Methylcyclohexane	C_7H_{14}	74.8	88.2
Methylbenzene (Toluene)	C_7H_8	103.2	111.8
n-Octane	C_8H_{18}	−19.0	25.0
2,2,4-Trimethylpentane (Isooctane)[1]	C_8H_{18}	**100.0**	115.5
2,4,4-Trimethyl-2-pentene	C_8H_{16}	103.5	105.7
Isopropylbenzene (Cumene)	C_9H_{12}	113.0	116.7
1,3,5-Trimethylbenzene (Mesitylene)	C_9H_{12}	>120	—

* Courtesy of Ethyl Corporation
** Tetraethyl lead
[1] 2,2,4-Trimethylpentane is sometimes incorrectly named "isooctane."

by preventing "knocking" caused by premature explosions. Among the first compounds to be employed as antiknock agents was tetraethyl lead (TEL), $(C_2H_5)_4Pb$ (Sec. 7.6). Note the improvement in the octane ratings of certain hydrocarbons that results when less than one part per thousand of tetraethyl lead (TEL) is added (Table 18.4). Tetramethyl lead (TML), $(CH_3)_4Pb$, a more recently produced lead alkyl, is used for the same purpose. In order to carry away the lead that otherwise would be deposited in the engine and cause fouling, ethylene bromide, $BrCH_2CH_2Br$, and ethylene chloride, $ClCH_2CH_2Cl$, are added to the lead alkyls as scavengers. The lead dihalides formed at the operating temperature of the engine are carried away with the exhaust. Currently, the air pollution problem which has increasingly plagued our large cities is being blamed to a large degree upon the amounts of automobile exhaust emitted into the atmosphere. In addition to the lead compounds described, automobile exhaust also contains unburned hydrocarbons, finely-divided solids called particulates, carbon monoxide, and certain of the oxides of nitrogen. The reduction of lead alkyls in automobile fuel, as now required by law, will necessitate the use of engines having lower compression ratios capable of using lower octane fuels, the inclusion of more aromatics, or the addition of new anti-knock agents. One new antiknock additive currently used in some unleaded gasolines is methylcyclopentadienylmanganese tricarbonyl, the structure of which is shown.

The pollution problem coupled with a scarcity of fuel supplies poses one of the greatest technological challenges of modern times.

18.6 *Methods Used in Gasoline Production*

The presence of branched-chain aliphatic, olefinic, and aromatic hydrocarbons in gasoline greatly increases its octane number (see Table 18.4). Accordingly, processes known as **reforming** or **isomerization** have been developed for the rearrangement of straight-chain molecules to branched-chain structures. Small branched-chain alkanes and alkenes then may be combined by a process called **alkylation** to produce larger molecules which fall within the gasoline range. Another process referred to as **aromatization** causes straight-chain saturated and unsaturated hydrocarbons of sufficient length to become cyclized and dehydrogenated into aromatic hydrocarbons. For example, 1-heptene, when formed into a ring compound and "aromatized," produces toluene. Other examples of how gasoline may be "tailor-made" by employing modern industrial techniques are shown by the following series of equations.

Cracking

$$C_{15}H_{32} \xrightarrow[\text{heat}]{\text{catalyst,}} C_8H_{18} + C_7H_{14}$$

$$C_8H_{18} \xrightarrow[\text{heat}]{\text{catalyst,}} \underset{\textit{n}\text{-Butane}}{CH_3CH_2CH_2CH_3} + \underset{\text{2-Butene}}{CH_3-\overset{H}{\underset{}{C}}=\overset{H}{\underset{}{C}}-CH_3} \quad \text{or} \quad \underset{\text{1-Butene}}{CH_3-CH_2-CH=CH_2}$$

Reforming

$$\textit{n}\text{-}C_4H_{10} \xrightarrow[\text{heat}]{\text{catalyst,}} \underset{\text{Isobutane}}{CH_3-\overset{H}{\underset{CH_3}{C}}-CH_3} \text{ (or Butene)} + H_2$$

$$\underset{\text{1-Butene}}{CH_3-CH_2-CH=CH_2} \xrightarrow[\text{heat}]{\text{catalyst,}} \underset{\text{Isobutene}}{CH_3-\overset{CH_3}{\underset{}{C}}=CH_2}$$

Alkylation

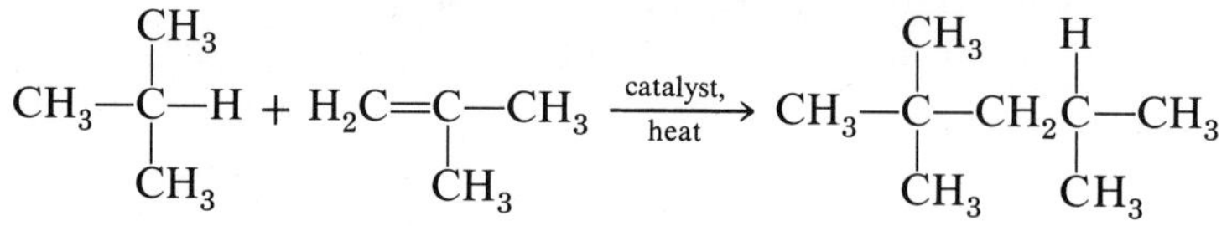

Isobutane   Isobutene   2,2,4-Trimethylpentane

Aromatization

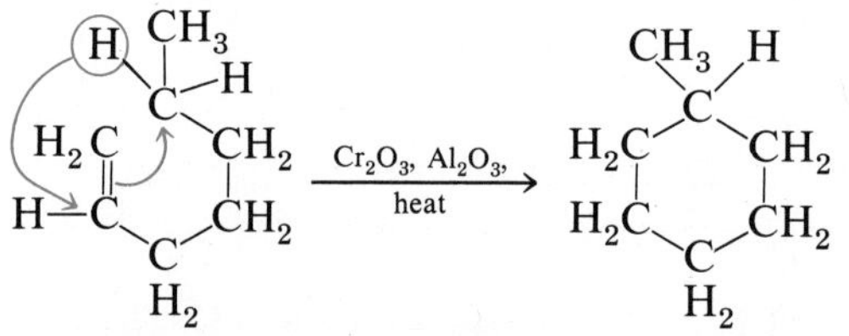

1-Heptene   Methylcyclohexane

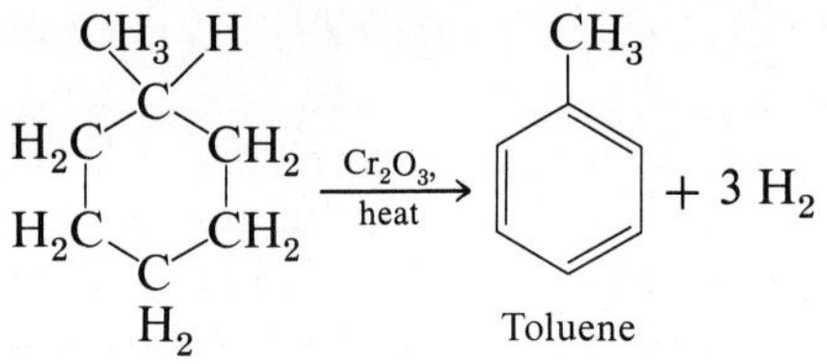

Toluene

18.7 *Synthetic Liquid Fuels*

The absence of petroleum deposits in Germany led to the development of two processes for the production of liquid fuels from coal. One of these, the **Fischer-Tropsch Process,** produces carbon monoxide and hydrogen from coke and steam. Hydrogen in excess then is combined with the carbon monoxide to yield hydrocarbons suitable as liquid fuels. The reaction is carried out at high temperatures and pressures and in the presence of catalysts.

$$C + H_2O \xrightarrow{300°C} CO + H_2$$

$$n\,CO + (2n + 1)H_2 \xrightarrow[250°C]{ThO_2,} C_nH_{2n+2} + n\,H_2O$$

Another method, known as the **Bergius Process,** hydrogenates coal directly by combining hydrogen and carbon under high pressures and temperatures. In practice, hydrogen under pressure is injected into a heavy asphaltic paste of pulverized coal. The hydrogen required for the reaction is obtained from methane and water by the following series of reactions.

$$CH_4 + H_2O \xrightarrow[750\,°C]{\text{catalyst,}} CO + 3\,H_2$$

$$CO + H_2O \xrightarrow[750\,°C]{\text{catalyst,}} CO_2 + H_2$$

Coal, pulverized in a ball mill, is mixed with the residuum of the previous run to provide the heavy viscous paste. A general equation for the process may be written.

$$n\,C + (n + 1)H_2 \xrightarrow[\substack{700 \text{ atmospheres,} \\ 400\,°C}]{Fe_2O_3,\ MoO_2,} C_n H_{2n+2}$$

The Bergius process is a versatile one which yields hydrocarbons that fall within the diesel fuel and jet fuel, as well as within the gasoline, range. During World War II Germany produced approximately 70% of its aviation fuel by this method. The U.S. Bureau of Mines in 1949, as an experiment, produced the first synthetic gasoline in this country by the Bergius process. The cost of gasoline when produced synthetically from coal is now not competitive with that produced from petroleum.

18.8   *Petrochemicals*

The name petrochemicals is one usually assigned to compounds, found in or derived from petroleum, that are used in the manufacture of commodities other than fuels. The hydrocarbons used as fuels and lubricants are mixtures or blends of many compounds. Although these products have a common origin with petrochemicals, the term usually is reserved to describe compounds isolated from petroleum in a high state of purity. Such compounds serve as indispensable raw materials in the plastics, rubber, and synthetic fibers industries. New developments in the manufacture of paints, pesticides, fertilizers, and detergents are made possible, in a large measure, by the ready availability of chemicals from the petroleum industry. Table 18.5 lists some of the aromatic hydrocarbons obtained from petroleum and the chemicals derived from them. Examples of everyday products made from these derived compounds are included in the table to illustrate the great commercial importance of these raw materials.

TABLE 18.5 *Petrochemicals Obtained from Petroleum*

Hydrocarbons	Derived Products and Uses

HNO₃, H₂SO₄ → Nitrobenzene → [H] → Aniline

Used in the preparation of drugs, dyestuffs.

Cl₂, FeCl₃ → Chlorobenzene → NaOH, heat, then H⁺ → Phenol

Used in the manufacture of phenolic resins for plastics, weed killers, wood preservatives, dyes, disinfectants, photographic developers.

3 Cl₂, uv → Benzene hexachloride

The γ isomer of $C_6H_6Cl_6$, called "Gammexane," is used as an insecticide.

H₂, Ni → Cyclohexane → [O] → HOOC—(CH₂)₄—COOH Adipic acid

Used in the manufacture of Nylon.

RX, AlCl₃ → An alkyl-benzene → H₂SO₄ → An alkyl-benzene-sulfonic acid (SO₃H) → NaOH → A sodium alkylbenzene-sulfonate (SO₃Na)

Used as alkylbenzesulfonate detergents.

H₂C=CH₂, AlCl₃ → Ethylbenzene (C₂H₅) → Fe₂O₃ 650° → Styrene (CH=CH₂)

Used in the manufacture of polystyrene and synthetic rubber.

Benzene

TABLE 18.5 *Continued*

Hydrocarbons	Derived Products and Uses

Toluene → (HNO₃, H₂SO₄) → 2,4-Dinitrotoluene → T.N.T.

Used as a military high explosive.

2,4-Dinitrotoluene → [H] → Benzene + 2,4-Diaminotoluene → 2,4-Toluene diisocyanate

Used in the manufacture of polyurethane foams for upholstery, insulation, and flotation material.

o-Xylene → [O] → Phthalic anhydride

Used in the manufacture of glyptal resins for surface coatings and auto finishes.

p-Xylene → [O] → Terephthalic acid

Used in the manufacture of polyester fibers —e.g., Dacron.

Summary

1. Petroleum is the principal source material from which we obtain fuels (gaseous and liquid), lubricants, and a large number of organic compounds.

2. Methane is the principal hydrocarbon constituent of a natural gas used as fuel; liquefied petroleum (L. P.), or "bottled" gas, is largely propane.

3. The composition of petroleum varies widely with the geographical location of the producing area.

4. The ability of a gasoline to perform well in an internal combustion engine is a measure of its "octane number."

5. The octane numbers of hydrocarbons increase with chain branching, unsaturation, and cyclization. Octane numbers of motor fuels may also be increased by the use of "additives."

6. High molecular weight hydrocarbons (above C_9) not suitable for use as gasoline are catalytically "cracked" or broken into smaller molecules. The latter are rearranged (isomerized), cyclized (aromatized), or combined (alkylated) to form compounds of high "octane number."

7. The Fischer-Tropsch synthesis and the Bergius process are two German developments for the production of liquid fuel from coal.

8. Petrochemicals are organic compounds found in or derived from petroleum and used in the production of commodities other than fuels.

New Terms

aromatization	motor fuel additives
catalytic cracking	octane number
liquefied petroleum gas (LPG)	petrochemical

Supplementary Exercises

EXERCISE 18.1 Explain why the inclusion of olefinic hydrocarbons in a gasoline blend improves its octane rating, but makes the gasoline less stable to storage.

EXERCISE 18.2 Write equations to show how each of the following petrochemicals could be converted to the industrially important products indicated.
    (a) propene $\longrightarrow$ 2-propanol
    (b) ethene $\longrightarrow$ ethylene glycol
    (c) *o*-xylene $\longrightarrow$ phthalic anhydride
    (d) toluene $\longrightarrow$ benzoic acid
    (e) benzene $\longrightarrow$ maleic anhydride

EXERCISE 18.3 Cylinders of the type used for L. P. gas storage at cottages and small homes contain, when filled, 100 pounds of fuel. If the liquefied gas is propane, how many liters of the gaseous hydrocarbon at STP must be compressed into each cylinder? (1 lb = 0.4536 kg)

EXERCISE 18.4 The following compounds are added as antioxidants to a gasoline to improve its storage characteristics. Draw the structure of each.
    (a) 2,6-di-*tert*-butylphenol
    (b) N-phenyl-β-naphthylamine
    (c) 2,4-dimethyl-6-*tert*-butylphenol

EXERCISE 18.5 Write equations for the preparation of hydrocarbon fuels from coal, water, and hydrogen by (a) the Fischer-Tropsch synthesis; (b) the Bergius process. Use $C_n H_{2n+2}$ as the type formula for your product.

19

Color in Organic Compounds, Dyes

Introduction

As was described in Chapter 6 organic molecules absorb electromagnetic radiation in the ultraviolet and visible regions of the spectrum and undergo a transition from one electronic state to another. When the transition is in the ultraviolet region, the organic compound will appear colorless, or white. However, when the transition is in the visible region (about 4000 to 7500 Å or 400 to 750 nm), the compound will appear to the human eye to be colored.

The proper mixture of all wavelengths within the visible region results in white light, but when radiation of a certain wavelength within the visible region is absent—that is, has been absorbed or removed by interference in any manner—the resultant light is colored. The color perceived by the eye is produced by a composite of all wavelengths not absorbed and is the complementary color of the particular wavelengths that were absorbed. For example, absorption of the longer wavelengths (red, orange, yellow) results in a blue or blue-green color; absorption of the shorter wavelengths (violet, blue) results in an orange or yellow color. Table 19.1 lists the complementary colors produced when various wavelengths within the visible region of the spectrum are absorbed.

TABLE 19.1 *Absorption of Visible Light and Complementary Colors*

Wavelength Absorbed (Å)	Color Absorbed	Complementary Color
4,000–4,250	violet	yellow-green
4,250–4,750	blue	yellow
4,750–4,900	blue-green	orange
4,900–5,000	green-blue	red
5,000–5,500	green	purple
5,500–5,750	yellow-green	violet
5,750–5,950	yellow	blue
5,950–6,050	orange	green-blue
6,050–6,800	red	blue-green
6,800–7,500	purple	green

19.1   Color and Molecular Structure

The amount and intensity of the color produced when a molecule absorbs in the visible region of the spectrum is a function of molecular structure. For organic molecules the electronic transitions involved in intensely colored substances are generally $\pi \rightarrow \pi^*$ or $n \rightarrow \pi^*$ transitions. Furthermore, the π-systems will tend to be rather extensive, involving not only a number of conjugated double bonds but also one or more characteristic structural units capable of undergoing the $\pi \rightarrow \pi^*$ or $n \rightarrow \pi^*$ transitions. Thus, color is largely a function of the π-system. If the absorption band is narrow and intense (large value of ε), the color will be brilliant and clear; however, if the band is broad or if there are several overlapping bands in the visible region, the color will be muddy or dull. In an attempt to correlate color with structure, early chemists noted that certain groups nearly always produced color in a molecule. Such groups were called **chromophores** (Gr., *chroma,* light; *phorein,* to bear). The following are examples of common chromophoric groups:

$-N=O$	$-N\overset{O}{\underset{O}{\diagdown}}$	$-N=N-$	$-C\overset{O}{\diagup}\left(\overset{H\ \ H}{\underset{C=C}{\mid\ \ \mid}}\right)_n C\overset{O}{\diagdown}-$
Nitroso	Nitro	Azo	Dicarbonyl
			($n = 0$ or some integer)

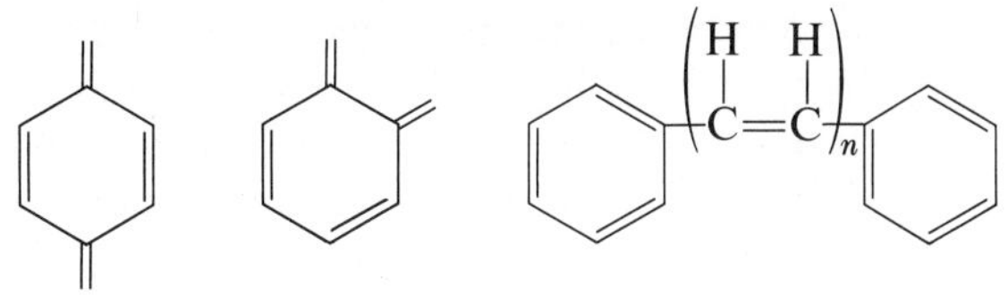

p-Quinoid   o-Quinoid   Diphenylpolyenes
                                   (n = 3 or more)

Certain electron-pair donor groups, when present in a molecule with a chromophore, appeared to intensify or augment the color. Such color-assisting groups were called **auxochromes** (Gr., *auxanein,* to increase). Auxochromes usually are acidic or basic groups which, when ortho or para to the chromophore, form with it an extended conjugated system. The hydroxyl group is the auxochrome in many colored organic compounds. Thus, in contrast to nitrobenzene, which is a very pale yellow, *o*- and *p*-nitrophenols are very bright yellow. The resonance structures possible for *o*- and *p*-nitrophenols are shown below.

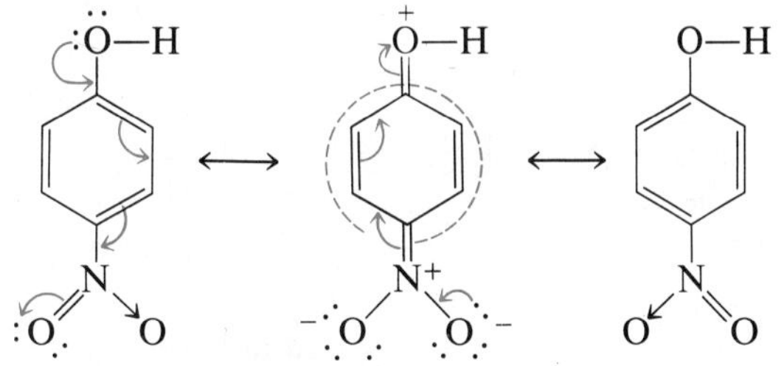

Resonance forms of *p*-Nitrophenol
(*p*-quinoid structure outlined by dotted line).

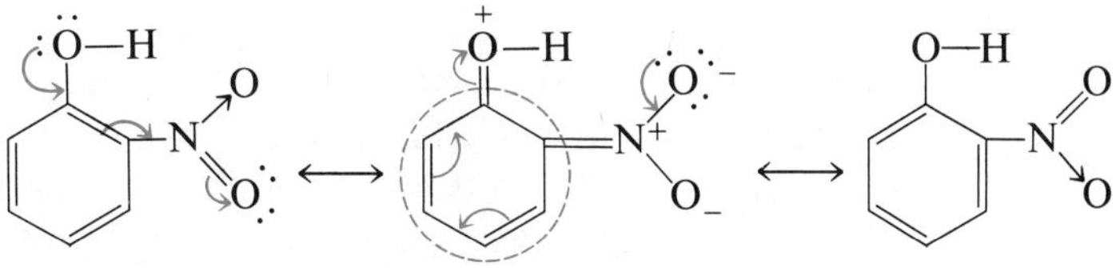

Resonance forms of *o*-Nitrophenol
(*o*-quinoid structure outlined by dotted line).

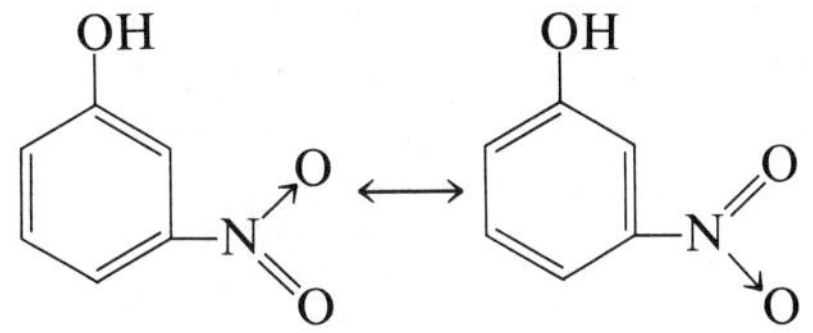

Resonance forms of *m*-Nitrophenol
(A conjugated system of alternate double and single bonds
between chromophore and auxochrome is not possible.)

Common auxochromes besides the hydroxyl group are —OR, —NH$_2$, —NHR, and —NR$_2$. In these color-assisting groups you will observe that each has one thing in common—namely, one or more pairs of electrons which may interact with and extend the conjugation already present. Although the chromophore-auxochrome theory has been a useful one, color in an organic compound is now recognized as the result of extensive conjugation, not the function of any particular group. If an organic compound is to be used as a dye, a third grouping, in addition to the chromophore and auxochrome, must be present. Such groups make possible salt formation and provide better water solubility. Sulfonic and carboxylic acid salts usually are employed for this purpose.

EXERCISE 19.1

Explain why the color of a *p*-nitrophenol solution changes from light yellow to orange when made basic with sodium hydroxide.

Color and Dyes

19.2 *Historical*

A compound to be used as a dye must possess, besides color, the ability to attach itself firmly to fabric, to leather, or to whatever is to be colored. Moreover, it must be fast to light and resistant to washing. Natural compounds which possess all the qualities cited above have never been plentiful or easily obtained, yet color has been so important in the life of man that it has been used since the beginning of civilization. One of the earliest dyes used by the ancients was indigo blue which occurs as a glucoside (Sec. 14.5) in the leaf of the indigo plant. Cloth samples taken from the tombs of the pharaohs revealed that indigo was known to the Egyptians as early as 2000 B.C., but the synthesis of indigo was not accomplished until 1880 by Baeyer. Another early dye, an object of commerce for thousands of years, was Tyrian purple. Although structurally similar to indigo, it is not of plant origin. It was obtained by the ancients from a snail (*Murex brandaris*) found on the small Mediterranean island of Tyre off the coast of present day Lebanon. Such large numbers of this small mollusk were required to produce but a little of the highly prized dye that the use of this color by any but the nobility was forbidden. Hence the expression "born to the purple." Another ancient dye is alizarin, which is found in the root of the madder plant. It was used to dye cotton and linen red. Its synthesis was accomplished in 1870. Unlike indigo, Tyrian purple and alizarin are no longer important as dyes. The structures of these three ancient dyes are given below.

You will be able to recognize in each the color-bearing and the color-assisting groups.

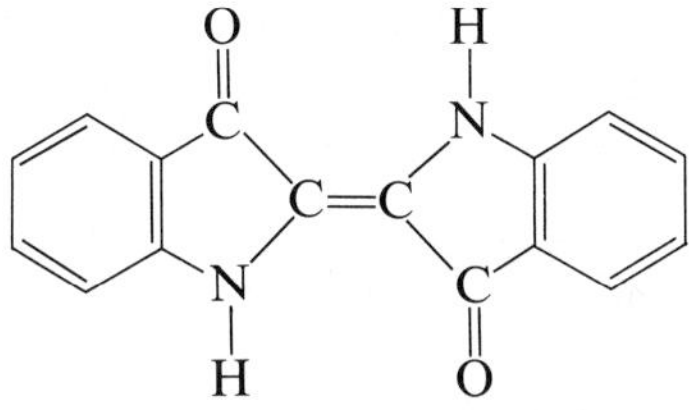

Indigo

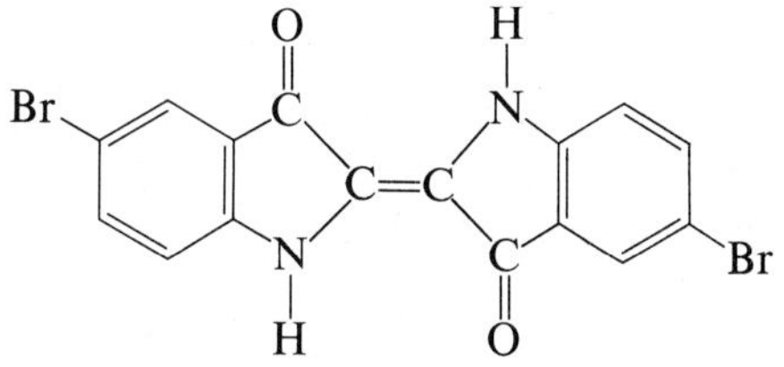

Tyrian purple (6,6'-dibromoindigo)[1]

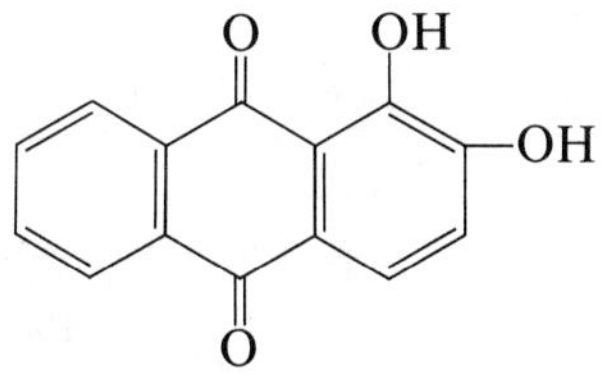

Alizarin

Progress in the production of synthetic dyes from coal-tar derivatives has been comparatively recent. The first synthetic dye to be produced commercially was mauve. It was accidentally discovered in 1856 by William Perkin, an eighteen-year-old assistant in Hofmann's (Sec. 13.3-C) laboratory. Perkin tried to oxidize an aniline derivative in the hope of obtaining quinine, but the black product, which the reaction yielded, contained a violet substance which had all the desirable properties of a dye. Perkin called the dye "mauveine" and, recognizing the commercial possibilities of his discovery, began to manufacture it in quantity. The aniline which Perkin used was not pure but contained some *o*- and *p*-toluidines. Mauveine since has been found to have the polycyclic structure shown here.

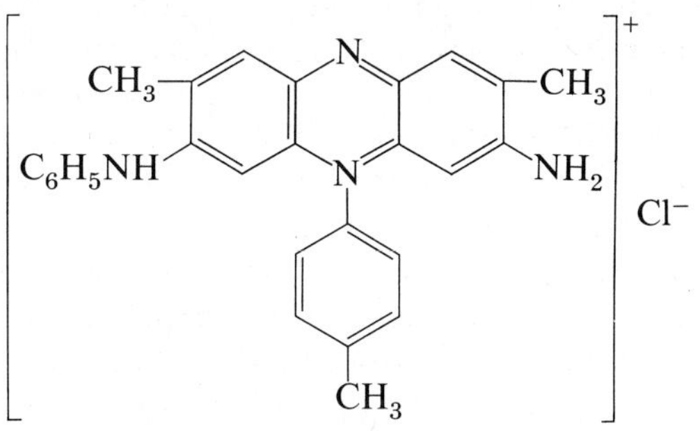

Mauveine

---

[1] The numbering of heterocyclic ring systems is illustrated in Table 17.1.

Perkin's discovery stimulated further research that led not only to the preparation of many other dyes, but, equally important, to the preparation of many other useful organic compounds. The impetus provided by the search for dyestuffs led to a period of intense chemical research so productive that it has been referred to as "the golden age of organic chemistry." Many of the synthetic dyes produced today are far superior to the naturally occurring ones used by the ancients. Not only our garments, but a host of other commodities throughout the market place, are now available in a wide variety of color—purple included.

19.3   *Classification of Dyes by Structure*

Dyes may be classified in two ways. The organic chemist classifies them according to a common parent structure. The dyer who is primarily interested in anchoring the dye to the fiber classifies them according to the method of application. The first, or chemical classification, includes the following principal types along with examples of each.

A. Nitro and Nitroso Dyes.   Dyes of this class are the *o*- or *p*-nitroso, nitrophenols, and the nitroso and nitronaphthols. One of the oldest nitro dyes is the high explosive, trinitrophenol (picric acid (Sec. 8.8-D)).

Names assigned to the hundreds of dyes in everyday use usually indicate a color and often the method of application, but seldom do the names of dyes reveal their chemical structures. Other examples of nitro and nitroso dyes are shown.

Martius yellow   Naphthol green Y

B. Azo Dyes.   This class represents the largest and most important group of dyes. The characteristic feature of each dye in this class is the chromophoric azo group ($-N{=}N-$), which forms part of the conjugated system and joins two or more aromatic rings. Azo dyes are prepared by coupling a diazotized aromatic amine with another aromatic amine or with a phenol or a naphthol (see Sec. 13.8-B). This simple reaction can be carried out directly within the fabric, and results in a product complete with chromophore, auxochrome, and conjugated system. One of the first azo dyes to be prepared was *para red,* obtained from diazotized *p*-nitroaniline and *β*-naphthol.

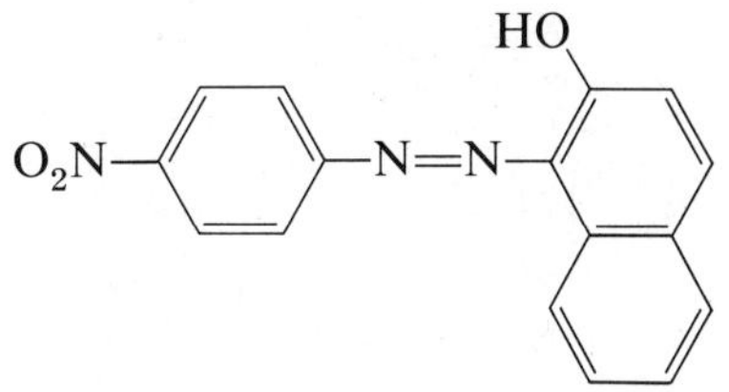

Para red

More than one aromatic ring, when included in the coupling reaction, extends the conjugated system and deepens the color of a dye. For example, diazotized benzidine (4,4′-diaminobiphenyl) will couple with 1-naphthyl-amine-4-sulfonic acid to produce a dye called *congo red*. Congo red also is used as a chemical indicator.

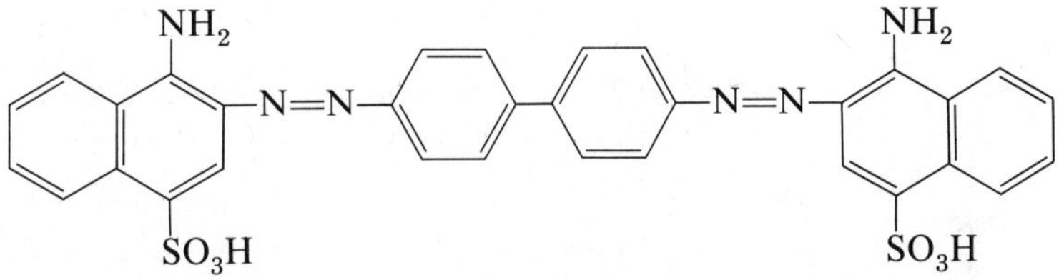

Congo red

C. Triphenylmethane Dyes. The triphenylmethane dyes can be identified from a common structural feature in which a central carbon atom is joined to two benzene rings and to a *p*-quinoid group.

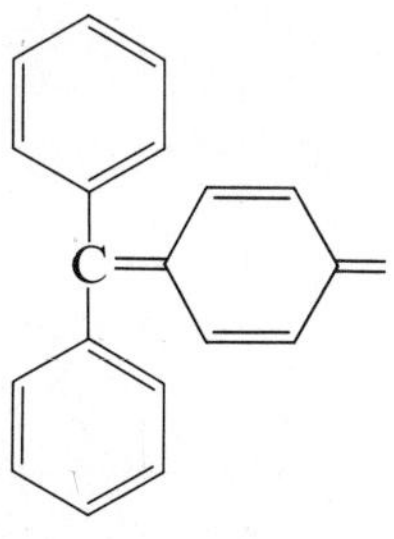

Malachite green is an example of a dye in this class. It is prepared by condensing benzaldehyde with dimethylaniline to give a colorless **leuco base** (Gr., *leuco,* white). Oxidation of the leuco base followed by acid treatment yields the dye.

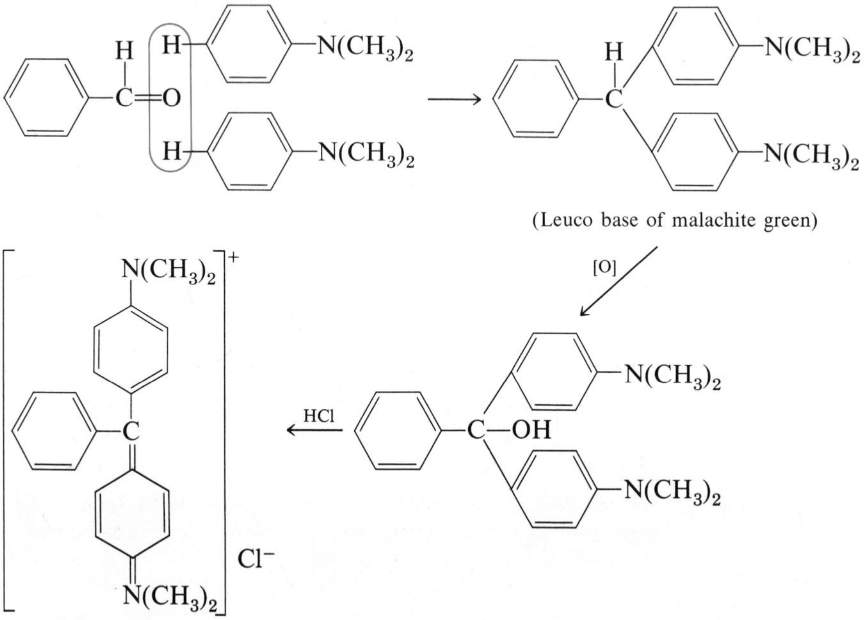

(Leuco base of malachite green)

Malachite green

Another triphenylmethane dye with a structure very similar to that of malachite green is *crystal violet*. Crystal violet is used as a biological stain and, until recently, has been widely used medicinally as an anthelmintic.

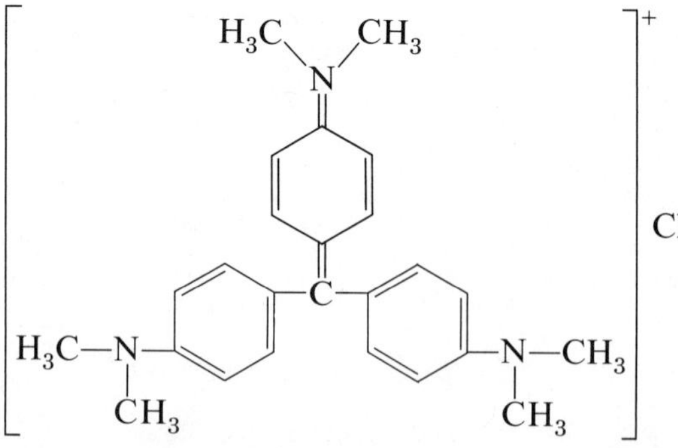

Crystal violet

The familiar phenolphthalein, an acid-base indicator rather than a dye, also has a triphenylmethane structure. It is prepared by condensing phthalic anhydride with phenol.

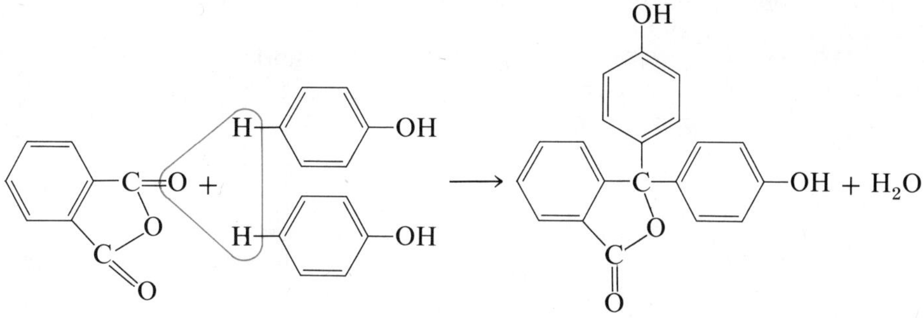

Phenolphthalein (colorless)

Indicators and some dyes are able to change color with a change in the hydrogen ion concentration of their solutions. Such changes in color are possible because an indicator is itself a weak acid or a weak base and enters into a salt-producing reaction. Phenolphthalein in acid solutions is colorless but turns red in alkaline solutions. A strong alkaline solution converts the indicator into a triphenylcarbinol derivative. Without an extended conjugated system the compound again is colorless.

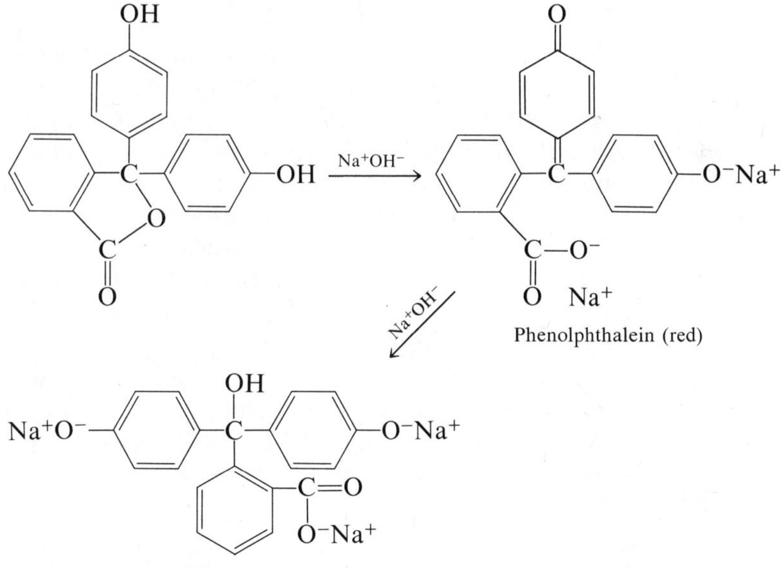

Phenolphthalein (red)

Phenolphthalein (colorless)

D. Indigoid Dyes. The indigoid dyes have as their identifying feature two conjugated carbonyl groups, $-\overset{\overset{\displaystyle O}{\|}}{C}-CH{=}CH-\overset{\overset{\displaystyle O}{\|}}{C}-$. Indigo is our best example of a dye of this class. It occurs as a colorless plant glucoside, *indican*. The latter yields glucose and *indoxyl* when hydrolyzed. Indoxyl, when exposed to air, oxidizes to indigo.

Indoxyl   Indigo

E. Anthraquinone Dyes. The anthraquinone dyes have a *p*-quinoid structure in common with two other benzene rings in a *fused ring* molecule. An example of an anthraquinone dye is alizarin (Sec. 19.2).

19.4   *Classification of Dyes by Method of Application*

The mechanics of dyeing vary with the nature of the material to be dyed. A dye process suitable for animal fibers such as wool or silk may be entirely unsatisfactory for dyeing cotton. Wool and silk are proteins and contain many basic and acidic groups. These groups serve as points of attachment for a dye because it also has acidic and basic functions. On the other hand, cotton is a carbohydrate and provides only neutral ether linkages and hydroxyl groups as points of attachment for hydrogen bonds. Affixing color to synthetic fibers such as the polyolefins—hydrocarbons entirely lacking in polar groups—requires still other techniques. One point is clear: a dye must do more than simply color the surface of a fiber. It should become a *part of the fiber,* and wear and wash with the fiber. A number of dyes fulfil these conditions very well when used on some materials but not when used on others. In all cases the nature of the material will determine the process and the dye to be employed in dyeing it. A number of different dyeing methods and dyes adaptable to each method are described in the following sections.

A. Direct or Substantive Dyes. Dyes of this class contain polar group-ings—acidic or basic—which are able to combine with polar groups in the fiber. Such dyes color a fabric directly when the latter is immersed in a hot, aqueous solution of the dye. Direct dyes are particularly well adapted to the dyeing of silk and wool. Picric acid and Martius yellow (Sec. 19.3-A) are two examples of direct dyes. Both compounds are acids and can combine with the free amino groups found in protein fibers. Nylon, a polyamide (Sec. 17.2), also may be dyed by this method.

B. Mordant Dyes.   This class of dyes includes those which can form an insoluble, colored salt (called a lake) with certain metallic oxides. Mordant dyeing is one of the earliest methods practiced for anchoring color to a fabric. Mordant dyes can be used on silk and wool or on cotton fibers. The fabric appears colored because the colored precipitate is bound to and envelops each fiber. The oxides of aluminum, chromium, and iron usually are used for mordanting colors. Alizarin is an example of a mordant dye.

Alizarin

C. Vat Dyes.   A vat dye is a substance which, in a reduced form, is water soluble and may be colorless. In this form it is introduced into the fabric. The fabric, after the dye is absorbed, is removed from the "vat" and exposed to air or treated with a chemical oxidizing agent. This step oxidizes the dye to a colored, insoluble form. The ancient dyes, indigo and Tyrian purple, are classic examples of vat dyes.

D. Ingrain Dyes.   Ingrain dyes, sometimes called developed dyes, are dyes which develop within the fabric itself. The azo dyes are good examples of ingrain dyes. The cloth to be dyed first is immersed in an alkaline solution of the compound to be coupled—usually a phenol or naphthol. A second immersion in a cold solution of a diazotized amine causes the coupling reaction to take place within the fabric. Dyes produced in this manner were referred to at one time as *ice colors* because of the low temperature requirement for stability on the part of the diazonium salt.

E. Disperse Dyes.   Dyes classified under this category are those which are soluble in the fiber but not in water. Disperse dyes are used in the coloring of many of the newer synthetic fibers. These sometimes are referred to as hydrophobic (water-fearing) fibers and usually are lacking in polar groups. The dye, in the form of a finely divided dispersion, is dissolved in some organic compound (often a phenolic), and the absorption into the fiber is carried out in a surface-active bath (soap solution) at high temperatures and pressures. Examples of disperse dyes used for acetate rayons, Dacron, Nylon and other synthetic fibers are Celliton Fast Pink B (1-amino-4-hydroxyanthraquinone) and Celliton Fast Blue B (1,4-N,N′-dimethylaminoanthraquinone).

Celliton Fast Pink   Celliton Fast Blue B

19.5 *Organic Pigments*

Our discussion of color in organic compounds would not be complete without a brief reference to colored pigments not used as dyes. A dye must be absorbed by the material to be colored, a pigment is applied to the surface. Pigments are not water soluble but, like the pigments in paints (Sec. 11.7), are suspended in the vehicle. Pigments are used mainly in decorative coatings, in printing inks, and for imparting color to plastics. If the chemical structure of a pigment can be modified to confer upon it water solubility, then it may serve as a dye.

The *phthalocyanines* comprise an important group of pigments. One of these, copper phthalocyanine, is prepared by heating phthalonitrile with copper at 200°.

Copper phthalocyanine

Summary

1. Absorption of light in the visible region produces color. Early chemists attributed color in organic molecules to the presence of multiple-bonded, electron-attracting groups which they called **chromophores.** Common chromophores are **nitroso, nitro, azo, dicarbonyl, quinoid,** and **diphenylpolyene** structures.

$$-N{=}O \qquad -N{=}O$$

Nitroso   Nitro

$$-C{-}(CH{=}CH)_n{-}C{-}$$

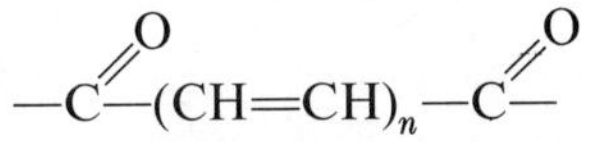

Dicarbonyl

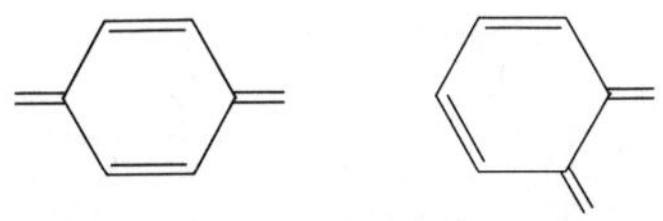

p- and *o*-quinoid 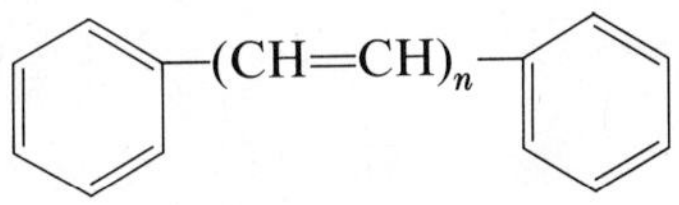   Diphenylpolyenes

2. Groups capable of extending the conjugation of a colored compound are electron-donating, color-intensifying groups called **auxochromes.** Common auxochromes are $-OH$, $-OR$, $-NH_2$, $-NHR$, $-NR_2$.

3. A colored organic compound may be used as a dye if it can be anchored to a fiber, is reasonably fast to light, and is resistant to washing.

4. Dyes can be classified according to structure as
   - (a) Nitroso or nitro dyes
   - (b) Azo dyes
   - (c) Triphenylmethane dyes
   - (d) Indigoid dyes
   - (e) Anthraquinone dyes

5. Dyes can be classified according to use as
   - (a) Direct dyes
   - (b) Mordant dyes
   - (c) Vat dyes
   - (d) Ingrain dyes

New Terms

auxochrome

azo dye

chromophore

complementary color

ingrain dye

leuco compound

mordant

a phthalocyanine

Supplementary Exercises

EXERCISE 19.2   A molecule absorbs strongly at 5.9 μ. Would any change resulting from this absorption be perceptible to the human eye?

EXERCISE 19.3   A drop of colorless mineral oil when allowed to spread on the surface of water produces a changing color pattern. Explain.

EXERCISE 19.4   An organic compound absorbs radiation of wavelengths 5,960–6,000 Å. What color region of the visible spectrum is this? What color is the compound likely to be?

EXERCISE 19.5   *β-Carotene*, $C_{40}H_{56}$, is a $(C=C)_8$ system that imparts a yellow color to carrots. Its open-chain isomer, *lycopene*, is a $(C=C)_{11}$ system responsible for the color of ripe tomatoes and watermelon. Explain why one is yellow and the other red. Consult Table 18.1 and suggest wavelength regions where absorption by each natural substance might take place.

EXERCISE 19.6   Draw structures for the following compounds.
    (a) 2,4-dinitrophenylhydrazine
    (b) diacetyl, $(CH_3-\underset{\underset{O}{\|}}{C}-)_2$

    (c) *p*-nitrobenzoic acid
    (d) quinone

Would these compounds be colored? Could any one be used satisfactorily as a dyestuff? Explain.

EXERCISE 19.7   In each of the dyes whose structures are given below, (a) outline the part of the structure considered to be the chromophore; (b) circle that part of each structure responsible for deepening the color—i.e., the auxochrome. Classify each according to structure and according to use.

(a)

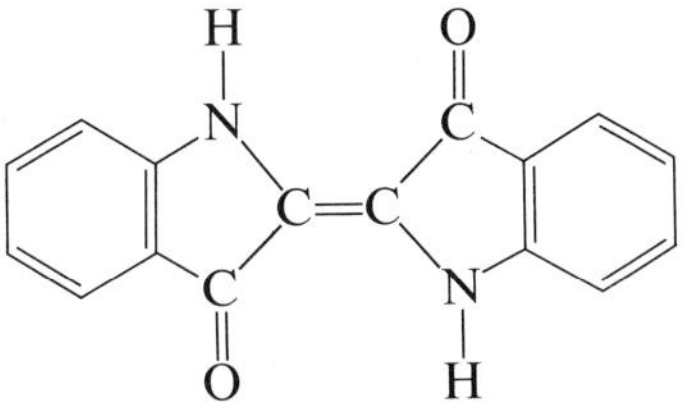

(b)

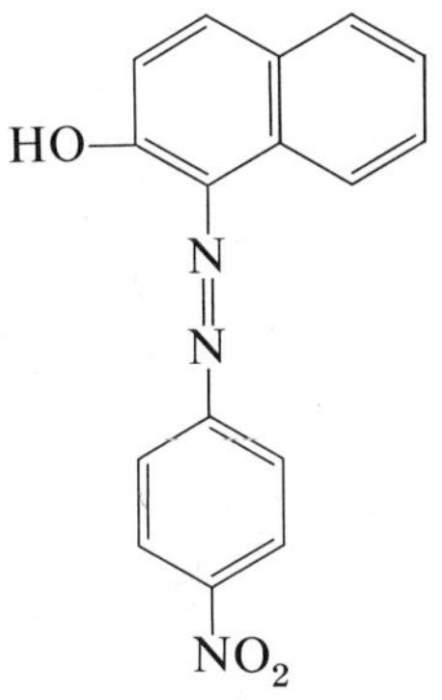

(c)

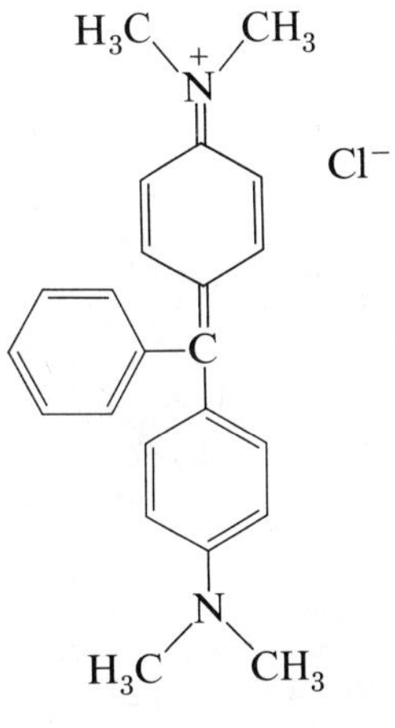

EXERCISE 19.8 Explain the following phenomena:
(a) phenolphthalein is colorless in acid solutions, red in weakly alkaline solutions, and again colorless in strongly alkaline solutions.
(b) an aqueous solution of *p*-nitrodimethylaniline is yellow but becomes colorless when the solution is made acidic.

EXERCISE 19.9 Congo Red is a useful indicator in the organic laboratory that is red in basic or neutral solution (pH 5.0 or above) and blue in *mineral* acid solution (pH 3.0 and below). The neutral (red) form of the dye is given in Sec. 19.3-B. Assume that two protons must be added to change the dye to its blue form and suggest a structure for the dicationic blue form. (*Hints:* (1) Protons will *not* be added to the —NH$_2$ or —SO$_3^-$Na$^+$ groups. (2) Resonance stabilization of the dication is essential.)

EXERCISE 19.10 Suggest a synthesis of Congo Red (Sec. 19.3-B) from 1-napththylamine (1-aminonaphthalene) and hydrazobenzene. (*Hint:* See Secs. 13.4-E and 13.5.)

Index

FATS   Solid or semisolid, naturally occurring, long chain carboxylic acid esters of glycerol.

FREE RADICAL   A highly reactive, short lived atom or group that has an odd, unpaired electron.

FUNCTIONAL GROUP   A structural feature that identifies a family of compounds and bestows upon its members a common property.

GLYCOLS   Compounds that contain two hydroxyl groups on adjacent carbon atoms.

HEMIACETAL   An unstable compound formed by the addition of an alcohol to the carbonyl group of an aldehyde.

HETEROCYCLIC   A ring compound that contains, in addition to carbon, one or more other kinds of atom.

HYBRIDIZED ORBITALS   Orbitals formed by blending orbitals in different subshells.

HYDROCARBONS   Compounds (mostly obtained from petroleum) that contain only hydrogen and carbon.

INVERT SUGAR   A levorotatory mixture of equal parts of glucose and fructose formed by hydrolyzing dextrorotatory sucrose.

ISOMERS   Compounds with the same molecular formula but with different structural formulas.

KETONES   Carbonyl compounds of the general formula $R_2C=O$.

LEVOROTATORY   A term applied to optically active compounds that rotate plane polarized light counterclockwise or to the left.

MOLECULAR ORBITAL   An orbital formed by the overlap of two or more atomic orbitals which encompasses more than one nucleus.

MUTAROTATION   A change with time in the rotation of a solution of an optically active compound.

NUCLEIC ACIDS   High molecular weight natural polymers found in the nuclei and cytoplasm of all living cells.

NUCLEOPHILIC REAGENT   A reagent with an unshared pair of electrons to donate, e.g., a base.